EXCLUSIVE REACTIONS

at High Momentum Transfer

EXCLUSIVE REACTIONS
at High Momentum Transfer

Proceedings of the International Workshop

21 – 24 May 2007
Jefferson Lab, Newport News, Virginia, USA

Editors

Anatoly Radyushkin
Old Dominion University & Jefferson Lab, USA

Paul Stoler
Rensselaer Polytechnic Institute, USA

World Scientific

NEW JERSEY • LONDON • SINGAPORE • BEIJING • SHANGHAI • HONG KONG • TAIPEI • CHENNAI

Published by

World Scientific Publishing Co. Pte. Ltd.

5 Toh Tuck Link, Singapore 596224

USA office: 27 Warren Street, Suite 401-402, Hackensack, NJ 07601

UK office: 57 Shelton Street, Covent Garden, London WC2H 9HE

British Library Cataloguing-in-Publication Data
A catalogue record for this book is available from the British Library.

EXCLUSIVE REACTIONS AT HIGH MOMENTUM TRANSFER
Proceedings of the International Workshop

ISBN-13 978-981-279-694-3
ISBN-10 981-279-694-0

Printed in Singapore by World Scientific Printers

ORGANIZING COMMITTEE

H. Avagyan (Jefferson Lab)
V. Burkert (Jefferson Lab)
R. Ent (Jefferson Lab)
C. Carlson (William & Mary)
A. Radyushkin (Old Dominion & Jefferson Lab)
M. Vanderhaeghen (William & Mary and Jefferson Lab)
P. Stoler (Rensselaer)
B. Wojtsekhowski (Jefferson Lab)

PREFACE

These are the proceedings of the Workshop on Exclusive Reactions at High Momentum Transfer, which was held 21–24 May, 2007 at the Thomas Jefferson National Accelerator Facility (Jefferson Lab) in Newport News, Virginia, USA. The workshop was sponsored by Jefferson Lab.

The workshop was open to all interested scientists, and embraced both theorists and experimentalists. The Organizing Committee thanks the speakers and participants at the conference for the stimulating atmosphere that prevailed. We trust the written versions of the talks presented in the volume will reproduce some of that feeling.

Thanks are also due to the people who worked hard to organize the conference and make it a success. The Conference group at Jefferson Lab was indispensible. Allow us to give special thanks to Ruth Bizot, Cynthia Lockwood, Marty Hightower, Mary Fox and Marti Bennett for all their efforts before, during and since the conference.

Paul Stoler
Anatoly Radyushkin
December 2007

CONTENTS

SESSION PRESENTATIONS

HARD EXCLUSIVE SCATTERING AT JLAB

P. KROLL

Fachbereich Physik, Universitaet Wuppertal,
D-42097 Wuppertal, Germany
E-mail: kroll@physik.uni-wuppertal.de

The various factorization schemes for hard exclusive processes and the status of their applications is briefly reviewed.

Keywords: Compton scattering, vector-meson electroproduction, form factors.

1. Introduction

In large momentum transfer exclusive processes the probe, say, a virtual photon has a wave length that is much shorter than the spacial extension of the hadronic target. This allows to look inside the hadrons and to study the interactions of their constituents, quarks and gluons. There is overwhelming evidence, mainly from inclusive reactions, that QCD is the correct theory for the interactions between quarks and gluons. QCD is a complicated theory. Quarks and gluons are confined, only their bound states - the hadrons - can be observed experimentally. The formation of hadrons from quarks and gluons occurs at soft scales where QCD perturbation theory is inapplicable. But, with the exception of lattice QCD, there is no analytical or numerical method known to solve QCD in the soft region. In any scattering process as hard as the external scale, for instance the virtuality, Q^2, of the probing photon, may be, soft hadronization is unavoidably involved too. Thus, one may wonder whether it is possible to calculate observables for hard processes. This is indeed possible in a number of cases thanks to the factorization properties of QCD: hard processes factorize into parton-level subprocesses amenable to perturbative QCD (and/or QED), and in soft hadronic matrix elements which embody the non-perturbative physics. For a number of processes there are rigorous proofs of factorization available, e.g. the pion electromagnetic form factor, deeply virtual lepton-nucleon scattering (DIS), deeply virtual Compton scattering (DVCS). For others factorization is a hypothesis with often good arguments for its validity. However, we have to be cautious in these cases. Collins and Qiu[1] found a counterexample, namely $h_1 h_2 \to h_3 h_4 X$ where h_i denotes a hadron, for which ($\mathbf{k}_\perp$) factorization breaks down. Given the theoretical complications involved in exclusive scattering and with regards to the large number of succesful tests of QCD properties accumulated over the last 30 years, the experimental and theoretical investigation of hard exclusive processes will not contribute towards the verification of QCD, rather we will learn about methods how to apply QCD.

In the following I will briefly review the factorization schemes used in exclusive scattering (Sects. 2 and 3). In Sect. 4 I will summarize our present knowledge on the generalized parton distributions (GPDs), the soft hadronic matrix elements occuring in the handbag factorization scheme. Next I will turn to applications of the handbag factorization to deeply virtual exclusive scattering (Sect. 5), discuss alternative theoretical approaches such as the Regge model (Sect. 6), and turn finally to wide-angle exclusive reaction (Sect. 7). Special emphasis is laid on the role of JLab in this physics - what has been achieved by JLab till now, what will be done in the future. In Sect. 8 I will present the summary.

2. The ERBL factorization scheme

A first factorization scheme for hard exclusive processes has been invented around 1980. Efremov and Radyushkin[2] as well as Brodsky and Lepage[3] showed that factorization holds for the pion form factor at large Q^2. This factorization scheme has been generalized later on to many other exclusive processes, lacking however proof in most cases [a]. Since the evolution equation for the associated soft matrix element, the so-called distribution amplitude (DA), is named after these authors, I take the liberty to give the full factorization scheme also this name - ERBL factorization. Other frequently used names for it are either misleading or lead to a clash of notation.

In order to sketch the ERBL scheme let me consider Compton scattering off protons at large Mandelstam variables $s, -t, -u$ as a typical and important example and let me consider only the at large scales dominant valence Fock state of the proton. The amplitudes of this process factorize into partonic subprocess $\gamma qqq \to \gamma qqq$ (see Fig. 1) and in proton DAs $\Phi_p(x_1, x_2, x_3)$ where the x_i are the usual momentum fractions. All partons

[a] Many authors have also contributed to the development of that field, e.g. Refs. 4,5.

of the valence Fock state participate in the subprocess, they are emitted or absorbed collinearly from their parent hadron and are quasi on-shell. This neccessitates the exchange of at least two hard gluons. The Compton amplitudes are given by convolutions of subprocess amplitudes and DAs

$$M \sim \Phi_p \otimes \mathcal{H} \otimes \Phi_p \,. \tag{1}$$

One may also consider higher Fock states of the involved hadrons but these contributions are suppressed by inverse powers of the hard scale as compared to the valence Fock state contribution.

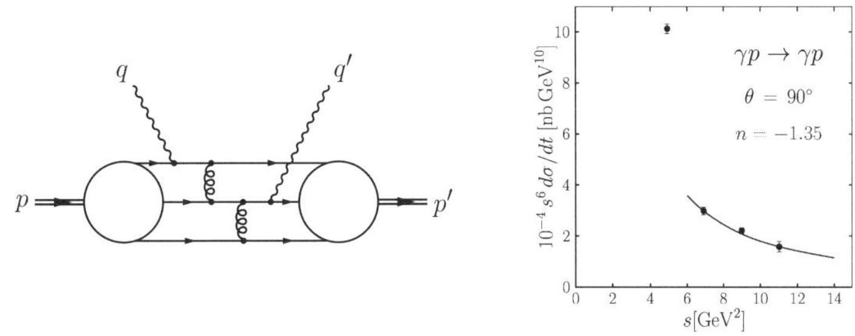

Fig. 1. Left: A typical graph for Compton scattering within the ERBL factorization scheme. Right: The Compton cross section, scaled by s^6, at a scattering angle of $90°$. Data taken from Ref. 8.

The ERBL factorization scheme implies dimensional counting[6,7] which means that at large momentum scales (or short distances) exclusive observables exhibit scaling, i.e. the fall off as a certain power of the hard scale asymptotically. The power laws are modified by perturbative logs generated by the running of α_s and the evolution of the DAs. Scaling often holds approximately in experiment although there seems to be no evidence for the perturbative logs. Recent precision data are often in conflict with dimensional counting. As an example the JLab Hall A data[8] on Compton scattering are shown in Fig. 1. Clearly the cross section does not drop as s^{-6} as predicted by dimensional counting. Violations of dimensional counting are also seen in the Pauli form factor[9] or in the precise BELLE data[10] on $\gamma\gamma \to p\bar{p}$. These counterexamples do not disprove dimensional counting. They merely indicate that the experimentally available scales for these data are not sufficiently large for applying dimensional counting and, hence, the ERBL factorization scheme.

The ERBL factorization scheme has been frequently applied to various exclusive processes, e.g. electromagnetic form factors, Compton scattering, photoproduction of mesons and various time-like processes. It turned out however that with very few exceptions the size of the ERBL contribution is too small, often by order of magnitude, in comparison with experiment. What does this mean? Are the scales available in present-day experiments, typically about 10 GeV^2, too low for applying ERBL factorization or is it possible to improve the results within that scheme? For instance, one may follow the suggestion of Chernyak and Zhitnitky[5] and use DAs which are concentrated in the end-point regions where one of the momentum fractions tends to zero. Such DAs provide much larger ERBL contributions, in some cases even agreement with experiment is achieved, e.g. for the pion form factor. It has been argued that the use of these CZ-type DAs lead to theoretical inconsistencies[11,12] since the bulk of the perturbative contribution is accumulated in the end-point regions where perturbation theory breaks down. One may also suspect that higher order pQCD corrections lead to a large K-factor but this has not yet been elaborated. However, the known NLO corrections for the pion form factor do not suffice for solving the difficulties with the size of the ERBL contribution if evaluated from DAs close to the asymptotic one, $\Phi_{AS} = 6x(1-x)$.

In order to cure some of the deficiencies of the the ERBL factorization scheme Sterman and Li[13] invented to so-called modified perturbative approach in which the quark transverse momenta are retained and Sudakov suppressions are taken into account. Configurations with large transverse separations of the quarks which occur in the end-point regions are suppressed and theoretically consistent results are obtained. For consistency the DAs are to be replaced by transverse momentum dependent light-cone wavefunctions.[14] In general the contributions obtained with the modified perturbative approach are also too small even if CZ-type wavefunctions are used, see for instance Ref. 15.

3. Handbag factorization in exclusive reactions

A new factorization scheme[16–18] became popular in 1996. In constrast to the ERBL scheme there is only one active parton that participates in the partonic subprocess, e.g. for Compton scattering $\gamma q \to \gamma q$, see Fig. 2. Similar to the ERBL scheme the active parton is emitted and reabsorbed by the hadron collinearly and is quasi on-shell. The soft hadronic matrix elements are now GPDs. The handbag factorization applies to two different kinematical regions of exclusive reactions. The deeply virtual region is characterized by large Q^2 but small Mandelstam $-t$. In the wide-angle region, on the other hand, Q^2 is assumed to be small while $-t$ and $-u$ are considered as large. Reactions studied in both the regions are Compton scattering and photo- and electroproduction of mesons in the space-like region as well as the crossed processes (e.g. $\gamma^{(*)}\gamma \leftrightarrow p\bar{p}$, $p\bar{p} \to \gamma^{(*)}M$) in the time-like region (see Fig. 2). Related to these processes is the photon-pseudoscalar-meson (P) transition form factor (see Fig. 2). The partonic subprocess is identical to that of Compton scattering but the hadronic matrix element for the $q\bar{q} \to P$ transition is just the DA appearing in the ERBL factorization scheme. The transition form factor is an exceptional case since the handbag and the ERBL factorization schemes fall together for it. The theoretical result for it, say, to NLO and evaluated from the asymptotic π DA is very close to experiment.[19] Only the origin of the remaining about 10% is still under debate. Suggested have been NNLO corrections, deviations from the asymptotic DA and/or power corrections.

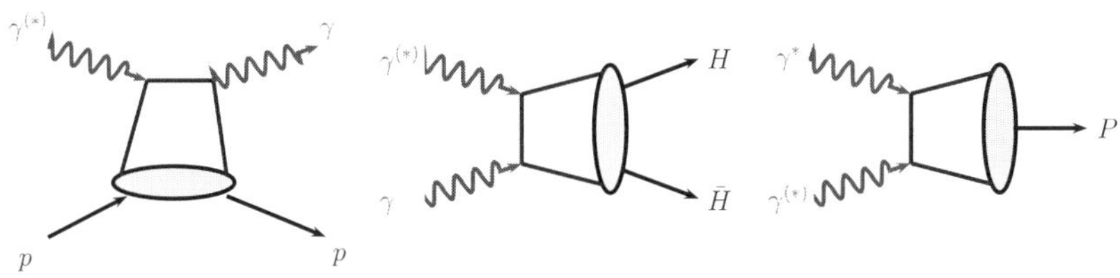

Fig. 2. Handbag factorization in the space- and time-like regions and the form factor for photon-pseudoscalar-meson transitions.

For parton helicity non-flip there are four GPDs for the proton in the space-like region denoted by H, $\widetilde{H}$, E and $\widetilde{E}$. They exist for each quark flavor and for the gluon and are functions of three variables, a momentum fraction x, the skewness ξ and t. For the GPDs a number of properties are known. Thus, H and $\widetilde{H}$ reduce to the ordinary unpolarized and polarized parton distribution functions (PDFs) in the forward limit $\xi, t \to 0$

$$\begin{aligned} H^q(x,0,0) &= q(x)\,, & \widetilde{H}^q(x,0,0) &= \Delta q(x)\,, \\ H^g(x,0,0) &= xg(x)\,, & \widetilde{H}^g(x,0,0) &= x\Delta g(x)\,. \end{aligned} \tag{2}$$

The forward limits of E and $\widetilde{E}$ are not accessible in DIS. The GPDs are related to the proton form factors by sum rules, e.g. for the Dirac form factor

$$F_1^a(t) = \int_{-1}^{1} dx H^a(x,\xi,t)\,, \qquad F_1(t) = \sum_a F_1^a(t)\,. \tag{3}$$

Analogous sum rules for E being related to the Pauli form factor, $\widetilde{H}$ (related to the axial form factor) and $\widetilde{E}$ (related to the pseudoscalar form factor). Other known properties of the GPDs are polynomiality, universality, evolution, Ji's sum rule and a couple of positivity constraints.

One may also consider parton helicity flip. These configurations define four more GPDs, termed H_T, $\widetilde{H}_T$, E_T and $\widetilde{E}_T$ for each quark flavor and for the gluon. These functions are practically unknown. They are very hard to access since parton helicity flip is frequently suppressed in partonic subprocesses. One may proceed and consider two (or more) active partons. It is straightforward to show that in order to match the requirement of collinear emission and absorption of quasi on-shell partons by the hadrons, at least one hard gluon is to be exchanged between the active partons. These contributions which have not yet been investigated, are therefore expected to be suppressed. It is interesting to note that, say, for Compton scattering off protons the case of three active partons is just the ERBL contribution if dominance of the valence Fock state is assumed.

4. What do we know about the GPDs?

A popular model which allows to construct the GPDs from the PDFs is the double distribution ansatz[20]

$$f_i(\beta,\alpha,t) = g_i(\beta,t)\,h_i(\beta)\,\frac{\Gamma(2n_i+2)}{2^{2n_i+1}\,\Gamma^2(n_i+1)}\,\frac{[(1-|\beta|)^2-\alpha^2]^{n_i}}{(1-|\beta|)^{2n_i+1}}\,. \tag{4}$$

The functions h_i represent the PDFs. In the case of H for instance

$$h_g = |\beta| g(|\beta|)\,,\; h^q_{\rm sea} = q_{\rm sea}(|\beta|)\,{\rm sign}(\beta)\,,\; h^q_{\rm val} = q_{\rm val}(\beta)\,\Theta(\beta)\,, \tag{5}$$

and $g_i(\beta, t=0)=1$, n_i either 1 or 2. The GPD is obtained from the double distribution by the following integral representation

$$H_i(\bar{x},\xi,t) = \int_{-1}^{1} d\beta \int_{-1+|\beta|}^{1-|\beta|} d\alpha\,\delta(\beta+\xi\alpha-\bar{x})\,f_i(\beta,\alpha,t)\; + {\rm D-term}\,. \tag{6}$$

For the t dependence of the GPDs, embodied in the function g_i, several ansaetze are to be found in the literature. The simplest idea is to assume that it represents a β independent kind of form factor but the implied $\beta - t$ factorization seems to be unrealistic.[21,22] Another idea is to generalize the Regge behaviour of the PDFs,[23] $q(\beta) \to \beta^{-\alpha(0)}$ for $\beta \to 0$, to non-zero values of t:

$$g_i(\beta,t) = {\rm e}^{b_i t}\;|\,\beta\,|^{-\alpha_i' t}\,, \tag{7}$$

Here a linear Regge trajectory, $\alpha_i = \alpha_i(0) + \alpha_i' t$, is assumed and an exponential t dependence of the corresponding residue (with a parameter b_i). There are many applications of the double distribution model, reggeized or not, for instance Refs. 24–29. The advantage of the double distribution model is that the reduction formulas (2) and polynomiality are automatically satisfied. The D-term[30] which is not related to the PDFs and hence a free function, provides the largest power of ξ in the moments. It only contributes to the real parts of the gluon and the flavor-singlet quark GPDs. Its quantitative role is not clear.

Alternatively, one may try to extract the GPDs from experimental data in analogy to the determination of the PDFs. First attempts to determine at least the zero-skewness GPDs this way have been published.[31,32] The idea is to exploit the sum rules (3) at zero skewness, e.g.

$$F_1^u = \int_0^1 dx\, H_v^u = 2F_1^p + F_1^n\,, \quad F_1^d = \int_0^1 dx\, H_v^d = 2F_1^n + F_1^p\,, \tag{8}$$

where the valence quark GPDs are defined by $H_v^q = H^q - H^{\bar{q}}$. A possible contribution from $H^s - H^{\bar{s}}$ has been neglected in (8). The measurements of the strangeness form factors[33–35] seem to indicate that this contribution is small although non-zero.[36] A weak evidence for $s(x) \neq \bar{s}(x)$ has been found by the CTEQ group.[37]

To determine the integrand from the integral is in a strict mathematical sense an ill-posed problem. But using an ansatz for the GPDs with a few free parameters adjusted to experiment, it is possible to extract the GPDs H, $\widetilde{H}$ and E for valence quarks. Admittedly the results on the GPDs depend on the ansatz which one may take as

$$H_v^q = q_v(x) \exp\left[f_q(x)t\right], \tag{9}$$

in which

$$f_q = \left[\alpha' \log(1/x) + B_q\right](1-x)^{n+1} + A_q x(1-x)^n\,, \tag{10}$$

and analogously for the other two GPDs. In Ref. 32 a standard slope for the Regge trajectory is assumed ($\alpha' = 0.9\ \mathrm{GeV}^2$), $n = 2$ taken and the CTEQ6 PDFs[38] are used as input. The parameters A_q and B_q are fitted to the form factor data. In Ref. 31, on the other hand, $A_q = B_q = 0$ is assumed as well as $n = 0$ while α' is fitted to the data. The ansatz (9), (10) is motivated by overlap of Gaussian light-cone wavefunctions at large $-t$ and large x and by Regge behavior at low $-t$ and small x (cf. the double distribution ansatz (4) and (6) in the limit $\xi \to 0$). It should be noted that there is a third somewhat different attempt to extract the zero-skewness GPDs from the form factors.[39]

The GPDs H and $\widetilde{H}$ extracted from the form factor data look similar to the corresponding PDFs at low $-t$ while, at larger $-t$ (beyond the zero of the Regge trajectory), all GPDs exhibit a pronounced peak which moves towards $x = 1$ with increasing $-t$. The GPDs H_v^u and H_v^d are both positive while $\widetilde{H}_v^u$ and $\widetilde{H}_v^d$ as well as E_v^u and E_v^d have opposite signs. The double distribution model (4), (6) possess also this property. The signs and sizes of the valence quark GPDs are fixed by the known lowest moments of the GPDs at $\xi = t = 0$ ($e_v^a(x) = E_v^a(x, \xi = 0, t = 0)$)

$$\begin{aligned} \int_0^1 dx\, u(x) = 2\,, \quad \int_0^1 dx\, \Delta u(x) &= 0.926\,, \quad \int_0^1 dx\, e_v^u(x) = 1.67\,, \\ \int_0^1 dx\, d(x) = 1\,, \quad \int_0^1 dx\, \Delta d(x) &= -0.341\,, \quad \int_0^1 dx\, e_v^d(x) = -2.03\,. \end{aligned} \tag{11}$$

The moments of $\widetilde{H}$ are known from β decays, those of E from the anomalous magnetic moments of proton ad neutron. Given that the GPDs are smooth functions without zeros they should reflect the properties of the moments at least at low ξ and low $-t$.

With the valence quark GPDs at hand one may evaluate Ji's sum rule[18] and determine the total and orbital angular momentum the valence quarks carry. Thus, for instance, from the GPDs derived in Ref. 32 one obtains

$$\begin{aligned} L_v^u &= -(0.24 \div 0.27)\,, & J_v^u &= 0.21 \div 0.24\,, \\ L_v^d &= 0.15 \div 0.19\,, & J_v^d &= -0.02 \div 0.02\,. \end{aligned} \tag{12}$$

The opposite signs of L_v^u and L_v^d but nearly the same magnitude are related to the corresponding property of E_v. Fourier transforming the zero-skewness GPDs with respect to the momentum transfer $\mathbf{\Delta}$ ($\mathbf{\Delta}^2 = -t$)[40] one learns about the transverse localization of partons, i.e. about their densities in the hybrid representation of longitudinal momentum fraction and transverse configuration space. One may also evaluate various moments of the GPDs and, with regard to their universality property, they provide the soft physics input for the calculation of hard wide-angle exclusive processes as for instance real Compton scattering, see Sect. 7.

Lattice QCD provides a method to calculate moments of the GPDs. In fact the lowest three moments of the GPDs have been worked out as yet[41] in scenarios with pion masses between 350 and 800 MeV. The extrapolation to the chiral limit has not yet been performed with a sufficient degree of accuracy. In so far the comparison of the lattice results with experiment or other theoretical or phenomenological results is to be done with reservation. Nevertheless, the t dependencies of ratios of moments either obtained from lattice QCD[41] or from phenomenology[32] are surprisingly close each other for $-t \leq 1.2$ GeV^2 and for a large range of pion masses in the lattice calculation. Also the lattice results[41] on the orbital angular momentum are in fair agreement with (12) given the uncertainties in both the approaches.

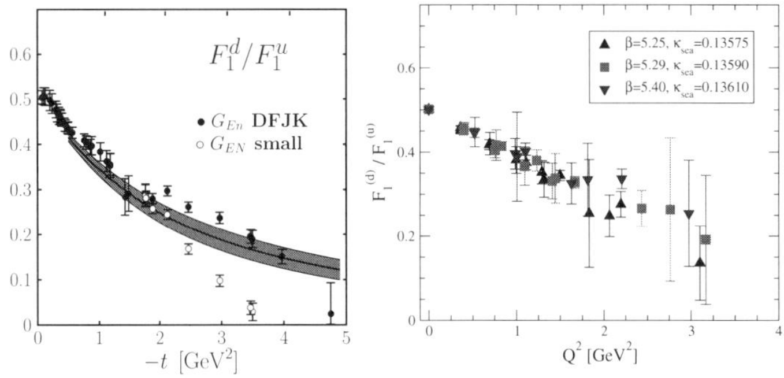

Fig. 3. The ratio of moments F_1^d/F_1^u from Refs. 32 (left) and 42 (right) versus $-t = Q^2$.

A particularly interesting feature of the GPD H is that the ratio of the lowest moments for d and u valence quarks, F_1^d/F_1^u, drops rapidly with increasing $-t$, see Fig. 3. This feature is seen in both the phenomenological[32] and the lattice[42] analysis. It seems also to be demanded by experiment although presently an extrapolation of the neutron's electric form factor is needed for the extraction of these moments from data. The JLab Hall A collaboration (E02-013) will provide data on G_E^n up to about 3.5 GeV^2 in the near future which will render an extrapolation unnecessary. Thus we have an indication that u quarks may dominate over d quarks in the proton form factor at large $-t$, a behavior that corresponds to that of the PDFs at large x:[38] $d_{\rm val}/u_{\rm val} \propto (1-x)^{1.6}$. This $t-x$ correlation of form factors and PDFs is a property of the ansatz (9), (10). Indeed one can show that the moments of the form factors drop as

$$F_1^q \propto \mid t \mid^{-(1+\beta_q)/2}, \tag{13}$$

where β_q is the power of $1-x$ with which the PDFs fall towards $x=1$ (CTEQ6M:[38] $\beta_u = 3.4$, $\beta_d = 5$). These results shed doubts on the assertion that the behavior of the Dirac form factor at intermediate values of momentum transfer is a consequence of dimensional counting.

5. Deeply virtual exclusive scattering

Hard electroproduction of photons, vector mesons (V) and pseudoscalar mesons constitute an important class of processes to which the handbag factorization scheme can be applied to. In fact for these processes rigorous proofs of factorization exist in the limit $Q^2 \to \infty$.[17,43,44] In Fig. 4 typical Feynman graphs are shown which contribute to these processes to leading-twist and LO pQCD accuracy. The dominant helicity amplitudes (ν, ν' (λ, λ') label the helicities of the incoming and outgoing proton (parton), explicite helicities refer to those of photons and mesons) read

$$\mathcal{M}^{\gamma}_{+\nu',+\nu} \sim \sum_a e_a^2 \int_{-1}^{1} d\bar{x} \left[\frac{\mathcal{F}^a_{\nu'\nu} + \widetilde{\mathcal{F}}^a_{\nu'\nu}}{\bar{x} - \xi + i\epsilon} + \frac{\mathcal{F}^a_{\nu'\nu} - \widetilde{\mathcal{F}}^a_{\nu'\nu}}{\bar{x} + \xi - i\epsilon} \right],$$

$$\mathcal{M}^{M(q)}_{0\nu',0\nu} \sim \sum_a \mathcal{C}_V \int_{-1}^{1} d\bar{x} \left[\sum_\lambda \mathcal{H}^{M(q)}_{0\lambda,0\lambda}\, \mathcal{F}^a_{\nu'\nu} + \sum_\lambda 2\lambda \mathcal{H}^{M(q)}_{0\lambda,0\lambda}\, \widetilde{\mathcal{F}}^a_{\nu'\nu} \right], \qquad (14)$$

where

$$\mathcal{F}^a_{\nu\nu} = H^a - \frac{\xi^2}{1-\xi^2}\, E^a\,, \qquad \mathcal{F}^a_{-\nu\nu} = 2\nu \frac{\sqrt{t_0 - t}}{2m(1-\xi^2)}\, E^a\,, \qquad (15)$$

and analogously for $\widetilde{\mathcal{F}}$. For vector-meson production there is an analogous contribution from the gluonic subprocess (see Fig. 4) to be added. Skewness is fixed in electroproduction by Bjorken-x: $\xi \simeq x_{\rm Bj}/(2 - x_{\rm Bj})$ at small $x_{\rm Bj}$. Since the interest lies in small $-t$, the $\gamma^* \to \gamma, V, P$ helicity non-flip transitions dominate. I.e. for the Compton process the transverse-transverse transition is leading while the longitudinal-longitudinal transition obviously dominates for the production of pseudoscalar mesons but also for vector mesons. This is so since the subprocesses shown in Fig. 4, suppress transversely polarized vector mesons. Parity conservation tells us furthermore that $\sum_\lambda \lambda \mathcal{H}_{0\lambda,0\lambda} = 0$ for vector mesons while, for pseudoscalar mesons, $\sum_\lambda \mathcal{H}_{0\lambda,0\lambda} = 0$. In other words, electroproduction of pseudoscalar mesons probes the GPDs $\widetilde{H}$ and $\widetilde{E}$, vector mesons the GPDs H and E to leading-twist order. To DVCS, on the other hand, all four GPDs contribute. The t dependence of the subprocess amplitudes is usually neglected in contrast to that of the GPDs since it provides corrections of order t/Q^2. With regard to flavor it is evident that DVCS probes the valence and sea quark GPDs to LO pQCD, ρ and ω production the gluon GPD in addition. The production of ϕ mesons is sensitive to the gluon and sea GPDs, J/Ψ production only to the gluon GPD since the charm content of the proton is tiny. The production of π^0, on the other hand, is only fed by the valence quark GPDs.

Different experiments probe different regions of ξ: $\simeq 10^{-3}$ by HERA, $\simeq 10^{-2}$ COMPASS, $\simeq 10^{-1}$ HERMES and $\simeq 0.2-0.6$ JLab. Guided by the double distribution model (4), (6), one expects that the role of gluons and sea quarks is diminishing with increasing skewness while that of the valence quarks is increasing. Thus, provided LO handbag physics is dominant at a given hard scale, the study of the mentioned processes over a wide range of $x_{\rm Bj}$ may allow to disentangle the various GPDs.

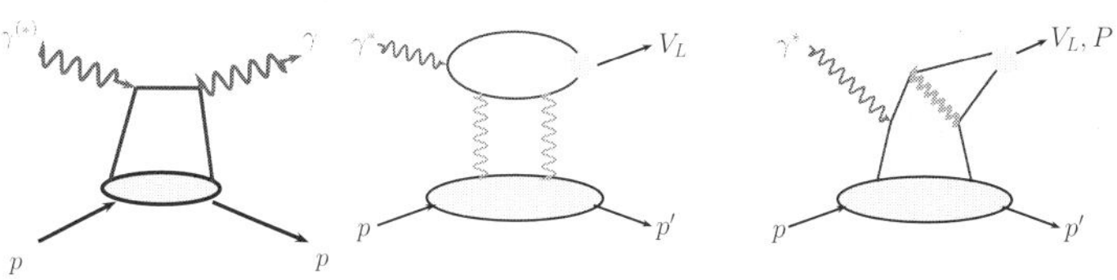

Fig. 4. Typical graphs for deeply virtual electroproduction for γ, V, P.

Vector-meson electroproduction is dominated by the GPD H, the others play a minor role. They are noticeable only in spin asymmetries like A_{LL} or A_{UT} measured with longitudinally polarized beam and target or a transversally polarized target, respectively. This is particularly the case for ρ production, for ω production these effects are much larger. Model estimates indicate that for $\widetilde{H}$ and E the valence quarks dominate for $\xi \geq 0.01$, sea quarks and gluons contributions seem to be small and cancel each other to some extent. Now, for valence quarks the following combinations occur

$$F_v^\rho = e_u F_v^u - e_d F_v^d \qquad F_v^\omega = e_u F_v^u + e_d F_v^d \,, \tag{16}$$

where $F_v = H_v, \widetilde{H}_v, E_v$. Given the signs of the GPDs discussed in Sec. 4, we see that H_v^ρ is large but H_v^ω is small while we have the opposite situation in the case of $\widetilde{H}_v$ and E_v. Thus, ω production is probably a very good case for studying $\widetilde{H}_v$ and E_v. This seems to be a rewarding task for JLab.

5.1. *Deeply virtual Compton scattering*

This process is considered to be the theoretical cleanest one and therefore a lot of theoretical and experimental work is devoted to its investigation. Still it is not a simple process. At NLO there are enhanced corrections from the gluonic GPDs which are particularly large at low ξ and overcompensate the suppression by α_s.[26,45] Another interesting feature of DVCS is the interference with the Bethe-Heitler process for which the final state photon is emitted from the lepton. Since the Bethe-Heitler amplitude is known for given nucleon form factors, the interference region of both the contributions allows to study DVCS at amplitude level. Measurements of $ep \to e\gamma p$ with a polarized beam or target allows to filter out the interference term.[46–48] In Fig. 5 a recent result from the Jlab Hall A collaboration[47] is shown. Clearly seen are the interference regions and a region where the DVCS contribution dominates. Whether the DVCS contribution seen in this experiment can be understood within the handbag frame work is still a pending issue, detailed phenomenological analyses of DVCS data have not yet been performed. Only the HERA DVCS cross section data[49,50] have already been analyzed[25] to NLO with GPDs obtained from the double distribution model (4), (6). Recently methods have been developed that provide fitting schemes to the data by using a kind of partial wave expansions of the DVCS amplitudes.[51] These methods have not yet probed in detail.

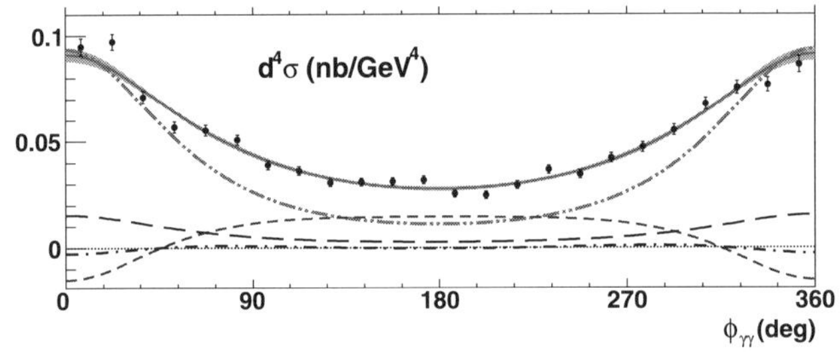

Fig. 5. Left: The cross section for $ep \to e\gamma p$. The dash-dot-dotted line represents the Bethe-Heitler contribution. Data taken from Ref. 47.

5.2. *Electroproduction of mesons*

The disadvantage of meson electroproduction as compared to DVCS is that a second soft hadronic matrix element is required, namely the meson wavefunction or DA. This is to be traded for the advantage of separating the GPDs H and E from $\widetilde{H}$ and $\widetilde{E}$ at leading-twist accuracy. While there is a large set of accurate data available for vector meson electroproduction, only a few data exist as yet for π production.[52] Here I will restrict myself to a few comments on vector-meson electroproduction.

There are several leading-twist, LO pQCD handbag calculations of vector-meson electroproduction[24,28,29] for which the basic graphs are shown in Fig. 4. It turned out that the handbag contribution overestimates the cross section for longitudinal photons ($\gamma_L^* p \to V_L p$) although with the tendency of approaching experiment with increasing Q^2, see Fig. 6. This an example of power corrections that persist up to very large scales.

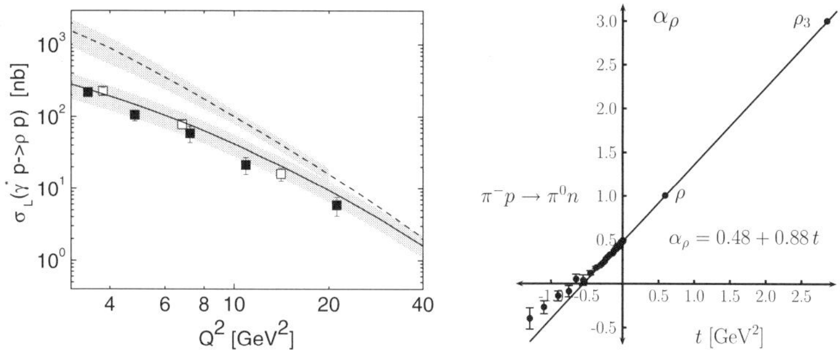

Fig. 6. Left: The longitudinal cross section for ρ production versus Q^2 at $W = 75$ GeV. The solid line represents the handbag predictions,[28] the dashed line the leading-twist contribution. The bands indicate the theoretical uncertainties. Data taken from Ref. 53,54. Right: The ρ Regge trajectory. Cross section data are taken from Ref. 70.

It has recently been shown that NLO corrections[55,56] are very large due to BFKL-type logarithms $\sim \ln 1/\xi$ and cancel to a large extent the LO term at low Q^2 and low $x_{\rm Bj}$. A recent attempt[57] to resum higher orders with methods known from DIS seems to indicate that the sum of all higher order corrections to the LO term is not large. Thus the issue of the size of higher order corrections is still unsettled.

A LO calculation that includes power corrections (modeled by quark transverse momenta) in order to suppress the leading-twist contribution to the $\gamma_L^* p \to V_L p$ amplitude and which also allows to calculate the $\gamma_T^* p \to V_T p$ amplitude is advocated for in Refs. 28,58. Only the subprocesses are caluclated within the modified perturbative approach while the partons are still emitted and reabsorbed by the proton collinearly. The results for $\sigma_L(\rho)$ obtained in Ref. 28 are shown in Fig. 6. With this approach good agreement with experiment is also achieved for the ratio $R = \sigma_L/\sigma_T$, some spin density matrix elements, A_{LL} and the target asymmetry A_{UT}.

Extension of this approach to other transitions is in principle possible. Interestingly, while to the longitudinal amplitude only H and E contribute, the other amplitudes are also fed by $\widetilde{H}$ and $\widetilde{E}$. As shown in Ref. 59 the two types of GPDs H, E and $\widetilde{H}, \widetilde{E}$ lead to special symmetry relations among the helicity amplitudes which are known from the exchange of particles with natural parity (N), and unnatural parity (U), respectively

$$\mathcal{M}^{N(U)}_{-\mu'\nu',-\mu\nu} = (-)(-1)^{\mu'-\mu}\,\mathcal{M}^{N(U)}_{\mu'\nu',\mu\nu}\,. \tag{17}$$

These symmetry relations prevent interferences between N and U type contributions in unpolarized vector-meson electroproduction. Such terms however appear for instance in double spin asymmetries like A_{LL}.

A final remark concerning vector-meson electroproduction is in order. The cross section data[53,54,60,61] reveal an asymmetric minimum at $W \simeq 3-4$ GeV and fixed Q^2. The mild increase of the cross section towards larger energies is well described by the handbag physics but not the sharp increase in the opposite direction. Whether a new dynamical mechanism sets in at low W or whether it is still handbag physics but with more complicated GPDs remains to be seen. The upcoming data on ρ electroproduction from CLAS may help in unravelling the physics in that kinematical region.

6. Alternative approaches to deeply virtual processes

Vector-meson electroproduction has a long history. Its main feature is that it behaves similar to elastic two-body reactions. At the beginning this diffractive nature was understood with the help of vector-meson dominance which views the photon as a superposition of vector mesons and, hence, the process as elastic vector-meson proton scattering. Pomeron exchange, supplemented by subleading Regge poles, lead to a fair description of vector-meson electroproduction at least at low photon virtualities. More complicated versions of the Pomeron (soft and hard ones, BFKL Pomeron) allowed for an extension of the Regge model to larger values of Q^2. Later on the Pomeron was viewed as two gluons[62] which couple perturbatively to the $q\bar{q}$ pair created by the virtual photon. Brodsky *et al*[63] discussed the limit of large Q^2 but small $x_{\rm Bj}$ and showed that the Pomeron-proton vertex is approximately given by the gluon PDF $g(x_{\rm Bj})$. This the so-called leading-log$(1/x_{\rm Bj})$ approximation which has frequently been applied.[64–66] Similar to that approach is the color-dipole model.[67] For the HERA setting of the kinematics, i.e. for $x_{\rm Bj}$ of the order of 10^{-3}, the leading-log approximation is close to the handbag approach, the latter is only enhanced by the skewness effect of about 20%. For larger values of $x_{\rm Bj}$ the leading-log approximation breaks down. It is also not clear how to generalize it to quarks.

There is a renewed interest in Regge ideas, not only for vector-meson electroproduction and the small x behaviour of the PDFs and GPDs but also for π production and even for DVCS. Complete Regge fits to data exist, e.g. Ref. 68,69. The spectrum of hadrons forms linear Regge trajectories (see Sect. 4) which means that $J_j = \alpha_i(t = m_j^2)$ for a hadronic resonance with mass m_j and spin J_j. The remarkable observation is that these Regge trajectories, continued to negative t, describe the energy dependence of the cross sections of soft two-body reactions at small $-t$. For instance, for $\pi^- p \to \pi^0 n$ to which only the ρ trajectory contributes, one finds

$$d\sigma/dt(\pi^- p \to \pi^0 n) \propto s^{2(\alpha_\rho(t)-1)}\,, \tag{18}$$

see Fig. 6. In other cases the cross section is subject to a superposition of several Regge trajectories. To each Regge trajectory a residuum is associated which is free function of t. In spite of this interesting connection between the particle spectrum and the energy dependence of cross sections, the predictiveness of the Regge model is low. It often fails with polarization observables but this can easily be cured by adding other Regge poles and cuts. The Regge model lacks an important property any good model should have - it cannot be disproved.

7. Wide-angle scattering

It has been argued[21,71] that at large $s, -t, -u$ the amplitude for real and virtual ($Q^2 < -t$) Compton scattering factorizes in analogy to DVCS (see Fig. 2). The cross section for real Compton scattering reads in this case

$$\begin{aligned}\frac{d\sigma}{dt} = \frac{d\hat{\sigma}}{dt}\Big\{&\frac{1}{2}\Big[R_V^2(t) + \frac{-t}{4m^2}R_T^2(t) + R_A^2(t)\Big] \\ &-\frac{us}{s^2+u^2}\Big[R_V^2(t) + \frac{-t}{4m^2}R_T^2(t) - R_A^2(t)\Big]\Big\}\,,\end{aligned} \tag{19}$$

($d\hat{\sigma}/dt$ is the Klein-Nishina cross section). Instead of a convolution as in (14) $1/x$ moments of zero-skewness GPDs occur now ($F_{vi} = H_v, \widetilde{H}_v, E_v$)

$$R_i(t) \simeq \sum_{a=u,d} e_a^2 \int_0^1 \frac{dx}{x}\, F_{vi}^a(x,0,t)\,, \tag{20}$$

The tensor form factor R_T describes proton helicity flip.[72] With the zero-skewness GPDs,[32] discussed in Sect. 4, at hand these Compton form factors can be evaluated and the Compton cross section predicted; there is no free parameter. A very good agreement with the recent JLab Hall A data[8] is achieved for sufficiently large Mandelstam variables. The handbag approach also predicts interesting spin effects. For instance, the helicity transfer from the initial photon to the outgoing proton reads

$$K_{LL} \simeq \frac{s^2-u^2}{s^2+u^2}\,\frac{R_A(t)}{R_V(t)}\,. \tag{21}$$

Also this result is in agreement with a JLab measurement.[73] The large positive value of K_{LL} found in Ref. 73, is very difficult to achieve in the ERBL factorization scheme.[74]

For the corresponding time-like process $\gamma\gamma \leftrightarrow p\bar{p}$ a cross section, similar to (19), can be derived[75,76] but the form factors are unknown and have to be extracted from experiment.[10] It turns out that these form factors are larger than the corresponding space-like form factors, a feature that is known from the electromagnetic form factors. The handbag approach accounts for all the features of the BELLE data.[10] It can be extended to other two-photon channels like pairs of hyperons or mesons. It also applies to photoproduction of mesons and to $p\bar{p} \to \gamma M$. As for deeply virtual meson electroprodcution (see Sect. 5.2) there are difficulties with the normalization of the cross sections[77] which have not yet been settled. Other features of these processes, as for instance the ratio of the $\gamma n \to \pi^- p$ and $\gamma p \to \pi^+ n$ measured at Jlab,[78] are quite well understood in the handbag approach.

8. Summary

In this talk I have sketched the factorization schemes in use for hard exclusive scattering processes and discussed their applications in some detail. The main interest has been focussed on the handbag approach since its prospects of becoming the standard description of both the deeply virtual and the wide-angle exclusive processes are best although a detailed comparison between theory and experiment is still pending. With the exception of vector-meson electroproduction for which already a vast amount of data exist, data for hard exclusive processes which cover a wide range of kinematics are still lacking but are excpected to become available in the near future from all pertinent experiments. The upgraded JLab will provide even more data on these processes in a few years. A definite judgement of the handbag approach cannot be given at present. In case that the handbag approach survives the detailed future tests we will learn much about the GPDs and the structure of the proton.

A special case are the valence quark GPDs at zero skewness which, with a few assumptions, can be accessed through the data on the nucleon form factors. JLab is in the position of providing more form factor data in the near future (G_M^n from CLAS, G_E^p from PR01-109, G_E^n from E02-013) which will lead to improved GPDs. From the upgraded JLab more form factor data can be expected that will allow for an extension of the t range in which the zero-skewness GPDs can be extracted. Lattice QCD results on moments of GPDs, provided these are reliably extrapolated to the limit of the physical mass of the pion, may diminish the dependence of these GPDs on the chosen parameterization.

References

1. J. Collins and J. W. Qiu, *Phys. Rev. D* **75**, 114014 (2007).
2. A. V. Efremov and A. V. Radyushkin, *Phys. Lett. B* **94**, 245 (1980).
3. G. P. Lepage and S. J. Brodsky, *Phys. Lett. B* **87**, 359 (1979).
4. A. Duncan and A. H. Mueller, *Phys. Rev. D* **21**, 1636 (1980).
5. V. L. Chernyak and A. R. Zhitnitsky, *Phys. Rept.* **112**, 173 (1984).
6. V. A. Matveev *et al.*, *Lett. Nuovo Cim.* **7**, 719 (1973).
7. S. J. Brodsky and G. R. Farrar, *Phys. Rev. D* **11**, 1309 (1975).
8. A. Danagoulian *et al.* [Hall A Collab.], *Phys. Rev. Lett.* **98**, 152001 (2007).
9. O. Gayou *et al.* [Hall A Collaboration], *Phys. Rev. C* **64**, 038202 (2001); *Phys. Rev. Lett.* **88**, 092301 (2002).
10. C. C. M. Kuo *et al.*, Phys. Lett. B **621**, 41 (2005).
11. N. Isgur and C. H. Llewellyn Smith, *Phys. Rev. Lett.* **52**, 1080 (1984).
12. A. V. Radyushkin, *Nucl. Phys. A* **527**, 153c (1991).
13. H. N. Li and G. Sterman, *Nucl. Phys. B* **381**, 129 (1992).
14. R. Jakob *et al.*, *Phys. Lett. B* **315**, 463 (1993) [*Err.-ibid. B* **319**, 545 (1993)].
15. J. Bolz *et al.*, *Z. Phys. C* **66**, 267 (1995).
16. D. Mueller *et al.*, *Fortsch. Phys.* **42**, 101 (1994).
17. A. V. Radyushkin, *Phys. Lett. B* **385**, 333 (1996).
18. X. D. Ji, *Phys. Rev. Lett.***78**, 610 (1997).
19. J. Gronberg *et al.* [CLEO Collaboration], *Phys. Rev. D* **57**, 33 (1998).
20. A. V. Radyushkin, *Phys. Lett. B* **449**, 81 (1999).
21. M. Diehl, T. Feldmann, R. Jakob and P. Kroll, *Eur. Phys. J. C* **8**, 409 (1999).
22. M. Penttinen, M. V. Polyakov and K. Goeke, *Phys. Rev. D* **62**, 014024 (2000).
23. P. V. Landshoff *et al.*, *Nucl. Phys. B* **28**, 225 (1971).
24. M. Vanderhaeghen *et al.*, *Phys. Rev. D* **60**, 094017 (1999).
25. A. Freund and M. McDermott, Eur. Phys. J. C **23**, 651 (2002).
26. A. V. Belitsky *et al.*, *Phys. Lett. B* **474**, 163 (2000).
27. V. Guzey and T. Teckentrup, *Phys. Rev. D* **74**, 054027 (2006).
28. S. V. Goloskokov and P. Kroll, *Eur. Phys. J. C* **50**, 829 (2007).
29. M. Diehl *et al.*, *Phys. Rev. D* **72**, 034034 (2005) [*Err.-ibid. D* **72**, 059902 (2005)].
30. M. V. Polyakov and C. Weiss, *Phys. Rev. D* **60**, 114017 (1999).
31. M. Guidal *et al.*, *Phys. Rev. D* **72**, 054013 (2005).
32. M. Diehl, T. Feldmann, R. Jakob and P. Kroll, *Eur. Phys. J. C* **39**, 1 (2005).
33. D. S. Armstrong *et al.* [G0 collaboration], *Phys. Rev. Lett.***95**, 092001 (2005).
34. A. Acha *et al.* [HAPPEX collaboration], *Phys. Rev. Lett.* **98**, 032301 (2007).
35. F. E. Maas *et al.* [A4 collaboration], *Phys. Rev. Lett.* **94**, 152001 (2005).
36. R. D. Young *et al.*, arXiv:0704.2618 [hep-ph].
37. H. L. Lai *et al.*, *JHEP* **0704**, 089 (2007).
38. J. Pumplin *et al.*, *JHEP* **0207**, 012 (2002).
39. S. Ahmad *et al.*, *Phys. Rev. D* **75**, 094003 (2007).

40. M. Burkardt, *Int. J. Mod. Phys. A* **18**, 173 (2003).
41. Ph. Hägler *et al.*, arXiv:0705.4295[hep-lat].
42. M. Gockeler *et al.*, PoS **LAT2006**, 120 (2006).
43. X. D. Ji, *Phys. Rev. D* **55**, 7114 (1997).
44. J. C. Collins, L. Frankfurt and M. Strikman, *Phys. Rev. D*56, 2982 (1997).
45. X. D. Ji and J. Osborne, Phys. Rev. D **58**, 094018 (1998).
46. S. Chen *et al.* [CLAS Collaboration], *Phys. Rev. Lett.* **97**, 072002 (2006).
47. C. Munoz Camacho *et al.* [Hall A Collab.], *Phys. Rev. Lett.* **97**, 262002 (2006).
48. A. Airapetian *et al.* [HERMES Collab.], *Phys. Rev. D* **75**, 011103 (2007).
49. C. Adloff *et al.* [H1 Collaboration], *Phys. Lett. B* **517**, 47 (2001).
50. S. Chekanov *et al.* [ZEUS Collaboration], *Phys. Lett. B* **573**, 46 (2003).
51. K. Kumericki, D. Muller and K. Passek-Kumericki, arXiv:hep-ph/0703179.
52. A. Airapetian *et al.* [HERMES collaboration], arXiv:0707.0222.
53. C. Adloff *et al.* [H1 collaboration], *Eur. Phys. J. C***13**, 371 (2000).
54. J. Breitweg *et al.* [ZEUS collaboration], *Eur. Phys. J. C***6**, 603 (1999).
55. D. Yu. Ivanov, L. Szymanowski and G. Krasnikov, JETP Lett. **80**, 226 (2004) [Pisma Zh. Eksp. Teor. Fiz. **80**, 255 (2004)].
56. M. Diehl and W. Kugler, arXiv:0708.1121 [hep-ph].
57. D. Yu. Ivanov, proceedings of the EDS07 conference, Hamburg (2007).
58. S. V. Goloskokov and P. Kroll, arXiv:0708.3569 [hep-ph].
59. S. V. Goloskokov and P. Kroll, *Eur. Phys. J. C* **42**, 281 (2005).
60. C. Hadjidakis *et al.* [CLAS Collaboration], *Phys. Lett. B***605**, 256 (2005).
61. A. Airapetian *et al.* [HERMES collaboration], *Eur. Phys. J. C* **17**, 389 (2000).
62. A. Donnachie and P. V. Landshoff, *Phys. Lett. B***185**,403 (1987).
63. S. J. Brodsky *et al.*, *Phys. Rev. D* **50**, 3134 (1994).
64. L. Frankfurt, W. Koepf and M. Strikman, *Phys. Rev. D* **54**, 3194 (1996).
65. A. D. Martin, M. G. Ryskin and T. Teubner, *Phys. Rev. D* **62**, 014022 (2000).
66. D. Y. Ivanov and R. Kirschner, *Phys. Rev. D* **58**, 114026 (1998).
67. I. P. Ivanov, N. N. Nikolaev and A. A. Savin, *Phys. Part. Nucl.* **37**, 1 (2006).
68. A. Donnachie and H. G. Dosch, *Phys. Lett. B* **502**, 74 (2001).
69. J. M. Laget, arXiv:0708.1250 [hep-ph].
70. V. N. Bolotov *et al.*, *Nucl. Phys. B* **73** (1974) 365.
71. A. V. Radyushkin, *Phys. Rev. D***58**,114008 (1998).
72. H. W. Huang *et al.*, *Eur. Phys. J. C*23, 301 (2002) [*Err.-ibid. C* **31**, 279 (2003)].
73. D. J. Hamilton *et al.* [Hall A Collab.], *Phys. Rev. Lett.***94**, 242001 (2005).
74. R. Thomson, A. Pang and C. R. Ji, *Phys. Rev. D* **73**, 054023 (2006).
75. M. Diehl, P. Kroll and A. Vogt, *Eur. Phys. J. C*26, 567 (2003).
76. A. Freund *et al.*, *Phys. Rev. Lett.***90**, 092001 (2003).
77. H. W. Huang and P. Kroll, *Eur. Phys. J. C*17, 423 (2000).
78. L. Y. Zhu *et al.* [Hall A Collaboration], *Phys. Rev. Lett.***91**, 022003 (2003).

ADS/CFT AND EXCLUSIVE PROCESSES IN QCD

STANLEY J. BRODSKY

Stanford Linear Accelerator Center
Stanford University, Stanford, California 94309
E-mail: sjbth@slac.stanford.edu

GUY F. DE TÉRAMOND

Universidad de Costa Rica
San José, Costa Rica, and
Stanford Linear Accelerator Center
Stanford University, Stanford, California 94309
E-mail: gdt@asterix.crnet.cr

The AdS/CFT correspondence between string theory in AdS space and conformal field theories in physical space-time leads to an analytic, semi-classical model for strongly-coupled QCD which has scale invariance and dimensional counting at short distances and color confinement at large distances. One can use holography to map the amplitude describing the hadronic state in the fifth dimension of Anti-de Sitter space AdS_5 to the light-front wavefunctions of hadrons in physical space-time, thus providing a relativistic description of hadrons in QCD at the amplitude level. In particular, we show that there is an exact correspondence between the fifth-dimensional coordinate of AdS space z and a specific impact variable ζ which measures the separation of the quark and gluonic constituents within the hadron in ordinary space-time. New relativistic light-front equations in ordinary space-time can then be derived which reproduce the results obtained using the 5-dimensional theory. The effective light-front equations possess elegant algebraic structures and integrability properties. This connection between the AdS and the light-front representations allows one to compute the analytic form of the frame-independent light-front wavefunctions, the fundamental entities which encode hadron properties and allow the computation of decay constants, form factors, deeply virtual Compton scattering, exclusive heavy hadron decays and other exclusive scattering amplitudes. As specific examples we compute the pion coupling constant f_π and study the behavior of the pion form factor $F_\pi(q^2)$ in the space and time-like regions. We also determine the Dirac nucleon form factors $F_1^p(q^2)$ and $F_1^n(q^2)$ in the space-like region.

1. Exclusive Processes in QCD

Exclusive processes provide essential tests of the quark and gluon structure of hadrons at the amplitude level and a window into the form of bound-state wavefunctions of QCD. For example, hadron form factors and the generalized distributions which control deeply virtual Compton scattering are expressible in terms of the light-front wavefunctions $\psi_n^H(x_i, \mathbf{k}_{\perp i}, \lambda_i)$, the bound-state eigensolutions of the QCD light-front Hamiltonian.

Light-front wavefunctions are the relativistic generalizations of the familiar Schrödinger wavefunctions of atomic physics, but they are determined at fixed light-front time $\tau = t + z/c$—the "front form" advocated by Dirac[1]—rather than at fixed ordinary time t. The light-front wavefunctions of a hadron are independent of the momentum of the hadron, and they are thus boost invariant; Wigner transformations and Melosh rotations are not required. The light-front formalism for gauge theories in light-cone gauge is particularly useful in that there are no ghosts and one has a direct physical interpretation of orbital angular momentum.

Given the light-front wavefunctions $\psi_{n/h}$ one can compute a large range of hadron observables. For example, the valence and sea quark and gluon distributions which are measured in deep inelastic lepton scattering are defined from the squares of the LFWFS summed over all Fock states n. Form factors, exclusive weak transition amplitudes[2] such as $B \to \ell\nu\pi$. and the generalized parton distributions[3] measured in deeply virtual Compton scattering $\gamma^* p \to \gamma p$ are (assuming the "handbag" approximation) overlaps of the initial and final LFWFS with $n = n'$ and $n = n' + 2$. The gauge-invariant distribution amplitude $\phi_H(x_i, Q)$ defined from the integral over the transverse momenta $\mathbf{k}_{\perp i}^2 \leq Q^2$ of the valence (smallest n) Fock state provides a fundamental measure of the hadron at the amplitude level;[4,5] they are the nonperturbative input to the factorized form of hard exclusive amplitudes and exclusive heavy hadron decays in perturbative QCD. The resulting distributions obey the DGLAP and ERBL evolution equations as a function of the maximal invariant mass, thus providing a physical factorization scheme.[6] In each case, the derived quantities satisfy the appropriate operator product expansions, sum rules, and evolution equations. At large x where the struck quark is far-off shell, DGLAP evolution is quenched,[7] so that the fall-off of the DIS cross sections in Q^2 satisfies Bloom-Gilman inclusive-exclusive duality at fixed W^2.

One of the most significant theoretical advances in recent years has been the application of the AdS/CFT correspondence[8] between string theories defined in 5-dimensional Anti–de Sitter (AdS) space-time and conformal field theories in physical space-time. QCD is not itself a conformal theory; however there are indications, both from theory[9,10] and phenomenology,[11,12] that the QCD coupling is slowly varying at small momentum transfer. In addition, one can argue that if the gluon has a maximum wavelength or an effective mass[13] due to confinement, that gluonic vacuum polarization corrections and the β-function must vanish in the infrared.

If there is a conformal window where the QCD coupling is large and approximately constant and quark masses can be neglected, then QCD resembles a conformal theory, thus motivating the application of AdS/CFT to QCD. Thus, even though QCD is not conformally invariant, one can use the mathematical representation of the conformal group in five-dimensional anti-de Sitter space to construct an analytic first approximation to the theory. The resulting AdS/QCD model gives accurate predictions for hadron spectroscopy and a description of the quark structure of mesons and baryons which has scale invariance and dimensional counting at short distances, together with color confinement at large distances.

As we shall discuss, one can use holography to map the amplitude $\Phi(z)$ describing the hadronic state in the fifth dimension of Anti-de Sitter space AdS_5 to the light-front wavefunctions $\psi_{n/h}$ of hadrons in physical space-time,[14] thus providing a relativistic description of hadrons in QCD at the amplitude level. In fact, there is an exact correspondence between the fifth-dimensional coordinate of anti-de Sitter space z and a specific impact variable ζ in the light-front formalism which measures the physical separation of the constituents within the hadron. One can derive this correspondence by noticing that the mapping of $z \to \zeta$ transforms the expression for the form factors in AdS/CFT to the exact QCD Drell-Yan-West expression in terms of light-front wavefunctions. The amplitude $\Phi(z)$ describing the hadronic state in AdS_5 can then be precisely mapped to the light-front wavefunctions $\psi_{n/h}$ of hadrons in physical space-time.[14] One thus obtains not only an accurate description of the hadron spectrum, but also a simple but realistic model of the valence light-front wavefunctions of mesons, baryons, and glueballs. Thus there is a remarkable mapping between the AdS description of hadrons and the Hamiltonian formulation of QCD in physical space-time quantized at fixed light front time $\tau = t + z/c$.

The AdS/QCD correspondence is particularly relevant for the description of hadronic form factors, since it incorporates the connection between the twist of the hadron to the fall-off of its current matrix elements, as well as essential aspects of vector meson dominance. It also provides a convenient framework for analytically continuing the space-like results to the time-like region. Recent applications to the form factors of mesons and nucleons [15–21] have followed from the pioneering work of refs. 22,23.

2. The AdS/CFT Duality

The essential principle underlying the AdS/CFT approach to conformal gauge theories is the isomorphism of the group of Poincare' and conformal transformations $SO(4,2)$ to the group of isometries of Anti-de Sitter space. The AdS metric is

$$ds^2 = \frac{R^2}{z^2}(\eta_{\mu\nu} dx^\mu dx^\nu - dz^2), \tag{1}$$

which is invariant under scale changes of the coordinate in the fifth dimension $z \to \lambda z$ and $x_\mu \to \lambda x_\mu$. Thus one can match scale transformations of the theory in $3+1$ physical space-time to scale transformations in the fifth dimension z. In the AdS/CFT duality, the amplitude $\Phi(z)$ represents the extension of the hadron into the additional fifth dimension. The behavior of $\Phi(z) \to z^\Delta$ at $z \to 0$ matches the twist-dimension of the hadron at short distances $x^2 \to 0$.

2.1. *The Hard-Wall Holographic Model*

As shown by Polchinski and Strassler,[24] one can simulate confinement by imposing boundary conditions in the holographic variable at $z = z_0 = 1/\Lambda_{\rm QCD}$. Confinement can also be introduced by modifying the AdS metric to mimic a confining potential. The resulting models, although *ad hoc*, provide a simple semi-classical approximation to QCD which incorporate both constituent counting rule behavior at short distances and confinement at large distances.

This simple approach, which has been described as a "bottom-up" approach, has been successful in obtaining general properties of scattering amplitudes of hadronic bound states at strong coupling,[24,25] the low-lying hadron spectra[26–29] and hadron couplings and chiral symmetry breaking.[30–33] The gauge theory/gravity duality also provides a convenient framework for the description of deep inelastic scattering structure functions at small x[34] and a unified description of hard and soft pomeron physics.[35] Recent applications to describe chiral symmetry breaking[36] and other meson and baryon properties, have also been carried out within the framework of a top-bottom approach to AdS/CFT using higher dimensional branes.[37]

2.2. *The Soft-Wall Holographic Model*

The predicted mass spectrum in the truncated space hard-wall (HW) model is linear $M \propto L$ at high orbital angular momentum L, in contrast to the quadratic dependence $M^2 \propto L$ in the usual Regge parameterization. It has been shown recently that by choosing a specific profile for a non-constant dilaton, the usual Regge dependence can be obtained.[38] The procedure allows one to retain conformal AdS metrics (1) and to introduce a smooth cutoff which depends on the profile of a dilaton background field φ

$$S = \int d^4x\, dz\, \sqrt{g}\, e^{-\varphi(z)} \mathcal{L}, \tag{2}$$

where φ is a function of the holographic coordinate z which vanishes in the ultraviolet limit $z \to 0$. The IR hard-wall or truncated space holographic model, discussed in the previous section, corresponds to a constant dilaton field in the confining region $\varphi(z) = \varphi_0$, $z \leq 1/\Lambda_{\rm QCD}$, and to very large values elsewhere: $\varphi(z) \to \infty$ for $z > 1/\Lambda_{\rm QCD}$. The introduction of a soft cutoff avoids the ambiguities in the choice of boundary conditions at the infrared wall. A convenient choice[38] for the background field with usual Regge behavior is $\varphi(z) = \kappa^2 z^2$. The resulting wave equations are equivalent to the radial equation of a two-dimensional oscillator previously found in the context of mode propagation on $AdS_5 \times S^5$ in the light-cone formulation of Type II supergravity.[39]

3. Light-Front Fock Representation

The light-front expansion is constructed by quantizing QCD at fixed light-cone time[1] $\tau = t + z/c$ and forming the invariant light-front Hamiltonian: $H_{LF}^{QCD} = P^+P^- - \vec{P}_\perp^2$ where $P^\pm = P^0 \pm P^z$.[40] The momentum generators P^+ and $\vec{P}_\perp$ are kinematical; *i.e.*, they are independent of the interactions. The generator $P^- = i\frac{d}{d\tau}$ generates light-cone time translations, and the eigen-spectrum of the Lorentz scalar H_{LF}^{QCD} gives the mass spectrum of the color-singlet hadron states in QCD: $H_{LC}\,|\psi_h\rangle = \mathcal{M}_h^2\,|\psi_h\rangle$. The state $|\psi_h\rangle$ is an expansion in multi-particle Fock eigenstates $\{\,|n\rangle\}$ of the free light-front Hamiltonian: $|\psi_h\rangle = \sum_n \psi_{n/h}|\psi_h\rangle$; the projection of the eigensolution on the free Fock basis gives the hadronic light-front wavefunctions. The coefficients of the Fock expansion $\psi_{n/h}(x_i, \mathbf{k}_{\perp i}, \lambda_i)$ are independent of the total momentum P^+ and $\mathbf{P}_\perp$ of the hadron and depend only on the relative partonic coordinates, the longitudinal momentum fraction $x_i = k_i^+/P^+$, the relative transverse momentum $\mathbf{k}_{\perp i}$ and λ_i, the constituent's spin along the z direction. Momentum conservation requires $\sum_{i=1}^n x_i = 1$ and $\sum_{i=1}^n \mathbf{k}_{\perp i} = 0$.

3.1. *Light-Front Wavefunctions in Impact Space*

The holographic mapping of hadronic LFWFs to AdS string modes is most transparent when one uses the impact parameter space representation.[41] The total position coordinate of a hadron or its transverse center of momentum $\mathbf{R}_\perp$, is defined in terms of the energy momentum tensor $T^{\mu\nu}$

$$\mathbf{R}_\perp = \frac{1}{P^+} \int dx^- \int d^2\mathbf{x}_\perp \, T^{++} \, \mathbf{x}_\perp . \tag{3}$$

In terms of partonic transverse coordinates

$$x_i \mathbf{r}_{\perp i} = x_i \mathbf{R}_\perp + \mathbf{b}_{\perp i}, \tag{4}$$

where the $\mathbf{r}_{\perp i}$ are the physical transverse position coordinates and $\mathbf{b}_{\perp i}$ frame independent internal coordinates, conjugate to the relative coordinates $\mathbf{k}_{\perp i}$. Thus, $\sum_{i=1}^{n} \mathbf{b}_{\perp i} = 0$ and $\mathbf{R}_\perp = \sum_{i=1}^{n} x_i \mathbf{r}_{\perp i}$. The LFWF $\psi_n(x_j, \mathbf{k}_{\perp j})$ can be expanded in terms of the $n-1$ independent coordinates $\mathbf{b}_{\perp j}$, $j = 1, 2, \ldots, n-1$

$$\psi_n(x_j, \mathbf{k}_{\perp j}) = (4\pi)^{\frac{(n-1)}{2}} \prod_{j=1}^{n-1} \int d^2\mathbf{b}_{\perp j} \exp\Big(i \sum_{j=1}^{n-1} \mathbf{b}_{\perp j} \cdot \mathbf{k}_{\perp j}\Big) \widetilde{\psi}_n(x_j, \mathbf{b}_{\perp j}). \tag{5}$$

The normalization is defined by

$$\sum_n \prod_{j=1}^{n-1} \int dx_j d^2\mathbf{b}_{\perp j} \left| \widetilde{\psi}_n(x_j, \mathbf{b}_{\perp j}) \right|^2 = 1. \tag{6}$$

One of the important advantages of the light-front formalism is that current matrix elements can be represented without approximation as overlaps of light-front wavefunctions. In the case of the elastic space-like form factors, the matrix element of the J^+ current only couples Fock states with the same number of constituents. If the charged parton n is the active constituent struck by the current, and the quarks $i = 1, 2, \ldots, n-1$ are spectators, then the Drell-Yan West formula[42–44] in impact space is

$$F(q^2) = \sum_n \prod_{j=1}^{n-1} \int dx_j d^2\mathbf{b}_{\perp j} \exp\Big(i\mathbf{q}_\perp \cdot \sum_{j=1}^{n-1} x_j \mathbf{b}_{\perp j}\Big) \left| \widetilde{\psi}_n(x_j, \mathbf{b}_{\perp j}) \right|^2, \tag{7}$$

corresponding to a change of transverse momenta $x_j \mathbf{q}_\perp$ for each of the $n-1$ spectators. This is a convenient form for comparison with AdS results, since the form factor is expressed in terms of the product of light-front wave functions with identical variables.

4. Light-Front Mapping of String Amplitudes

We can now establish an explicit connection between the AdS/CFT and the LF formulae. To make more transparent the holographic connection between AdS_5 and the conformal quantum field theory defined at its asymptotic $z \to 0$ boundary, it is convenient to write the AdS metric (1) in terms of light front coordinates $x^\pm = x^0 \pm x^3$

$$ds^2 = \frac{R^2}{z^2}\left(dx^+dx^- - d\mathbf{x}_\perp^2 - dz^2\right). \tag{8}$$

It is also useful to express (7) in terms of an effective single particle transverse distribution $\widetilde{\rho}$ [14]

$$F(q^2) = 2\pi \int_0^1 dx \frac{(1-x)}{x} \int \zeta d\zeta \, J_0\left(\zeta q \sqrt{\frac{1-x}{x}}\right) \tilde{\rho}(x,\zeta), \tag{9}$$

where we have introduced the variable

$$\zeta = \sqrt{\frac{x}{1-x}} \left|\sum_{j=1}^{n-1} x_j \mathbf{b}_{\perp j}\right|, \tag{10}$$

representing the x-weighted transverse impact coordinate of the spectator system. On the other hand, the expression for the form factor in AdS space is represented as the overlap in the fifth dimension coordinate z of the normalizable modes dual to the incoming and outgoing hadrons, Φ_P and $\Phi_{P'}$, with the non-normalizable mode, $J(Q,z) = zQK_1(zQ)$, dual to the external source[34]

$$F(Q^2) = R^3 \int \frac{dz}{z^3} \, \Phi_{P'}(z) J(Q,z) \Phi_P(z). \tag{11}$$

If we compare (9) in impact space with the expression for the form factor in AdS space (11) for arbitrary values of Q using the identity

$$\int_0^1 dx \, J_0\left(\zeta Q \sqrt{\frac{1-x}{x}}\right) = \zeta Q K_1(\zeta Q), \tag{12}$$

then we can identify the spectator density function appearing in the light-front formalism with the corresponding AdS density

$$\tilde{\rho}(x,\zeta) = \frac{R^3}{2\pi} \frac{x}{1-x} \frac{\left|\Phi(\zeta)\right|^2}{\zeta^4}. \tag{13}$$

Equation (13) gives a precise relation between string modes $\Phi(\zeta)$ in AdS_5 and the QCD transverse charge density $\tilde{\rho}(x,\zeta)$. The variable ζ represents a measure of the transverse separation between point-like constituents, and it is also the holographic variable z characterizing the string scale in AdS. Consequently the AdS string mode $\Phi(z)$ can be regarded as the probability amplitude to find n partons at transverse impact separation $\zeta = z$. Furthermore, its eigenmodes determine the hadronic spectrum.[14]

In the case of a two-parton constituent bound state, the correspondence between the string amplitude $\Phi(z)$ and the light-front wave function $\widetilde{\psi}(x, \mathbf{b})$ is expressed in closed form[14]

$$\left|\widetilde{\psi}(x,\zeta)\right|^2 = \frac{R^3}{2\pi}\, x(1-x)\, \frac{|\Phi(\zeta)|^2}{\zeta^4}, \tag{14}$$

where $\zeta^2 = x(1-x)\mathbf{b}_\perp^2$. Here $\mathbf{b}_\perp$ is the impact separation and Fourier conjugate to $\mathbf{k}_\perp$.

In presence of the dilaton background $\varphi = \kappa^2 z^2$ the form factor in AdS space has the additional term $e^{-\kappa^2 z^2}$ in the metric

$$F(Q^2) = R^3 \int \frac{dz}{z^3}\, e^{-\kappa^2 z^2}\, \Phi(z) J_\kappa(Q,z) \Phi(z), \tag{15}$$

to be properly normalized to the charge operator at $Q = 0$. The non-normalizable electromagnetic mode $J_\kappa(Q,z)$ in (15)

$$J_\kappa(Q,z) = \Gamma\left(1 + \frac{Q^2}{4\kappa^2}\right) U\left(\frac{Q^2}{4\kappa^2}, 0, \kappa^2 z^2\right), \tag{16}$$

couples to the dilaton field as it propagates in AdS space and $U(a,b,c)$ is the confluent hypergeometric function with the integral representation

$$\Gamma(a) U(a,b,z) = \int_0^\infty e^{-zt} t^{a-1} (1+t)^{b-a-1} dt. \tag{17}$$

In the large Q^2 limit, $Q^2 \gg 4\kappa^2$ we find that $J_\kappa(Q,z) \to zQK_1(zQ)$. Thus, for large transverse momentum the current decouples from the dilaton background, and we recover our previous results for the ultraviolet behavior of matrix elements. When summed over all Fock states the Drell-Yan-West (DYW) formula gives an exact result. The formula describes the coupling of the free electromagnetic current to the elementary constituents in the interaction representation. In the presence of a dilaton field in AdS space, the electromagnetic AdS mode is no longer dual to a the free quark current, but dual to a dressed current, *i.e.*, a hadronic electromagnetic current including virtual $\overline{q}q$ pairs and thus confined. We thus expect that the modified mapping corresponds to the presence of higher Fock states in the hadron.

4.1. *Holographic Light-Front Representation*

The mapping of z from AdS space to ζ in the LF space allows the equations of motion in AdS space to be recast in the form of a light-front Hamiltonian equation[40]

$$H_{LF} \mid \phi\rangle = \mathcal{M}^2 \mid \phi\rangle, \tag{18}$$

a remarkable result which allows the discussion of the AdS/CFT solutions in terms of light-front equations in physical 3+1 space-time. By substituting $\phi(\zeta) = \left(\frac{\zeta}{R}\right)^{-3/2} \Phi(\zeta)$, in the AdS wave equation describing the propagation of scalar modes in AdS space

$$\left[z^2 \partial_z^2 - (d-1) z\, \partial_z + z^2 \mathcal{M}^2 - (\mu R)^2\right] \Phi(z) = 0, \tag{19}$$

we find an effective Schrödinger equation as a function of the weighted impact variable ζ

$$\left[-\frac{d^2}{d\zeta^2} + V(\zeta)\right] \phi(\zeta) = \mathcal{M}^2 \phi(\zeta), \tag{20}$$

with the effective potential $V(\zeta) \to -(1-4L^2)/4\zeta^2$ in the conformal limit, where we identity ζ with the fifth dimension z of AdS space: $\zeta = z$. We have written above $(\mu R)^2 = -4 + L^2$. The solution to (20) is $\phi(z) = z^{-\frac{3}{2}} \Phi(z) = C z^{\frac{1}{2}} J_L(z\mathcal{M})$. This equation reproduces the AdS/CFT solutions for mesons with relative orbital angular momentum L. The holographic hadronic light-front wave functions $\phi(\zeta) = \langle \zeta | \phi \rangle$ are normalized according to

$$\langle \phi | \phi \rangle = \int d\zeta \, |\langle \zeta | \phi \rangle|^2 = 1, \tag{21}$$

and represent the probability amplitude to find n-partons at transverse impact separation $\zeta = z$. Its eigenmodes determine the hadronic mass spectrum.

The effective wave equation (20) is a relativistic light-front equation defined at $x^+ = 0$. The AdS metric ds^2 (8) is invariant if $\mathbf{x}_\perp^2 \to \lambda^2 \mathbf{x}_\perp^2$ and $z \to \lambda z$ at equal light-front time. The Casimir operator for the rotation group $SO(2)$ in the transverse light-front plane is L^2. This shows the natural holographic connection to the light front.

The lowest stable state $L = 0$ is determined by the Breitenlohner-Freedman bound.[45] Its eigenvalues are set by the boundary conditions at $\phi(z = 1/\Lambda_{\rm QCD}) = 0$ and are given in terms of the roots of Bessel functions: $\mathcal{M}_{L,k} = \beta_{L,k} \Lambda_{\rm QCD}$. Normalized LFWFs $\widetilde{\psi}_{L,k}$ follow from (14)

$$\widetilde{\psi}_{L,k}(x, \zeta) = B_{L,k} \sqrt{x(1-x)} J_L\left(\zeta \beta_{L,k} \Lambda_{\rm QCD}\right) \theta\big(z \le \Lambda_{\rm QCD}^{-1}\big), \tag{22}$$

where $B_{L,k} = \Lambda_{\rm QCD}/\sqrt{\pi} J_{1+L}(\beta_{L,k})$. The resulting wavefunctions depicted in Fig. 1 display confinement at large interquark separation and conformal symmetry at short distances, reproducing dimensional counting rules for hard exclusive processes and the scaling and conformal properties of the LFWFs at high relative momenta in agreement with perturbative QCD.

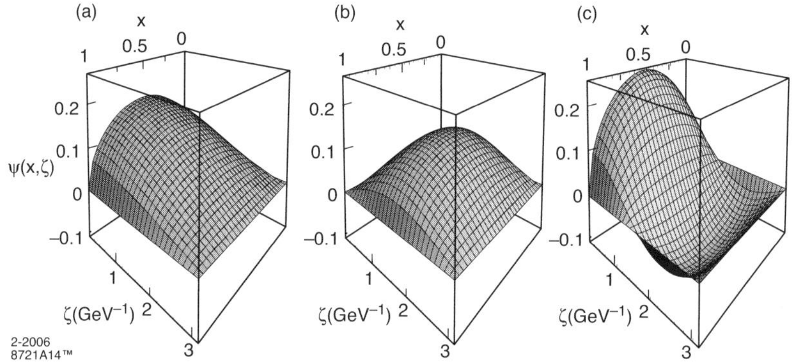

Fig. 1. AdS/QCD Predictions for the light-front wavefunctions of a meson in the hard-wall model: (a) $n = 0, L = 0$, (b) $n = 0, L = 1$, (c) $n = 1, L = 0$.

Since they are complete and orthonormal, these AdS/CFT model wavefunctions can be used as an initial ansatz for a variational treatment or as a basis for the diagonalization of the light-front QCD Hamiltonian. We are now in fact investigating this possibility with J. Vary and A. Harinandrath. Alternatively, one can introduce confinement by adding a two-dimensional harmonic oscillator potential $\kappa^4\zeta^2$ in the LF transverse plane to the conformal kernel in Eq. (20). One can also introduce nonzero quark masses for the meson. The procedure is straightforward in the $\mathbf{k}_\perp$ representation by using the substitution $\frac{\mathbf{k}_\perp^2}{x(1-x)} \to \frac{\mathbf{k}_\perp^2+m_1^2}{x} + \frac{\mathbf{k}_\perp^2+m_2^2}{1-x}$.

5. Integrability of AdS/CFT Equations

The integrability methods of Ref. [46] find a remarkable application in the AdS/CFT correspondence. Integrability follows if the equations describing a physical model can be factorized in terms of linear operators. These ladder operators generate all the eigenfunctions once the lowest mass eigenfunction is known. In holographic QCD, the conformally invariant $3+1$ light-front differential equations can be expressed in terms of ladder operators and their solutions can then be expressed in terms of analytical functions. In the conformal limit the ladder algebra for bosonic (B) or fermionic (F) modes is given in terms of the operator ($\Gamma^B = 1, \;\; \Gamma^F = \gamma_5$)

$$\Pi_\nu^{B,F}(\zeta) = -i\left(\frac{d}{d\zeta} - \frac{\nu+\frac{1}{2}}{\zeta}\,\Gamma^{B,F}\right), \tag{23}$$

and its adjoint

$$\Pi_\nu^{B,F}(\zeta)^\dagger = -i\left(\frac{d}{d\zeta} + \frac{\nu+\frac{1}{2}}{\zeta}\,\Gamma^{B,F}\right), \tag{24}$$

with commutation relations

$$\left[\Pi_\nu^{B,F}(\zeta), \Pi_\nu^{B,F}(\zeta)^\dagger\right] = \frac{2\nu+1}{\zeta^2}\,\Gamma^{B,F}. \tag{25}$$

For $\nu \geq 0$ the Hamiltonian is written as a bilinear form $H_{LC}^{B,F} = \Pi_\nu^{B,F\,\dagger}\Pi_\nu^{B,F}$. In the fermionic case the eigenmodes also satisfy a first order LF Dirac equation. For bosonic modes, the lowest stable state $\nu = 0$ corresponds to the Breitenlohner-Freedman bound. Higher orbital states are constructed from the L-th application of the raising operator $a^\dagger = -i\Pi^B$ on the ground state.

6. Hadronic Spectra in AdS/QCD

The holographic model based on truncated AdS space can be used to obtain the hadronic spectrum of light quark $q\bar{q}, qqq$ and gg bound states. Specific hadrons are identified by the correspondence of the amplitude in the fifth dimension with the twist dimension of the interpolating operator for the hadron's valence Fock state, including its orbital angular momentum excitations. Bosonic modes with conformal dimension $2 + L$ are dual to the interpolating operator $\mathcal{O}_{\tau+L}$ with $\tau = 2$. For fermionic modes $\tau = 3$. For example, the set of three-quark baryons with spin $1/2$ and higher is described by the light-front Dirac equation

$$\left(\alpha\,\Pi^F(\zeta) - \mathcal{M}\right)\psi(\zeta) = 0, \tag{26}$$

where $i\alpha = \begin{pmatrix} 0 & I \\ -I & 0 \end{pmatrix}$ in the Weyl representation. The solution is

$$\psi(\zeta) = C\sqrt{\zeta}\left[J_{L+1}\left(\zeta\mathcal{M}\right)u_+ + J_{L+2}\left(z\mathcal{M}\right)u_-\right], \tag{27}$$

with $\gamma_5 u_\pm = u_\pm$. A discrete four-dimensional spectrum follows when we impose the boundary condition $\psi_\pm(\zeta = 1/\Lambda_{\rm QCD}) = 0$: $\mathcal{M}_{\alpha,k}^+ = \beta_{\alpha,k}\Lambda_{\rm QCD}$, $\mathcal{M}_{\alpha,k}^- = \beta_{\alpha+1,k}\Lambda_{\rm QCD}$, with a scale-independent mass ratio.[28]

Figure 2(a) shows the predicted orbital spectrum of the nucleon states and Fig. 2(b) the Δ orbital resonances. The spin-3/2 trajectories are determined from the corresponding Rarita-Schwinger equation. The solution of the spin-3/2 for polarization along Minkowski coordinates, ψ_μ, is similar to the spin-1/2 solution. The data for the baryon spectra are from [47]. The internal parity of states is determined from the SU(6) spin-flavor symmetry. Since only one parameter, the QCD mass scale Λ_{QCD}, is introduced, the

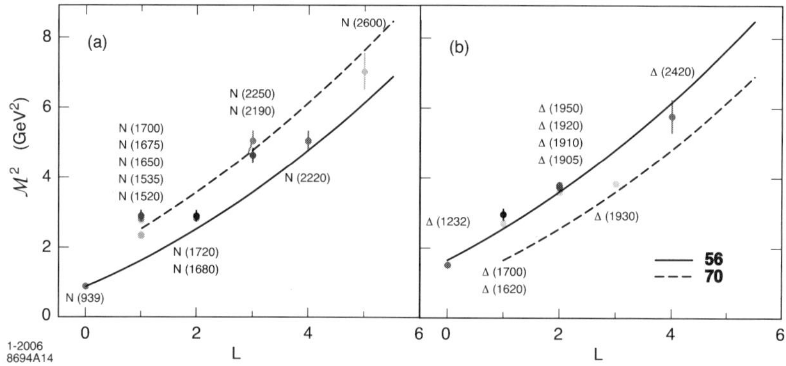

Fig. 2. Predictions for the light baryon orbital spectrum for $\Lambda_{QCD} = 0.25$ GeV. The **56** trajectory corresponds to L even $P = +$ states, and the **70** to L odd $P = -$ states.

agreement with the pattern of physical states is remarkable. In particular, the ratio of Δ to nucleon trajectories is determined by the ratio of zeros of Bessel functions. The predicted mass spectrum in the truncated space model is linear $M \propto L$ at high orbital angular momentum, in contrast to the quadratic dependence $M^2 \propto L$ in the usual Regge parameterization. One can obtain $M^2 \propto (L+n)$ dependence in the holographic model by the introduction of a harmonic potential $\kappa^2 z^2$ in the AdS wave equations.[38] This result can also be obtained by extending the conformal algebra written above. An account of the extended algebraic holographic model and a possible supersymmetric connection between the bosonic and fermionic operators used in the holographic construction will be described elsewhere.

7. Analytic Results for Hadronic Observables in AdS/QCD

7.1. *Hadronic Form Factors of Arbitrary Twist*

A string mode ϕ_τ which couples to a local hadronic interpolating operator of twist τ defined at the asymptotic boundary of AdS space has scaling dimension τ, $\Phi_\tau \sim z^\tau$, as $z \to 0$. In the soft-wall model a normalized string mode Φ_τ is given by

$$\Phi_\tau(z) = \frac{1}{R^{3/2}} \sqrt{\frac{2}{\Gamma(\tau - 1)}}\, \kappa^{\tau-1} z^\tau. \tag{28}$$

for $n = 0$. It is convenient to perform the change of variable $t = \frac{x}{1-x}$ in the integral representation for the bulk-to-boundary propagator (16). Upon integration by parts the resulting expression, there follows

$$J_\kappa(Q, z) = \kappa^2 z^2 \int_0^1 \frac{dx}{(1-x)^2}\, x^{\frac{Q^2}{4\kappa^2}}\, e^{-\kappa^2 z^2 x/(1-x)}, \tag{29}$$

the result found by Grigoryan and Radyushkin in Ref. 17. We can then find an analytical expression for the form factor of a hadronic state of arbitrary twist τ by substituting the integral representation (29) for $J_\kappa(q, z)$ in (15). We find the result

$$F(Q^2) = \Gamma(\tau) \frac{\Gamma\left(1 + \frac{Q^2}{4\kappa^2}\right)}{\Gamma\left(\tau + \frac{Q^2}{4\kappa^2}\right)}. \tag{30}$$

In the absence of anomalous dimensions, the twist is an integer, $\tau = N$, and we can simplify (30) by using the recurrence formula

$$\Gamma(N + z) = (N - 1 + z)(N - 2 + z) \ldots (1 + z)\Gamma(1 + z). \tag{31}$$

We find

$$F(Q^2) = \frac{1}{1 + \frac{Q^2}{4\kappa^2}}, \quad N = 2, \tag{32}$$

$$F(Q^2) = \frac{2}{\left(1 + \frac{Q^2}{4\kappa^2}\right)\left(2 + \frac{Q^2}{4\kappa^2}\right)}, \quad N = 3, \tag{33}$$

$$\cdots$$

$$F(Q^2) = \frac{(N-1)!}{\left(1 + \frac{Q^2}{4\kappa^2}\right)\left(2 + \frac{Q^2}{4\kappa^2}\right) \cdots \left(N - 1 + \frac{Q^2}{4\kappa^2}\right)}, \quad N, \tag{34}$$

which is expressed as a product of $N - 1$ poles, corresponding to the first $N - 1$ states along the vector meson radial trajectory. For large Q^2 it follows that

$$F(Q^2) \to (N-1)! \left[\frac{4\kappa^2}{Q^2}\right]^{(N-1)}, \tag{35}$$

and we recover the conformal power-law counting rules for hard scattering.[48–50]

7.2. *Meson Form Factors*

As we have discussed, hadron form factors can be predicted in AdS from the overlap of the normalizable modes dual to the incoming and outgoing hadrons with the bulk-to-boundary propagator, $J(Q, z)$, dual to the external source, or equivalently by using the DYW formula in physical space-time. For example, we can compute the pion form factor from the AdS expressions (11) and (15) for the hadronic string modes Φ_π in the hard-wall (HW)

$$\Phi_\pi^{HW}(z) = \frac{\sqrt{2}\Lambda_{QCD}}{R^{3/2}J_1(\beta_{0,1})} z^2 J_0\left(z\beta_{0,1}\Lambda_{QCD}\right), \tag{36}$$

and soft-wall (SW) model

$$\Phi_\pi^{SW}(z) = \frac{\sqrt{2}\kappa}{R^{3/2}} z^2, \tag{37}$$

respectively. Since the pion mode couples to a twist-two boundary interpolating operator which creates a two-component hadronic bound state, the form factor is given in the SW model by the simple monopole form (32) corresponding to $n = 2$

$$F_\pi(Q^2) = \frac{4\kappa^2}{4\kappa^2 + Q^2}. \tag{38}$$

The hadronic scale is evaluated by fitting the space-like data for the form factor as shown in Figure 3, where we plot the product $Q^2 F_\pi(Q^2)$ for the soft and hard-wall holographic models. Both models would seem to describe the overall behavior of the space-like data; however, when the low energy data is examined in detail, the SW model gives a noticeable better description as shown in Figure 4. When the results for the pion form factor are analytically continued to the time-like region, $q^2 \to -q^2$ we obtain the results shown in Figure 5 for $\log\left(|F_\pi(q^2)|\right)$. The monopole form of the SW model exhibits a pole at the ρ mass and reproduces well the ρ peak with $M_\rho = 4\kappa^2 = 750$ MeV. In the strongly coupled semiclassical gauge/gravity limit hadrons have zero widths and are stable. The form factor accounts for the scaling behavior in the space-like region, but does not give rise to the additional structure found in the time-like region, since the ρ pole saturates 100% the monopole form (38).

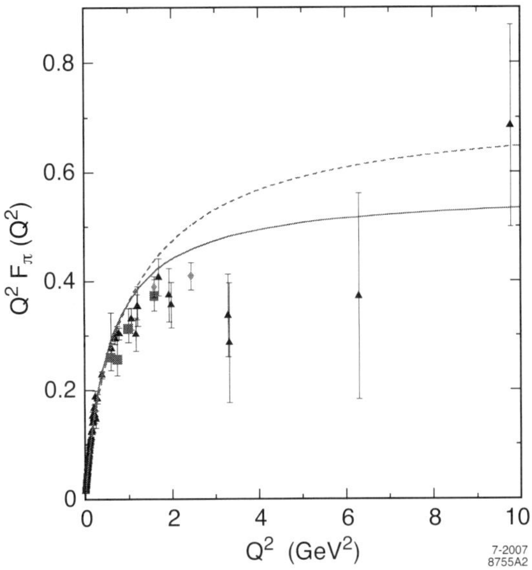

Fig. 3. Space-like scaling behavior for $Q^2F_\pi(Q^2)$ as a function of $Q^2 = -q^2$. The continuous line is the prediction of the soft-wall model for $\kappa = 0.375$ GeV. The dashed line is the prediction of the hard-model for $\Lambda_{QCD} = 0.22$ GeV. The triangles are the data compilation from Baldini *et al.*,[51] the boxes are JLAB 1 data[52] and diamonds are JLAB 2 data.[53]

7.3. *Contributions to Meson Form Factors and Structure Functions at Large Momentum Transfer in AdS/QCD*

The form factor of a hadron at large Q^2 arises from the small z kinematic domain in AdS space. According to the AdS/CFT duality, this corresponds to small distances $x^\mu x_\mu \sim 1/Q^2$ in physical space-time, the domain where the current matrix elements are controlled by the conformal twist-dimension, Δ, of the hadron's interpolating operator. In the case of the front form, where $x^+ = 0$, this corresponds to small transverse separation $x^\mu x_\mu = -\mathbf{x}_\perp^2$.

As we have shown,[14] one can use holography to map the functional from of the string modes $\Phi(z)$ in AdS space to the light front wavefunctions in physical space time by identifying z with the transverse variable $\zeta = \sqrt{\frac{x}{1-x}}|\vec{\eta}_\perp|$. Here $\vec{\eta}_\perp = \sum_{i=1}^{n-1} x_i \mathbf{b}_{\perp i}$ is the weighted impact separation, summed over the impact separation of the spectator constituents. The leading large-Q^2 behavior of form factors in AdS/QCD arises from small $\zeta \sim 1/Q$, corresponding to small transverse separation.

For the case of a meson with two constituents the form factor can be

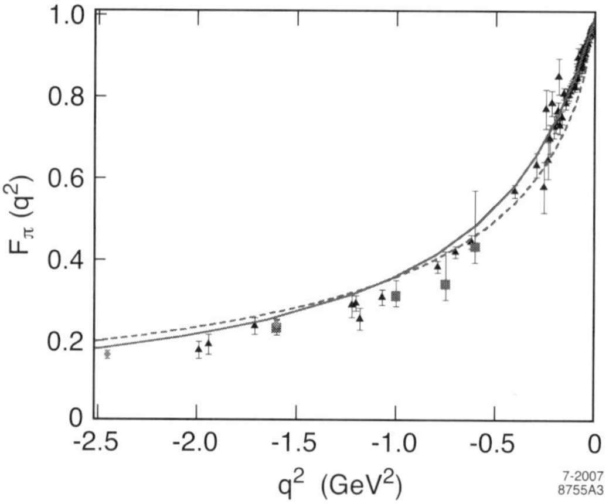

Fig. 4. Space-like behavior of the pion form factor $F_\pi(q^2)$ as a function of q^2 for $\kappa = 0.375$ GeV and $\Lambda_{QCD} = 0.22$ GeV. Continuous line: soft-wall model, dashed line: hard-wall model. Triangles are the data compilation from Baldini *et al.*,[51] boxes are JLAB 1[52] and diamonds are JLAB 2.[53]

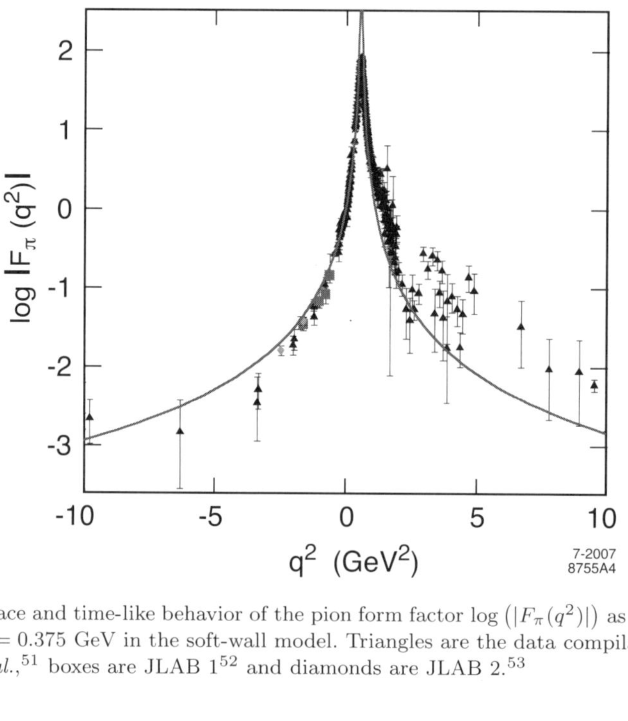

Fig. 5. Space and time-like behavior of the pion form factor $\log\left(|F_\pi(q^2)|\right)$ as a function of q^2 for $\kappa = 0.375$ GeV in the soft-wall model. Triangles are the data compilation from Baldini *et al.*,[51] boxes are JLAB 1[52] and diamonds are JLAB 2.[53]

written in terms of an effective light-front transverse density in impact space

$$F(q^2) = \int_0^1 dx \int db^2\, \widetilde{\rho}(x, b, Q), \tag{39}$$

with $\widetilde{\rho}(x, b, Q) = \pi J_0\left(b\, Q(1-x)\right) |\widetilde{\psi}(x, b)|^2$ and $b = |\mathbf{b}_\perp|$. The kinematics are illustrated in figure 6 for the case of a meson with two constituents in the soft-wall model

$$\widetilde{\psi}_{q\overline{q}/\pi}(x, \mathbf{b}_\perp) = \frac{\kappa}{\sqrt{\pi}} \sqrt{x(1-x)}\; e^{-\frac{1}{2}\kappa^2 x(1-x)\mathbf{b}_\perp^2}, \tag{40}$$

where the Gaussian form of the LFWF at large Q^2 favors short-distance configurations with small $\zeta^2 = b_\perp^2 x(1-x) \sim 1/Q^2$. Since we are mainly interested in studying the contribution from different regions to the form factor at large Q^2, we have replaced the modified bulk-to-boundary propagator $J_\kappa(Q, z)$ (16) by its large Q^2 form $zQK_1(zQ)$. One sees a shift of the integrand $\tilde{\rho}(x, b, Q)$ toward small $|\mathbf{b}_\perp|$ and small $1-x$ at high Q^2. A similar behavior is observed for the LFWF obtained from the hard wall model.

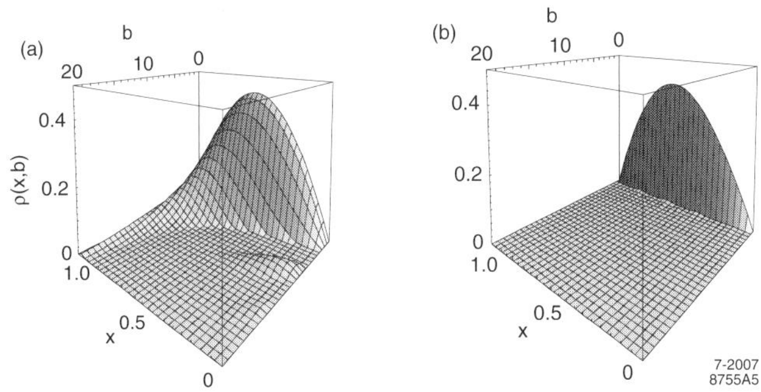

Fig. 6. Effective partonic density $\rho(x, b, Q)$ in terms of the longitudinal momentum fraction x, the transverse relative impact variable $b = |\mathbf{b}_\perp|$ and momentum transfer Q for the soft wall model. As Q increases the distribution becomes increasingly important near $x = 1$ and $\mathbf{b}_\perp = 0$. This is illustrated in (a) for $Q = 1$ GeV/c. At very large Q (figure (b)), the distribution is peaked towards $\mathbf{b}_\perp = 0$. The value of κ is 0.375 GeV.

7.4. *The Meson Structure Function*

The pion structure function $q_\pi(x, Q^2)$ is computed by integrating the square of the pion light-front wave function up to the scale Q^2

$$q_\pi(x, Q^2) = \int^{Q^2} \frac{d^2\mathbf{k}_\perp}{16\pi^3} \left|\psi_{\overline{q}q/\pi}(x, \mathbf{k}_\perp)\right|^2, \tag{41}$$

where the LFWF in $\mathbf{k}_\perp$ space

$$\psi_{\overline{q}q/\pi}(x, \mathbf{k}_\perp) = \frac{4\pi}{\kappa\sqrt{x(1-x)}}\, e^{-\frac{\mathbf{k}_\perp^2}{2\kappa^2 x(1-x)}}. \tag{42}$$

is the Fourier transform (5) for a two-parton bound state. We find

$$q_\pi(x, Q^2) = 1 - e^{-\frac{Q^2}{\kappa^2 x(1-x)}}, \tag{43}$$

In the large Q^2 limit $q_\pi(x, Q^2 \to \infty) \equiv q_\pi(x) = 1$, which is the behavior of a strongly coupled theory found in QCD(1+1).[54] Identical results[55] are obtained for the pion in the hard wall model.

7.5. *The Pion Decay Constant*

The pion decay constant is given by the matrix element of the axial isospin current $J^{\mu 5a}$ between a physical pion and the vacuum state[56] $\langle 0 | J_W^+(0) | \pi^-(P^+, \vec{P}_\perp)\rangle$, where J_W^+ is the flavor changing weak current. Only the valence state with $L_z = 0$, $S_z = 0$, contributes to the decay of the $\pi^\pm$. Expanding the hadronic initial state in the decay amplitude into its Fock components we find

$$f_\pi = 2\sqrt{N_C}\int_0^1 dx \int \frac{d^2\vec{k}_\perp}{16\pi^3}\, \psi_{\overline{q}q/\pi}(x, k_\perp). \tag{44}$$

This light-cone equation allows the exact computation of the pion decay constant in terms of the valence pion light-front wave function.[6]

7.6. *The Meson Distribution Amplitude*

The meson distribution amplitude $\phi(x, Q)$ is defined as[4]

$$\phi(x, Q) = \int^{Q^2} \frac{d^2\mathbf{k}_\perp}{16\pi^3}\, \psi(x, \mathbf{k}_\perp). \tag{45}$$

It follows that

$$\phi_\pi(x, Q \to \infty) = \frac{4}{\sqrt{3}\pi} f_\pi \sqrt{x(1-x)}, \tag{46}$$

with

$$f_\pi = \frac{1}{8}\sqrt{\frac{3}{2}}\, R^{3/2} \lim_{\zeta \to 0} \frac{\Phi(\zeta)}{\zeta^2}, \tag{47}$$

since $\phi(x, Q \to \infty) \to \widetilde{\psi}(x, \mathbf{b}_\perp \to 0)/\sqrt{4\pi}$ and $\Phi_\pi \sim \zeta^2$ as $\zeta \to 0$. The pion decay constant depends only on the behavior of the AdS string mode near the asymptotic boundary, $\zeta = z = 0$ and the mode normalization. For the truncated-space (TS) pion mode we find $f_\pi^{TS} = \frac{\sqrt{3}}{8 J_1(\beta_{0,k})} \Lambda_{\rm QCD} = 83.4$ MeV, for $\Lambda_{QCD} = 0.22$ MeV. The corresponding result for the transverse harmonic oscillator (HO) pion mode is $f_\pi^{HO} = \frac{\sqrt{3}}{8} \kappa = 86.6$ MeV, for $\kappa = 0.375$ GeV. The values of Λ_{QCD} and κ are determined from the space-like form factor data as discussed above. The experimental result for f_π is extracted from the rate of weak π decay and has the value $f_\pi = 92.4$ MeV.[47]

It is interesting to note that the pion distribution amplitude predicted by AdS/QCD (46) has a quite different x-behavior than the asymptotic distribution amplitude predicted from the PQCD evolution[4] of the pion distribution amplitude $\phi_\pi(x, Q \to \infty) = \sqrt{3} f_\pi x(1-x)$. The broader shape of the pion distribution increases the magnitude of the leading twist perturbative QCD prediction for the pion form factor by a factor of 16/9 compared to the prediction based on the asymptotic form, bringing the NLO PQCD prediction close to the empirical pion form factor.[57] The Fermilab E791 measurements[58] of the x and $k_\perp$ fall-off of the diffractive dijet cross section $\pi A \to Jet\ Jet\ A$ also suggest a broad pion distribution amplitude for jet transverse momenta $k_\perp^2 < 2$ GeV^2.

7.7. *The Nucleon Dirac Form Factors*

As a final example we compute the spin non-flip nucleon form factor in the soft wall model. Consider the spin non-flip form factors

$$F_+(Q^2) = g_+ \int \frac{dz}{z}\, e^{-\kappa^2 z^2} J_\kappa(Q, z) |\Psi_+(z)|^2, \tag{48}$$

$$F_-(Q^2) = g_- \int \frac{dz}{z}\, e^{-\kappa^2 z^2} J_\kappa(Q, z) |\Psi_-(z)|^2, \tag{49}$$

where the effective charges g_+ and g_- are determined from the spin-flavor structure of the theory. We choose the struck quark to have $S^z = +1/2$. The two AdS solutions Ψ_+ and Ψ_- correspond to nucleons with total angular momentum $J^z = +1/2$ and $-1/2$. For the $SU(6)$ spin-flavor symmetry

$$F_1^p(Q^2) = \int \frac{dz}{z}\, e^{-\kappa^2 z^2} J_\kappa(Q, z) |\Psi_+(\zeta)|^2, \tag{50}$$

$$F_1^n(Q^2) = -\frac{1}{3} \int \frac{dz}{z}\, e^{-\kappa^2 z^2} J_\kappa(Q, z) \left[|\Psi_+(z)|^2 - |\Psi_-(z)|^2 \right], \tag{51}$$

where $F_1^p(0) = 1$, $F_1^n(0) = 0$. The bulk-to-boundary propagator $J_\kappa(Q, z)$ is the solution (16) of the AdS wave equation for the external electromagnetic current, and the plus and minus components of the twist 3 nucleon mode in the soft wall model are

$$\Psi_+(z) = \frac{\sqrt{2}\kappa^2}{R^{3/2}}\, z^3, \tag{52}$$

$$\Psi_-(z) = \frac{\kappa^3}{R^{3/2}}\, z^4. \tag{53}$$

The results for $Q^4 F_1^p(Q^2)$ and $Q^4 F_1^n(Q^2)$ follow from the analytic form for the form factors for any τ given in Section 7.1 and are shown in Figure 7.

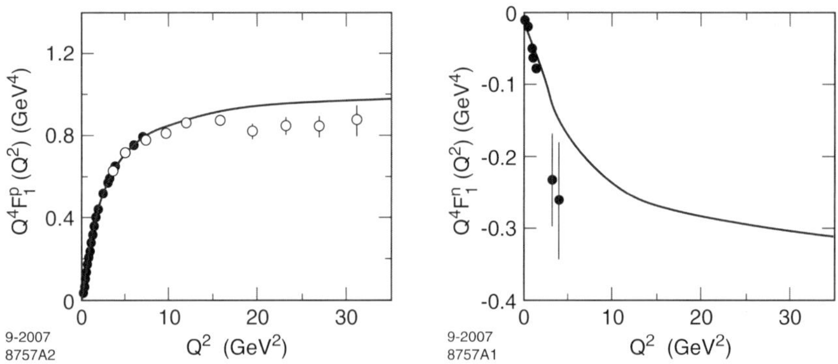

Fig. 7. Predictions for $Q^4 F_1^p(Q^2)$ and $Q^4 F_1^n(Q^2)$ in the soft wall model for $\kappa = 0.424$ GeV. The data compilation is from Diehl.[59]

8. Overview of Hard Exclusive Processes

The leading power fall-off of the hard scattering amplitude follows from the conformal scaling of the underlying hard-scattering amplitude: $T_H \sim 1/Q^{n-4}$, where n is the total number of fields (quarks, leptons, or gauge fields) participating in the hard scattering.[48,49] Thus the reaction is dominated by subprocesses and Fock states involving the minimum number of interacting fields. In the case of $2 \to 2$ scattering processes, this implies $d\sigma/dt(AB \to CD) = F_{AB\to CD}(t/s)/s^{n-2}$, where $n = N_A + N_B + N_C + N_D$ and n_H is the minimum number of constituents of H. These dimensional counting rules are also a key feature of AdS/CFT.[24] The near-constancy of the effective QCD coupling at small scales helps explain the general empirical success of the dimensional counting rules for the near-conformal power law fall-off of form factors and fixed-angle scaling.[60] For example, measurements of pion photoproduction are consistent with dimensional counting $s^7 d\sigma/dt(\gamma p \to \pi^+ n) \sim$ constant at fixed CM angle for $s > 7$ GeV. The angular distributions seen in hard large CM angle scattering reactions are consistent with quark interchange, a result predicted by the hard wall AdS/QCD model. Reviews are given in refs. 61 and 62. One sees the onset of perturbative QCD scaling behavior even for exclusive nuclear amplitudes such as deuteron photodisintegration (Here $n = 1 + 6 + 3 + 3 = 13$) $s^{11} d\sigma/dt(\gamma d \to pn) \sim$ constant at fixed CM angle.[63–65] The measured deuteron form factor[66] also appears to follow the leading-twist QCD predictions[67] at large momentum transfers in the few GeV region. A measurement of $d\sigma/dt(\gamma d \to \Delta^{++}\Delta^{)}$ in the scaling region can establish the role of "hidden-color" degrees of freedom[68] of the nuclear wavefunction in hard deuteron reactions.

Recently the Hall A collaboration[69] at JLab has reported a significant exception to the general empirical success of dimensional counting in fixed CM angle Compton scattering $d\sigma/dt(\gamma p \to \gamma p) \sim F(\theta_{CM})/s^8$ instead of the predicted $1/s^6$ scaling. The deviations from fixed-angle conformal scaling may be due to corrections from resonance contributions in the JLab energy range. It is interesting that the hadron form factor $R_V(t)$,[70] which multiplies the $\gamma q \to \gamma q$ amplitude is found by Hall-A to scale as $1/t^2$, in agreement with the PQCD and AdS/CFT prediction. In addition the Belle measurement[71] of the timelike two-photon cross section $d\sigma/dt(\gamma\gamma \to p\bar{p})$ is consistent with $1/s^6$ scaling.

Although large-angle proton-proton elastic scattering is well described by dimensional scaling $s^{10} d\sigma/dt(pp \to pp) \sim$ constant at fixed CM angle, extraordinarily large spin-spin correlations are observed.[72] The ratio of scattering cross sections for spin-parallel and normal to the scattering plane versus spin-antiparallel reaches $R_{NN} \simeq 4$ in large angle $pp \to pp$ at $\sqrt{s} \simeq 5$ GeV; this is a remarkable example of "exclusive transversity". Color transparency is observed at lower energies but it fails[73] at the same energy where R_{NN} becomes large. In fact, these anomalies have a natural explanation[74] as a resonance effect related to the charm threshold in pp scattering. Alternative explanations of the large spin correlation are discussed and reviewed in ref. 75. Resonance formation is a natural phenomenon when all constituents are relatively at rest. For example, a resonance effect can occur due to the intermediate state $uuduudc\bar{c}$ at the charm threshold $\sqrt{s} = 5$ GeV in pp collisions. Since the c and $\bar{c}$ have opposite intrinsic parity, the resonance appears in the $L = J = S = 1$ partial wave for $pp \to pp$ which is only allowed for spin-parallel and normal scattering $A_{NN} = 1$.[74] Resonance formation at the charm threshold also explains the dramatic quenching of color transparency seen in quasielastic pn scattering by the EVA BNL experiment[73] in the same kinematic region. The reason why these effects are so dramatic in $pp \to pp$ scattering is that the amplitude for the formation of an $uuduudc\bar{c}$ s-channel resonance in the intermediate state is of the same magnitude as the fast-falling background $pp \to pp$ PQCD amplitude from quark interchange at large CM angles: $M(pp \to pp) \sim 1/u^2t^2$. We have predicted that the open charm cross section in pp scattering will be of order of 1 μb at threshold.[74]

Color transparency[76,77] is a key property of color gauge theory, and it thus stands at the foundations of QCD. Color transparency has been confirmed in diffractive dijet production,[78] pion photoproduction[79] and vector meson electroproduction,[80] but it is very important to also systematically validate it in large angle hadron scattering processes. One also expects similar novel QCD phenomena in large-angle photoproduction $\gamma p \to \pi N$ near the charm threshold, including the breakdown of color transparency and strong spin-spin correlations. These effects can be tested by measurements at the new JLab 12 GeV facility, which would confirm resonance formation in a low partial wave in $\gamma p \to \pi N$ at $\sqrt{s} \simeq 4$ GeV due to attractive forces in the $uud\bar{c}c$ channel.

9. Conclusions

We have shown how the AdS/CFT correspondence between Anti-de Sitter space and conformal gauge provides an analytically tractable approximation to QCD in the regime where the QCD coupling is large and constant. In particular, there is an exact correspondence between the fifth-dimension coordinate z of AdS space and an impact variable ζ which measures the invariant separation of the quark constituents within the hadron in ordinary space-time. This connection allows one to compute the analytic form of the frame-independent light-front wavefunctions of mesons and baryons, the fundamental entities which encode hadron properties and allow the computation of exclusive scattering amplitudes.

The phenomenology of the AdS/QCD model is just beginning, but it can be anticipated that it will have many applications to hadron dynamics. For example, the model LFWFs obtained from AdS/QCD provide a basis for understanding hadron structure functions and fragmentation functions at the amplitude level; the same wavefunctions also describe hadron formation from the coalescence of co-moving quarks. The spin correlations which underly single and double spin correlations are also described by the AdS/QCD eigensolutions. The AdS/QCD hadronic wavefunctions provide predictions for the generalized parton distributions of hadrons and their weak decay amplitudes from first principles. The amplitudes relevant to diffractive reactions could also be computed. We also anticipate that the extension of the AdS/QCD formalism to heavy quarks will allow a great variety of heavy hadron phenomena to be analyzed from first principles.

Acknowledgments

This research was supported by the Department of Energy contract DE–AC02–76SF00515. The research of GdT is supported in part by an Aportes grant from Florida Ice & Farm.

References

1. P. A. M. Dirac, Rev. Mod. Phys. **21**, 392 (1949).
2. S. J. Brodsky and D. S. Hwang, Nucl. Phys. B **543**, 239 (1999) [arXiv:hep-ph/9806358].
3. S. J. Brodsky, M. Diehl and D. S. Hwang, Nucl. Phys. B **596**, 99 (2001) [arXiv:hep-ph/0009254].
4. G. P. Lepage and S. J. Brodsky, Phys. Lett. B **87**, 359 (1979).
5. A. V. Efremov and A. V. Radyushkin, Phys. Lett. B **94**, 245 (1980).
6. G. P. Lepage and S. J. Brodsky, Phys. Rev. D **22**, 2157 (1980).
7. S. J. Brodsky and G. P. Lepage, SLAC-PUB-2294 (1979).
8. J. M. Maldacena, Adv. Theor. Math. Phys. **2**, 231 (1998) [Int. J. Theor. Phys. **38**, 1113 (1999)] [arXiv:hep-th/9711200]; S. S. Gubser, I. R. Klebanov and A. M. Polyakov, Phys. Lett. B **428**, 105 (1998) [arXiv:hep-th/9802109]; E. Witten, Adv. Theor. Math. Phys. **2**, 253 (1998) [arXiv:hep-th/9802150].
9. L. von Smekal, R. Alkofer and A. Hauck, Phys. Rev. Lett. **79**, 3591 (1997) [arXiv:hep-ph/9705242].
10. S. Furui and H. Nakajima, arXiv:hep-lat/0612009; S. Furui and H. Nakajima, arXiv:0708.1421 [hep-lat].
11. S. J. Brodsky, S. Menke, C. Merino and J. Rathsman, Phys. Rev. D **67**, 055008 (2003) [arXiv:hep-ph/0212078].
12. A. Deur, V. Burkert, J. P. Chen and W. Korsch, Phys. Lett. B **650**, 244 (2007) [arXiv:hep-ph/0509113].
13. J. M. Cornwall, Phys. Rev. D **26**, 1453 (1982).
14. S. J. Brodsky and G. F. de Teramond, Phys. Rev. Lett. **96**, 201601 (2006) [arXiv:hep-ph/0602252].
15. G. F. de Teramond, arXiv:hep-ph/0606143.
16. H. R. Grigoryan and A. V. Radyushkin, Phys. Lett. B **650**, 421 (2007) [arXiv:hep-ph/0703069]; arXiv:0709.0500 [hep-ph]; H. R. Grigoryan, arXiv:0709.0939 [hep-ph].
17. H. R. Grigoryan and A. V. Radyushkin, arXiv:0706.1543 [hep-ph].

18. S. J. Brodsky and G. F. de Teramond, arXiv:hep-th/0702205.
19. S. J. Brodsky and G. F. de Teramond, arXiv:0707.3859 [hep-ph].
20. T. Huang and F. Zuo, arXiv:0708.0936 [hep-ph].
21. H. J. Kwee and R. F. Lebed, arXiv:0708.4054 [hep-ph].
22. J. Polchinski and L. Susskind, arXiv:hep-th/0112204.
23. S. Hong, S. Yoon and M. J. Strassler, JHEP **0604**, 003 (2006) [arXiv:hep-th/0409118].
24. J. Polchinski and M. J. Strassler, Phys. Rev. Lett. **88**, 031601 (2002) [arXiv:hep-th/0109174].
25. S. J. Brodsky and G. F. de Teramond, Phys. Lett. B **582**, 211 (2004) [arXiv:hep-th/0310227].
26. H. Boschi-Filho and N. R. F. Braga, JHEP **0305**, 009 (2003) [arXiv:hep-th/0212207].
27. G. F. de Teramond and S. J. Brodsky, arXiv:hep-th/0409074.
28. G. F. de Teramond and S. J. Brodsky, Phys. Rev. Lett. **94**, 201601 (2005) [arXiv:hep-th/0501022].
29. D. K. Hong, T. Inami and H. U. Yee, Phys. Lett. B **646**, 165 (2007) [arXiv:hep-ph/0609270].
30. J. Erlich, E. Katz, D. T. Son and M. A. Stephanov, Phys. Rev. Lett. **95**, 261602 (2005) [arXiv:hep-ph/0501128].
31. L. Da Rold and A. Pomarol, Nucl. Phys. B **721**, 79 (2005) [arXiv:hep-ph/0501218]; JHEP **0601**, 157 (2006) [arXiv:hep-ph/0510268].
32. J. Hirn and V. Sanz, JHEP **0512**, 030 (2005) [arXiv:hep-ph/0507049]; J. Hirn, N. Rius and V. Sanz, Phys. Rev. D **73**, 085005 (2006) [arXiv:hep-ph/0512240].
33. K. Ghoroku, N. Maru, M. Tachibana and M. Yahiro, Phys. Lett. B **633**, 602 (2006) [arXiv:hep-ph/0510334].
34. J. Polchinski and M. J. Strassler, JHEP **0305**, 012 (2003) [arXiv:hep-th/0209211].
35. R. C. Brower, J. Polchinski, M. J. Strassler and C. I. Tan, arXiv:hep-th/0603115.
36. J. Babington, J. Erdmenger, N. J. Evans, Z. Guralnik and I. Kirsch, Phys. Rev. D **69**, 066007 (2004) [arXiv:hep-th/0306018].
37. For a recent review of the top-bottom approach to the string/gauge correspondence, including applications to finite temperature and phase transitions, see: K. Peeters and M. Zamaklar, arXiv:0708.1502 [hep-ph] and references therein.
38. A. Karch, E. Katz, D. T. Son and M. A. Stephanov, Phys. Rev. D **74**, 015005 (2006) [arXiv:hep-ph/0602229].
39. R. R. Metsaev, arXiv:hep-th/0002008.
40. S. J. Brodsky, H. C. Pauli and S. S. Pinsky, Phys. Rept. **301**, 299 (1998) [arXiv:hep-ph/9705477].
41. D. E. Soper, Phys. Rev. D **15**, 1141 (1977).
42. S. D. Drell and T. M. Yan, Phys. Rev. Lett. **24**, 181 (1970).
43. G. B. West, Phys. Rev. Lett. **24**, 1206 (1970).
44. S. J. Brodsky and S. D. Drell, Phys. Rev. D **22**, 2236 (1980).

45. P. Breitenlohner and D. Z. Freedman, Annals Phys. **144**, 249 (1982).
46. L. Infeld, Phys. Rev. **59**, 737 (1941).
47. S. Eidelman *et al.* [Particle Data Group], Phys. Lett. B **592**, 1 (2004).
48. S. J. Brodsky and G. R. Farrar, Phys. Rev. D **11**, 1309 (1975).
49. V. A. Matveev, R. M. Muradian and A. N. Tavkhelidze, Lett. Nuovo Cim. **7**, 719 (1973).
50. S. Ferrara, A. F. Grillo and G. Parisi, Nuovo Cim. A **12**, 952 (1972).
51. R. Baldini, S. Dubnicka, P. Gauzzi, S. Pacetti, E. Pasqualucci and Y. Srivastava, Eur. Phys. J. C **11**, 709 (1999).
52. V. Tadevosyan *et al.* [Jefferson Lab F(pi) Collaboration], Phys. Rev. C **75**, 055205 (2007) [arXiv:nucl-ex/0607007].
53. T. Horn *et al.* [Fpi2 Collaboration], Phys. Rev. Lett. **97**, 192001 (2006) [arXiv:nucl-ex/0607005].
54. H. C. Pauli and S. J. Brodsky, Phys. Rev. D **32**, 2001 (1985).
55. A. V. Radyushkin, Phys. Lett. B **642**, 459 (2006) [arXiv:hep-ph/0605116].
56. M. E. Peskin and D. V. Schroeder, *Reading, USA: Addison-Wesley (1995).*
57. H. M. Choi and C. R. Ji, Phys. Rev. D **74**, 093010 (2006) [arXiv:hep-ph/0608148].
58. E. M. Aitala *et al.* [E791 Collaboration], Phys. Rev. Lett. **86**, 4768 (2001) [arXiv:hep-ex/0010043].
59. M. Diehl, Nucl. Phys. Proc. Suppl. **161**, 49 (2006) [arXiv:hep-ph/0510221].
60. S. J. Brodsky and G. P. Lepage, Adv. Ser. Direct. High Energy Phys. **5**, 93 (1989).
61. D. W. Sivers, S. J. Brodsky and R. Blankenbecler, Phys. Rept. **23**, 1 (1976).
62. C. G. White *et al.*, Phys. Rev. D **49**, 58 (1994).
63. R. J. Holt, Phys. Rev. C **41**, 2400 (1990).
64. C. Bochna *et al.* [E89-012 Collaboration], Phys. Rev. Lett. **81**, 4576 (1998) [arXiv:nucl-ex/9808001].
65. P. Rossi *et al.* [CLAS Collaboration], arXiv:hep-ph/0405207.
66. S. Rock *et al.*, Phys. Rev. D **46**, 24 (1992).
67. S. J. Brodsky and B. T. Chertok, Phys. Rev. D **14**, 3003 (1976).
68. S. J. Brodsky, C. R. Ji and G. P. Lepage, Phys. Rev. Lett. **51**, 83 (1983).
69. A. Danagoulian *et al.* [Hall A Collaboration], Phys. Rev. Lett. **98**, 152001 (2007) [arXiv:nucl-ex/0701068].
70. M. Diehl, T. Feldmann, R. Jakob and P. Kroll, Eur. Phys. J. C **8**, 409 (1999) [arXiv:hep-ph/9811253].
71. A. Chen, *Photon 2001, Ascona, Switzerland, 2-7 Sep 2001*
72. G. R. Court *et al.*, Phys. Rev. Lett. **57**, 507 (1986).
73. I. Mardor *et al.*, Phys. Rev. Lett. **81**, 5085 (1998).
74. S. J. Brodsky and G. F. de Teramond, Phys. Rev. Lett. **60**, 1924 (1988).
75. D. Dutta and H. Gao, Phys. Rev. C **71**, 032201 (2005) [arXiv:hep-ph/0411267].
76. S. J. Brodsky and A. H. Mueller, Phys. Lett. B **206**, 685 (1988).
77. G. Bertsch, S. J. Brodsky, A. S. Goldhaber and J. F. Gunion, Phys. Rev. Lett. **47**, 297 (1981).

78. E. M. Aitala *et al.* [E791 Collaboration], Phys. Rev. Lett. **86**, 4773 (2001) [arXiv:hep-ex/0010044].
79. B. Clasie *et al.*, arXiv:0707.1481 [nucl-ex].
80. A. Airapetian *et al.* [HERMES Collaboration], Phys. Rev. Lett. **90**, 052501 (2003) [arXiv:hep-ex/0209072].

HADRON STRUCTURE MATTERS IN COLLISIONS AT HIGH ENERGY AND MOMENTUM

A. W. THOMAS

Jefferson Lab, 12000 Jefferson Ave.,
Newport News, VA 23185, USA
and
College of William and Mary, Williamsburg VA 23187, USA
E-mail: awthomas@jlab.org

We review key predictions of the description of the long range structure of the nucleon in terms of a pion cloud. We explain how this, together with other aspects of nucleon structure, such as the relativistic motion of the valence quarks and the one-gluon-exchange hyperfine interaction, provides a natural explanation of the proton spin crisis.

Keywords: Style file; LaTeX; Proceedings; World Scientific Publishing.

1. Introduction

There are many cases where the description of high energy reactions in terms of the leading Fock components of the participating hadron wave functions provide a very clear and effective understanding of the physics. On the other hand, phenomena such as the famous excess of $\bar{d}$ over $\bar{u}$ in the sea of the proton can be easily understood in terms of the pion cloud of the proton. While the latter is associated with the spontaneous breaking of chiral symmetry and the flavor asymmetry of the sea was anticipated on this basis, such phenomena cannot be easily described in terms of the naive Fock space expansion of the hadron wave function.

We review a number of examples where the pion cloud provides a simple and intuitive explanation of important properties of the nucleon. We see that it also plays a very important role in providing a bridge between modern lattice QCD calculations at large quark mass and the experimental observables, yielding new insights into hadron structure. Finally, we apply the lessons coming out of the initial discussions to a very important challenge for modern nuclear and particle physics, namely the well-known proton "spin crisis", originally reported by the European Muon Collaboration based on their high energy measurements of spin dependent deep inelastic scattering. We shall see that well known non-perturbative physics, including the pion cloud, provide a natural explanation of the curent data on the spin structure function.

2. Role of the Pion Cloud

Amongst the many models developed to incorporate the constraints of chiral symmetry into a quantitative picture of hadron structure the cloudy bag model (CBM) was unique.[2–4] It did not attempt to describe the short distance structure in terms of pions, since the natural degrees of freedom in QCD are the quarks and gluons. It simply introduced the pion field on top of a successful model of quark confinement (the MIT bag model), in the simplest way consistent with chiral symmetry. It was possible to prove rigorously that, provided the size of the bag (read confinement region) was sufficiently large (say 0.8fm or more), the pion cloud could be treated perturbatively[5] – the size of the quark core provided a natural mechanism for suppressing the emission of more than one or two pions at a time.

Although the CBM began with the minimal coupling of the pion field to the surface of the bag confining the valence quarks necessary to restore chiral symmetry, it was soon transformed into a form with pseudo-vector coupling throughout the bag volume,[8] which made it much easier to derive some crucial low energy theorems, such as the Weinberg-Tomozawa formula. (It was also discovered that, at least for massless u and d quarks, the results derived using surface coupling were valid in the volume coupled version. For example, the single pion exchange force between two nucleons was identical to that between point-like nucleons for separations greater than twice the bag radius ($r > 2R$), with the form factor arising from finite nucleon size only modifying it when the bags overlapped, in *both* formulations.) The chiral quark model of Manohar and Georgi,[9] which was published a few years later, was very similar to that version except that it did not carry out the projection onto colorless, bare hadronic states (the P-space projection).[4] Only a decade later was it realized[10] that this projection is essential to ensure the correct infrared limit and hence the correct leading and next-to-leading non-analytic (LNA and NLNA) behaviour of various hadronic properties.[11] Unfortunately, this lesson is still not widely appreciated and many calculations within so-called chiral quark models are still carried out in a manner which actually violates chiral symmetry!

We now review a number of important physical results which follow directly from the perturbative nature of the pion cloud found naturally within the cloudy bag model.

2.1. *Neutron charge distribution*

Within the CBM one had at once a very natural and beautiful way to understand the charge distribution of the neutron.[3] It had of course been known that the long range negative tail of the neutron was explained in terms of pion emission. But within the Chew-Low model the expansion in terms of the number of pions was divergent and one could say nothing about the short distance structure (inside 1 fm). In the CBM it was suddenly simple; the neutron could be thought of as predominantly a bare neutron bag (with zero charge distribution), but occasionally as a π^- and a bare proton bag whose charge distribution was trivial to calculate. Furthermore, the structure of the pion cloud was such that it peaked in the bag surface, so that the peak of the negative charge distribution of the neutron had a natural interpretation in terms of the bag surface (or the size of the volume within which the valence quarks were confined). On a personal note, I first realized this on New Year's Eve 1980 and had to wait in considerable anticipation for many years before experimental measurements were able to pin down the position of this peak. Modern recoil polarization measurements have established that this peak is indeed around 0.8 fm,[15] as anticipated in the CBM.

The studies of many different phenomena within the CBM are consistent with a picture in which the total probability of a physical nucleon consisting of a single pion and a bare nucleon is approximately 20% (and a pion and a bare Δ around 8-10%).[16] For the proton (rounding 20% to 21% to simplify the algebra) this means roughly 14% π^+-bare n and 7% π^0-bare p. Similarly for the neutron it's 14% π^--bare p and 7% π^0-bare n. Taking the charge density at the centre of a bare p to be x and at the center of a bare n to be 0, this gives for the ratio of the central charge densities of the physical n to p to be approximately $0.14x/(0.79x + 0.07x) = 0.14/0.86 \sim 1/6$, which is very close to the experimental ratio extracted from modern studies of the neutron electric form factor at Nikhef, Mainz, MIT-Bates and JLab.[12–15] Small corrections may be expected from the $\pi\Delta$ component of the wave function and from hyperfine effects, which may result in a slightly non-zero charge distribution in the bare n, but the result just derived in this simple and physically transparent manner remains essentially correct.

2.2. *Flavor symmetry violation*

Feynman's much appreciated physical insight led to calculations in the early 1970s of a possible pion contribution to the nucleon sea.[17] However, this was largely ignored until the discovery of the famous EMC effect generated enormous interest in the possible role of an excess of pions associated with nuclear binding on the structure functions of nuclei.[18,19] At this time it was realized that the presence of a pion cloud around the nucleon would have profound consequences for the flavor structure of the proton sea. In particular, with the biggest pionic component of the proton wave function being π^+n and the π^+ containing only down anti-quarks, it was clear that one expected an excess of anti-down over anti-up quarks in the nucleon sea.[20] Using the probabilities noted earlier one expected an excess of about 0.14 anti-down quarks.

At the time this novel property of the nucleon sea was predicted it was extremely unusual to think about deep inelastic scattering in terms of hadronic Fock components and certainly not in terms of a pion cloud. It was not until the experimental confirmation of a violation of the Gottfried sum-rule almost 8 years later[22] that this began to change. Early discussions of the violation of the Gottfried sum-rule often talked about flavor symmetry violation and appeared to confuse that with a violation of isospin, or charge independence. In fact, the simple explanation of how this excess of anti-down quarks was predicted[20,23] makes it clear that it has *nothing* to do with isospin violation. The 2:1 ratio of the π^+n to π^0p Fock components is a direct consequence of isospin symmetry. In contrast, the term "flavor symmetry" was at best vague and at worst misleading.

In summary, the concept of a pion cloud surrounding the proton led to a simple and natural prediction of an excess of anti-down quarks in the proton sea. The prediction was not only qualitatively but quantitatively in agreement with experiments performed many years later. This discovery has completely changed the standard fits to parton distribution functions.

Perhaps surprisingly, the original paper which predicted the excess of anti-down sea quarks was not primarily concerned with that problem. Its main subject was actually to point out the considerable difference between the strength of the strange and non-strange sea predicted by the meson cloud picture. Indeed, with the hadron size providing a natural high momentum cut-off on the meson-baryon dynamics, it was observed that the strange sea arising from the fluctuation p $\rightarrow K^+\Lambda$ would necessarily be much smaller than the non-strange sea associated with the pion cloud.[20] This was proposed as a natural explanation of the 2:1 ratio that had already been seen in neutrino deep inelastic scattering.[21] Of course, as well as this sea arising from the meson clouds, which are non-perturbative in the usual sense in which one discusses QCD, there is necessarily a perturbative sea generated through the process $g \to q + \bar{q}$ which would be approximately flavor blind (at least for u, d and s).

The consequences for the strange sea of the nucleon were explored further by Signal and Thomas just a few years later.[24] It was noted that while the number of s and $\bar{s}$ quarks must be equal in the proton, their momentum distributions will in general not be the same.[25] This is essentially because of the unequal sharing of light-cone momentum between the K^+ and Λ in the corresponding piece of the proton wave function. Although we still have no experimental guidance on this issue, it has turned out to be very important in the context of using neutrino deep inelastic scattering to test the Standard Model. It is clearly very important to reliably and accurately determine at least the integral of $x(s(x) - \bar{s}(x))$ as soon as possible.

2.3. *Decuplet contribution to octet magnetic moments*

The first systematic inclusion of the decuplet baryons in a calculation of octet magnetic moments was in the study by Théberge and Thomas.[26] On physical grounds it was clear that any spin flip quantity would be sensitive to the inclusion of the decuplet. The very convergence of the expansion in pion number for the nucleon was directly related to the role of the Δ in the vertex renormalization of the pion-nucleon coupling constant.[4,5] The results for the octet magnetic moments were in very reasonable agreement with experiment. Within the context of formal χPT it was a decade before decuplet contributions were incorporated in the analysis and even then it took a while to realize that the correct LNA structure required a projection onto bare-baryon pion configurations.

3. Discoveries in Modern Lattice QCD

One of the unexpected but very positive consequences of our *lack* of supercomputing power is the fact that it has not been possible to compute physical hadron properties in lattice QCD. In fact, with computation time scaling like m_π^{-9} (if we include the larger lattice size needed), calculations have covered the pion mass range from 0.3 to 1.0 GeV (or higher). Far from being a disappointment, this has given us a wealth of unexpected insight into how QCD behaves as the light quark masses are varied.[27] In terms of the insight this has given us into hadron structure it is both truly invaluable and thus far under-utilized.

The most striking feature of the lattice data is that in the region $m_\pi > 0.4$ GeV, in fact for almost all of the simulations made so far, all baryon properties show a smooth dependence on quark mass, totally consistent with a constituent quark model. The rapid, non-linear dependence on m_π required by the LNA and NLNA behavior of χPT are notably absent from the data!

The conventional view of χPT has no explanation for this simple, universal observation. Worse, in seeking to apply χPT to extrapolate the data back to the physical pion mass, it has been necessary to rely on ad hoc cancellations between the high order terms in the usual power series expansion (supplemented by the required non-analytic behavior). In fact, there is strong evidence that such series expansions have been applied well beyond their region of convergence[28] and that as a result the extrapolations are largely unreliable.

On the other hand, the picture of the pion cloud that we have presented here yields an extremely natural explanation of the universal, constituent quark model behaviour of hadron properties found in the lattice simulations for $m_\pi > 0.4$ GeV. The natural high momentum cut-off on the momentum of the emitted pion, which is associated with the finite size (typically $R \sim 1$ fm) of the bare baryon (i.e., the bag in the CBM), strongly suppresses pion loop contributions as m_π increases. The natural mass scale which sets the boundary between rapid chiral variation and constituent quark type behavior is $1/R \sim 0.2\,\text{to}\,0.4$ GeV. Indeed, when in the early investigation of the quark mass dependence of nucleon properties the CBM was compared directly with lattice data, the agreement was remarkably good.[30] (Similar results have been obtained recently within the chiral quark soliton model.[31]) The results were equally as impressive for the N and Δ masses and magnetic moments, the proton charge radius and the moments of its parton distribution functions.[32] The key features necessary to reproduce the behaviour found at large quark mass in lattice QCD *and* to reproduce the experimentally measured data at the physical mass seem to be that:

- The treatment of the pion cloud (chiral) corrections ensures the correct LNA (and NLNA, although in practice this seems less important in many applications) behaviour of QCD
- The pion cloud contribution is suppressed for m_π beyond 0.4 GeV, and
- the underlying quark model exhibits constituent quark like behaviour for the corresponding range of current quark masses.

The CBM satisfies all of these properties.

In practice, of course, in analyzing lattice QCD data one does not want to rely on any particular quark model. However, one does need to suppress the pion cloud as m_π goes up, and the simple use of a finite range regulator (FRR) in the evaluation of the pion loops that yield the LNA and NLNA behaviour ensures this at the cost of one additional parameter, the cut-off mass Λ. If the data are good enough one can use this as a fitting parameter but in general it is sufficient[33] to choose a value consistent with the physical arguments presented above (e.g., Λ is ~ 0.8 GeV for a dipole regulator, 0.6 GeV for a monopole and 0.4 GeV using a θ-function). The sensitivity of the extrapolation to the choice of the functional form of the FRR is then an additional source of systematic error in the final quoted result. In the case of the nucleon mass the corresponding systematic error was[34] of the order of a mere 0.1%.

One of the most remarkable results of this physical understanding of the role of the pion cloud and, in particular, its suppression at large pion mass has been the unexpected discovery of a connection between lattice simulations based upon quenched QCD (QQCD) and full QCD.[35] In a study of the quark mass dependence of the N and Δ masses,[36] it was discovered that if the self-energies appropriate to either QQCD or full QCD were regulated using the same dipole form for the FRR (the dipole being the most natural physical choice given that the axial form factor of the nucleon has a dipole form) with mass parameter $\Lambda = 0.8$ GeV (the preferred value, as noted above), then the residual expansions for the nucleon mass in QQCD and QCD (and also for the Δ in QQCD and QCD) were the same within the errors of the fit! This is a remarkable result which a posteriori gives enormous support to the physical picture of the baryons consisting of confined valence quarks surrounded by a perturbative pion cloud. The baryon core is basically determined by the confinement mechanism and provided the choice of lattice scale reproduces the physically known confining force (either through the string tension or the Sommer parameter,[37] derived from the heavy quark potential) it makes little difference whether one uses QQCD or full QCD to describe that core. What *does* matter is the change in the chiral coefficients as one goes from QQCD to full QCD.

Until recently, the most significant application of this discovery was the application to the calculation of the octet magnetic moments and charge radii based on accurate QQCD simulations that extend to rather low quark mass. Using the constraints of charge symmetry this has led to some extremely accurate calculations of the strange quark contributions to the magnetic moment[38] and charge radius[39] of the proton. Indeed, those calculations are in excellent agreement with the current world data[40] but, in a unique example in modern strong interaction physics, they are an order of magnitude more accurate.

Very recently, as we briefly explain in the next section, this analysis of quenched and full QCD data for Δ and N masses has proven crucial to a satisfactory resolution of the spin problem in terms of well known hadron structure physics.

4. Where is the Spin of the Proton?

A critical consequence of the success of the analysis of quenched and full QCD data, described in the previous section, is that, for the first time, we can be confident of having control of the pion self-energy contributions for the N and Δ. This is extremely important for hadron structure, because the hyperfine interaction associated with one-gluon-exchange (OGE) has traditionally been taken as the source of the Δ-N mass difference. Indeed, it is used to fix the effective strong coupling strength, α_s. This was confused for some time by the suggestion that the pion cloud could make a substantial contribution to $m_\Delta - m_N$. However, the analysis of Ref. [36] led to the conclusion that the pion loops give less than 50 MeV of this 300 MeV mass difference. Thus we can be confident, when modeling hadron structure, in including *both* pion cloud (chiral) effects and OGE corrections without fear of double counting.

This a crucial observation when it comes to a modern interpretation of the proton"spin crisis". The other observations necessary to understand this problem now are:

- The fraction of the proton spin carried by its quarks is now known to be $33 \pm 3 \pm 5\%$ – significantly larger than the original EMC result, which was of order 14% and consistent with zero.
- Extensive measurements of spin structure functions and high-p_T hadron production has established that there is little or no polarized glue in the proton, with the consequence that the axial anomaly is most likely to play little or no role in resolving the spin crisis.

On the other hand, three simple pieces of structure physics, suggested immediately after the spin crisis was announced but lost in the rush to follow the U(1) axial anomaly bandwagon, do suffice to explain the modern data.

We refer to Ref. [41] for details but briefly: the relativistic motion of the confined valence quarks reduces the quark spin to around 65%; an exchange current correction associated with OGE (in which one has an intermediate p-wave anti-quark) reduces this to 50%; and finally the pion cloud converts spin to pion orbital angular momentum, reducing the 50% by a factor $Z - P_{N\pi}/3 + 5/3P_{\Delta\pi} \sim (0.7, 0.8)$, thus yielding $\Sigma \sim$ 35–40%, in excellent agreement with the experimental value noted above. Each of these simple physical effects has the effect of converting quark spin to either valence quark or anti-quark orbital angular momentum. This observation has important consequences for the study of Generalized Parton Distributions (GPDs), which will be a major component of the physics program following the 12 GeV Upgrade at Jefferson Lab.

5. Concluding Remarks

We reviewed the role of the pion cloud in nucleon structure. Time and again this picture has led to new discoveries and predictions. It is extremely unlikely that it will cease to be either useful or inspiring in the near future,[42] as major new facilities, such as the 12 GeV Upgrade at Jefferson Lab, enable us to probe hadron and nuclear structure in completely new ways. In particular, in connection with the famous proton "spin problem", we have seen that even in deep inelastic scattering features of non-perturbative hadron structure such as the one-gluon-exchange hyperfine interaction and the pion cloud play a crucial role. Furthermore, the studies of GPDs at JLab, after the Upgrade, will be essential in establishing the distribution of orbital angular momentum on the valence quarks and anti-quarks in the proton predicted by this explanation.

Acknowledgements

This work was supported by the U. S. DOE through contract No. DE-AC-060R23177, under which Jefferson Science Associates operates Jefferson Lab.

References

1. H. Pagels, Phys. Rept. **16**, 219 (1975).
2. S. Theberge *et al.*, Phys. Rev. D **22**, 2838 (1980) [Erratum-ibid. D **23**, 2106 (1981)].
3. A. W. Thomas, S. Theberge and G. A. Miller, Phys. Rev. D **24** (1981) 216.
4. A. W. Thomas, Adv. Nucl. Phys. **13**, 1 (1984).

5. L. R. Dodd, A. W. Thomas and R. F. Alvarez-Estrada, Phys. Rev. D **24**, 1961 (1981).
6. D. H. Lu, A. W. Thomas and A. G. Williams, Phys. Rev. C **57**, 2628 (1998)
7. A. W. Thomas *et al.*, J. Phys. Conf. Ser. **9**, 321 (2005) [arXiv:hep-lat/0502002].
8. A. W. Thomas, J. Phys. G **7**, L283 (1981).
9. A. Manohar and H. Georgi, Nucl. Phys. B **234**, 189 (1984).
10. E. Jenkins and A. V. Manohar, Phys. Lett. B **259**, 353 (1991).
11. A. W. Thomas and G. Krein, Phys. Lett. B **456**, 5 (1999) [arXiv:nucl-th/9902013].
12. I. Passchier *et al.*, Phys. Rev. Lett. **82**, 4988 (1999) [arXiv:nucl-ex/9907012].
13. J. Bermuth *et al.*, Phys. Lett. B **564**, 199 (2003) [arXiv:nucl-ex/0303015].
14. H. Zhu *et al.* [E93026 Collaboration], Phys. Rev. Lett. **87**, 081801 (2001)
15. J. J. Kelly, AIP Conf. Proc. **698**, 393 (2004).
16. J. Speth and A. W. Thomas, Adv. Nucl. Phys. **24**, 83 (1997).
17. J. D. Sullivan, Phys. Rev. D **5**, 1732 (1972).
18. M. Ericson and A. W. Thomas, Phys. Lett. B **128**, 112 (1983).
19. C. H. Llewellyn Smith, Phys. Lett. B **128**, 107 (1983).
20. A. W. Thomas, Phys. Lett. B **126**, 97 (1983).
21. H. Abramowicz *et al.*, Z. Phys. C **15**, 19 (1982).
22. P. Amaudruz *et al.* [NMC], Phys. Rev. Lett. **66**, 2712 (1991).
23. A. I. Signal *et al.*, Mod. Phys. Lett. A **6**, 271 (1991).
24. A. I. Signal and A. W. Thomas, Phys. Lett. B **191**, 205 (1987).
25. W. Melnitchouk and M. Malheiro, Phys. Lett. B **451**, 224 (1999)
26. S. Theberge and A. W. Thomas, Nucl. Phys. A **393**, 252 (1983).
27. A. W. Thomas, Nucl. Phys. Proc. Suppl. **119**, 50 (2003)
28. D. B. Leinweber *et al.*, Nucl. Phys. A **755**, 59 (2005)
29. D. B. Leinweber *et al.*, Phys. Rev. D **60**, 034014 (1999)
30. D. B. Leinweber *et al.*, Phys. Rev. D **61**, 074502 (2000)
31. K. Goeke *et al.*, Eur. Phys. J. A **27**, 77 (2006);
    K. Goeke *et al.*, arXiv:hep-ph/0702031.
32. W. Detmold *et al.*, Phys. Rev. D **66**, 054501 (2002)
33. G. P. Lepage, arXiv:nucl-th/9706029.
34. D. B. Leinweber *et al.*, Phys. Rev. Lett. **92**, 242002 (2004)
35. S. Aoki *et al.* [CP-PACS Collaboration], Phys. Rev. Lett. **84**, 238 (2000)
36. R. D. Young *et al.*, Phys. Rev. D **66**, 094507 (2002)
37. R. Sommer, Nucl. Phys. B **411**, 839 (1994) [arXiv:hep-lat/9310022].
38. D. B. Leinweber *et al.*, Phys. Rev. Lett. **94**, 212001 (2005)
39. D. B. Leinweber *et al.*, Phys. Rev. Lett. **97**, 022001 (2006)
40. R. D. Young *et al.*, Phys. Rev. Lett. **97**, 102002 (2006)
41. F. Myhrer and A. W. Thomas, "Resolution of the Proton Spin Problem", to be published (2007)
42. M. Strikman and C. Weiss, Phys. Rev. D **69**, 054012 (2004)

INCLUSIVE PERSPECTIVES

PAUL HOYER

Department of Physical Sciences, University of Helsinki, P.O.B. 64, FIN-00014 University of Helsinki, Finland
www.helsinki.fi/~hoyer

I discuss the relation between inclusive and exclusive dynamics suggested by Bloom-Gilman duality. Duality implies the simultaneous applicability of two distinct limits, the standard DIS limit of hard inclusive processes taken at fixed x_B and a limit where the hadronic mass is held fixed. I review experimental evidence for the relevance of the fixed mass limit in inclusive processes at high x_F. Semi-local duality suggests that inclusive and exclusive processes occur on the same target Fock states. DIS scaling then implies that the Fock states contributing to hard exclusive processes have a large transverse size, *i.e.*, that the hard scattering occurs off a single parton which carries a large fraction of the hadron momentum.

The inclusive-exclusive connection

I shall take an inclusive perspective on exclusive dynamics, which is the focus of this workshop. Following Drell, Yan and West[1] we expect, as illustrated in Fig. 1, to recover exclusive form factors from Deep Inelastic Scattering (DIS) in the limit of $x_B \to 1$, where $x_B = Q^2/2m\nu$ is the Bjorken variable.

As first observed by Bloom and Gilman,[2] and more recently confirmed also for nuclear target and spin dependence by data from Jefferson Lab and DESY,[3] the inclusive-exclusive connection works much better than anyone could have expected. In Fig. 2 the $ep \to eN^*$ cross sections for $N^* = \Delta(1232)$ and $S_{11}(1535)$, which determine the corresponding $p \to N^*$ transition form factors, are compared to the scaling (high Q^2) DIS cross section at common values of the Nachtmann variable $\xi \simeq x_B$,

$$\xi = \frac{2x_B}{1+\sqrt{1+4x_B^2 m^2/Q^2}} = x_B\left(1 - x_B^2\frac{m^2}{Q^2} + \ldots\right) \tag{1}$$

The x_B (and ξ) of a given N^* tends to unity, $x_B \to 1$, with increasing

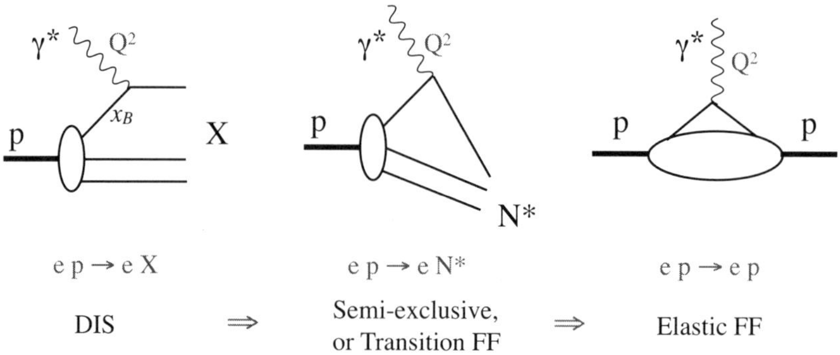

Fig. 1. The inclusive – exclusive connection in $ep \to eX$. As $x_B = Q^2/2m\nu \to 1$ the mass W of the hadronic final state X decreases towards the target nucleon mass. There is a continuous connection between inclusive DIS, $p \to N^*$ transition form factors and the nucleon elastic form factor.

virtuality Q^2 of the photon,

$$M_{N^*}^2 = m_N^2 + \frac{1 - x_B}{x_B} Q^2 \tag{2}$$

Bloom-Gilman duality refers to the remarkable fact that the N^* cross sections at low Q^2 equal the (smooth, scaling) DIS cross section measured at high Q^2 but at the same value of ξ.

The duality between exclusive resonances and inclusive DIS is all the more impressive since exclusive scattering is *coherent* on the entire target, whereas DIS scaling is obtained from *incoherent* scattering on single partons. According to (2) the high Q^2 limit of exclusive processes is taken with $x_B \to 1$:

$$Q^2 \to \infty \text{ at fixed } (1 - x_B)Q^2 \qquad \text{(BB limit)} \tag{3}$$

Berger and Brodsky[5] first pointed out novel coherence effects in this BB limit of the Drell-Yan process $\pi N \to \mu^+\mu^- X$[a]. The BB limit (3) may be contrasted with the usual Bjorken limit of hard inclusive processes,

$$Q^2 \to \infty \text{ at fixed } x_B \qquad \text{(Bj limit)} \tag{4}$$

[a]Berger and Brodsky found multiparton coherence even when $Q(1-x)$ was held fixed. This earlier onset of coherent effects is apparently due to the leading twist quark distribution of the pion, $f_{q/\pi}(x)$, being suppressed for $x \to 1$ due to the helicity mismatch between the pion and the quark.

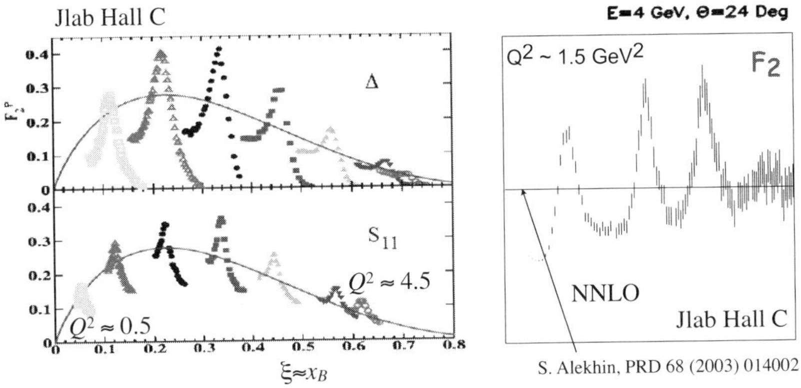

Fig. 2. Bloom-Gilman duality.[3] Left: The proton structure function $F_2(\xi)$ for $ep \to e + X$ is plotted in the $X = \Delta(1232)$ and $X = S_{11}(1535)$ resonance regions for several values of Q^2 in the range $0.5 \ldots 4.5$ GeV^2. The Nachtmann variable ξ defined in (1) equals x_B up to target mass corrections. The smooth curve is a fit to the Jlab data at high Q^2. Right: The F_2 structure function in the resonance region at $Q^2 = 1.5$ GeV^2 compared to the high Q^2 NNLO fit from Ref. [4].

which is the basis of the twist expansion. The twist expansion does not apply in the BB limit (3), where (as we shall presently discuss) scattering on several partons remains coherent.

The relation between hard inclusive and exclusive processes demonstrated by duality suggests that DIS is not as incoherent as often thought. In fact, it has for some time been realized that even in the Bj limit the hard photon interaction of DIS remains coherent with soft rescattering of the struck quark on spectators in the target.[6] Recently, this has led to doubts about the validity of factorization in hard hadronic processes.[7]

Conversely, duality shows that hard exclusive scattering has features in common with inclusive dynamics. Unfortunately, little is known about the general properties of scattering in the BB limit (3), *e.g.*, how the hard subprocess of scale Q may be factored from incoherent soft processes in this limit. Higher twist corrections to DIS have been observed to increase at large x_B,[8] and are expected to be governed by the scale $Q^2(1 - x_B)$ (which is fixed in the BB limit). The dominant production mechanism of heavy quarks was shown to be multiple soft scattering in the target (rather than hard scattering on a single target parton) when the quark pair mass $M_{Q\bar{Q}}$ increases with the momentum fraction x_F carried by the quarks such that $(1 - x_F)M_{Q\bar{Q}}^2$ is fixed.[9]

In the following I shall recall some general features of coherence in the BB limit, consider experimental evidence that this limit is relevant, and argue that it merits further study.

Hard-soft coherence in large x Fock states

Partons are mutually coherent when their lifetimes (inverse energies) are commensurate. Interactions of coherent partons are added at the amplitude level, *i.e.*, interference effects cannot be neglected. The Light-Front (LF) energy[b] of a Fock state with total momentum P is

$$P^- \equiv P^0 - P^z = \sum_i \frac{p_{i\perp}^2 + m_i^2}{x_i P^+} \qquad \left(\sum_i x_i = 1\right) \tag{5}$$

where x_i, $\boldsymbol{p}_{i\perp}$ and m_i denote the momentum fraction, transverse momentum and mass of parton i, respectively. Contributions to P^-P^+ of order Q^2 can thus arise in two distinct ways:

- From *hard* partons with $p_{i\perp}^2 \sim Q^2$ or $m_i^2 \sim Q^2$
- From *soft* partons with $p_{i\perp}^2 \sim m_i^2 \sim \Lambda_{QCD}^2$ but $x_i \sim \Lambda_{QCD}^2/Q^2$

Thus hard partons with large x_i can be coherent with soft partons of small x_i. If, as in the BB limit, the hard parton takes nearly all the momentum,

$$x \to 1 \quad \text{with} \quad (1-x)p_\perp^2 \sim \Lambda_{QCD}^2 \quad \text{fixed} \tag{6}$$

then all the other partons have small $x_i \sim \Lambda_{QCD}^2/p_\perp^2$ and give contributions of $\mathcal{O}\left(p_\perp^2\right)$ to P^-P^+ in (5). Thus the full Fock state interacts coherently in the BB limit.

Asymmetric Fock states where one quark carries nearly all the momentum are not as unusual as they may seem at first. A good example is provided by DIS itself (Figs. 3 and 4). In the target rest frame, where the virtual photon has large positive longitudinal momentum ($q^+ \simeq 2\nu$), the photon typically fluctuates into a $\mathrm{q\bar{q}}$ pair before interacting in the target. The momentum fraction z carried by the quark is distributed according to the splitting function

$$\frac{dP(\gamma \to \mathrm{q\bar{q}})}{dz} \propto \left[z^2 + (1-z)^2\right] \tag{7}$$

[b]When the longitudinal momentum components are large compared to masses and transverse momenta coherence measured by the LF and ordinary energies give the same result, since $E = \sqrt{p_\parallel^2 + p_\perp^2 + m^2} \simeq |p_\parallel| + (p_\perp^2 + m^2)/2|p_\parallel|$. The $|p_\parallel|$ terms cancel in energy differences due to momentum conservation.

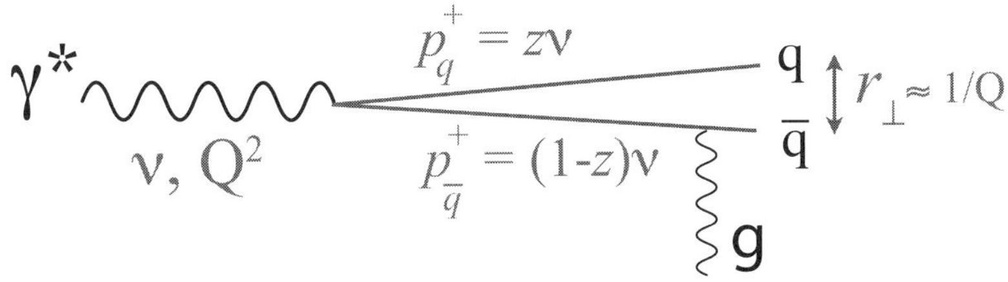

Fig. 3. DIS as viewed in the target rest frame. The virtual photon fluctuates into a $q\bar{q}$ pair which interacts with a target gluon. When the momentum fractions z and $1-z$ of the quarks are similar the pair has a small transverse size $r_\perp \sim 1/Q$ and the gluon is hard.

The quark and antiquark typically carry commensurate momentum fractions $z = \mathcal{O}\left(\frac{1}{2}\right)$. However, only $q\bar{q}$ Fock states which have lifetimes of the same order as the virtual photon are coherent with (and hence can contribute to) the hard process. $P^-P^+ \sim \mathcal{O}\left(Q^2\right)$ in (5) requires $p_{q\perp} \sim Q$. Both quarks are 'hard' and the size of the Fock state is $r_\perp \sim 1/p_{q\perp} \sim 1/Q$. The quark pair is a color singlet and its cross section in the target is $\sigma(q\bar{q}) \sim 1/Q^2$. Color transparency forces the target gluon to be hard and perturbative. The corresponding DIS subprocess $\gamma^* g \to q\bar{q}$ is therefore of higher order in α_s.

The lowest order (parton model) subprocess $\gamma^* q \to q$ appears as an *endpoint* contribution where $1 - z \sim \Lambda_{QCD}^2/Q^2$ (Fig. 4). The antiquark then has finite momentum in the target[c] even in the Bj limit (4), $p_{\bar{q}}^+ = (1-z)q^+ \sim \Lambda_{QCD}^2/mx_B$. Because the photon splitting function (7) is finite for $z \to 1$ the probability for the asymmetric splitting is given by the width of the z interval: $P(\gamma \to q\bar{q}) \sim \Delta z \sim \Lambda_{QCD}^2/Q^2$. The asymmetric Fock state is coherent with the virtual photon since the product P^-P^+ in (5) is of $\mathcal{O}\left(Q^2\right)$ due to the 'soft' antiquark contribution ($x_{\bar{q}} \sim \Lambda_{QCD}^2/Q^2$). With $p_{q\perp} \sim \Lambda_{QCD}$ the transverse size of the Fock state is large and it interacts in the target with a normal hadronic cross section. The resulting DIS cross section thus scales dimensionally, $\sigma_{DIS} \sim P(\gamma \to q\bar{q})/\Lambda_{QCD}^2 \sim 1/Q^2$. The gluons in Fig. 4 are nonperturbative and represent the soft scattering of the antiquark in the target.

[c]This is obviously required in order that the antiquark entering the target can equivalently be interpreted as a quark emerging from the target.

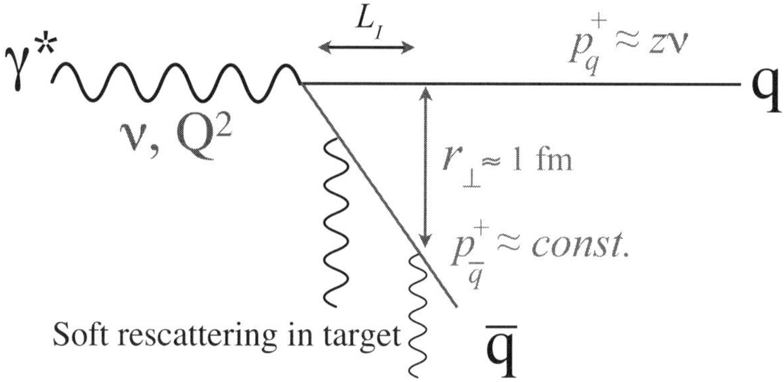

Fig. 4. If the virtual photon fluctuates into an asymmetric q$\bar{q}$ pair, such that the momentum of the antiquark in the target does not grow with ν and Q^2, the pair has a large transverse size $r_\perp \sim 1$ fm. The nonperturbative scattering of the pair in the target (which occurs within the Ioffe length L_I of the photon vertex) determines the DIS cross section.

For our present discussion it is essential to note that the soft interactions of the antiquark are coherent with the virtual photon interaction as long as they occur within the Ioffe length $L_I \simeq 1/2m_N x_B$. The photon coherence length is Lorentz dilated in the target rest frame,

$$L_I \simeq \frac{1}{Q} \cdot \frac{\nu}{Q} = \frac{\nu}{Q^2} \tag{8}$$

and thus remains finite in the Bj limit. Soft, coherent interactions of the antiquark are unsuppressed and in fact *required* for the fast quark to materialize as a jet in the final state (in $A^+ = 0$ gauge the fast quark does not interact).

The possibility that soft interactions of partons with small momentum fractions affect the scattering of hard partons with high x may be relevant also for the dynamics of exclusive processes. The magnitude of such effects depends on the behaviour of hadron wave functions in the limit where one parton carries nearly all the hadron momentum. This is a non-perturbative issue which at present must be decided by experiment. I next review some data involving hadron wave functions which points to the relevance of these coherence effects.

Experimental hints

Lepton pair production

The BB limit (3) of the Drell-Yan amplitude is sketched in Fig. 5. As the virtuality Q^2 of the lepton pair increases with its momentum fraction x_F such that $(1 - x_F)Q^2$ is fixed, the stopped quark with a small fraction $1 - x \sim 1 - x_F$ remains coherent with the virtual photon interaction. Also the time-scale of the gluon which transfers longitudinal momentum between the quarks in the pion is commensurate with that of the virtual photon. Hence the produced leptons are coherent with the full wave function of the pion. It turns out[5] that the virtual photon is *longitudinally* polarized in this kinematic limit, in contrast to the transverse polarization obtained for $Q^2 \to \infty$ at fixed x_F. This reflects the conservation of helicity from the pion to the lepton pair.

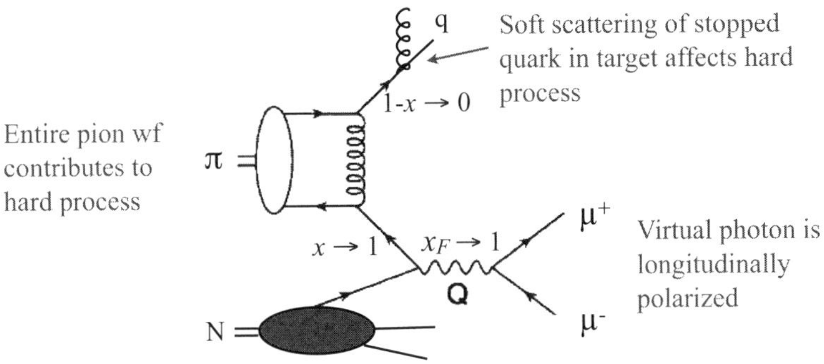

Fig. 5. The Drell-Yan process $\pi N \to \mu^+\mu^- X$ in the BB limit (3) where the momentum fraction of the lepton pair $x_F \to 1$ such that $(1 - x_F)Q^2$ is fixed. The virtuality of the gluon which transfers longitudinal momentum onto the annihilating antiquark scales as Q^2. The gluon exchange, as well as the subsequent soft rescattering of the stopped quark, are coherent with the photon interaction. The lepton pair senses the entire pion wave function and carries zero helicity in the BB limit.

The E615 data[10] (Fig. 6), while not agreeing with the details of the model calculation, does indeed show a trend for the photon to become longitudinally polarized for $x_F \gtrsim 0.8$ at $Q^2 \simeq 20$ GeV^2. It is interesting to note that this effect sets in already at a sizeable $(1 - x_F)Q^2 \simeq 4$ GeV^2. There is evidence[11] for a similar effect in J/ψ production at high x_F.

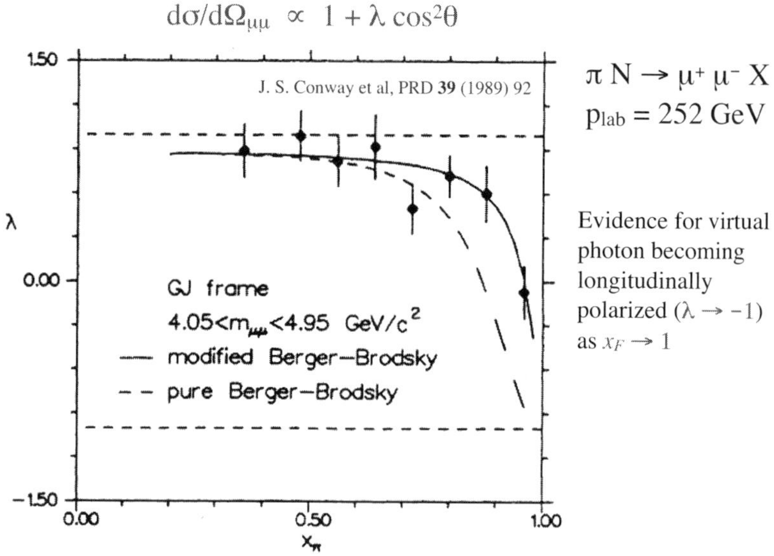

Fig. 6. E615 data[10] on the angular distribution of the muon pair in $\pi N \to \mu^+\mu^- X$ at $Q^2 \simeq 20\ \mathrm{GeV}^2$. The polarization of the pair changes from transverse towards longitudinal when the pair carries a high momentum fraction of the pion.

Single spin asymmetry

In an inclusive process $a + b \to c + X$ parity allows the cross section to depend on the spin component orthogonal to the reaction plane of one of the particles. This single spin asymmetry (SSA) of a spin $\frac{1}{2}$ particle is given by

$$A_N = \frac{d\sigma^\uparrow - d\sigma^\downarrow}{d\sigma^\uparrow + d\sigma^\downarrow} = \frac{2\Sigma_{\{\sigma\}}\mathrm{Im}\left[\mathcal{M}^*_{\leftarrow,\{\sigma\}}\mathcal{M}_{\rightarrow,\{\sigma\}}\right]}{\Sigma_{\{\sigma\}}\left[\left|\mathcal{M}_{\rightarrow,\{\sigma\}}\right|^2 + \left|\mathcal{M}_{\leftarrow,\{\sigma\}}\right|^2\right]} \tag{9}$$

where the $\mathcal{M}$ are helicity amplitudes, and the helicities $\{\sigma\}$ of all particles except the polarized one are summed over. From this expression it is clear that $A_N \neq 0$ requires

- Helicity flip
- A dynamical phase (absorptive part)

When the produced particle c has large transverse momentum these features must occur in interactions which are *coherent* with the hard subprocess for a sizeable asymmetry to occur. However, both helicity flip and absorptive parts are strongly suppressed in perturbative processes, being proportional to current quark masses and α_s, respectively. Thus it was long ago noted[12] that A_N should be small in large $p_\perp$ processes.

Data does not agree with these theoretical expectations. The Lambda polarization in $pp \to \Lambda + X$ increases strongly[13] with $x_F(\Lambda)$, reaching values $A_N \gtrsim 0.3$ at $x_F \simeq 0.8$. A_N furthermore shows no sign of decreasing with transverse momentum, in the measured range $p_\perp(\Lambda) \lesssim 3$ GeV. Similar results (Fig. 7) were later obtained for $p^\uparrow p \to \pi + X$ by E704[14] and recently by STAR.[15]

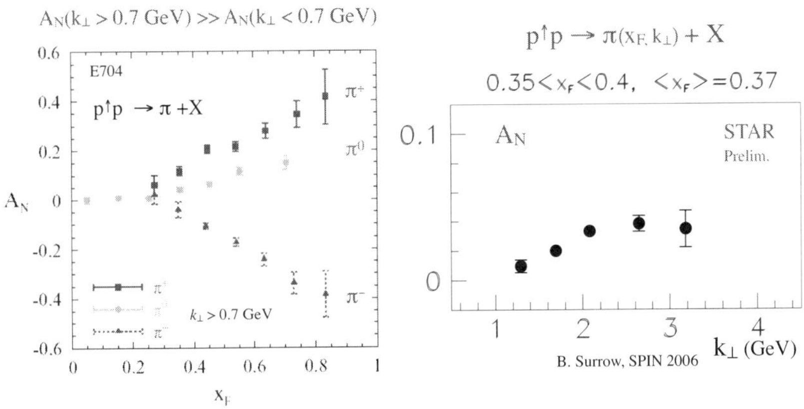

Fig. 7. The single spin asymmetry (9) in $p^\uparrow p \to \pi + X$. Left: Fermilab E704 data[14] as a function of x_F, for pions with transverse momenta $k_\perp > 0.7$ GeV. Right: The $k_\perp$ dependence from STAR[15] at $x_F \simeq 0.37$.

The fact that A_N does not decrease with transverse momentum up to several GeV is hard to understand in a standard perturbative QCD analysis. A_N is antisymmetric in the transverse momentum $p_{c\perp}$ of particle c, and is normalized by the total reaction rate which is even in $p_{c\perp}$. Hence it is inevitable that $A_N \propto 1/p_{c\perp}$ for $p_{c\perp} \to \infty$. It has also formally been shown[16] that the SSA is a 'twist-3' effect.

For a single quark to produce a pion with $x_F \simeq 0.8$ (as in the $pp \to \pi + X$ E704 data) it would have to carry $x \gtrsim 0.9$ of the proton momentum, and then deliver a fraction $z \gtrsim 0.9$ of its momentum to the pion. At such large values of x and z the quark distribution and fragmentation functions are very small. It was in fact shown[18] that the leading twist cross section is an order of magnitude below the E704 cross section.

A possible way out of this dilemma is that the BB limit (3) is more relevant for describing the data than the twist expansion, at the measured values of x_F and $p_\perp$. This is suggested by the increase of A_N for $x_F \to 1$ (Fig. 7) and by the moderate values of $(1-x_F)p_\perp^2$ as compared to the onset of the longitudinal virtual photon polarization in the E615 data on lepton pair production (Fig. 6).

The coherence of soft and hard processes in the BB limit allows the helicity flip and absorptive part required for $A_N \neq 0$ to be generated in a *soft part of the amplitude* (Fig. 8), thus avoiding the suppression noted in Ref.[12] This was explicitly verified in a perturbative analysis.[17] The large values $A_N \simeq 0.4$ seen experimentally[d] are also easier to understand when the hard process is coherent over the whole wave function of the polarized hadron.

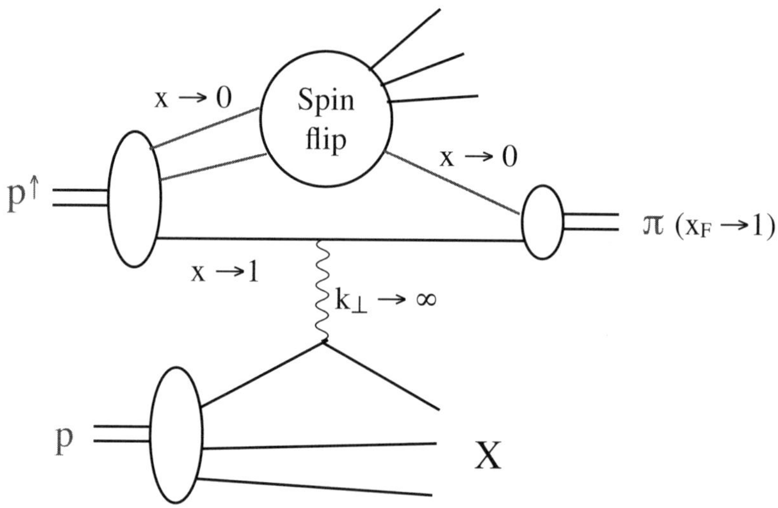

Fig. 8. Dynamics[17] of the single spin asymmetry in the BB limit (3). The hard scattering occurs off the parton with high momentum fraction x (the energy of the projectile is assumed to much larger than the transverse momentum, $E_p \gg k_\perp$). The soft interactions of the partons with low x are coherent with the hard scattering. Hence a helicity flip and absorptive part in the soft scattering suffices to correlate the pion angular distribution with the projectile spin.

[d]The A_N measuresd in DIS is an order of magnitude smaller.[19]

Bloom-Gilman duality

Finally I return to Bloom-Gilman duality (Fig. 2), and to what it may teach us about the dynamics of exclusive processes. Consider the target hadron as a superposition of Fock states. The least 'miraculous' explanation of duality is that, for a given x_B, inclusive and exclusive final states are produced *by the same (mixture of) target Fock states*, in the whole range of Q^2 at which duality applies.[20] The precocious scaling in inclusive DIS implies that the photon wavelength is small compared to the interparton distance in the Fock state, so that the hard scattering (for exclusive as well as inclusive processes) occurs on a single quark. Coherence between the struck quark and soft partons in the same Fock state is still possible, according to our discussion above of the BB limit (3).

At low values of Q^2 the mass W of the hadronic final state is in the resonance region (*cf.* (2)). The formation of resonances occurs, however, at a late time, $t_R > L_I$ in (8), and is thus incoherent with the hard photon scattering. The scattering probability has already been 'set', and may only be distributed within the mass uncertainty prevailing at the resonance formation time, $\Delta W \sim 1/t_R$. The semilocal duality of Fig. 2 indicates that ΔW is of the order of the resonance spacing. Thus unitarity ensures that the resonance bumps average the smooth (high Q^2) scaling curve.

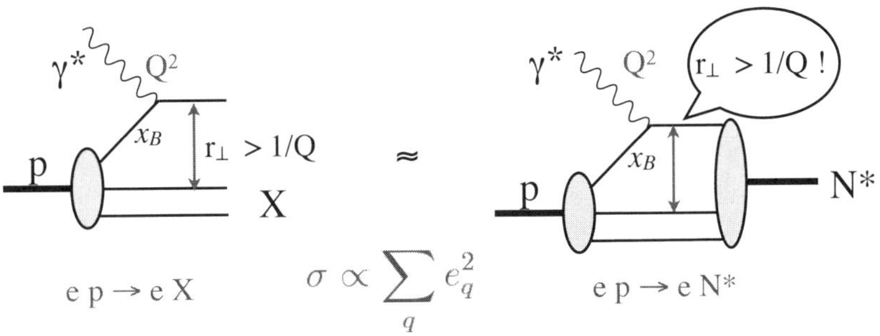

Fig. 9. Bloom-Gilman duality follows naturally if[20] the virtual photon scatters off the same (superposition of) target Fock states at all Q^2. At low Q^2 (*rhs.*) the formation of the N^* resonance in the final state is incoherent with the photon interaction and thus does not affect the cross section. Scaling at high Q^2 (*lhs.*) implies that the photon strikes a single quark in the target. Duality then requires this to be the case also for the $p \longrightarrow N^*$ transition form factor.

The above scenario requires that target Fock states contributing to exclusive form factors have large transverse size compared to the photon resolution, $r_\perp > 1/Q$. This is the 'end-point' contribution in the formalism developed by Brodsky and Lepage (BL).[21] The struck quark carries nearly all the hadron momentum, as in the BB limit (3). While there is no doubt about the BL analysis for compact Fock states with $r_\perp \sim 1/Q$, the importance of the end-points has been debated for a long time.[22] The above interpretation of duality suggests that the situation is analogous to DIS, where photon splitting into compact $\mathrm{q\bar{q}}$ states corresponds to the higher order subprocess $\gamma^* g \to \mathrm{q\bar{q}}$, whereas the dominant parton model contribution $\gamma^*\mathrm{q} \to \mathrm{q}$ originates from asymmetric states of large transverse extent.

The importance of the endpoint contribution in exclusive processes depends on the behaviour of hadron wave functions in the limit where one parton takes nearly all the momentum, $x \to 1$. It is sometimes assumed that this behaviour may be calculated perturbatively, starting from a non-perturbative wave function with no support at high x. However, the E615 data[10,23] indicates that $f_{q/\pi}(x) \propto (1-x)^{1.2}$ for $x \to 1$, disfavouring the perturbative result $f_{q/\pi}(x) \propto (1-x)^2$ for the quark distribution in the pion. Furthermore, in the non-perturbative AdS/CFT approach[24] the pion distribution amplitude $\phi_\pi(x) \sim \sqrt{x(1-x)}$ falls off slower at the endpoints than the perturbative result $\phi_\pi(x) \sim x(1-x)$.

Measurements of color transparency can in principle determine the transverse size of Fock states contributing to, say, $ep \to ep$ at high Q^2. In scattering on nuclei, $eA \to ep(A-1)$, compact proton Fock states emerging from the hard process would be transparent to the nucleus. Present data[25] shows little evidence for color transparency in these processes, suggesting a large transverse size for the relevant proton Fock states.

The properties of subprocess amplitudes in QCD allow endpoint contributions to be important. For example, the effective size of the photoproduction subprocess $\gamma u \to \pi^+ d$ at high momentum transfer t is measured[26] by its sensitivity to a small virtuality Q^2 of the (transverse) photon (Fig. 10). The derivative of the differential cross section $d\sigma/dt$ *wrt.* Q^2 turns out to diverge logarithmically at $Q^2 = 0$, formally implying an infinite size. The effective size would be even larger for a pion distribution amplitude which vanishes more slowly than the perturbative one at the endpoints. Helicity flip and rescattering effects also enhance endpoint contributions.

Concluding remarks

Bloom-Gilman duality expresses a remarkable relation between inclusive and exclusive processes. It requires the simultaneous and precocious validity of two distinct high Q^2 limits: The Bj limit (4) where x_B is held fixed and

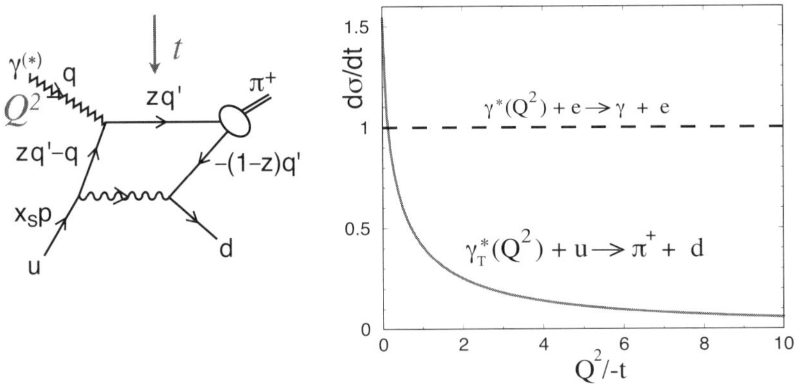

Fig. 10. The effective size of the QCD subprocess $\gamma + u \to \pi^+ + d$ at large momentum transfer (left) can be 'measured' by making the photon slightly virtual.[26] The Q^2 derivative of the cross section is logarithmically infinite at $Q^2 = 0$ (right), assuming a pion distribution amplitude of asymptotic form. This indicates that the QCD subprocess is not transversally compact, in contrast to QED Compton scattering (dashed line).

what we have called the BB limit (3). In the latter limit the mass W of the hadronic system is fixed, and it is thus appropriate for exclusive form factors ($M_{N^*} = W$).

The dynamics of the two limits is very different. All partons in a Fock state where one parton carries nearly all the momentum, $x \sim 1-\Lambda^2_{QCD}/Q^2$, remain coherent in the $Q^2 \to \infty$ limit. I argued that there is experimental evidence of such contributions in the Drell-Yan process and in single spin asymmetries at high x_F. This kind of Fock states may also be relevant for exclusive processes, where they appear as endpoint contributions in the standard exclusive framework developed by Brodsky and Lepage.

The multiparton coherence of the BB limit implies that the standard twist expansion is inapplicable. It appears desirable to improve our understanding of this limit, including the factorization of coherent from incoherent scattering dynamics.

Acknowledgments

I wish to commend the Organizers for this interesting workshop and thank them for their hospitality.

References

1. S. D. Drell and T. M. Yan, Phys. Rev. Lett. **24**, 181 (1970);
   G. B. West, Phys. Rev. Lett. **24**, 1206 (1970).
2. E. D. Bloom and F. J. Gilman, Phys. Rev. Lett. **25**, 1140 (1970);
   Phys. Rev. D **4**, 2901 (1971).
3. W. Melnitchouk, R. Ent and C. Keppel, Phys. Rept. **406**, 127 (2005) [arXiv:hep-ph/0501217].
4. S. Alekhin, Phys. Rev. D **68**, 014002 (2003) [arXiv:hep-ph/0211096].
5. E. L. Berger and S. J. Brodsky, Phys. Rev. Lett. **42**, 940 (1979).
6. S. J. Brodsky, P. Hoyer, N. Marchal, S. Peigne and F. Sannino, Phys. Rev. D **65**, 114025 (2002) [arXiv:hep-ph/0104291];
   S. J. Brodsky, D. S. Hwang and I. Schmidt, Phys. Lett. B **530**, 99 (2002) [arXiv:hep-ph/0201296]. J. C. Collins, Phys. Lett. B **536**, 43 (2002) [arXiv:hep-ph/0204004].
7. J. Collins and J. W. Qiu, Phys. Rev. D **75**, 114014 (2007) [arXiv:0705.2141 [hep-ph]].
8. M. Virchaux and A. Milsztajn, Phys. Lett. B **274**, 221 (1992);
   S. I. Alekhin and A. L. Kataev, Phys. Lett. B **452**, 402 (1999) [arXiv:hep-ph/9812348].
9. S. J. Brodsky, P. Hoyer, A. H. Mueller and W. K. Tang, Nucl. Phys. B **369**, 519 (1992).
10. J. S. Conway *et al.*, Phys. Rev. D **39**, 92 (1989).
11. C. Biino *et al.*, Phys. Rev. Lett. **58**, 2523 (1987).
12. G. L. Kane, J. Pumplin and W. Repko, Phys. Rev. Lett. **41**, 1689 (1978).
13. B. Lundberg *et al.*, Phys. Rev. D **40**, 3557 (1989).
14. D. L. Adams *et al.* [FNAL-E704 Collaboration], Phys. Lett. B **264**, 462 (1991).
15. B. Surrow [STAR Collaboration], AIP Conf. Proc. **915**, 293 (2007) [arXiv:0705.3483 [hep-ex]].
16. J. W. Qiu and G. Sterman, Phys. Rev. Lett. **67**, 2264 (1991); Nucl. Phys. B **378**, 52 (1992); Phys. Rev. D **59**, 014004 (1999) [arXiv:hep-ph/9806356].
17. P. Hoyer and M. Järvinen, JHEP **0702**, 039 (2007) [arXiv:hep-ph/0611293].
18. C. Bourrely and J. Soffer, Eur. Phys. J. C **36**, 371 (2004) [arXiv:hep-ph/0311110].
19. A. Airapetian *et al.* [HERMES Collaboration], Phys. Rev. Lett. **94**, 012002 (2005) [arXiv:hep-ex/0408013].
20. P. Hoyer, Proc. Frascati 2005, *Quark-hadron duality and the transition to pQCD,* pp. 263-270, [arXiv:hep-ph/0509139].
21. G. P. Lepage and S. J. Brodsky, Phys. Rev. D **22**, 2157 (1980).
22. N. Isgur and C. H. Llewellyn Smith, Phys. Rev. Lett. **52**, 1080 (1984) and Nucl. Phys. **B317**, 526 (1989);
    A.V. Radyushkin Nucl. Phys. **A527**, 153c (1991) and Nucl. Phys. **A532**, 141c (1991);
    S. J. Brodsky and G. P. Lepage, In *Perturbative Quantum Chromodynamics,* (A. H. Mueller, Ed.) World Scientific (1989), p. 23-240 (SLAC-PUB-4947).
23. W. Melnitchouk, Eur. Phys. J. A **17**, 223 (2003) [arXiv:hep-ph/0208258].

24. S. J. Brodsky and G. F. de Teramond, Phys. Rev. Lett. **96**, 201601 (2006) [arXiv:hep-ph/0602252].
25. K. Garrow *et al.*, Phys. Rev. C **66**, 044613 (2002) [arXiv:hep-ex/0109027].
26. P. Hoyer, J. T. Lenaghan, K. Tuominen and C. Vogt, Phys. Rev. D **70**, 014001 (2004) [arXiv:hep-ph/0210124].

FITTING DVCS AT NLO AND BEYOND

K. KUMERIČKI

Department of Physics, Faculty of Science, University of Zagreb
HR-10002 Zagreb, Croatia

D. MÜLLER

Institut für Theoretische Physik II, Ruhr-Universität Bochum
D-44780 Bochum, Germany

K. PASSEK-KUMERIČKI

Theoretical Physics Division, Rudjer Bošković Institute
HR-10002 Zagreb, Croatia

We outline the twist-two analysis of deeply virtual Compton scattering within the conformal partial wave expansion of the amplitude, represented as a Mellin–Barnes integral. The complete next-to-leading order results, including evolution, are obtained in the $\overline{\mathrm{MS}}$ and a conformal factorization scheme. Within the latter, exploiting conformal symmetry, the radiative corrections are evaluated up to next-to-next-to-leading order. Using a new proposed parameterization for GPDs, we study the convergence of perturbation theory and demonstrate that our formalism is suitable for a fitting procedure of DVCS observables.

1. Introduction

The proton structure has been widely explored in inclusive measurements, mainly, in deeply inelastic lepton-proton scattering (DIS). Here the scattering essentially occurs due to the exchange of a virtual boson (photon) on a single parton, and so one can access *parton distribution functions* (PDFs). This universal, however, convention-dependent functions $q_a(x)$ are interpreted as probabilities that partons of certain flavour a will be found with given longitudinal momentum fraction x. Since the PDFs are naturally defined in a translation invariant manner, they do not carry information about the transversal distribution of partons. Some information about transversal degrees of freedom can be obtained from elastic lepton-proton scattering. Namely, the electromagnetic form factors $F_{1,2}(t)$ are Fourier transforms of the electric and magnetic charge distribution in nucleon, and can be, e.g., in the infinite momentum frame, interpreted as probability that partons are found at some transversal distance **b** from the center-of-mass.

However, it is anticipated that longitudinal and transversal degrees of freedom have a cross talk.[1] The three dimensional distribution of partons in the nucleon can be addressed within *generalized parton distributions* (GPDs).[2,3] The usefulness of the GPDs has also been widely realized in connection with the spin puzzle, since they encode the angular momentum carried by the individual parton species.[3] These intrinsically non-perturbative quantities can be revealed by analyzing hard exclusive leptoproduction of mesons or photon. The former processes are perhaps theoretically more problematic to describe but offer a direct view into individual flavour GPDs. To the latter one the *deeply virtual Compton scattering* (DVCS) process contributes. DVCS is theoretically considered as the cleanest probe of GPDs, however, only certain flavour combinations of GPDs appear. Comprehensive reviews on GPDs can be found in Ref. [**?**,4].

The non-forward Compton scattering process is described by the tensor

$$T_{\mu\nu}(q,P,\Delta) = \frac{i}{e^2}\int d^4x\, e^{ix\cdot q}\langle P_2,S_2|Tj_\mu(x/2)j_\nu(-x/2)|P_1,S_1\rangle, \qquad (1)$$

where $q=(q_1+q_2)/2$, $P=P_1+P_2$ and $\Delta=P_2-P_1$. The generalized Bjorken limit corresponds to $Q^2=-q^2\to\infty$ with the scaling variables

$$\xi = \frac{Q^2}{P\cdot q}, \qquad \eta = -\frac{\Delta\cdot q}{P\cdot q}, \qquad (2)$$

and the momentum transfer squared Δ^2 being fixed. Note that in the forward case the hadronic DIS tensor $W_{\mu\nu}$ arises from the optical theorem: $W_{\mu\nu}=\Im\mathrm{m}T_{\mu\nu}(q,P=2p,\Delta=0)/(2\pi)$, where $p=P_1=P_2$ and $\xi\to x_{Bj}$.

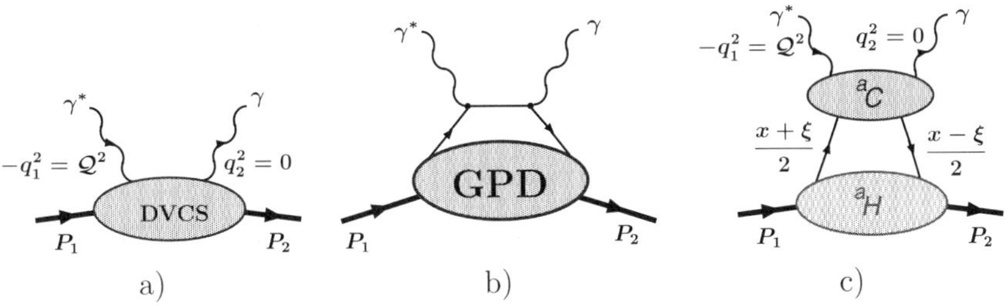

Fig. 1. a) DVCS. b) LO perturbative contribution to DVCS. c) Factorization (to all orders in α_s) on an example of special parity-even, helicity conserving contribution (5).

In DVCS – Fig.1a – the virtuality of the incoming photon $\mathcal{Q}^2 = -q_1^2$ is large while the final photon is on-shell. The skewness parameter η and the Bjorken-like scaling parameter ξ are then equal to twist-two accuracy, i.e., $\eta = \xi + \mathcal{O}(1/\mathcal{Q}^2)$. In the generalized Bjorken limit, similarly as for DIS structure functions, the amplitude factorizes into long- and short-distance contributions (Fig.1b and c): short-distance physics is perturbatively calculable Compton scattering on a parton, while long-distance physics is encoded in a non-perturbative amplitude for a parton being emitted and later reabsorbed by the nucleon. The latter *amplitude* is called GPD.

The GPDs with even parity, we are considering here, are defined as

$$ {}^{q}F(x,\eta,\Delta^2) = \int \frac{dz^-}{2\pi}\, e^{ixP^+z^-} \langle P_2|\bar{q}(-z)\gamma^+ q(z)|P_1\rangle\Big|_{z^+=0,\,\mathbf{z}_\perp=\mathbf{0}} \qquad (3) $$

$$ {}^{G}F(x,\eta,\Delta^2) = \frac{4}{P^+}\int \frac{dz^-}{2\pi}\, e^{ixP^+z^-} \langle P_2|G_a^{+\mu}(-z)G_{a\mu}^{\ \ +}(z)|P_1\rangle\Big|_{z^+=0,\,\mathbf{z}_\perp=\mathbf{0}}, $$

and similarly for odd parity. Moreover, it is convenient to decompose GPDs,

$$ {}^{a}F = \frac{\bar{u}(P_2)\gamma^+ u(P_1)}{P^+}\, {}^{a}H + \frac{\bar{u}(P_2)i\sigma^{+\nu}u(P_1)\Delta_\nu}{2MP^+}\, {}^{a}E\,, \qquad a = q, G\,, \qquad (4) $$

into helicity conserving and non-conserving generalized form factors. The Compton tensor (1) is analogously decomposed into Compton form factors (CFFs) with corresponding parity and helicity properties. In the forward limit ($\Delta \to 0$) GPDs reduce to PDFs. Together with sum rules, e.g., relating GPDs to electromagnetic form factors, this provides constraints that are important for GPD modelling. But as they do not constrain much the skewness (η) dependence, modelling is still a difficult problem. One guiding feature is a polynomiality property of GPDs: n-th (Mellin) moment of GPD is even polynomial in η of order n or $n \pm 1$.

In the following we outline the perturbative QCD approach to DVCS that is based on the conformal partial wave expansion, represented as Mellin-Barnes integral.[6,7] For shortness, we concentrate on the singlet part of the parity-even and helicity-conserving CFF $\mathcal{H}$, i.e., GPD H, which is the dominant contribution in collider experiments.

2. Deeply virtual Compton scattering beyond NLO

The separation of short- and long-distance dynamics in $\mathcal{H}$ can be written as a convolution (Fig.1c) over the longitudinal momentum fraction x,

$$ {}^{a}\mathcal{H}(\xi,\Delta^2,\mathcal{Q}^2) = \int_{-1}^{1} dx\, {}^{a}C(x,\xi,\mathcal{Q}^2/\mu^2)\, {}^{a}H(x,\eta=\xi,\Delta^2,\mu^2)\,, \qquad (5) $$

where the factorization scale μ^2 is often taken $\mu^2 = \mathcal{Q}^2$. Here the index $a \in \{\text{NS},\text{S}[\Sigma, G]\}$ denotes either non-singlet or singlet (containing quarks $[\Sigma]$ and gluons $[G]$) part. The coefficient function C^a is a perturbative quantity, known to NLO, which describes $q\gamma^* \to q\gamma$ and $g\gamma^* \to g\gamma$ subprocesses. Note that the well known leading-order (LO) contribution to C^a is actually a pure QED process (Fig. 1b). Obviously, to stabilize the perturbation series and investigate its convergence one needs the second order in α_s, i.e., next-to-next-to-leading order (NNLO) contributions. The importance of NNLO in singlet case is amplified by the fact that at LO photons scatter only off charged partons, whereas gluons start contributing at NLO.

Instead of the convolution (5), one might sum conformal GPD moments. The resulting series only converges in the unphysical region and it is necessary to resum this series, e.g., by the Mellin-Barnes integral[7]

$$ {}^{\mathrm{S}}\mathcal{H} = \frac{1}{2i}\int_{c-i\infty}^{c+i\infty} dj\, \xi^{-j-1}\left[i + \tan\left(\frac{\pi j}{2}\right)\right] \boldsymbol{C}_j(\mathcal{Q}^2/\mu^2, \alpha_s(\mu))\boldsymbol{H}_j(\xi, \Delta^2, \mu^2)\,. \tag{6} $$

where $\boldsymbol{C}_j = \left({}^{\Sigma}C_j, {}^{G}C_j\right)$ and $\boldsymbol{H}_j = \left({}^{\Sigma}H_j, {}^{G}H_j\right)$ are conformal moments. They are analogous to common Mellin moments used in DIS but the integral kernel x^j is replaced by Gegenbauer functions $C_j^{\nu}(x)$, which are eigenfunctions of LO evolution kernel for quarks ($\nu = 3/2$) and gluons ($\nu = 5/2$).

The advantages of using conformal moments, i.e., Mellin-Barnes representation, are manifold. It allows for an efficient and stable numerical treatment, it enables a simple inclusion of evolution, and it opens a new door for interesting modelling of GPDs. Finally, utilizing the conformal operator product expansion (COPE) and known NNLO DIS results, it enables the assessment of NNLO contributions. Note that neither Wilson coefficients nor anomalous dimensions are calculated (only so-called quark bubble insertions were evaluated) in non-forward kinematics at NNLO.

While massless QCD is conformally invariant at tree level, this invariance is broken at the loop level since renormalization introduces a mass scale, leading to the running of the coupling constant ($\beta \neq 0$). Assuming the existence of a non-trivial fixed point α_s^*, i.e., $\beta(\alpha_s^*) = 0$, the COPE prediction for Wilson coefficients in general kinematics reads

$$ C_j(\alpha_s^*) = c_j(\alpha_s^*)\, {}_2F_1\left({\frac{2+j+\gamma_j(\alpha_s^*)}{4}, \frac{4+2j+\gamma_j(\alpha_s^*)}{4} \atop \frac{5+2j+\gamma_j(\alpha_s^*)}{2}} \middle| \frac{\eta^2}{\xi^2}\right) \left(\frac{\mu^2}{\mathcal{Q}^2}\right)^{\gamma_j(\alpha_s^*)/2} . \tag{7} $$

For $\eta = 0$ Eq. (7) reduces to the DIS Wilson coefficients $C_j \to c_j$ and thus fixes the normalization c_j. The choice $\eta = \xi$ corresponds to DVCS. The anomalous dimensions γ_j, governing the evolution, are the same as in DIS.

The restoration of the running coupling is governed by the renormalization group invariance. It is supplemented by an ambiguous initial condition. We take a prescription ($\overline{\text{CS}}$ scheme) in which the form of the Wilson-coefficients is not altered and where mixing under evolution is "pushed" to NNLO:

$$C_j^{\overline{\text{CS}}} = \sum_{k=j}^{\infty} C_k(\alpha_s(\mathcal{Q}))\mathcal{P}\exp\left\{\int_{\mathcal{Q}}^{\mu}\frac{d\mu'}{\mu'}\left[\gamma_j(\alpha_s(\mu'))\delta_{kj} + \frac{\beta}{g}\Delta_{kj}(\alpha_s(\mu'))\right]\right\} \quad (8)$$

with $C_k(\alpha_s(\mathcal{Q}))$ obtained from the $\eta = \xi$ limit of Eq. (7). The mixing Δ_{kj} matrix, appearing at NNLO, is neglected. We take c_j and γ_j to NNLO accuracy[8] and obtain the DVCS prediction to NNLO in the $\overline{\text{CS}}$ scheme.

For a general factorization scheme, e.g., the $\overline{\text{MS}}$ scheme, conformal symmetry breaking superficially occurs also due to the renormalization prescription for the composite operators. Then conformal symmetry is also spoiled by terms which are not proportional to β. Utilizing an appropriate rotation of Eq. (8), known to NLO, one arrives at the corresponding $\overline{\text{MS}}$ result, where evolution induces a mixing of GPD conformal moments.

3. Results

We have used the formalism described in the preceding section to investigate the size of NNLO corrections to non-singlet and singlet CFFs, to obtain complete (non-diagonal evolution included) $\overline{\text{MS}}$ NLO predictions,[7] and to perform fits, in both $\overline{\text{MS}}$ and $\overline{\text{CS}}$ schemes, to DVCS and DIS data.[7]

One can use a simple Regge-inspired ansatz for GPDs

$$\boldsymbol{H}_j(\eta, \Delta^2, \mathcal{Q}_0^2) = \begin{pmatrix} N'_\Sigma F_\Sigma(\Delta^2)\, B\big(1+j-\alpha_\Sigma(0), 8\big) \\ N'_G F_G(\Delta^2)\, B\big(1+j-\alpha_G(0), 6\big) \end{pmatrix} + \mathcal{O}(\eta^2)\,, \quad (9)$$

with (p_a is a flavour dependent integer)

$$\alpha_a(\Delta^2) = \alpha_a(0) + 0.15\Delta^2\,, \qquad F_a(\Delta^2) = \frac{j+1-\alpha_a(0)}{j+1-\alpha_a(\Delta^2)}\left(1 - \frac{\Delta^2}{M_0^{a\,2}}\right)^{-p_a}. \quad (10)$$

In the forward case ($\Delta = 0$) this ansatz reduces to the standard building blocks for PDFs: $N\, x^{-\alpha(0)}\,(1-x)^{\beta}$. We have performed the analysis of radiative corrections with generic parameters and made fits of parameters N_Σ, $\alpha_\Sigma(0)$, M_0^Σ, N_G, $\alpha_G(0)$, M_0^G. The work on a more realistic η-dependent ansatz is in progress.

The quantities that we utilize as a measures of the scheme dependence and, foremostly, as indicators for the convergence of the perturbation series

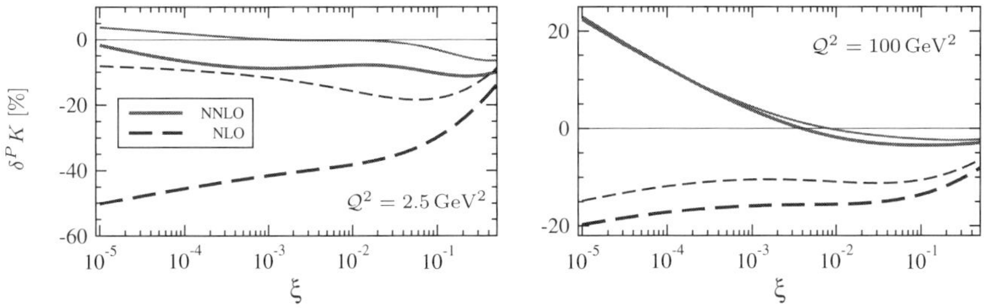

Fig. 2. Relative NLO and NNLO corrections (11) in the $\overline{\text{CS}}$ scheme ($\Delta^2 = 0.25\text{GeV}^2$, input scale $\mathcal{Q}_0^2 = 2.5\text{GeV}^2$). Thick [thin] lines – "hard" ["soft"] gluon: $N_G = 0.4$, $\alpha_G(0) = \alpha_\Sigma(0) + 0.05$ [$N_G = 0.3$, $\alpha_G(0) = \alpha_\Sigma(0) - 0.02$].

are naturally the ratios of CFFs. We compare the modulus and phase at order N^PLO to the one at order N^{P-1}LO, where $P = \{0, 1, 2\}$ stands for LO, NLO, and NNLO order, respectively:

$$\delta^P K = \frac{\left|\mathcal{H}^{\text{N}^P\text{LO}}\right|}{\left|\mathcal{H}^{\text{N}^{P-1}\text{LO}}\right|} - 1\,, \qquad \delta^P \varphi = \arg\left(\frac{\mathcal{H}^{\text{N}^P\text{LO}}}{\mathcal{H}^{\text{N}^{P-1}\text{LO}}}\right). \tag{11}$$

The phase differences are small, and we will not comment on them here further. The NLO corrections to the moduli in $\overline{\text{MS}}$ and $\overline{\text{CS}}$ schemes have a similar ξ-shape, where $\overline{\text{MS}}$ corrections are generally larger. The relative NLO and NNLO corrections in $\overline{\text{CS}}$ scheme are depicted in Fig. 2. From the left panel, showing corrections at the input scale, we realize that the large negative NLO corrections to the modulus (thick dashed) in the 'hard' gluon scenario are shrunk at NNLO to less than 10% (thick solid), in particular in the small ξ region. In the 'soft' gluon case the NNLO corrections (thin solid) are $\pm 5\%$. For $\xi \sim 0.5$, the corrections are reduced only unessentially and are around 5% and 10% at both NLO and NNLO level. If evolution is switched on (right panel), our findings drastically change. For $\xi \gtrsim 5 \cdot 10^{-2}$ NNLO corrections are stabilized on the level of about 3% at $\mathcal{Q}^2 = 100$ GeV^2. But they start to grow with decreasing ξ and reach at $\xi \approx 10^{-5}$ the 20% level. As in DIS, this breakdown of perturbation theory at small ξ obviously stems from evolution and is thus universal, i.e., process independent.

The Mellin-Barnes integral approach offers the possibility for a fast and numerically stable analysis. Sofar our numerical routine is designed for the purpose of fitting DVCS (and DIS) observables[9] and testing various GPD ansaetze in the small ξ-region; an example is presented in Fig. 3. Note that the large change of the scaling prediction within the considered order does not influence the quality of NLO and NNLO fits.

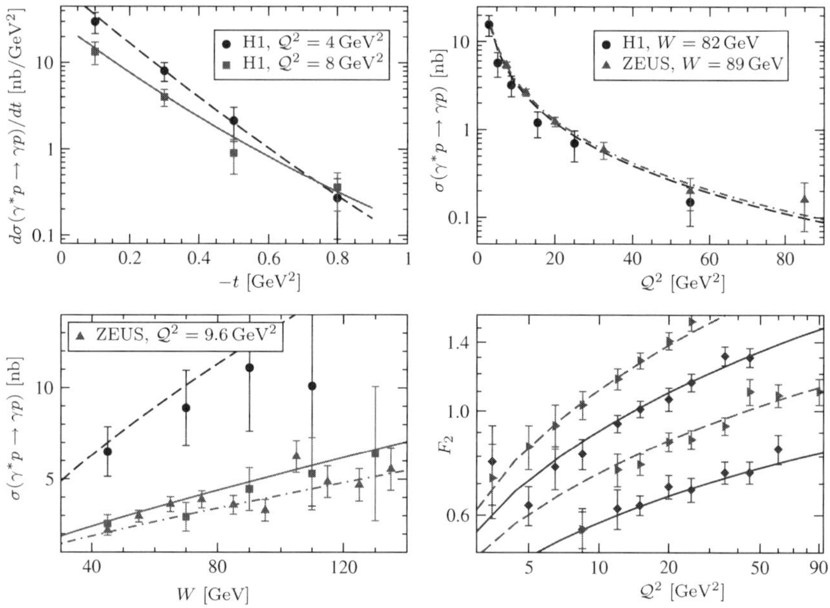

Fig. 3. Differential DVCS cross section versus Δ^2 (up,left), Δ^2-integrated one versus $\mathcal{Q}^2$ (up,right) and W (down,left) as well as DIS structure function F_2 (down,right).

Fourier transform of resulting GPDs for $\eta = 0$,

$$H(x, \mathbf{b}) = \int \frac{d^2\mathbf{\Delta}}{(2\pi)^2}\, e^{-i\mathbf{b}\cdot\mathbf{\Delta}} H(x, \eta = 0, \Delta^2 = -\mathbf{\Delta}^2)\,, \tag{12}$$

can be interpreted in the infinite momentum frame as probability density,[1] see Fig. 4b. The average transversal parton distance squared $\langle \mathbf{b}^2 \rangle$ is given by the GPD slope $B = \langle \mathbf{b}^2 \rangle / 4$, shown in Fig. 4a. The results confirm the picture, mentioned in the introduction, about the correlation of transversal and longitudinal degrees of freedom: harder partons are closer to the center.

4. Summary

We have shown that the representation of CFFs as Mellin-Barnes integrals offers a useful tool in analyzing DVCS: the inclusion of evolution is simple, numerical treatment is stable and fast. Also this Mellin–Barnes integral representation can be used within the standard $\overline{\mathrm{MS}}$ scheme beyond LO. Such a representation can straightforwardly be obtained from the momentum fraction representation and, therefore, also the hard electroproduction

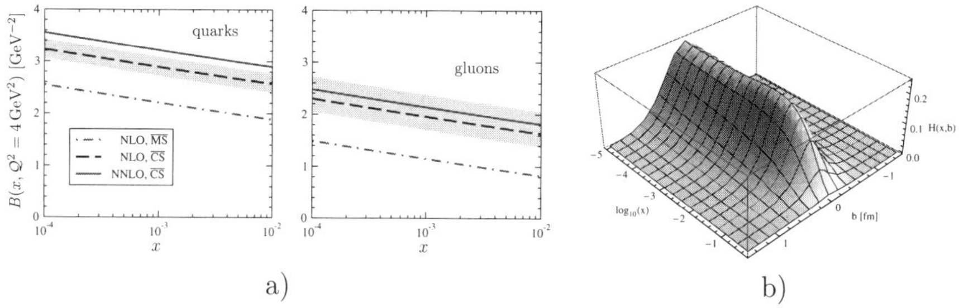

Fig. 4. a) Resulting GPD slope $B(\mathcal{Q}^2 = 4\,\mathrm{GeV}^2)$ and b) 3D picture of gluon GPD (12).

of mesons can be given in terms of Mellin–Barnes integrals. This opens the road for the 'global' analysis of experimental data within the perturbative GPD formalism to NLO accuracy. Furthermore, the use of conformal symmetry enables elegant approach to higher-order radiative corrections. We have shown that although NLO corrections to DVCS can be sizable, and are strongly dependent on the gluonic input, the NNLO corrections are small to moderate, supporting the perturbative framework. The observed change in the scale dependence is not so conclusive: similarly as in DIS we encounter large NNLO effects for $\xi < 10^{-3}$, which signal a breakdown of naive perturbation series in the evolution operator. The problem of treatment or resummation of these large corrections is relevant primarily to our partonic interpretation of the nucleon content. As long as we precisely define the treatment of the evolution operator, perturbative QCD can be employed as a tool for analyzing data also in the small ξ region. Finally, fits to available DVCS and DIS reveal the transversal distribution of partons.

References

1. M. Burkardt, *Phys. Rev.* **D62**, p. 071503 (2000);
   Int. J. Mod. Phys. **A18**, 173 (2003).
2. D. Müller and *et al.*, *Fortschr. Phys.* **42**, p. 101 (1994).
   A. Radyushkin, *Phys. Lett.* **B380**, p. 417 (1996).
3. X.-D. Ji, *Phys. Rev. Lett.* **78**, 610 (1997).
4. M. Diehl, *Phys. Rept.* **388**, 41 (2003).
5. A. V. Belitsky and A. V. Radyushkin, *Phys. Rept.* **418**, 1 (2005).
6. D. Müller and A. Schäfer, *Nucl. Phys.* **B739**, 1 (2006).
7. K. Kumerički, D. Müller and K. Passek-Kumerički (2007), hep-ph/0703179.
8. E. B. Zijlstra and W. L. van Neerven, *Nucl. Phys.* **B383**, 525 (1992).
   A. Vogt, S. Moch and J. A. M. Vermaseren, *Nucl. Phys.* **B691**, 129 (2004).
9. C. Adloff *et al.*, *Phys. Lett.* **B517**, p. 47 (2001).
   S. Chekanov *et al.*, *Phys. Lett.* **B573**, 46 (2003).
   A. Aktas *et al.*, *Eur. Phys. J.* **C44**, 1 (2005).

SPIN-ORBIT CORRELATIONS AND SINGLE-SPIN ASYMMETRIES

M. BURKARDT

Dept. of Physics, New Mexico State University, Las Cruces, NM 88003, USA
*E-mail: burkardt@nmsu.edu**

Several examples for the role of orbital angular momentum and spin-orbit correlations in hadron structure are discussed.

1. Introduction

In nonrelativistic quantum mechanics, Fourier transforms of the form factors yield charge distributions in the center of mass frame. In general, the concept of a center of mass has no analog in relativistic theories, and thus the position space interpretation of form factors is frame dependent. This is different in the Infinite Momentum Frame (IMF) or light-cone framework, where a Galilean subgroup of transverse boosts allows introducing the transverse center of longitudinal momentum as the weighted average of transverse (i.e. $\perp$ to the boost direction) positions of all partons, weighted by their momentum fractions. This center of $\perp$ momentum is the reference point in the interpretation of the (two dimensional) Fourier transform of Generalized Parton Distributions (GPDs) as Impact Parameter dependent parton Distributions (IPDs).[1] A similar interpretation exists for the 2-d Fourier transform of form factors[2] as the latter can also be obtained by integrating GPDs over the momentum fraction.

The distribution of partons in impact parameter also plays a role for Single-Spin Asymmetries (SSAs): as one expects the final state interactions on the ejected quark in a Semi-Inclusive Deep-Inelastic Scattering (SIDIS) experiment to be on average attractive, any sideward deformation of IPDs is expected to result in an enhancement of the transverse momentum distribution of the ejected quark in the opposite direction. This observation forms the basis for a qualitative link between GPDs and SSAs.[3]

2. Impact Parameter Dependent PDFs and the Sivers Effect

The Fourier transform of the GPD $H_q(x,0,t)$ yields the distribution $q(x,\mathbf{b}_\perp)$ of unpolarized quarks, for an unpolarized target, in impact parameter space

$$q(x,\mathbf{b}_\perp) = \int \frac{d^2\mathbf{\Delta}_\perp}{(2\pi)^2} H_q(x,0,-\mathbf{\Delta}_\perp^2)\, e^{-i\mathbf{b}_\perp\cdot\mathbf{\Delta}_\perp}, \tag{1}$$

with $\mathbf{\Delta}_\perp = \mathbf{p}'_\perp - \mathbf{p}_\perp$. For a transversely polarized target (e.g. when polarized in the $+\hat{x}$-direction) the impact parameter dependent PDF (IPD) $q_{+\hat{x}}(x,\mathbf{b}_\perp)$ is no longer axially symmetric and the transverse deformation is described by the gradient of the Fourier transform of the GPD $E_q(x,0,t)$

$$q_{+\hat{x}}(x,\mathbf{b}_\perp) = q(x,\mathbf{b}_\perp) - \frac{1}{2M}\frac{\partial}{\partial b_y}\int \frac{d^2\mathbf{\Delta}_\perp}{(2\pi)^2} E_q(x,0,-\mathbf{\Delta}_\perp^2)\, e^{-i\mathbf{b}_\perp\cdot\mathbf{\Delta}_\perp} \tag{2}$$

$E_q(x,0,t)$ and hence the details of this deformation are not very well known, but its x-integral, the Pauli form factor F_2, is. This allows to relate the average transverse deformation resulting from Eq. (2) to the contribution from the corresponding quark flavor to the anomalous magnetic moment. This observation is important in understanding the sign of the Sivers function.

In a target that is polarized transversely (*e.g.* vertically), the quarks in the target nucleon can exhibit a (left/right) asymmetry of the distribution $f_{q/p^\uparrow}(x_{\mathrm{B}},\boldsymbol{k}_T)$ in their transverse momentum $\boldsymbol{k}_T$[4,5]

$$f_{q/p^\uparrow}(x_{\mathrm{B}},\boldsymbol{k}_T) = f_1^q(x_{\mathrm{B}},k_T^2) - f_{1T}^{\perp q}(x_{\mathrm{B}},k_T^2)\frac{(\hat{\boldsymbol{P}}\times\boldsymbol{k}_T)\cdot\boldsymbol{S}}{M}, \tag{3}$$

where $\boldsymbol{S}$ is the spin of the target nucleon and $\hat{\boldsymbol{P}}$ is a unit vector opposite to the direction of the virtual photon momentum. The fact that such a term may be present in (3) is known as the Sivers effect and the function $f_{1T}^{\perp q}(x_{\mathrm{B}},k_T^2)$ is known as the Sivers function. The latter vanishes in a naive parton picture since $(\hat{\boldsymbol{P}}\times\boldsymbol{k}_T)\cdot\boldsymbol{S}$ is odd under naive time reversal (a property known as naive-T-odd), where one merely reverses the direction of all momenta and spins without interchanging the initial and final states. The momentum fraction x, which is equal to x_{B} in DIS experiments, represents the longitudinal momentum of the quark *before* it absorbs the virtual photon, as it is determined solely from the kinematic properties of the virtual photon and the target nucleon. In contradistinction, the transverse momentum $\boldsymbol{k}_T$ is defined in terms of the kinematics of the final state and hence it represents the asymptotic transverse momentum of the active quark *after* it has left the target and before it fragments into hadrons. Thus the Sivers function for semi-inclusive DIS includes the final state interaction between struck quark and target remnant, and time reversal invariance no longer requires that it vanishes. Indeed, as time reversal not only reverses the

Fig. 1. In SIDIS (a) the ejected (say, "red") quark is attracted by the ("anti-red") spectators. In contradistinction, in DY (b), before annihilating with the ("red") active quark, the approaching ("anti-red") antiquark is repelled by the ("anti-red") spectators.

signs of all spins and momenta, but also transforms final state interactions (FSI) into initial state interactions (ISI), it has been shown that the Sivers function relevant for SIDIS and that relevant for Drell-Yan (DY) processes must have opposite signs,[6]

$$f_{1T}^{\perp}(x_{\mathrm{B}}, k_T^2)_{SIDIS} = -f_{1T}^{\perp}(x_{\mathrm{B}}, k_T^2)_{DY}, \tag{4}$$

where the asymmetry in DY arises from the ISI between the incoming antiquark and the target. The experimental verification of this relation would provide a test of the current understanding of the Sivers effect within QCD. It is instructive to elucidate its physical origin in the context of a perturbative picture: for instance, when the virtual photon in a DIS process hits a red quark, the spectators must be collectively anti-red in order to form a color-neutral bound state, and thus attract the struck quark (Fig. 1). In DY, when an anti-red antiquark annihilates with a target quark, the target quark must be red in order to merge into a photon, which carries no color. Since the proton was colorless before the scattering, the spectators must be anti-red and thus repel the approaching antiquark.

The significant distortion of parton distributions in impact parameter space (2) provides a natural mechanism for a Sivers effect. In semi-inclusive DIS, when the virtual photon strikes a u quark in a $\perp$ polarized proton, the u quark distribution is enhanced on the left side of the target (for a proton with spin pointing up when viewed from the virtual photon perspective). Although in general the final state interaction (FSI) is very complicated, we expect it to be on average attractive thus translating a position space distortion to the left into a momentum space asymmetry to the right and vice

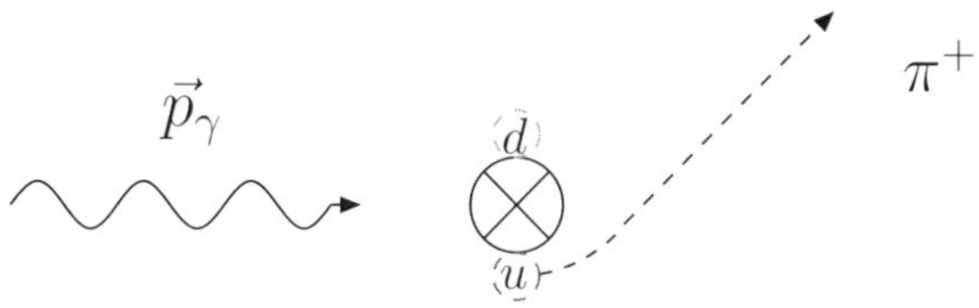

Fig. 2. The transverse distortion of the parton cloud for a proton that is polarized into the plane, in combination with attractive FSI, gives rise to a Sivers effect for u (d) quarks with a $\perp$ momentum that is on the average up (down).

versa (Fig. 2). Since this picture is very intuitive, a few words of caution are in order. First of all, such a reasoning is strictly valid only in mean field models for the FSI as well as in simple spectator models.[7] Furthermore, even in such mean field models there is no one-to-one correspondence between quark distributions in impact parameter space and unintegrated parton densities (e.g. Sivers function). While both are connected by a Wigner distribution,[8] they are not Fourier transforms of each other. Nevertheless, since the primordial momentum distribution of the quarks (without FSI) must be symmetric we find a qualitative connection between the primordial position space asymmetry and the momentum space asymmetry (with FSI). Another issue concerns the x-dependence of the Sivers function. The x-dependence of the position space asymmetry is described by the GPD $E(x, 0, -\Delta_\perp^2)$. Therefore, within the above mechanism, the x dependence of the Sivers function should be related to the x dependence of $E(x, 0, -\Delta_\perp^2)$. However, the x dependence of E is not known yet and we only know the Pauli form factor $F_2 = \int dx E$. Nevertheless, if one makes the additional assumption that E does not fluctuate as a function of x then the contribution from each quark flavor q to the anomalous magnetic moment κ determines the sign of $E^q(x, 0, 0)$ and hence of the Sivers function. Making these assumptions, as well as the very plausible assumption that the FSI is on average attractive, one finds that $f_{1T}^{\perp u} < 0$, while $f_{1T}^{\perp d} > 0$. Both signs have been confirmed by a flavor analysis based on pions produced in a SIDIS experiment by the HERMES collaboration.[9]

3. Charge Density in the Center of the Neutron

As integrating the GPD H^q over the momentum fraction x of the active quark yields the Dirac form factor F_1^q, integrating Eq.(1) over x also provides an interpretation of the Dirac form factor F_1^q as the 2d Fourier transform of the charge density (from quarks with flavor q) in impact parameter space

$$\rho(\mathbf{b}_\perp) = \int \frac{d^2\mathbf{\Delta}_\perp}{(2\pi)^2} e^{i\mathbf{b}_\perp\cdot\mathbf{\Delta}_\perp} F_1^q(t = -\mathbf{\Delta}_\perp^2). \tag{5}$$

The main advantage of Eq. (5) compared to the Fourier transform of the Sachs form factors is that $\rho(\mathbf{b}_\perp)$ has a density interpretation.[2,10]

Application of Eq.(5) to F_1 for the neutron,[11] yields a charge density $\rho(\mathbf{b}_\perp)$ that is negative not only at very large $\mathbf{b}_\perp$ but also near $\mathbf{b}_\perp = 0$. The negative charge density at large distances $\mathbf{b}_\perp$ has the well known interpretation in terms of the pion cloud through the virtual process $n \to p\pi^-$, but the negative charge density near the origin appears to be mysterious. The key for intuitively understanding the negative charge density in the center of the neutron seems to be Orbital Angular Momentum (OAM).

With the benefit of hindsight, the first evidence for the presence of OAM in the nucleon wave function came from the existence of a large anomalous magnetic moment $\kappa_p = 1.79$ and $\kappa_n = -1.91$. Indeed, in a relativistic theory, an anomalous magnetic moment necessarily implies the presence of wave function components with nonzero OAM (for a detailed discussion, see e.g. Refs. 12 and 13). Further insight can be gained by performing a flavor decomposition of the anomalous magnetic moment. Neglecting the small contribution from strange and heavier quarks and making use of charge symmetry, one finds for the contribution from u and d quarks

$$\kappa_p^u = \kappa_n^d = 1.67 \qquad \kappa_p^d = \kappa_n^u = -1.91 \tag{6}$$

respectively. Here charge factors have been taken out such that for example $\kappa_p = 1.79 = \frac{2}{3}\kappa_p^u - \frac{1}{3}\kappa_p^d$. For the purpose of this paper, we observe that not only are κ_N^q large but that the magnitude of the contribution from the minority flavor (d in the proton and u in the neutron) is even slightly larger than that of the corresponding majority flavor. Given that there are less down quarks in the proton, in combination with the fact that a nonzero anomalous magnetic moment requires wave function components with OAM, this result suggests that a d quark in a proton has a significantly higher probability to be found with nonzero OAM than a u quark.

The second piece of evidence comes from studies of the Sivers function $f_{1T}^\perp$. A recent flavor analysis based on pions produced in SIDIS suggests a nonzero Sivers function for both u and d quarks with approximately equal magnitude and opposite sign $f_{u/p^\uparrow} \approx -f_{d/p^\uparrow}$.[9] Again we note that even though the proton contains more u than d quarks, the Sivers function for d quarks is comparable in magnitude with those for u quarks, again indicating that d quark wave function components have a larger p wave component.

Finally, we turn our attention to recent lattice calculations. Using the Ji relation $J_q = \frac{1}{2}\int dx\, x\, [H_q(x,0,0) + E_q(x,0,0)]$ to determine the contribution J_q of quark flavor q to the nucleon spin from the GPDs H_q and E_q,[14] and after subtracting the quark spin contribution, one finds[15]

$$L_u \approx -L_d \approx 0.15 \tag{7}$$

i.e. about equal in magnitude and with opposite sign. While it is not entirely clear how to relate the OAM obtained through the Ji relation, to the OAM in light-cone wave function (the latter being relevant for the anomalous magnetic moment and the Sivers function) this result confirms our observation that the smaller number of d quarks yields the same magnitude for the OAM, i.e. again a larger contribution from each d quark (in the proton).

Despite the fact that there are less d than u quarks in the proton, they contribute with about the same magnitude to the anomalous magnetic moment, the Sivers function, and the quark OAM from the Ji-relation. These observations indicate that the wave function for d quarks (in a proton) has a larger p-wave component than the one for u quarks. Charge symmetry implies that in a neutron u quarks have a larger p wave components than d quarks. Since p wave function components are suppressed at the origin, this naturally suppresses u quarks in a neutron for small $\mathbf{b}_\perp$ compared to an $SU(2)$ symmetric solution thus providing a qualitative explanation for the surprising result from Ref. [11].

4. Tensor Correlations

Another set of observables that are sensitive to spin-orbit correlations are Transverse Momentum dependent parton Distributions (TMDs). Projecting out quarks with transverse spin $\mathbf{s}$, the most general expression for the $\mathbf{k}_T^2$-dependence of parton distributions reads[16]

$$\begin{aligned} q(x,\mathbf{k}_T,\mathbf{s},\mathbf{S}) = \frac{1}{2}\Big[& f_1 + s^i S^i h_1 + \frac{1}{M} S^i \varepsilon^{ij} k^j f_{1T}^\perp + \frac{1}{M} s^i \varepsilon^{ij} k^j h_1^\perp \\ & + \frac{1}{M}\Lambda s^i k^i h_{1L}^\perp + \frac{1}{2M^2} s_i S_j \left(2k^i k^j - \mathbf{k}_T^2 \delta^{ij}\right) h_{1T}^\perp \Big], \end{aligned} \tag{8}$$

where Λ is the longitudinal nucleon polarization and $\mathbf{S}$ its transverse spin. Two more terms appear when one also considers longitudinally polarized quarks.

In the following, we will focus on the chirally odd tensor correlation $h_{1T}^\perp$, which contributes to matrix elements with a double spin asymmetry in orthogonal transverse directions. In a helicity basis, this implies that it contributes to matrix elements where both quark and nucleon helicities flip — but in opposite directions — resulting in a helicity mismatch by two units. A specific example is the transition from a nucleon state with $S_z = +\frac{1}{2}$ to $S_z = -\frac{1}{2}$ while the spin of the active quark flips from $s_z = -\frac{1}{2}$ to $s_z = +\frac{1}{2}$. The active quark thus has to absorb $L_z = -2$. The fact that L_z changes by two units requires either the presence of wave function components with $L_z = \pm 2$ (s-d interference), or matrix elements that are quadratic in the p wave component. In either case, applying the power-counting techniques from Ref. [17], this implies that $h_{1T}^{\perp} \stackrel{x\to 1}{\longrightarrow} (1-x)^5$.

We will focus on contributions quadratic in the p wave component and neglect s-d interference. Consider a nucleon that is polarized in the $+\hat{x}$ direction. When the active quark has $l_x = \pm 1$, its distribution is enhanced in the $\hat{y} - \hat{z}$ plane. When viewed from the $\hat{z}$ direction, the distribution is thus enhanced along the $\hat{y}$ axis, but suppressed along the $\hat{x}$ axis (imagine a bagel in the $\hat{y}-\hat{z}$ plane viewed from the side). On the other hand, for $l_x = 0$, the quark distribution is enhanced along the $\hat{x}$ axis (peanut aligned along with the $\hat{x}$ axis). As this deformation is described by $h_{1T}^{\perp}$, this TMD thus appears the ideal tool to decide whether (and in which spin configuration) the polarized quark density looks more like a bagel or a peanut, or perhaps even a pretzel.[18]

In order to better understand the specific implications for the nucleon's angular momentum structure, let us consider for example the case $h_{1T}^{\perp} > 0$. According to the above discussion, this case corresponds to $l_x = 0$ when quark and nucleon transverse spin are parallel, and $l_x = \pm 1$ when they are anti-parallel. In a bag model or potential model, this type of correlation arises naturally when j_q is in the same direction as the nucleon spin, as the p wave component arises from the lower component of the quark wave function, and is largest when j_q and s_q are anti-parallel with l_q parallel to j_q. In most models, one would thus expect $h_{1T}^{\perp u} > 0$. The case $h_{1T}^{\perp} < 0$ corresponds to $l_x = \pm 1$ when quark and nucleon transverse spin are parallel, and $l_x = 0$ when they are anti-parallel. This naturally arises in quark states with l_q and s_q coupled to a net j_q that is oriented opposite to the nucleon spin. We thus expect $h_{1T}^{\perp d} < 0$ with smaller absolute magnitude than $h_{1T}^{\perp u} > 0$.

In contradistinction to the Sivers and Boer-Mulders functions, $h_{1T}^{\perp}$ is (naive) T-even and does not *require* the presence of nontrivial ISI/FSI phases. Nevertheless, measurements of $h_{1T}^{\perp}$ in SIDIS or DY experiments may still be affected by ISI/FSI. While such effects may be small compared to the intrinsic $h_{1T}^{\perp}$, it is not *a priori* clear how to separate ISI/FSI from intrinsic effects.

In Ref. [19] it has been shown that a tensor correlation analogous to the one described by $h_{1T}^{\perp}$ in momentum space is described by the Fourier transform of the GPD $\tilde{H}_T''$ in impact parameter space. There the issue of ISI/FSI does not arise, as the GPDs are defined through matrix elements of local operators and directly probe the intrinsic quark densities. While it is not possible to directly map a density in position space onto a density in momentum, (single particle) p-orbits have the same angular distribution in momentum space as they have in position space. Therefore, a bagel in momentum space corresponds to a bagel in position space as well and one would thus expect that $h_{1T}^{\perp}$ has the same sign as $\tilde{H}_T''$. Generalizing the approach from Ref. 20, more concrete relations between tensor correlation in momentum and impact-parameter space have been derived in Ref. 21 for the specific case of diquark models.

References

1. M.Burkardt, Phys. Rev. D **62**, 071503 (2000), Erratum-ibid. D **66**, 119903 (2002).
2. D. Soper. Phys. Rev. D **5**, 1956 (1972).
3. M. Burkardt, Phys. Rev. D **66**, 114005 (2002); Phys. Rev. D **69**, 057501 (2004).
4. D.W. Sivers, Phys. Rev. D **43**, 261 (1991).
5. A. Bacchetta et al., Phys. Rev. D **70**, 117504 (2004).
6. J.C. Collins, Phys. Lett. B **536**, 43 (2002).
7. S.J. Brodsky, D.S. Hwang, and I. Schmidt, Nucl. Phys. B **642**, 344 (2002); M. Burkardt and D.S. Hwang, Phys. Rev. **D69**, 074032 (2004); L.P. Gamberg et al., Phys. Rev. D **67**, 071504 (2003); D. Boer, S.J. Brodsky, and D.S. Hwang, Phys. Rev. D **67**, 054003 (2003); A. Bacchetta et al., Phys. Lett **B578**, 109 (2004); M. Radici et al., hep-ph/0708.0232; L.P. Gamberg et al. hep-ph/0708.0324; D. Boer et al., Phys. Rev. D **67**, 054003 (2003); L.P. Gamberg et al., Phys. Rev. D **67** (2003) 071504.
8. A.V. Belitsky, X. Ji, and F. Yuan, Phys. Rev. D **69**, 074014 (2004).
9. A. Airapetian et al. (HERMES collaboration), Phys. Rev. Lett. **94**, 012002 (2005).
10. M. Burkardt, proc. Workshop on Lepton Scattering, Hadrons and QCD, Adelaide, Australia, March 2001, hep-ph/0105324; P.V. Pobylitsa, Phys. Rev. D **66**, 094002 (2002).
11. G.A. Miller, arXiv:0705.2409
12. M. Burkardt and G. Schnell, Phys. Rev. D **74**, 013002 (2006).
13. S.J. Brodsky and S. Gardner, Phys. Lett. B **643**, 22 (2006).
14. X. Ji, Phys. Rev. Lett. **78**, 610 (1997).

15. M. Göckeler et al. (QCDSF collaboration), Phys. Rev. Lett. **98**, 222001 (2007); Ph. Hägler et al. (LHPC collaboration), hep-lat/0705.4295.
16. A. Bacchetta et al., Phys. Rev. Lett. **85**, 712 (2000).
17. H. Avakian et al., hep-ph/0705.1553
18. G.A. Miller, Phys. Rev. C **68**, 022201 (2003); G.A. Miller, arXiv:0708.2297
19. M. Diehl and P.Hägler, Eur. Phys. J. **C44**, 87 (2005).
20. Z. Lu and I. Schmidt, Phys. Rev. D **75**, 073008 (2007).
21. S. Meissner, A. Metz, and K. Goeke, Phys. Rev. D **76**, 034 (2007).

ELECTROPRODUCTION OF SOFT PIONS AT LARGE MOMENTUM TRANSFERS

V. M. BRAUN[1], D. Yu. IVANOV[2] and A. PETERS[1]

[1] *Institut für Theoretische Physik, Universität Regensburg, D-93040 Regensburg, Germany*

[2] *Sobolev Institute of Mathematics, 630090 Novosibirsk, Russia*

We consider pion electroproduction on a proton target close to threshold for Q^2 in the region 1 − 10 GeV^2. The momentum transfer dependence of the S-wave multipoles at threshold, E_{0+} and L_{0+}, is calculated using light-cone sum rules.

Keywords: threshold electroproduction, chiral symmetry

1. Threshold Pion Production

Pion electroproduction at threshold from a proton target

$$
\begin{aligned}
e(l) + p(P) &\to e(l') + \pi^+(k) + n(P')\,, \\
e(l) + p(P) &\to e(l') + \pi^0(k) + p(P')
\end{aligned}
\tag{1}
$$

can be described in terms of two generalised form factors defined as[1]

$$
\begin{aligned}
&\langle N(P')\pi(k)|j_\mu^{em}(0)|p(P)\rangle = \qquad\qquad (2)\\
&= -\frac{i}{f_\pi}\bar{N}(P')\gamma_5\left\{\left(\gamma_\mu q^2 - q_\mu \not{q}\right)\frac{1}{m^2}G_1^{\pi N}(Q^2) - \frac{i\sigma_{\mu\nu}q^\nu}{2m}G_2^{\pi N}(Q^2)\right\}N(P)\,,
\end{aligned}
$$

which can be related to the S-wave transverse E_{0+} and longitudinal L_{0+} multipoles:

$$
\begin{aligned}
E_{0+}^{\pi N} &= \frac{\sqrt{4\pi\alpha_{\rm em}}}{8\pi f_\pi}\sqrt{\frac{(2m+m_\pi)^2+Q^2}{m^3(m+m_\pi)^3}}\left(Q^2G_1^{\pi N} - \frac{1}{2}mm_\pi G_2^{\pi N}\right), \\
L_{0+}^{\pi N} &= \frac{\sqrt{4\pi\alpha_{\rm em}}}{8\pi f_\pi}\frac{m|\omega_\gamma^{\rm th}|}{2}\sqrt{\frac{(2m+m_\pi)^2+Q^2}{m^3(m+m_\pi)^3}}\left(G_2^{\pi N} + \frac{2m_\pi}{m}G_1^{\pi N}\right).
\end{aligned}
\tag{3}
$$

The differential cross section at threshold is given by

$$\frac{d\sigma_{\gamma^*}}{d\Omega_\pi}\Big|_{\rm th} = \frac{2|\vec{k}_f|W}{W^2-m^2}\Big[(E_{0+}^{\pi N})^2 + \epsilon\frac{Q^2}{(\omega_\gamma^{\rm th})^2}(L_{0+}^{\pi N})^2\Big]. \tag{4}$$

Here and below $m = 939$ MeV is the nucleon mass, $W^2 = (k+P')^2$ is the invariant energy, $\vec{k}_f$ and $\omega_\gamma^{\rm th}$ are the pion three-momentum and the photon energy in the c.m. frame. The generalised form factors in (2) are real functions of the momentum transfer Q^2 at the threshold $W = m + m_\pi$. For generic W the definition in (2) can be extended to specify two of the existing six invariant amplitudes, $G_{1,2}(Q^2) \to G_{1,2}(Q^2, W)$, which become complex functions.

The celebrated low-energy theorem (LET)[2–4] relates the S-wave multipoles or, equivalently, the form factors G_1, G_2 at threshold, to the nucleon electromagnetic and axial form factors for vanishing pion mass $m_\pi = 0$

$$\frac{Q^2}{m^2}G_1^{\pi^0 p} = \frac{g_A}{2}\frac{Q^2}{(Q^2+2m^2)}G_M^p\,, \quad G_2^{\pi^0 p} = \frac{2g_A m^2}{(Q^2+2m^2)}G_E^p\,, \tag{5}$$
$$\frac{Q^2}{m^2}G_1^{\pi^+ n} = \frac{g_A}{\sqrt{2}}\frac{Q^2}{(Q^2+2m^2)}G_M^n + \frac{1}{\sqrt{2}}G_A\,, \quad G_2^{\pi^+ n} = \frac{2\sqrt{2}g_A m^2}{(Q^2+2m^2)}G_E^n\,.$$

Here the terms in $G_{M,E}$ are due to pion emission off the initial proton state, whereas for charged pion in addition there is a contribution corresponding to the chiral rotation of the electromagnetic current.

The subsequent discussion concentrated mainly on the corrections to (5) due to finite pion mass.[5,6] More recently, the threshold pion production for small Q^2 was reconsidered and the low-energy theorems re-derived in the framework of the chiral perturbation theory (CHPT), see [7] for a review. The new insight gained from CHPT calculations[8] is that the expansion at small Q^2 has to be done with care as the limits $m_\pi \to 0$ and $Q^2 \to 0$ do not commute, in general. The LET predictions seem to be in good agreement with experimental data on pion photoproduction,[9] However, it appears[7,10] that the S-wave electroproduction cross section (4) for already $Q^2 \sim 0.1$ GeV^2 cannot be explained without taking into account chiral loops.

For larger momentum transfers the situation is much less studied as the power counting of CHPT cannot be applied. The traditional derivation of LET using PCAC and current algebra does not seem to be affected as long as the emitted pion is 'soft' with respect to the initial and final state nucleons simultaneously. The corresponding condition is, parametrically, $Q^2 \ll \Lambda^3/m_\pi$ (see, e.g. [5]) where Λ is some hadronic scale, and might be satisfied for $Q^2 \sim 1$ GeV^2 or even higher. We are not aware of any dedicated analysis of the threshold production in the $Q^2 \sim 1$ GeV^2 region, however.

It was suggested in Ref. [11] that in the opposite limit of very large momentum transfers the standard pQCD collinear factorisation approach[12,13] becomes applicable and the helicity-conserving $G_1^{\pi N}$ form factor can be calculated for $m_\pi = 0$ in terms of chirally rotated nucleon distribution amplitudes. In practice one expects that the onset of the pQCD regime is postponed to very large momentum transfers because the factorisable contribution involves a small factor $\alpha_s^2(Q)/\pi^2$ and has to win over nonperturbative "soft" contributions that are suppressed by an extra power of Q^2 but do not involve small coefficients.

The purpose of this study is to suggest a realistic QCD-motivated model for the Q^2 dependence of the $G_{1,2}$ form factors alias S-wave multipoles at threshold in the region $Q^2 \sim 1 - 10$ GeV^2 that can be accessible in current and future experiments in Jefferson Laboratory and elsewhere (HERMES, MAMI).

2. Light-Cone-Sum Rules

In Ref. [14] we have developed a technique to calculate baryon form factors for moderately large Q^2 using light-cone sum rules (LCSR).[15,16] This approach is attractive because in LCSRs "soft" contributions to the form factors are calculated in terms of the same nucleon distribution amplitudes (DAs) that enter the pQCD calculation and there is no double counting. Thus, the LCSRs provide one with the most direct relation of the hadron form factors and distribution amplitudes that is available at present, with no other nonperturbative parameters.

The same technique can be applied to pion electroproduction. In Ref. [1] the G_1 and G_2 form factors were estimated in the LCSR approach for the range of momentum transfers $Q^2 \sim 5 - 10$ GeV^2. For this work, we have reanalysed the sum rules derived in [1] taking into account the semi-disconnected pion-nucleon contributions in the intermediate state. We demonstrate that, with this addition, the applicability of the sum rules can be extended to the lower Q^2 region and the LET results in (5) are indeed reproduced at $Q^2 \sim 1$ GeV^2 to the required accuracy $\mathcal{O}(m_\pi)$. The results presented below essentially interpolate between the large-Q^2 limit considered in [1] and the standard LET predictions at low momentum transfers.

Accurate quantitative predictions are difficult for several reasons, e,g, because the nucleon distribution amplitudes are poorly known. In order to minimise the dependence of various parameters in this work we only use

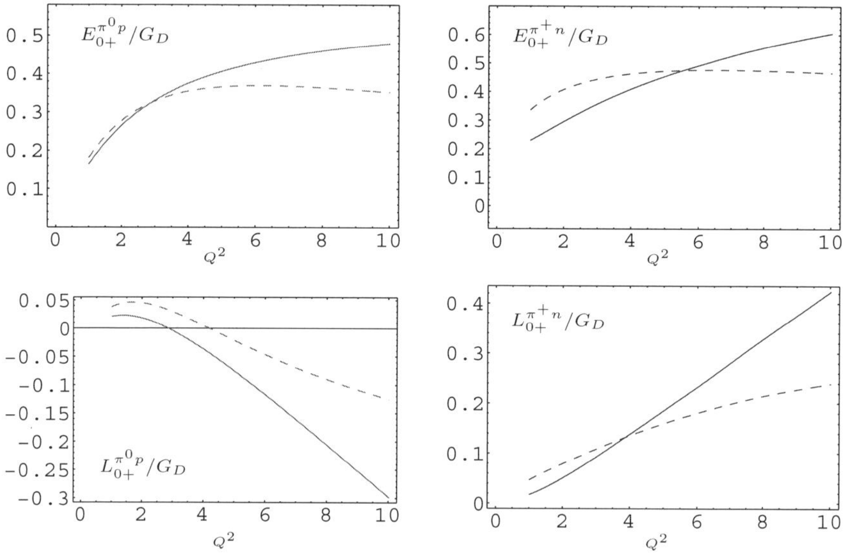

Fig. 1. The LCSR-based model (solid curves) for the Q^2 dependence of the electric and longitudinal partial waves at threshold E_{0+} and L_{0+}, (3), in units of GeV^{-1}, normalised to the dipole formula (7).

the LCSRs to predict certain form factor ratios and then normalise to the electromagnetic nucleon form factors as measured in the experiment, see [17] for the details. In particular we use the parametrisation of the proton magnetic form factor from [18] and for the neutron magnetic form factor from [19]. For the proton electric form factor we use the fit[18,20] to the combined JLab data in the $0.5 < Q^2 < 5.6$ GeV^2 range

$$\mu_p \frac{G_E^p}{G_M^p} = 1 - 0.13(Q^2 - 0.04) \tag{6}$$

and put the neutron electric form factor to zero, which should be good to our accuracy. Note that using (6) for larger values of Q^2 up to 10 GeV^2 is an extrapolation which may be not justified.

The resulting LCSR-based model is shown by the solid curves in Fig. 1, where the four partial waves at threshold that are related to the generalised form factors through the Eq. (3) are plotted as a function of Q^2, normalised to the dipole formula

$$G_D(Q^2) = 1/(1 + Q^2/\mu_0^2)^2 \tag{7}$$

where $\mu_0^2 = 0.71$ GeV^2. This model is used in the numerical analysis presented below. We expect that its accuracy is about 50%. It can be improved in future by the calculation of radiative corrections to the LCSRs, especially if sufficiently accurate lattice calculations of the moments of nucleon distribution amplitudes become available. To give a rough idea about possible uncertainties, the "pure" LCSR predictions (all form factors and other input taken from the sum rules) are shown by dashed curves for comparison.

3. Moving Away From Threshold

As a simple approximation, we suggest to calculate pion production near threshold in terms of the generalised form factors (2) and taking into account pion emission from the final state which dominates the P-wave contribution in the chiral limit (cf. [11]). In particular, we use the following expression:

$$\begin{aligned}
&\langle N(P')\pi(k)|j^{em}_\mu(0)|p(P)\rangle \\
&= -\frac{i}{f_\pi}\bar{N}(P')\gamma_5\left\{\left(\gamma_\mu q^2 - q_\mu \not{q}\right)\frac{1}{m^2}G_1^{\pi N}(Q^2) - \frac{i\sigma_{\mu\nu}q^\nu}{2m}G_2^{\pi N}(Q^2)\right\}N(P) \\
&\quad + \frac{ic_\pi g_A}{2f_\pi[(P'+k)^2) - m^2]}\bar{N}(P')\not{k}\gamma_5(\not{P}'+m)\left\{F_1^p(Q^2)\left(\gamma_\mu - \frac{q_\mu \not{q}}{q^2}\right)\right. \\
&\quad \left. + \frac{i\sigma_{\mu\nu}q^\nu}{2m}F_2^p(Q^2)\right\}N(P)\,. \qquad (8)
\end{aligned}$$

Here $F_1^p(Q^2)$ and $F_2^p(Q^2)$ are the Dirac and Pauli electromagnetic form factors of the proton, $c_{\pi^0} = 1$ and $c_{\pi^+} = \sqrt{2}$ is the isospin factor, $g_A = 1.267$ and $f_\pi = 93$ MeV.

The separation of the generalised form factor contribution and the final state emission in (8) can be justified in the chiral limit $m_\pi \to 0$ but involves ambiguities in contributions $\sim \mathcal{O}(m_\pi)$. We have chosen not to include the term $\sim \not{k}$ in the numerator of the proton propagator in the third line in (8) so that this contribution strictly vanishes at the threshold. In addition, we found it convenient to include the term $\sim q_\mu \not{q}/q^2$ in the Lorentz structure that accompanies the F_1 form factor in order to make the amplitude formally gauge invariant. To avoid misunderstanding, note that our expression is not suitable for making a transition to the photoproduction limit $Q^2 = 0$ in which case, e.g. pion radiation from the initial state has to be taken in the same approximation to maintain gauge invariance.

The virtual photon cross section can be written as a sum of terms

$$d\sigma_{\gamma^*} = \frac{\alpha_{\rm em}}{8\pi}\frac{k_f}{W}\frac{d\Omega_\pi}{W^2 - m^2}|\mathcal{M}_{\gamma^*}|^2 \qquad (9)$$

with

$$|\mathcal{M}_{\gamma^*}|^2 = M_T + \epsilon\, M_L + \sqrt{2\epsilon(1+\epsilon)}\, M_{LT}\, \cos(\phi_\pi) \\ + \epsilon M_{TT}\, \cos(2\phi_\pi) + \lambda\sqrt{2\epsilon(1-\epsilon)}\, M'_{LT}\, \sin(\phi_\pi)\,; \tag{10}$$

in the last term λ is the beam helicity.

The complete expressions for the invariant functions are rather cumbersome but are simplified significantly in the chiral limit $m_\pi \to 0$ and assuming $k_f = \mathcal{O}(m_\pi)$. We obtain

$$\begin{aligned} f_\pi^2 M_T &= \frac{4\vec{k}_i^2 Q^2}{m^2} |G_1^{\pi N}|^2 + \frac{c_\pi^2 g_A^2 \vec{k}_f^2}{(W^2-m^2)^2} Q^2 m^2 G_M^2 \\ &\quad + \cos\theta \frac{c_\pi g_A |k_i||k_f|}{W^2-m^2} 4Q^2 G_M \mathrm{Re}\, G_1^{\pi N}\,, \\ f_\pi^2 M_L &= \vec{k}_i^2 |G_2^{\pi N}|^2 + \frac{4c_\pi^2 g_A^2 \vec{k}_f^2}{(W^2-m^2)^2} m_N^4 G_E^2 \\ &\quad - \cos\theta \frac{c_\pi g_A |k_i||k_f|}{W^2-m^2} 4m^2 G_E \mathrm{Re}\, G_2^{\pi N}\,, \\ f_\pi^2 M_{LT} &= -\sin\theta \frac{c_\pi g_A |k_i||k_f|}{W^2-m^2} Qm \Big[G_M \mathrm{Re}\, G_2^{\pi N} + 4G_E \mathrm{Re}\, G_1^{\pi N} \Big]\,, \\ f_\pi^2 M_{TT} &= 0\,, \\ f_\pi^2 M'_{LT} &= -\sin\theta \frac{c_\pi g_A |k_i||k_f|}{W^2-m^2} Qm \Big[G_M \mathrm{Im}\, G_2^{\pi N} - 4G_E \mathrm{Im}\, G_1^{\pi N} \Big]\,, \end{aligned} \tag{11}$$

where $\vec{k}_i$ is the c.m.s. momentum in the initial state. Note that the single spin asymmetry contribution $\sim M'_{LT}$ involves imaginary parts of the generalised form factors that arise because of the final state interaction. In our approximation $M_{TT} = 0$ which is because we do not take into account the D- and higher partial waves. Consequently, the $\sim \cos(2\phi)$ contribution to the cross section is absent.

We find that the integrated cross sections scale like $\sigma_{\gamma^* p \to \pi N} \sim 1/Q^6$, which is in agreement with the structure function measurements in the threshold region by E136.[21] The S-wave contribution appears to be larger than P-wave up to $W \simeq 1.16$ GeV. The ratio of $\pi^0 p$ and $\pi^+ n$ final states is approximately $1:2$ and almost Q^2-independent.

The comparison of our calculation for the structure function $F_2^p(W,Q^2)$ in the threshold region $W^2 < 1.4$ GeV^2 to the SLAC E136 data[21] at the average value $Q^2 = 7.14$ GeV^2 and $Q^2 = 9.43$ GeV^2 is shown in Fig. 3. The predictions are generally somewhat below these data ($\sim 50\%$), apart from the last data point at $W^2 = 1.4$ GeV^2 which is significantly higher.

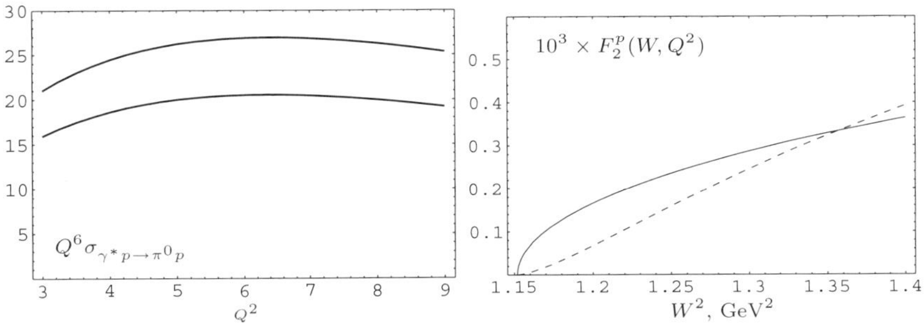

Fig. 2. *Left panel:* The integrated cross section $Q^6\sigma_{\gamma^* p\to\pi^0 p}$ (in units of $\mu b\times\text{GeV}^6$) as a function of Q^2 for $W = 1.11$ GeV (lower curve) and $W = 1.15$ GeV (upper curve). *Right panel:* The S-wave (solid) vs. the P-wave (dashed) contribution to the structure function $F_2^p(W,Q^2)$ as a function of W^2 for $Q^2 = 7.14$ GeV^2.

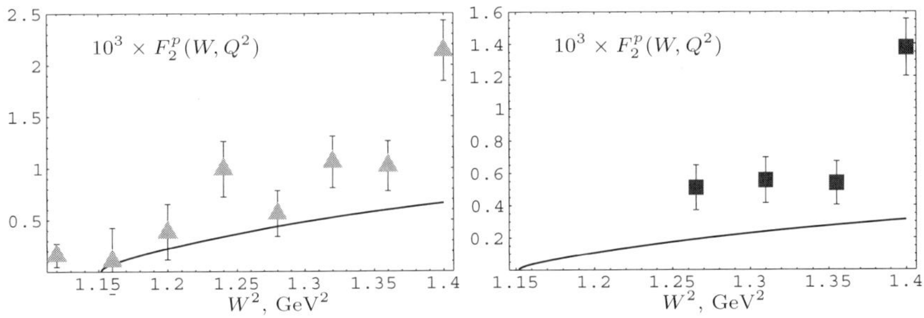

Fig. 3. The structure function $F_2^p(W,Q^2)$ as a function of W^2 scaled by a factor 10^3 compared to the SLAC E136 data[21] at the average value $Q^2 = 7.14$ GeV^2 (left panel) and $Q^2 = 9.43$ GeV^2 (right panel).

Note that in our approximation there is no D-wave contribution, and the final state interaction is not included. Both effects can increase the cross section so that we consider the agreement as satisfactory. We believe that the structure function at $W^2 = 1.4$ GeV^2 already contains a considerable D-wave contribution and also one from the tail of the Δ-resonance and thus cannot be compared with our model, at least in its present form.

To avoid misunderstanding we stress that the estimates of the cross sections presented here are not state-of-the-art and are only meant to provide one with the order-of-magnitude estimates of the threshold cross sections that are to our opinion most interesting. These estimates can be improved in many ways, for example taking into account the energy dependence of the generalised form factors generated by the FSI and adding a model for the D-wave contributions. The model can also be tuned to reproduce the existing lower Q^2 and/or larger W experimental data.

Acknowledgements

We gratefully acknowledge discussions with A. Afanasev, V. Kubarovsky, A. Lenz, A. Schäfer, P. Stoler and I. Strakovsky on various aspects of this project. V.B. thanks U. Meissner for bringing Ref. [8] to his attention and useful comments. The work of D.I. was partially supported by grants from RFBR-05-02-16211, NSh-5362.2006.2 and BMBF(06RY258). The work by A.P. was supported by the Studienstiftung des deutschen Volkes.

References

1. V. M. Braun, D. Yu. Ivanov, A. Lenz and A. Peters, Phys. Rev. D **75**, 014021 (2007).
2. N. M. Kroll and M. A. Ruderman, Phys. Rev. **93**, 233 (1954).
3. Y. Nambu and D. Lurie, Phys. Rev. **125**, 1429 (1962).
4. Y. Nambu and E. Shrauner, Phys. Rev. **128**, 862 (1962).
5. A. I. Vainshtein and V. I. Zakharov, Nucl. Phys. B **36**, 589 (1972).
6. S. Scherer and J. H. Koch, Nucl. Phys. A **534**, 461 (1991).
7. V. Bernard, N. Kaiser and U. G. Meissner, Int. J. Mod. Phys. E **4**, 193 (1995).
8. V. Bernard, N. Kaiser and U. G. Meissner, Phys. Rev. Lett. **69**, 1877 (1992).
9. D. Drechsel and L. Tiator, J. Phys. G **18**, 449 (1992).
10. V. Bernard, N. Kaiser, T. S. H. Lee and U. G. Meissner, Phys. Rev. Lett. **70**, 387 (1993).
11. P. V. Pobylitsa, M. V. Polyakov and M. Strikman, Phys. Rev. Lett. **87** (2001) 022001.
12. A. V. Efremov and A. V. Radyushkin, Phys. Lett. B **94**, 245 (1980).
13. G. P. Lepage and S. J. Brodsky, Phys. Rev. D **22**, 2157 (1980).
14. V. M. Braun, A. Lenz, N. Mahnke and E. Stein, Phys. Rev. D **65**, 074011 (2002); V. M. Braun, A. Lenz and M. Wittmann, Phys. Rev. D **73**, 094019 (2006).
15. I. I. Balitsky, V. M. Braun and A. V. Kolesnichenko, Nucl. Phys. B **312**, 509 (1989).
16. V. L. Chernyak and I. R. Zhitnitsky, Nucl. Phys. B **345**, 137 (1990).
17. V. M. Braun, D. Yu. Ivanov and A. Peters, paper in preparation
18. E. J. Brash, A. Kozlov, S. Li and G. M. Huber, Phys. Rev. C **65**, 051001 (2002).
19. P. E. Bosted, Phys. Rev. C **51** (1995) 409; E. Tomasi-Gustafsson, F. Lacroix, C. Duterte and G. I. Gakh, Eur. Phys. J. A **24** (2005) 419.
20. O. Gayou *et al.* [Jefferson Lab Hall A Collaboration], Phys. Rev. Lett. **88**, 092301 (2002).
21. P. E. Bosted *et al.*, Phys. Rev. D **49**, 3091 (1994).

COLOR TRANSPARENCY: 33 YEARS AND STILL RUNNING

M. STRIKMAN

Dept. of Physics, Pennsylvania State University, University Park, PA 16802, USA

I review history of the color transparency (CT) which started with discovery of the J/ψ meson, discovery of high energy CT phenomena and the recent progress in the investigations of CT at intermediate energies.

1. Historical introduction

One of the distinctive properties of QCD is the suppression of the interaction of small size color singlet wave packets with hadrons. It plays a key role in ensuring approximate Bjorken scaling in deep inelastic scattering, in proving QCD factorization theorems for high energy hard exclusive processes, etc. It leads to a number of color transparency (CT) phenomena in the hard coherent / quasielastic interactions with nuclei at high energies. Also, the CT phenomenon allows to probe minimal small size components in the hadrons. In addition, at intermediate energies CT phenomena provide unique probes of the space time evolution of wave packets which is relevant for interpretation of the RHIC heavy ion collision data.

For me the story of CT goes back to the discovery of J/ψ. It was impossible to explain within the concepts of the pre-QCD theory of strong interactions why the decay width of J/ψ is so small, and (this was learned soon after its discovery) why the photoproduction cross section is so small. These issues were subject of numerous discussions between Leonya Frankfurt and Volodya Gribov during the winter of 74-75 with VG trying to reconcile J/ψ properties with the soft Pomeron logic and LF arguing that for a system consisting of heavy quarks the radius should be significantly smaller than one given by radius of pion emission (this was in contrast to the widely accepted idea at that time due to Fermi, that the radius of a hadron is determined by the pion cloud and therefore should be approximately universal). More generally LF argued that all matrix elements involving heavy quarks should be suppressed, leading to a strong reduction of the cross section of J/ψ - nucleon interaction ($\propto 1/M^2_{J/\psi}$) and "an unusual conclusion that nucleon becomes transparent to the hadrons built of heavy quarks".[1] This was a clear break with the strong interaction picture with one soft scale which was discussed before J/ψ.

A perturbative model for the interaction of hadrons via two gluon exchange was applied to $J/\psi - N$ interaction by Gunion and Soper[2] who demonstrated that within the model the smallness of the J/ψ-nucleon interaction is related to the spacial small size of J/ψ. Arguments that the suppression should be present also in the nonperturbative domain were given in [3] where it was argued also that small $J/\psi(\psi')$ nucleon cross section extracted from the photoproduction data using the vector dominance model underestimates the genuine $J/\psi - N$ and especially $\psi' - N$ cross section by a large factor.

An independent development was the discussion of the hard exclusive processes like nucleon form factor, large angle hadron-hadron scattering in the large Q^2 limit. A debate was going on whether the minimal Fock space components highly localized in space give the dominant contribution in the kinematic range studied experimentally, or the process is dominated by the end point contributions corresponding to quark - gluon configurations of average size. For a recent review see [4] .

A. Mueller has suggested to use exclusive processes off nuclei, namely large angle reaction $pA \to pp(A-1)$ in order to discriminate between the two mechanisms,[5] while S.Brodsky[6] made a prediction that the cross of the process $\pi A \to \pi p(A-1)$ should be proportional to the number of protons in the target. It is feasible to study these processes as well as quasielastic electron - nucleus scattering only in the kinematics where at least one hadron in the final state has relatively small momentum leading to a need to take into account space time evolution of the quark-gluon wave packets involved in the collision which greatly reduces the CT effect [7] .

This called for finding high energy processes which are dominated by the interaction of hadrons in small size configurations which could be legitimately calculated in pQCD and which are not affected by the space-time evolution of small wave packets. A key observation was that, due to the possibility of treating configurations as frozen during the collision process one can introduce a notion of the cross section of scattering of a small dipole configuration (say $q\bar{q}$) of transverse size d on the nucleon [8,9] which in the leading log approximation is given by [10]

$$\sigma(d,x) = \frac{\pi^2}{3}\alpha_s(Q^2_{eff})d^2\left[x_N G_N(x,Q^2_{eff}) + 2/3x_N S_N(x_N,Q^2_{eff})\right], \quad (1)$$

where $Q^2_{eff} = \lambda/d^2, \lambda = 4 \div 10$, and S is the sea quark distribution for quarks making up the dipole. Here, in difference from the original estimate, we include also the contribution of quark exchanges which is important for the interactions at intermediate energies. Note that Eq.(1) predicts a rapid increase of the dipole -hadron cross section with increase of energy which is qualitatively different from the expectation of the two gluon exchange model [2] where cross section does not depend on energy.

First, we will consider the case a more simple case of high energy CT where only two conditions are required: dominance of small size configurations and smallness of $q\bar{q} - N$ interaction. Next we will consider a more complicated case of CT in the intermediate energy processes where it is masked to large extent by the expansion effects.

2. Discovery of high energy CT

To observe CT in a high energy process one needs to find a trigger which selects small size configurations in the projectile. One idea is to select a special final state: diffraction of a pion into two high transverse momentum jets. Qualitatively one expects in this case $d \sim 1/p_t(jet)$. Another idea is to select a small initial state - diffraction of a longitudinally polarized virtual photon into a meson. It employs the decrease of the transverse separation between q and $\bar{q}$ in the wave function of $\gamma^*_L, d \propto 1/Q$. The pQCD results for these processes where first derived in [8,11], with the proofs of the QCD factorization for these processes given for dijet production in [10] and for meson production in [12] (where in addition to production of vector mesons a general case of arbitrary mesons was considered).

2.1. *Pion dissociation into two jets*

The space time picture of the process is as follows - long before the target pion fluctuates into $q\bar{q}$ configuration with transverse separation d, which elastically scatters off the target with an amplitude which for $t = 0$ is given by Eq.(1) (up to small corrections due to different off shellness of $q\bar{q}$ pair in the initial and final states, followed by the transformation of the pair into two jets. A slightly simplified final answer is

$$A(\pi N \rightarrow 2\,jets + N)(z, p_t, t=0) \propto \int d^2 d\psi_\pi^{q\bar{q}}(d,z)\sigma_{q\bar{q}-N(A)}(d,s)e^{ip_t d}, \quad (2)$$

where z in the light-cone fraction of the pion momentum carried by a quark, $\psi_\pi^{q\bar{q}}(z,d) \propto z(1-z)_{d\rightarrow 0}$ is the quark-antiquark Fock component of the meson light cone wave function. Presence of the plane wave factor in the final state leads to an expectation of an earlier onset of scaling than in the case of the vector meson production where vector meson wave function enters.

The FNAL experiment[13] confirmed key CT predictions of [8] : a) a strong increase of the cross section of the $\pi + A \rightarrow$ "*two jets*" $+ A$ process with A(A=carbon, and platinum): $\sigma \propto A^{1.61\pm 0.08}$ as compared to the prediction $\sigma \propto A^{1.54}$ *, b) the $z^2(1-z)^2$ dependence of the cross section on the fraction of energy z carried by the jet, c) the k_t dependence of the cross section. Note that the CT prediction for the A-dependence was a factor of seven different from the A-dependence for the soft diffraction.

In the recent update of the analysis Ashery reported[14] a fit to the z distribution using Gegenbauer polynomials for different ranges of p_t. For $1.25 \leq p_t \leq 1.5\,\mathrm{GeV/c}$ higher order polynomials appear to be important. Since the CT is observed for this p_t range as well this indicates that squeezing occurs already before the leading term $(1-z)z$ dominates.

2.2. *Vector meson production at HERA*

Exclusive vector meson production was extensively studied at HERA. The leading twist picture of the process[11] is, in a sense, a mirror image of the dijet production - virtual longitudinally polarized photon first transformed to a small transverse size pair which interacts elastically with a target and next transforms to a vector meson. Hence the process is described by the same equation (2) as for pion case with a substitution of the plane wave $q\bar{q}$ wave function by the $q\bar{q}$ wave function of the longitudinally polarized virtual photon.

The extensive studies of the vector meson production were performed at HERA. Several of the theoretical predictions were confirmed including fast x-dependence of the process at large Q^2, consistent with the x-dependence of $G_N^2(x, Q_{eff}^2)$, and convergence of the t-dependence to the universal one

*In QCD a naive expectation of the CT that the amplitude is proportional to A is modified [8,11] due to the leading twist gluon shadowing which should be present at sufficiently small x. This effect is not important for the x range of the experiment [13] .

at large Q^2_{eff} where it is given by the two gluon form factor. At the same time the data confirm a conclusion of the model studies [15] that in a wide range of virtualities one needs to take into account a higher twist effect of the finite transverse size of γ_L to explain the absolute cross section and t-dependence of the data. The leading twist dominance for the absolute cross section for all mesons and for the t-dependence for light mesons requires very large Q^2 since only in this case one can neglect the transverse size of the $q\bar{q}$ pair in γ_L as compared to that in the meson wave function. The same mechanism leads to $Q^2_{eff}/Q^2 \ll 1$ even at large Q^2.

To summarize this section. The presence of small size $q\bar{q}$ Fock components in light mesons is unambiguously established. At transverse separations d ≤ 0.3 fm pQCD reasonably describes small "$q\bar{q}$ - dipole" - nucleon interactions for $10^{-4} < x < 10^{-2}$. Color transparency is established for the small dipole interaction with nuclei for $x \sim 10^{-2}$. Further studies of high energy CT and onset of color opacity will be performed at LHC in the ultraperipheral heavy ion collisions, see [16] for a review.

3. Color transparency for intermediate energies

3.1. *Expansion effects*

In this section we discuss searches for CT at Jlab and BNL which correspond to the kinematics where the expansion / contraction of the interacting small system is very important (essential longitudinal distances are not large enough for using of the frozen approximation) and strongly suppresses color transparency effect[7,17] .

The maximal longitudinal distance for which coherence effects are still present is determined by the minimal characteristic internal excitation energies of the hadron h. The estimates[7,17] show that for the case of a nucleon ejectile coherence is completely lost at the distances $l_c \sim 0.3 \div 0.4\text{fm} \cdot p_h$, where p_h is measured in GeV/c [†]

To describe the effect of the loss of coherence two complementary languages were suggested. In Ref. [7] based on the quark-gluon representation of point-like configuration (PLC) wave function it was argued that the main effect is quantum diffusion of the wave packet so that

$$\sigma^{PLC}(Z) = \left(\sigma_{hard} + \frac{Z}{l_c}[\sigma - \sigma_{hard}]\right)\theta(l_c - Z) + \sigma\theta\,(Z - l_c)\,. \qquad (3)$$

[†]It is of interest that a much larger value of l_c/p_h is assumed in modeling of heavy ion collisions at RHIC.

This equation is justified for an early stage of time development in the leading logarithmic approximation when perturbative QCD can be applied. Also, one can expect that Eq. (3) smoothly interpolates between the hard and soft regimes. A sudden change of σ^{PLC} would be inconsistent with the observation of an early (relatively low Q^2) Bjorken scaling [18] . Eq.(3) implicitly incorporates the geometric scaling for the PLC - nucleon interactions which for the discussed energy range includes nonperturbative effects.

The time development of the *PLC* can also be obtained by its interaction with a nucleus using a baryonic basis for the wave function of PLC:

$$|\Psi_{PLC}(t)\rangle = \Sigma_{i=1}^{\infty} a_i \exp(iE_i t)\,|\Psi_i\rangle = \exp(iE_1 t)\Sigma_{i=1}^{\infty} a_i \exp\left(\frac{i(m_i^2 - m_1^2)t}{2P}\right)|\Psi_i\rangle\,, \qquad (4)$$

where $|\Psi_i\rangle$ are the eigenstates of the Hamiltonian with masses m_i, and p is the momentum of PLC which satisfies $E_i \gg m_i$. As soon as the relative phases of the different hadronic components become large (of the order of one) the coherence is likely to be lost.

Numerical results of the quantum diffusion model[7,18] and the model based on the expansion over hadronic basis with sufficiently large number of intermediate states [17c] give similar numerical results. However though both approaches model certain aspects of dynamics of expansion, a complete treatment of this phenomenon in QCD is so far missing. In particular, the phenomenon of spontaneously broken chiral symmetry may lead to presence of two scales in the rate of expansion, one corresponding to regime where quarks can be treated as massless, and another where virtualities become small enough and quark acquire effective masses of the order of 300 MeV.

3.2. *Large angle quasielastic A(p,2p) process*

First data on the CT reaction $A(p, 2p)$ were obtained at BNL. They were followed by the dedicated experiment EVA. The final results of EVA[19] can be summarized as follows. Our calculation within the eikonal approximation with proper normalization of the wave function agrees well the p_p=5.9 GeV/c data. The transparency increases significantly for p_p= 9 GeV/c where l_c= 2.7 fm. Hence momenta of the incoming proton $\sim$ 10 GeV are sufficient to rather significantly suppress expansion effects. Hence one can use proton projectiles with energies above $\sim$10 GeV to study other aspects of the strong interaction dynamics. At the same time Glauber level transparency for p_p=11.5 $\div$ 14.2 GeV/c a problem for all current models including those which were specifically suggested to explain initial indications of the non-monotonous energy dependence of the transparency. This is because the drop of the transparency occurs over a large range of s': 24 GeV$^2 \leq s' \leq 30$ GeV^2 which is too broad for a resonance or for interference of quark exchange and Landshoff mechanisms.

In any case the trend, if confirmed by future data, would strongly suggest that the leading power quark exchange mechanism of elastic scattering dominates only at very large energies. This is consistent with the recent data from Jlab studies of the large angle Compton scattering. These data are not described by the minimal Fock space quark counting rule mechanism, while they agree well with predictions based on dominance of the box diagram contribution [4,20] .

3.3. *Color transparency in meson production*

It is natural to expect that it is easier to reach CT regime for the interaction/production of mesons than for baryons since only two quarks have to come close together.

The J/ψ coherent and quasielastic photoproduction experiments did find a weak absorption of J/ψ indicating presence of CT. There was also evidence for CT in the ρ-meson production. However these experiments did not have good enough resolution in the missing mass to suppress hadron production in the nucleus vertex, making interpretation of these experiments somewhat ambiguous.

A high resolution experiment of pion production recently reported evidence for the onset of CT [21] in the process $eA \to e\pi^+A^*$. In the chosen kinematics $\vec{p}_\pi \| \vec{q}$ which minimizes contribution of the elastic rescattering. The coherent length defined as the distance between the point where γ^* converted to a $q\bar{q}$ and where $q\bar{q}$ interacts with a nucleon - $l_{in} = (Q^2 + M^2_{q\bar{q}}/2q_0)$ is small for the kinematics of [21] and varies weakly with Q^2. This simplifies interpretation of the Q^2 dependence of the transparency as compared to the case of small x where l_{in} becomes comparable to the nucleus size. The experimental results agree well with predictions of [22] where CT was calculated based on the quantum diffusion model - Eq. (3).

It is worth emphasizing also, that in the Jlab kinematics one probes large x processes, which are dominated for the pion case (and probably also for the ρ-meson case) in the pQCD limit by the contribution of the ERBL region. In this case l_{in} has a different meaning than for small x processes where the DGLAP region dominates. It corresponds to the longitudinal distance between the point where γ^* knocks out a $q\bar{q}$ pair from the nucleon and the nucleon center. This distance can be both positive and negative, and hence its variation does not lead to a change of the rate of the absorption of the produced pair by the other nucleons.

Results for the ρ-meson production where also reported at this workshop.[23] To interpret this experiment one needs to take into account the effect of absorption due to decays of ρ^0 to two pions inside the nucleus, and the elastic rescattering contribution which is more important in this case than in the pion experiment since the data are integrated over a large range of the transverse momenta of the ρ meson .[24] Up to these effects, we expect similar transparency for this reaction and for π-meson production.

4. Directions for the future studies at Jlab

There are already approved plans for extending CT studies of the A(e,e'p), A(e,e'π) reactions to much higher energies at 12 GeV. This will finally allow to reach kinematics where l_c is larger than the interaction length for a nucleon/pion in the nuclear media.

A complementary strategy is to use processes where multiple rescatterings dominate in light nuclei (^{2}H,^{3}He) which allows to suppress the expansion effects. An additional advantage of these processes is that one can use for the calculations generalized eikonal approximation, see review in [25]. In particular, these reactions are well suited to search for a precursor of CT - suppression of the configurations in nucleons with pion cloud in the hard processes like the nucleon form factors at relatively small $Q^2 \geq 1\text{GeV}^2$ - chiral transparency.[26] The simplest reaction of this kind is production of a slow Δ isobar in the process e^{2}H $\rightarrow$ e+p+Δ^0 which should be suppressed in the chiral transparency regime.

Two other examples are (i) large angle $\gamma + N \rightarrow "meson" + N$ reaction in nuclei where one should first look for a change of A-dependence from $\propto A^{1/3}$ to $\propto A^{2/3}$ already in the region where expansion effects are large due to transition from the vector dominance regime to the regime of point-like photon interaction in which photon penetrates to any point in the nucleus, (ii) A-dependence of virtual compton scattering, namely at what Q^2 transition from vector dominance regime to the CT regime occurs. HERMES data are consistent with our prediction based on CT and closure - but accuracy of the data is moderate.

To summarize, the high energy CT is well established and will be further studied at LHC and EIC. It is likely that Jlab experiments at 12 GeV will observe significant CT effects for the processes with meson production and will provide allow a decisive test of whether nucleon form factors at $Q^2 \sim 15\text{GeV}^2$ are dominated by PLC or mean field configurations. CT will allow also to establish interplay between soft and hard physics for many other exclusive large momentum transfer processes at Jlab, EIC, LHC as well as at hadronic factories J-PARC, FAIR.

I thank my collaborators on the studies of CT phenomena for numerous discussions.

References

1. L. L. Frankfurt and V. A. Khoze, in Proceedings of 10th LNPI Winter School, Leningrad, USSR, 1975, v2,pp 196-408; *Yad.Fiz.* **23** 926 (1976).
2. J. F. Gunion and D. E. Soper, *Phys. Rev.* **D15**, 2617 (1977).
3. L. L. Frankfurt and M. I. Strikman, *Nucl. Phys.* **B250**, 143 (1985).
4. A. V. Radyushkin, arXiv:hep-ph/0409215.
5. A.H. Mueller in Proceedings of 17th rencontre de Moriond, Moriond, 1982 ed. J Tran Thanh Van (Editions Frontieres, Gif-sur-Yvette, France, 1982) Vol. I, p.13.
6. S.J. Brodsky in Proceedings of the 13th Int'l Symposium on Multiparticle Dynamics, ed. W. Kittel, W. Metzger and A. Stergiou (World Scientific, Singapore 1982,) p.963.
7. G.R. Farrar, H. Liu, L.L. Frankfurt & M.I. Strikman, *Phys. Rev. Lett.* **61** 686 (1988).
8. L. Frankfurt, G. A. Miller and M. Strikman, *Phys. Lett.* **B304**, 1 (1993).
9. B. Blaettel, G. Baym, L. L. Frankfurt and M. Strikman, *Phys. Rev. Lett.* **70**, 896 (1993).
10. L. Frankfurt, G. A. Miller and M. Strikman, *Phys. Rev.* **D65**, 094015 (2002).
11. S. J. Brodsky, L. Frankfurt, J. F. Gunion, A. H. Mueller and M. Strikman, *Phys. Rev.* **D50**, 3134 (1994).
12. J. C. Collins, L. Frankfurt and M. Strikman, *Phys. Rev.* **D56**, 2982 (1997).
13. E. M. Aitala *et al.* [E791 Collaboration], *Phys. Rev. Lett.* **86**, 4768 (2001), ibid **86**, 4773 (2001).
14. D. Ashery, *Nucl. Phys. Proc. Suppl.* **161**, 8 (2006).
15. L. Frankfurt, W. Koepf and M. Strikman, *Phys. Rev.* **D54**, 3194 (1996), ibid D **57**, 512 (1998).
16. K. Hencken *et al.*, arXiv:0706.3356 [nucl-ex].
17. B.K. Jennings and G.A. Miller, *Phys. Lett.* **B236** 209 (1990), *Phys. Rev.* **D44** 692 (1991), *Phys. Rev. Lett.* **70** 3619 (1992).
18. L. L. Frankfurt and M. I. Strikman, *Phys. Rept.* **160**, 235 (1988).
19. J. L. S. Aclander *et al.*, *Phys. Rev.* **C70**, 015208 (2004).
20. A. Danagoulian *et al.*, Phys. Rev. Lett. **98**, 152001 (2007).
21. B. Clasie *et al.*, arXiv:0707.1481 [nucl-ex].
22. A. Larson, G. A. Miller and M. Strikman, *Phys. Rev.* **C74**, 018201 (2006).
23. K.Hafidi, contribution to the proceeding of the workshop.
24. L. Frankfurt, G. A. Miller and M. Strikman, in preparation.
25. M. M. Sargsian, *Int. J. Mod. Phys.* **E10**, 405 (2001).
26. L. Frankfurt, T. S. H. Lee, G. A. Miller and M. Strikman, *Phys. Rev.* **C55**, 909 (1997).

MESON CLOUDS AND NUCLEON ELECTROMAGNETIC FORM FACTORS

GERALD A. MILLER

Department of Physics, University of Washington
Seattle, Washington 98195-1560, USA
E-mail: miller@phys.washington.edu

In contrast with common non-relativistic lore, the usual Sachs form factors are not the Fourier transforms of charge or magnetization densities. Instead, the two-dimensional Fourier transform of the electromagnetic F_1 form factor is the charge charge density of partons in the transverse plane. An analysis of the available data for neutron form factors leads to the result that the neutron charge density is negative at the center, and that the square of the transverse charge radius is positive. This contrasts with many expectations. Additionally, the use of measured proton form factors leads to the result that the proton's central d quark charge density is larger than that of the u quark by about 30%. The proton (neutron) charge density has a long range positively (negatively) charged component indicative of a pion cloud.

Keywords: Generalized Parton Distributions; Form Factors, Quark Densities

1. Introduction

This talk is mainly concerned with the question, "What do form factors really measure?", and secondarily with the question, "What is the relation between form factors and the orbital angular momentum of the objects that make up the neutron or proton?". The article [1], which contains more details, is the source the present presentation.

A way to focus the discussion is to ask yourself the question, "What is the charge density at the center of the nucleon?" The neutron has no net charge, but the charge density need not vanish. So we can ask, "Is the central charge density negative or positive?". This talk provides a surprising answer.

There are long-standing existing answers based on models [2,3]. The neutron can make a spontaneous quantum transition to a state consisting of a proton and a π^-.[2] The relatively light pion can spread out over a larger region of space than the proton. Hence one expects that the charge density is negative at the edge of the neutron and positive at the center. The same result is obtained with very different logic from the one-gluon exchange mechanism [3], which acts repulsively between two negatively charged d-quarks. But enough talk about models! Here we shall be concerned with model independent information.

2. Outline

We shall begin with a discussion of electromagnetic form factors. This will be followed by a discussion of light cone coordinates and the relevant kinematic subgroup of the Poincaré group. Then the definitions of generalized parton distributions GPDs and a bit of math lead to the key result[4] that the two-dimensional Fourier transform of the electromagnetic form factor F_1 gives the charge density $\rho(b)$ where b is the distance from the transverse center of mass, and no information about the longitudinal momentum or position is available. Then the data analysis will be discussed, and an attempt at providing an interpretation will be made. But really, finding an accurate interpretation is anyone's game.

3. Definitions

The electromagnetic form factors are matrix elements of the current operator, $J^\mu(x)$, between nucleon states of different momentum:

$$\langle p',\lambda'|J^\mu(0)|p,\lambda\rangle = \bar{u}(p',\lambda')\left(\gamma^\mu F_1(Q^2) + i\frac{\sigma^{\mu\alpha}}{2M}q_\alpha F_2(Q^2)\right)u(p,\lambda), \quad (1)$$

where the momentum transfer $q_\alpha = p'_\alpha - p_\alpha$ is taken as space-like, so that $Q^2 \equiv -q^2 > 0$, and M is the nucleon mass. The nucleon polarization states are chosen to be those of definite light-cone helicities λ, λ'.[5] The charge (Dirac) form factor is F_1, normalized such that $F_1(0)$ is the nucleon charge, and the magnetic (Pauli) form factor is F_2, normalized such that $F_2(0)$ is the anomalous magnetic moment. the Sachs form factors[6]

$$G_E(Q^2) \equiv F_1(Q^2) - \frac{Q^2}{4M^2}F_2(Q^2),\ G_M(Q^2) \equiv F_1(Q^2) + F_2(Q^2), \quad (2)$$

were introduced to provide an expression for the electron-nucleon cross section (in the one photon exchange approximation) that depends on the quantities G_E^2 and G_M^2 but not the product $G_E G_M$. In the Breit frame, in which $\mathbf{p} = -\mathbf{p}'$, G_E is the nucleon helicity flip matrix element of J^0. Furthermore, the scattering of neutrons from the electron cloud of atoms measures the derivative $-dG_E(Q^2)/dQ^2$ at $Q^2 = 0$, widely interpreted as six times the mean-square charge radius of the neutron. However, any probability or density interpretation of G_E is spoiled by a non-zero value of Q^2, no matter how small.[1] This is because the initial and final states have different momentum, and therefore relativistically have different wave functions. The factorization of relative and center of mass wave functions that is obtained from the non-relativistic Galilean invariance is not obtained relativistically. The internal wave function depends upon the total momentum of the nucleon. Any attempt to analytically correct for the total momentum by incorporating relativistic corrections in a p^2/m_q^2 type of expansion would be doomed, by the presence of the very light current quark mass, m_q, to be model-dependent. That is, at small values of Q^2, one finds

$$G_E^n \sim Q^2 (\int d^3r \left(r^2|\psi|^2 + \frac{C}{m_q^2} \right), \tag{3}$$

where the first term represents the traditional effect depending on the square of the wave function and the unknown coefficient C represents the correction due to the total momentum of the system.

4. Light cone coordinates

These useful coordinates involve the use of a "time"

$$x^+ = (ct + z)/\sqrt{2} = (x^0 + x^3)/\sqrt{2}. \tag{4}$$

The corresponding evolution operator is the not the Hamiltonian, p^0, but instead

$$p^- = (P^0 - p^3)/\sqrt{2}. \tag{5}$$

The orthogonal spatial coordinate is

$$x^- = (x^0 - x^3)/\sqrt{2}. \tag{6}$$

If one quantizes at $x^+ = 0$, then $x^- = \sqrt{2}z$, and this why x^- is thought of as the spatial variable. The canonically conjugate momentum is given by

$$p^+ = (p^0 + p^3)/\sqrt{2}. \tag{7}$$

We note that

$$p_\mu x^\mu = p^- x^+ + p^+ x^- - \mathbf{p} \cdot \mathbf{b}. \tag{8}$$

The transverse coordinates perpendicular to the 0 and 3 directions are denoted as $\mathbf{b}$ and $\mathbf{p}$.

5. Relativistic formalism–kinematic subgroup of the Poincaré group

The Lorentz transformation defined by a transverse velocity $\mathbf{v}$ has properties very similar to that of Galilean transformations. Under these transformations

$$k^+ \to k^+ \tag{9}$$

$$\mathbf{k} \to \mathbf{k} - k^+\mathbf{v}, \tag{10}$$

and k^- transforms so that $k^2 = k^+k^- - \mathbf{k}^2$ is not changed. Transverse boosts are like non-relativistic boosts according to Eq. (10).

This means that one may use[4,7,8] nucleon states that are transversely localized. The state with transverse center of mass $\mathbf{R}$ set to 0 is formed by taking a linear superposition of states of transverse momentum:

$$\left|p^+, \mathbf{R} = \mathbf{0}, \lambda\right\rangle \equiv \mathcal{N} \int \frac{d^2\mathbf{p}}{(2\pi)^2} \left|p^+, \mathbf{p}, \lambda\right\rangle . \tag{11}$$

where $\left|p^+, \mathbf{p}, \lambda\right\rangle$ are light-cone helicity eigenstates[5] and $\mathcal{N}$ is a normalization factor. The relevant range of integration in Eq. (11) must be restricted to $|\mathbf{p}| \ll p^+$ to maintain the interpretation of a nucleon moving with well-defined longitudinal momentum.[9] Thus we use the infinite momentum frame, for which the nucleon may accurately be regarded as a set of a large number of partons.

6. The main result

Using Eq. (11) sets the transverse center of momentum of a state of total very large momentum p^+ to zero, so that transverse distance $\mathbf{b}$ relative to $\mathbf{R}$. can be defined. Thus we may define a useful combination of quark-field operators:

$$\hat{O}_q(x, \mathbf{b}) \equiv \int \frac{dx^-}{4\pi} q_+^\dagger\left(-\frac{x^-}{2}, \mathbf{b}\right) q_+\left(\frac{x^-}{2}, \mathbf{b}\right) e^{ixp^+x^-}, \tag{12}$$

where the subscript $+$ denotes the use of only independent quark field operators. The impact parameter dependent PDF is defined[9] as the matrix element of this operator in the state of Eq. (11):

$$q(x, \mathbf{b}) \equiv \left\langle p^+, \mathbf{R} = \mathbf{0}, \lambda\right| \hat{O}_q(x, \mathbf{b}) \left|p^+, \mathbf{R} = \mathbf{0}, \lambda\right\rangle . \tag{13}$$

The use of Eq. (11) in Eq. (13) allows one to show[1] that $q(x, \mathbf{b})$ is the two-dimensional Fourier transform of the GPD H_q:

$$q(x, \mathbf{b}) = \int \frac{d^2q}{(2\pi)^2} e^{i\,\mathbf{q}\cdot\mathbf{b}} H_q(\xi = 0, x, t = -\mathbf{q}^2), \tag{14}$$

with H_q appearing because the initial and final helicities are each λ.

One finds a probability interpretation[4] by integrating $q(x, \mathbf{b})$ over all values of x. This sets the value of x^- to 0, so that

$$\int dx\; q(x, \mathbf{b}) \equiv \left\langle p^+, \mathbf{R} = \mathbf{0}, \lambda \right| q_+^\dagger(0, \mathbf{b}) q_+(0, \mathbf{b}) \left| p^+, \mathbf{R} = \mathbf{0}, \lambda \right\rangle, \tag{15}$$

and a density appears in the matrix elelment. If one multiplies the above relation by the quark charge e_q (in units of e), sums over quark flavors, uses Eq. (11) with $\hat{O}_q(x, \mathbf{b}) = e^{-i\hat{\mathbf{p}}\cdot\mathbf{b}} \hat{O}_q(x, \mathbf{0}) e^{i\hat{\mathbf{p}}\cdot\mathbf{b}}$ along with the sum rule relating the GPD to the form factor, the resulting infinite-momentum-frame IMF parton charge density in transverse space is

$$\rho(b) \equiv \sum_q e_q \int dx\; q(x, \mathbf{b}) = \int \frac{d^2q}{(2\pi)^2} F_1(Q^2 = \mathbf{q}^2) e^{i\,\mathbf{q}\cdot\mathbf{b}}. \tag{16}$$

7. Data analysis and results

We exploit Eq. (16) by using measured form factors to determine $\rho(b)$. Recent parameterizations[11–13] of G_E and G_M are very useful so we express F_1 in terms of G_E, G_M. Then $\rho(b)$ can be expressed as an integral involving known functions:

$$\rho(b) = \int_0^\infty \frac{dQ\;Q}{2\pi} J_0(Qb) \frac{G_E(Q^2) + \tau G_M(Q^2)}{1 + \tau}, \tag{17}$$

with $\tau = Q^2/4M^2$ and J_0 a cylindrical Bessel function. mass.

The charge density of the proton is shown in Fig. 1, and that for the neutron in Fig. 2. The proton density seems to be well determined, using the entire range of the parameterization,[12] which greatly overestimates the errors, leads to little variation. The surprising feature is the negative central value of the neutron charge density. This results from the negative definite nature of F_1.[1] The neutron density is sensitive to unknown values of F_1 at high Q^2. Cutting off the integral appearing in Eq. (17) at $Q = 2\sqrt{2}M$ leads to big changes, as shown in Fig. 2.

The neutron charge density has interesting features, as shown in Fig. 3 which displays the quantity $b\rho(b)$. It is the integral of this quantity that integrates to 0. The neutron charge density is negative at the center, positive

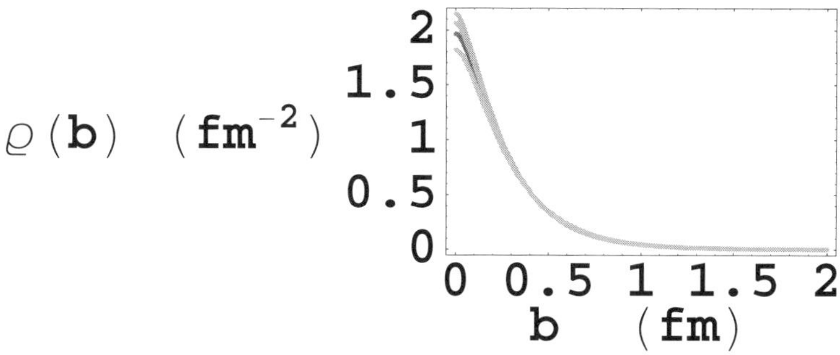

Fig. 1. The proton charge density $\rho(b)$ using the full spread of the parameters of Kelly's[12] parameterization.

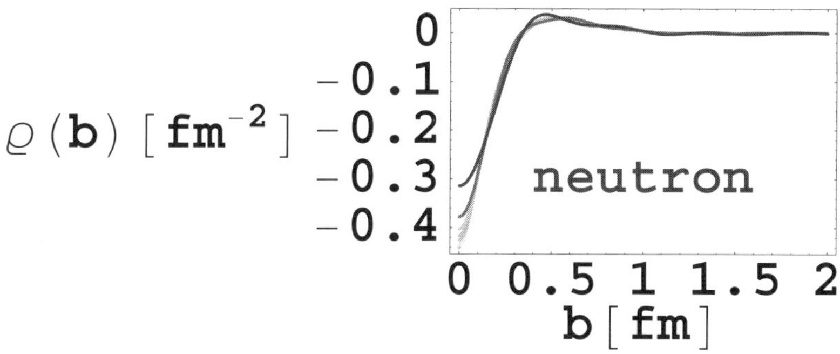

Fig. 2. The neutron charge density $\rho(b)$. The upper limit of the integral over Q in the Fourier transformation Eq. (17), ranges from $Q^2 = 8M^2$ to 20 M^2, with the largest (least negative) value at $b = 0$ obtained using $8M^2$.

in the middle, and again negative at the outer edge. The medium-ranged positive charge density is sandwiched by inner and outer regions of negative charge.

One can gain information about the individual u and d quark densities by invoking charge symmetry (invariance under a rotation by π about the z (charge) axis in isospin space)[14] so that the u, d densities in the proton are the same as the d, u densities in the neutron. We also neglect the effects of $s\bar{s}$[15] or heavier pairs of quarks. In this case $\rho_u(b) = \rho_p(b) - \rho_n(b)/2,\ \rho_d(b) = \rho_p(b) - 2\rho_n(b)$. Fig. 4, and are that the central down quark density is larger than that of the up quark by about 30%.

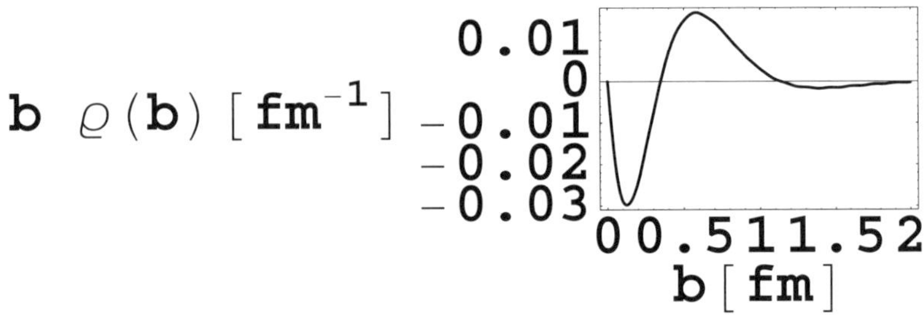

Fig. 3. The effective neutron charge density $b\rho(b)$, obtained using Kelly's[12] parameterization.

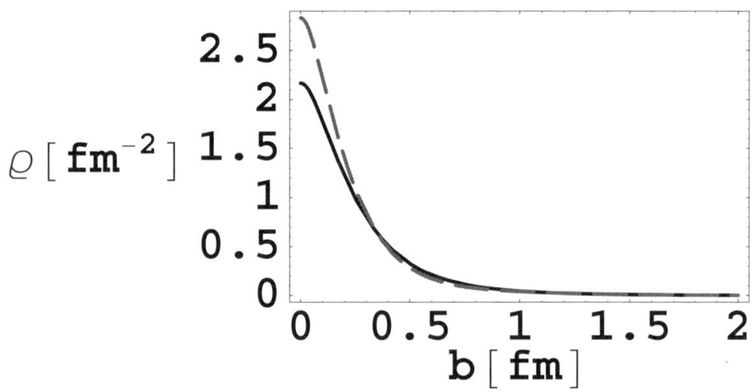

Fig. 4. The densities for d (dashed) and u (solid) quarks in the proton.

8. Summary and Interpretation

Model independent information about parton distributions has been obtained.[1] In particular, the central density of the neutron is negative. One possible interpretation is that quarks with high orbital angular momentum do not penetrate the interior, but the negatively charged pion, with only one unit of orbital angular momentum can both penetrate the interior and exist at long ranges. If this is the case, the negatively charged pions would be suppressed at medium range, allowing the related density to be positive.

Future measurements of neutron electromagnetic form factors could render the present results more precise, or modify them considerably.

Obtaining a qualitative and intuitive understanding of our results presents a challenge to lattice QCD and to builders of phenomenological models.

Acknowledgments

This work is partially supported by the USDOE.

References

1. G. A. Miller, *"Charge Density of the Neutron* arXiv:0705.2409 [nucl-th], to be published *Phys. Rev. Lett.*.
2. A. W. Thomas, S. Théberge and G. A. Miller, *Phys. Rev. D* **24**, 216 (1981).
3. J. L. Friar, *Part. Nucl.* **4**, 153 (1972); R. D. Carlitz, S. D. Ellis and R. Savit, *Phys. Lett. B* **68**, 443 (1977); N. Isgur, G. Karl and D. W. L. Sprung, *Phys. Rev. D* **23**, 163 (1981).
4. D. E. Soper, *Phys. Rev. D* **15**, 1141 (1977).
5. D.E. Soper, *Phys. Rev. D* **5**, 1956 (1972).
6. R. G. Sachs, *Phys. Rev.* **126**, 2256 (1962)
7. M. Burkardt, *Int. J. Mod. Phys. A* **18**, 173 (2003).
8. M. Diehl, *Eur. Phys. J. C* **25**, 223 (2002) *[Erratum-ibid. C* **31**, 277 (2003)].
9. M. Burkardt, *Phys. Rev. D* **62**, 071503 (R) (2000).
10. M. Diehl *et al.*, Nucl. Phys. B **596**, 33 (2001).
11. R. Bradford, A. Bodek, H. Budd and J. Arrington, *Nucl. Phys. Proc. Suppl.* **159**, 127 (2006).
12. J. J. Kelly, *Phys. Rev. C* **70**, 068202 (2004).
13. J. Arrington, *Phys. Rev. C* **69**, 022201 (R) (2004).
14. G. A. Miller, A. K. Opper and E. J. Stephenson, *Ann. Rev. Nucl. Part. Sci.* **56**, 253 (2006) G. A. Miller, *Phys. Rev. C* **57**, 1492 (1998); G. A. Miller, A. W. Thomas and A. G. Williams, *Phys. Rev. Lett.* **56**, 2567 (1986).
15. A. Acha *et al.*, *Phys. Rev. Lett.* **98**, 032301 (2007).

COVARIANCE, DYNAMICS AND SYMMETRIES, AND HADRON FORM FACTORS

M. S. BHAGWAT, I. C. CLOËT and C. D. ROBERTS

Physics Division
Argonne National Laboratory
Argonne, IL 60439-4843, U.S.A.

We summarise applications of Dyson-Schwinger equations to the theory and phenomenology of hadrons. Some exact results for pseudoscalar mesons are highlighted with details relating to the $U_A(1)$ problem. We describe inferences from the gap equation relating to the radius of convergence for expansions of observables in the current-quark mass. We recapitulate upon studies of nucleon electromagnetic form factors, providing a comparison of the ln-weighted ratios of Pauli and Dirac form factors for the neutron and proton.

Keywords: Dyson-Schwinger equations; Hadron electromagnetic form factors

1. Introduction

The Dyson-Schwinger equations (DSEs) are a nonperturbative means of studying QCD in the continuum. This is illustrated in the overview provided by Ref. [1]. A strength of the approach is that the chiral limit; viz., the domain of physical pion masses, is directly accessible. Indeed, with the established existence of a systematic, nonperturbative and symmetry-preserving truncation scheme[2,3] the DSEs were used to prove that the pion is simultaneously a Goldstone mode and a bound-state of effectively massive constituents.[4] Moreover, the exact nature of those dressed-quark and dressed-antiquark constituents is explained by the DSEs.[5,6] Namely, the quark-parton of QCD acquires a momentum-dependent mass function that at infrared momenta is larger by two orders-of-magnitude than the current-quark mass, an effect which owes primarily to a heavy cloud of gluons that clothes a low-momentum quark.[a] This is illustrated in Fig. 1.

[a]The 2-point Schwinger functions for ghosts and gluons experience analogous nonperturbative modification in the infrared.[7]

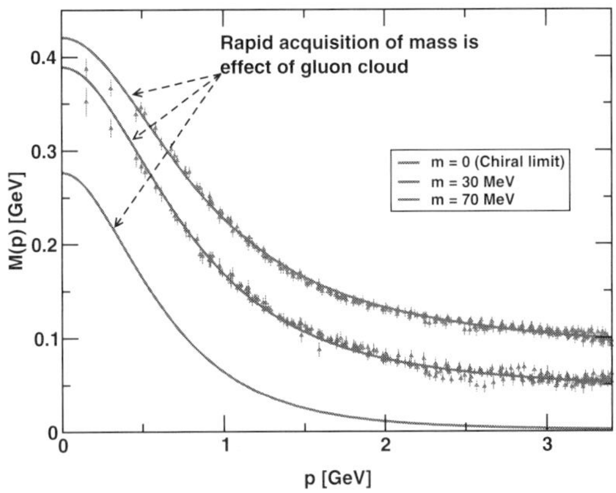

Fig. 1. Dressed-quark mass function, $M(p)$: solid curves – DSE results, obtained as explained in Refs. [5,6], "data" – numerical simulations of unquenched lattice-QCD.[8] In this figure one observes the current-quark of perturbative QCD evolving into a constituent-quark as its momentum becomes smaller. The constituent-quark mass arises from a cloud of low-momentum gluons attaching themselves to the current-quark. This is dynamical chiral symmetry breaking: an essentially nonperturbative effect that generates a quark mass *from nothing*; viz., it occurs even in the chiral limit.

Since a weak coupling expansion of the DSEs generates every diagram in perturbation theory, it is straightforward to ensure that model-dependent assumptions are restricted to infrared momenta. One can therefore turn a comparison of DSE predictions with data into a probe of the long-range interaction between light-quarks in QCD; namely, a unique means of exploring light-quark confinement. In large part this study is the same as drawing a map of the infrared behaviour of the QCD β-function. It is a fact too often ignored that the potential between infinitely heavy quarks measured in numerical simulations of quenched-QCD – the static potential – is not related in any known way to the confinement of light-quarks.

2. Gap equation

Since the gap equation is so fundamental to understanding hadron physics we reproduce it here:

$$S(p)^{-1} = Z_2\,(i\gamma\cdot p + m^{\rm bm}) + Z_1 \int_q^\Lambda g^2 D_{\mu\nu}(p-q)\frac{\lambda^a}{2}\gamma_\mu S(q)\Gamma^a_\nu(q,p), \quad (1)$$

where $\int_q^\Lambda$ represents a Poincaré invariant regularisation of the integral, with Λ the regularisation mass-scale,[4] $D_{\mu\nu}$ is the dressed-gluon propagator, Γ_ν is the dressed-quark-gluon vertex, and $m^{\rm bm}$ is the quark's Λ-dependent bare current-mass. The vertex and quark wave-function renormalisation constants, $Z_{1,2}(\zeta^2, \Lambda^2)$, depend on the gauge parameter.

The solution; namely, the dressed-quark propagator, can be written

$$S(p) = \frac{1}{i\gamma \cdot p\, A(p^2, \zeta^2) + B(p^2, \zeta^2)} = \frac{Z(p^2, \zeta^2)}{i\gamma \cdot p + M(p^2)} \tag{2}$$

wherein the mass function, $M(p^2)$, illustrated in Fig. 1, is independent of the renormalisation point, ζ. The solution is obtained from Eq. (1) augmented by the renormalisation condition $S(p)^{-1}\big|_{p^2=\zeta^2} = i\gamma \cdot p + m(\zeta^2)$, where $m(\zeta^2)$ is the running mass: $Z_2(\zeta^2, \Lambda^2)\, m^{\rm bm}(\Lambda) = Z_4(\zeta^2, \Lambda^2)\, m(\zeta^2)$, with Z_4 the Lagrangian-mass renormalisation constant. In QCD the chiral limit is strictly defined by:[4] $Z_2(\zeta^2, \Lambda^2)\, m^{\rm bm}(\Lambda) \equiv 0\,, \forall \Lambda^2 \gg \zeta^2$, which states that the renormalisation-point-invariant current-quark mass $\hat{m} = 0$. In this limit it is possible to unambiguously define the gauge invariant vacuum quark condensate in terms of $S(p)$.[4,9,10] This fact emphasises that gauge covariant quantities contain gauge invariant information.

The question of whether $M(p^2)$ has an expansion in current-quark mass around its chiral-limit value bears upon the radius of convergence for chiral perturbation theory. With this question one is asking whether it is possible to write

$$M(p^2; \hat{m}) = M(p^2; \hat{m} = 0) + \sum_{n=1}^{\infty} a_n \hat{m}^n \tag{3}$$

on a measurable domain of current-quark mass. It was found[10] that such an expansion exists for $\hat{m} < \hat{m}_{\rm rc}$. The value of $\hat{m}_{\rm rc}$ can be reported as follows: for a pseudoscalar meson constituted from a quark, f, with mass $\hat{m}_{\rm rc}$ and its antiparticle, $\bar{f}$, $m^{0^-}_{\bar{f}f} = 0.45\,{\rm GeV}$. Since physical observables, such as the leptonic decay constant of a pseudoscalar meson, are expressed in terms of $M(p^2)$, it follows that a chiral expansion is meaningful only for $(m^{0^-}_{\bar{f}f})^2 \lesssim 0.2\,{\rm GeV}^2$.[b] This entails, e.g., that it is only valid to employ chiral perturbation theory to fit and extrapolate results from numerical simulations of lattice-regularised QCD when the simulation parameters provide for $m_\pi^2 \lesssim 0.2\,{\rm GeV}^2$. Lattice results at larger pion masses are not within the domain of convergence of chiral perturbation theory.

[b]NB. Irrespective of the current-mass of the other constituent, a pseudoscalar meson containing one current-quark whose mass exceeds $\hat{m}_{\rm cr}$ is never within the domain of uniform convergence.

3. Mesons and the axial-vector Ward-Takahashi identity

Dynamical chiral symmetry breaking is a fact in QCD and this amplifies the importance of the axial-vector Ward-Takahashi identity. The existence of a sensible truncation of the DSEs has enabled proof via that identity of a body of exact results for pseudoscalar mesons. They relate even to radial excitations and/or hybrids,[11–13] and heavy-light[14] and heavy-heavy mesons.[15] The results have been illustrated using a renormalisation-group-improved rainbow-ladder truncation,[16,17] which also provided a prediction of the electromagnetic pion form factor.[18]

Implications for neutral pseudoscalar mesons have been elucidated.[19] In the general case the axial-vector Ward-Takahashi identity is written

$$P_\mu \Gamma^a_{5\mu}(k;P) = \mathcal{S}^{-1}(k_+) i\gamma_5 \mathcal{F}^a + i\gamma_5 \mathcal{F}^a \mathcal{S}^{-1}(k_-) - 2i\mathcal{M}^{ab}\Gamma^b_5(k;P) - \mathcal{A}^a(k;P)\,, \tag{4}$$

wherein: $\{\mathcal{F}^a \,|\, a = 0,\ldots,N_f^2-1\}$ are the generators of $U(N_f)$; the dressed-quark propagator $\mathcal{S} = \text{diag}[S_u, S_d, S_s, S_c, S_b, \ldots]$ is matrix-valued; $\mathcal{M}(\zeta)$ is the matrix of renormalised (running) current-quark masses and $\mathcal{M}^{ab} = \text{tr}_F\left[\{\mathcal{F}^a,\mathcal{M}\}\mathcal{F}^b\right]$, where the trace is over flavour indices. The inhomogeneous axial-vector vertex in Eq. (4), $\Gamma^a_{5\mu}(k;P)$, where P is the total momentum of the quark-antiquark pair and k the relative momentum, satisfies a Bethe-Salpeter equation, and likewise the pseudoscalar vertex, $\Gamma^b_5(k;P)$.

The final term in Eq. (4) expresses the non-Abelian axial anomaly. It involves

$$\mathcal{A}_U(k;P) = \int d^4x d^4y \, e^{i(k_+\cdot x - k_-\cdot y)} N_f \left\langle \mathcal{F}^0 \, q(x)\, \mathcal{Q}(0)\, \bar{q}(y)\right\rangle, \tag{5}$$

wherein the matrix element represents an operator expectation value in full QCD and $\mathcal{Q}(x) = i\frac{\alpha_s}{8\pi}\,\epsilon_{\mu\nu\rho\sigma}F^a_{\mu\nu}F^a_{\rho\sigma}(x) = \partial_\mu K_\mu(x)$ is the topological charge density operator, where $F^a_{\mu\nu}$ is the gluon field strength tensor. It is plain and important that only $\mathcal{A}^{a=0}$ is nonzero.

It is fundamentally important that while $\mathcal{Q}(x)$ is gauge invariant, the associated Chern-Simons current, K_μ, is not. It follows that in QCD no physical state can couple to K_μ and hence that physical states cannot provide a resolution of the so-called $U_A(1)$-problem; namely, they cannot play any role in ensuring that the η' is not a Goldstone mode.

In thinking about the $U_A(1)$-problem it is only necessary to focus on the case $\mathcal{A}^{a=0} \neq 0$ because if that is not so, then following Ref. [4] it is clear that the η' is certainly a Goldstone mode. $\mathcal{A}^0(k;P)$ is a pseudoscalar vertex and can therefore be expressed

$$\mathcal{A}^0(k;P) = \mathcal{F}^0\gamma_5\left[i\mathcal{E}_{\mathcal{A}}(k;P) + \gamma\cdot P\mathcal{F}_{\mathcal{A}}(k;P)\right.$$
$$\left. + \gamma\cdot k k\cdot P\mathcal{G}_{\mathcal{A}}(k;P) + \sigma_{\mu\nu}k_\mu P_\nu \mathcal{H}_{\mathcal{A}}(k;P)\right]. \qquad (6)$$

Equation (4) can now be used to derive a collection of chiral-limit, pointwise Goldberger-Treiman relations, important amongst which is the identity

$$2f_{\eta'}E_{\eta'}(k;0) = 2B_0(k^2) - \mathcal{E}_{\mathcal{A}}(k;0), \qquad (7)$$

where $B_0(k^2)$ is obtained in solving the chiral-limit gap equation.

It is now plain that if

$$\mathcal{E}_{\mathcal{A}}(k;0) = 2B_0(k^2), \qquad (8)$$

then $f_{\eta'}E_{\eta'}(k;0) \equiv 0$. This being true, then the homogeneous Bethe-Salpeter equation for the η' does not possess a massless solution in the chiral limit. The converse is also true; namely, the absence of such a solution requires Eq. (8). Hence, Eq. (8) is a necessary and sufficient condition for the absence of a massless η' bound-state. It is the chiral limit that is being discussed, in which case $B_0(k^2) \neq 0$ if, and only if, chiral symmetry is dynamically broken. Hence, the absence of a massless η' bound-state is only assured through the existence of an intimate connection between DCSB and an expectation value of the topological charge density.

Reference [19] derives a range of corollaries, amongst which are mass formulae for neutral pseudoscalar mesons, and presents an *Ansatz* for the Bethe-Salpeter kernel that enables their illustration. Despite its simplicity, the model is elucidative and phenomenologically efficacious; e.g., it predicts η–η' mixing angles of $\sim -15°$, π^0–η angles of $\sim 1°$, and a strong neutron-proton mass difference of $0.75\,(\hat{m}_d - \hat{m}_u)$.

4. Baryons

Partly because attention to this sector has only recently increased, the current level of expertise in studying baryons is roughly the same as it was with mesons more than ten years ago; viz., model building and phenomenology. We are a little ahead of that game, however, because much has been learnt in meson applications; e.g., as outlined above, a veracious understanding of the structure of dressed-quarks and -gluons has been acquired.

In quantum field theory a nucleon appears as a pole in a six-point quark Green function. The pole's residue is proportional to the nucleon's Faddeev amplitude, which is obtained from a Poincaré covariant Faddeev equation that adds-up all possible quantum field theoretical exchanges and interactions that can take place between three dressed-quarks. This is important because modern, high-luminosity experimental facilities employ large momentum transfer reactions; viz., $Q^2 > M_N^2$ where M_N is the nucleon's mass, and hence a veracious understanding of contemporary data requires a Poincaré covariant description of the nucleon.

A tractable truncation of the Faddeev equation is based[20] on the observation that an interaction which describes mesons also generates diquark correlations in the colour-$\bar{3}$ channel.[21] The dominant correlations for ground state octet and decuplet baryons are 0^+ and 1^+ diquarks because, e.g.: the associated mass-scales are smaller than the baryons' masses,[22,23] namely (in GeV) – $m_{[ud]_{0^+}} = 0.74 - 0.82$, $m_{(uu)_{1^+}} = m_{(ud)_{1^+}} = m_{(dd)_{1^+}} = 0.95 - 1.02$; and the electromagnetic size of these correlations is less than that of the proton[24] – $r_{[ud]_{0^+}} \approx 0.7\,\text{fm}$, which implies $r_{(ud)_{1^+}} \sim 0.8\,\text{fm}$ based on the ρ-meson/π-meson radius-ratio.[18,25]

The Faddeev equation's kernel is completed by specifying that the quarks are dressed, with two of the three dressed-quarks correlated always as a colour-$\bar{3}$ diquark. Binding is then effected by the iterated exchange of roles between the bystander and diquark-participant quarks. A Ward-Takahashi-identity-preserving electromagnetic current for the baryon thus constituted is subsequently derived.[26] It depends on the electromagnetic properties of the axial-vector diquark correlation.

A study of the nucleon's mass and the effect on this of a pseudoscalar meson cloud are detailed in [27]. Lessons learnt were employed in a series of studies of nucleon properties, including form factors.[28–31] The calculated ratio $\mu_p G_E^p(Q^2)/G_M^p(Q^2)$ passes through zero at $Q^2 \approx 6.5\,\text{GeV}^2$.[29] The analogous ratio for the neutron was calculated.[30] In the neighbourhood of $Q^2 = 0$, $\mu_n\, G_E^n(Q^2)/G_M^n(Q^2) = -\frac{r_n^2}{6}\, Q^2$, where r_n is the neutron's electric radius. The evolution of $\mu_p G_E^p(Q^2)/G_M^p(Q^2)$ and $\mu_n G_E^n(Q^2)/G_M^n(Q^2)$ on $Q^2 \gtrsim 2\,\text{GeV}^2$ are both primarily determined by the quark-core of the nucleon. While the proton ratio decreases uniformly on this domain,[28,29] the neutron ratio increases steadily until $Q^2 \simeq 8\,\text{GeV}^2$.[30]

A form factor ratio motivated by ideas from perturbative QCD[32] is depicted in Fig. 2. The parameter Λ is interpreted as a mass-scale that defines the upper-bound on the domain of so-called soft momenta in the perturbative analysis. A plausible value for such a quantity is $\Lambda \sim M_N$.[28c]

---

[c]NB. A value of $\Lambda \sim 0.3\,\text{GeV}$ corresponds to a length-scale $r_\Lambda \sim 1\,\text{fm}$ and it is not credible that perturbative QCD is applicable at ranges greater than the proton's radius.

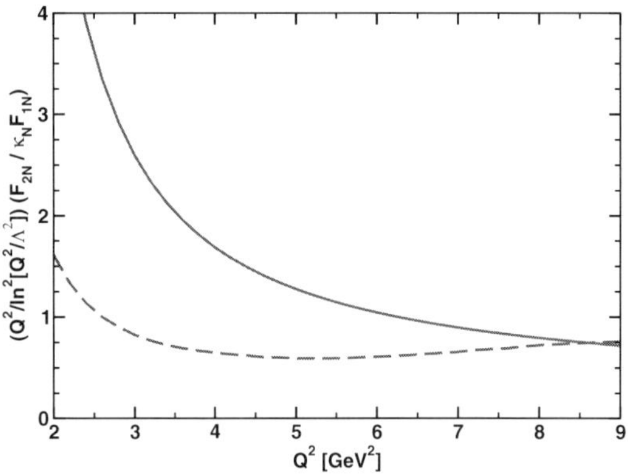

Fig. 2. Weighted nucleon Pauli/Dirac form factor ratio, calculated using the framework of Ref. [28] and presented with $\Lambda = 0.94$GeV: solid curve – neutron; dashed curve – proton.

It is interesting that the model of Ref. [28] yields neutron and proton ratios which cross at $Q^2 \simeq 8.5\,\text{GeV}^2$. The model's prediction for truly asymptotic momenta is currently being explored.

Stimulated by the possibility that Nature's fundamental "constants" might actually exhibit spatial and temporal variation,[33] we have begun to explore the current-quark-mass-dependence of the nucleon magnetic moments. This complements work on hadron σ-terms.[34–36] Preliminary results for the quark-core contribution to this variation can be expressed through the following ratios evaluated at the physical current-quark mass:

N	p	n
$-\frac{\delta\mu_N}{\mu_N}/\frac{\delta m}{m}$	0.016	0.0042

(9)

It is likely that pseudoscalar meson contributions will increase these values by a factor of $\gtrsim 10$.[37]

5. Epilogue

Confinement and dynamical chiral symmetry breaking (DCSB) can only be veraciously understood in relativistic quantum field theory. The DSEs provide a natural vehicle for the exploration of these phenomena.

DCSB is a singularly effective mass generating mechanism. For light-quarks it far outweighs the Higgs mechanism. It is understood via QCD's gap equation, which delivers a quark mass function with a momentum-dependence that connects the perturbative domain with the nonperturbative, constituent-quark domain.

The existence of a sensible truncation scheme enables the proof of exact results using the DSEs. That the truncation scheme is also tractable means the results can be illustrated, and furnishes a practical tool for the prediction of observables. The consequent opportunities for rapid feedback between experiment and theory brings within reach an intuitive understanding of nonperturbative strong interaction phenomena.

It can be argued that confinement is expressed in the analyticity properties of dressed Schwinger functions. To build understanding it is essential to work toward an accurate map of the confinement force between light-quarks and elucidate if/how that evolves from the potential between two static quarks. Among the rewards are a clear connection between confinement and DCSB, an accounting of the distribution of mass within hadrons, and a realistic picture of hybrids and exotics.

It is important to understand the relationship between parton properties on the light-front and the rest frame structure of hadrons. This is a challenge because, e.g., it is difficult to see how DCSB, a keystone of low-energy QCD, can be realised on the light-front. Parton distribution functions must be calculated in order to learn their content. Parametrisation is insufficient. It would be very interesting to know how, if at all, the distribution functions of a Goldstone mode differ from those of other hadrons. Answers to these and kindred questions are being sought using the DSEs.[38,39]

Acknowledgments

We acknowledge useful communications with V. V. Flambaum and A. Höll. This work was supported by the Department of Energy, Office of Nuclear Physics, contract no. DE-AC02-06CH11357; and benefited from the facilities of the ANL Computing Resource Center.

References

1. C. D. Roberts, *et al.*, Eur. Phys. J. *Special Topics* **140**, 53 (2007).
2. H. J. Munczek, Phys. Rev. D **52**, 4736 (1995).
3. A. Bender, C. D. Roberts and L. Von Smekal, Phys. Lett. B **380**, 7 (1996).
4. P. Maris, C. D. Roberts and P. C. Tandy, Phys. Lett. B **420**, 267 (1998).
5. M. S. Bhagwat, *et al.*, Phys. Rev. C **68**, 015203 (2003).
6. M. S. Bhagwat and P. C. Tandy, AIP Conf. Proc. **842**, 225 (2006).
7. L. von Smekal, R. Alkofer and A. Hauck, Phys. Rev. Lett. **79**, 3591 (1997).
8. P. O. Bowman, *et al.*, Phys. Rev. D **71**, 054507 (2005).
9. K. Langfeld, *et al.*, Phys. Rev. C **67**, 065206 (2003).
10. L. Chang, *et al.*, Phys. Rev. C **75**, 015201 (2007).
11. A. Höll, A. Krassnigg and C. D. Roberts, Phys. Rev. C **70**, 042203(R) (2004).

12. A. Höll, *et al.*, Phys. Rev. C **71**, 065204 (2005).
13. C. McNeile and C. Michael [UKQCD Collaboration], Phys. Lett. B **642**, 244 (2006).
14. M. A. Ivanov, Yu. L. Kalinovsky and C. D. Roberts, Phys. Rev. D **60**, 034018 (1999).
15. M. S. Bhagwat, *et al.*, Eur. Phys. J. A **31**, 630 (2007).
16. P. Maris and C. D. Roberts, Phys. Rev. C **56**, 3369 (1997).
17. P. Maris and P. C. Tandy, Phys. Rev. C **60**, 055214 (1999).
18. P. Maris and P. C. Tandy, Phys. Rev. C **62**, 055204 (2000).
19. M. S. Bhagwat, *et al.*, "Flavour symmetry breaking and meson masses," arXiv:0708.1118 [nucl-th].
20. R. T. Cahill, C. D. Roberts and J. Praschifka, Austral. J. Phys. **42**, 129 (1989).
21. R. T. Cahill, C. D. Roberts and J. Praschifka, Phys. Rev. D **36**, 2804 (1987).
22. C. J. Burden, *et al.*, Phys. Rev. C **55**, 2649 (1997).
23. P. Maris, Few Body Syst. **32**, 41 (2002).
24. P. Maris, Few Body Syst. **35**, 117 (2004).
25. M. S. Bhagwat and P. Maris, "Vector meson form factors and their quark-mass dependence," arXiv:nucl-th/0612069.
26. M. Oettel, M. Pichowsky and L. von Smekal, Eur. Phys. J. A **8**, 251 (2000).
27. M. B. Hecht, *et al.*, Phys. Rev. C **65**, 055204 (2002).
28. R. Alkofer, *et al.*, Few Body Syst. **37**, 1 (2005).
29. A. Höll, *et al.*, Nucl. Phys. A **755**, 298 (2005).
30. M. S. Bhagwat, *et al.*, Nucl. Phys. A **790**, 10 (2007).
31. A. Höll, C. D. Roberts and S. V. Wright, AIP Conf. Proc. **857**, 46 (2006).
32. A. V. Belitsky, X. d. Ji and F. Yuan, Phys. Rev. Lett. **91**, 092003 (2003).
33. J. P. Uzan, Rev. Mod. Phys. **75**, 403 (2003).
34. V. V. Flambaum, *et al.*, Few Body Syst. **38**, 31 (2006).
35. A. Höll, *et al.*, Nucl. Phys. Proc. Suppl. **161**, 87 (2006).
36. V. V. Flambaum and R. B. Wiringa, "Dependence of nuclear binding on hadronic mass variation," arXiv:0709.0077 [nucl-th].
37. V. V. Flambaum, *et al.*, Phys. Rev. D **69**, 115006 (2004).
38. M. B. Hecht, C. D. Roberts and S. M. Schmidt, Phys. Rev. C **63**, 025213 (2001).
39. I. C. Cloet, W. Bentz and A. W. Thomas, "Transversity quark distributions in a covariant quark-diquark model," arXiv:0708.3246 [hep-ph].

N to Δ ELECTROMAGNETIC AND AXIAL FORM FACTORS IN FULL QCD

C. ALEXANDROU

Department of Physics, University of Cyprus, P.O. Box 20537, 1678, Nicosia, Cyprus
E-mail: alexand@ucy.ac.cy, www.ucy.ac.cy

Lattice results on the N to Δ electromagnetic, axial-vector and pseudoscalar form factors are evaluated using dynamical staggered sea quarks and domain wall valence quarks for pion masses in the range of 580-350 MeV, as well as, dynamical and quenched Wilson fermions for similar pion masses.

Keywords: Lattice QCD, Form factors, Nucleon Resonances

1. Introduction

State-of-the-art lattice QCD calculations can yield model independent results on N to Δ transition form factors, thereby providing direct comparison with experiment. One such example is the N to Δ quadrupole form factors that have been accurately measured in a series of recent experiments at low[1,2] and high momentum transfers.[3] They encode information on the deformation of the nucleon and Δ. We present results on these N to Δ electromagnetic form factors, as well as on the dominant axial-vector N to Δ transition form factors $C_5^A(q^2)$ and $C_6^A(q^2)$. Experiments using electroproduction of the Δ resonance are in the progress[4] to measure the parity violating asymmetry in N to Δ, which, to leading order, is connected to $C_5^A(q^2)$. Evaluation of the pseudoscalar $\pi N\Delta$ form factor, $G_{\pi N\Delta}(q^2)$, follows once the N to Δ sequential propagators are computed. In addition, we evaluate the nucleon axial-vector form factors and the πNN form factor, $G_{\pi NN}(q^2)$. Having both the nucleon and the N to Δ form factors allows us to discuss ratios of form factors that are expected to show weaker quark mass dependence and be less sensitive to other lattice artifacts. Furthermore, knowledge of the axial-vector form factors and the πNN and $\pi N\Delta$ form factors allows us to check the Goldberger-Treiman relations.

The light quark regime is studied in two ways: Besides using configurations with two degenerate flavors of dynamical Wilson fermions we use a hybrid combination of domain wall valence quarks, which have chiral symmetry on the lattice, and MILC configurations generated with three flavors of staggered sea quarks using the Asqtad improved action.[5]

2. Lattice Techniques

Observables in lattice QCD are given by the vacuum expectation value of gauge invariant operators in Euclidean time:

$$< \Omega|\hat{O}|\Omega >= \frac{1}{Z}\int d[U]d[\bar{\psi}]d[\psi]\ O[U,\bar{\psi},\psi]e^{-S_g[U]-S_F[U,\bar{\psi},\psi]} \tag{1}$$

Integrating over the fermionic degrees of freedom we obtain

$$< \Omega|\hat{O}|\Omega >= \frac{1}{Z}\int d[U]\ \det(D[U])O[U,D^{-1}[U]]e^{-S_g[U]} \tag{2}$$

where $D^{-1}_{jn}[U]$ substitutes each appearance of $-\bar{\psi}_n\psi_j$ in the operator and describes valence quarks whereas $\det(D[U])$ corresponds to sea quarks. The path integral over the gauge fields is done numerically by stochastically generating a representative ensemble of gauge configurations according to the probability

$$P[U] = \frac{1}{Z}\exp\left\{-S_g[U] + \ln\left(\det(D[U])\right)\right\}. \tag{3}$$

In this work, besides Wilson fermions for the sea and valence quarks, we use staggered sea quarks ($\det(D_{\text{staggered}}[U])$) and domain wall valence fermions ($D^{-1}_{\text{DW}}[U]$). The expectation values are obtained by summing over the U-ensemble: $< \Omega|\hat{O}|\Omega >= \lim_{N\to\infty}\frac{1}{N}\sum_{k=1}^{N} O[U^k, D^{-1}[U^k]]$.

The evaluation of form factors involves taking numerically the Fourier transform of two- and three-point functions with respect to momentum transfer which, on a finite box of spatial length L, takes discrete values in units of $2\pi/L$. For large values of momentum transfer the results become noisy and therefore we are limited up to $Q^2 \equiv -q^2 \sim 2$ GeV^2. To ensure that finite volume effects are kept small we take box sizes such that $Lm_\pi \gtrsim 4.5$, where m_π is the pion mass[a]. In addition, discretization errors due to the finite lattice spacing a must be checked. Wilson fermions have $\mathcal{O}(a)$ discretization errors and staggered fermions with Asqtad action and domain wall fermions (hybrid approach) have $\mathcal{O}(a^2)$ errors. Therefore agreement between results in these two approaches provides an indication that cut-off effects are under control. Finally, we use larger bare u- and d -quark masses than physical and extrapolation to the chiral limit must be considered.

[a]One exception is in the case of dynamical Wilson fermions at the smallest pion mass for which $Lm_\pi = 3.6$ as marked in the Table.

Form factors are extracted from three-point functions, $G^{\Delta JN}(t_2, t_1; \mathbf{q}) = < \Omega| \sum_{\mathbf{x_1},\mathbf{x_2}} e^{i\mathbf{q}\cdot\mathbf{x_1}} \hat{T} \hat{J}_{\tilde{h}}(\mathbf{x_2}, t_2) \hat{J}(\mathbf{x_1}, t_1) \hat{J}_h^\dagger(0)|\Omega >$, shown in the diagram below:

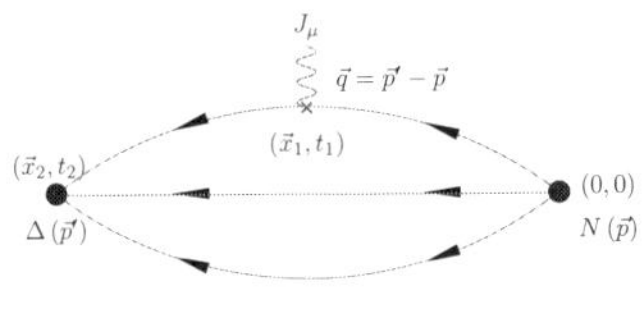

Interpolating fields for N and Δ are:

$$J^p(x) = \epsilon^{abc}[u^{Ta}(x) C\gamma_5 d^b(x)] u^c(x),$$

$$J_\sigma^{\Delta^+}(x) = \frac{1}{\sqrt{3}}\epsilon^{abc}\{2[u^{Ta}(x) C\gamma_\sigma d^b(x)] u^c(x) + [u^{Ta}(x) C\gamma_\sigma u^b(x)] d^c(x)\}$$

In all cases we apply Gaussian smearing at the source and sink. In the case of unquenched Wilson fermions HYP-smearing is applied to the gauge links used in the Gaussian smearing of the source and sink. In the case of domain wall fermions we use HYP-smeared MILC configurations in all computations. We carry out sequential inversions by fixing the quantum numbers at the sink and source, which means that the sink time t_2 is fixed, whereas the insertion time t_1 can vary and any operator can be inserted at t_1. In this work we consider the vector current, $j_\mu^a = \bar{\psi}\gamma_\mu \frac{\tau^a}{2}\psi$, the axial-vector current, $A_\mu^a = \bar{\psi}\gamma_\mu\gamma_5 \frac{\tau^a}{2}\psi$ and the pseudoscalar current, $P^a = \bar{\psi} i\gamma_5 \frac{\tau^a}{2}\psi$, where τ^a are Pauli matrices acting in flavor space. All $\vec{x}_1$ and $\vec{x}_2$ are summed over and we vary t_1 in search for a plateau. The exponential time dependence and unknown overlaps of the interpolating fields with the physical states cancel by dividing the three-point function with appropriate combinations of two-point functions.[6]

The lattice parameters that we use are given in the Table.

Wilson fermions			
number of confs	κ	m_π (GeV)	m_N (GeV)
Quenched $32^3 \times 64$, $\beta = 6.0$, $a^{-1} = 2.14(6)$ GeV ($a = 0.09$ fm) from nucleon mass at chiral limit			
200	0.1554	0.563(4)	1.267(11)
200	0.1558	0.490(4)	1.190(13)
200	0.1562	0.411(4)	1.109(13)
	κ_c =0.1571	0.	0.938(9)
Unquenched[7] $24^3 \times 40$,$\beta = 5.6$, $a^{-1} = 2.56(10)$ GeV ($a = 0.08$ fm)			
185	0.1575	0.691(8)	1.485(18)
157	0.1580	0.509(8)	1.280(26)
Unquenched[8] $24^3 \times 32$,$\beta = 5.6$, $a^{-1} = 2.56(10)$ GeV			
200	0.15825	0.384(8)$\leftarrow Lm_\pi = 3.6$	1.083(18)
	$\kappa_c = 0.1585$	0.	0.938(33)

Hybrid scheme $a^{-1} = 1.58$ GeV ($a = 0.125$ fm) from MILC collaboration						
number of confs	Volume	$(am_{u,d})^{\text{sea}}$	$(am_s)^{\text{sea}}$	$(am_q)^{DW}$	m_π^{DW} (GeV)	m_N (GeV)
150	$20^3 \times 64$	0.03	0.05	0.0478	0.589(2)	1.392(9)
198	$20^3 \times 64$	0.02	0.05	0.0313	0.501(4)	1.255(19)
100	$20^3 \times 64$	0.01	0.05	0.0138	0.362(5)	1.138(25)
150(300 for CMR)	$28^3 \times 64$	0.01	0.05	0.0138	0.354(2)	1.210(24)

3. *N* to Δ Electromagnetic Form Factors

The N to Δ matrix element of the electromagnetic current can be decomposed into a dominant magnetic dipole, G_{M1}, and two suppressed electric and Coulomb quadrupole form factors, G_{E2} and G_{C2}. A non-zero G_{E2} and G_{C2} signal a deformation in the nucleon and/or Δ. Precise experimental data on the quadrupole to dipole ratios, $R_{EM}(\text{EMR}) = -\frac{G_{E2}(q^2)}{G_{M1}(q^2)}$, and $R_{SM}(\text{CMR}) = -\frac{|\vec{q}|}{2m_\Delta}\frac{G_{C2}(q^2)}{G_{M1}(q^2)}$, suggest deformation of the nucleon/Δ.[9]

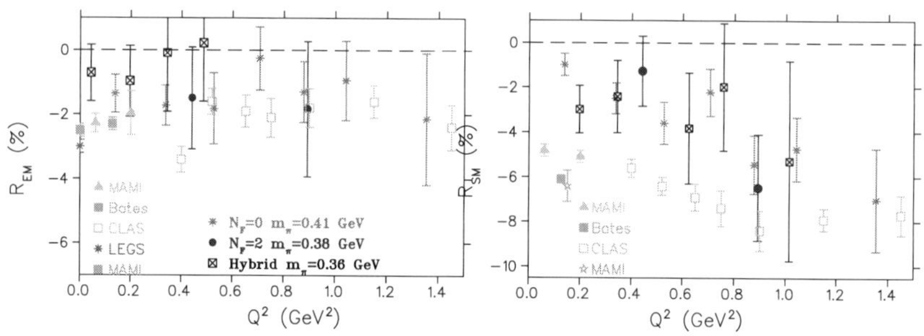

Fig. 1. The EMR (left) and CMR (right) for the lightest pion mass in our three type of simulations.

In Fig. 1 we show the EMR and CMR ratios for the smallest pion mass in the quenched case, for two dynamical flavors of Wilson fermions and in the hybrid approach. For the first time in full QCD, we achieve good enough accuracy to exclude a zero value for these ratios. Furthermore, at low Q^2, unquenched results become more negative bringing lattice results closer to experiment and showing the importance of the pion cloud effects at small Q^2.

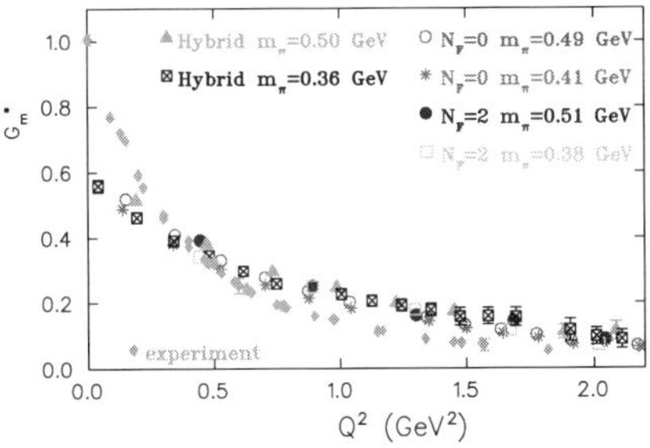

Fig. 2. Magnetic dipole form factor G_m^*, in the Ash parameterization: $G_m^* = \frac{1}{\sqrt{1+\frac{Q^2}{(m_N+m_\Delta)^2}}} G_{M1}$.

In Fig. 2 we show lattice results for G_m^* for Wilson fermions and in the hybrid approach as well as experimental results. At low Q^2 lattice results are below experiment for the pion masses considered here. It remains an open question whether extrapolation to the physical pion mass can bridge the gap between lattice and experiment.

4. Nucleon and N to Δ Axial-Vector Form Factors

In the case of Wilson fermions, besides N to Δ we also calculate the nucleon axial-vector form factors. The LHP collaboration[10] has evaluated these form factors in the hybrid approach with the same parameters as those used in our N to Δ study and therefore, in this case, we use their results to compare. The nucleon axial- vector form factors G_A and G_p are given by

$$\langle N(p')|A^3_\mu|N(p)\rangle = i\sqrt{\frac{m_N^2}{E_N(\mathbf{p}')E_N(\mathbf{p})}}\bar{u}(p')\left[G_A(q^2)\gamma_\mu\gamma_5 + \frac{q_\mu\gamma_5}{2m_N}G_p(q^2)\right]\frac{\tau^3}{2}u(p) \tag{4}$$

Since the final state is no longer the Δ a new set of sequential inversions is needed. The decomposition of the N to Δ matrix element of the axial-vector current can be written in terms of four transition form factors:[11]

$$< \Delta(p',s')|A^3_\mu|N(p,s) >= i\sqrt{\frac{2}{3}}\sqrt{\frac{m_N m_\Delta}{E_\Delta(\mathbf{p}')E_N(\mathbf{p})}}\ \bar{u}^\lambda_{\Delta^+}(p',s')\left[C_5^A(q^2)g_{\lambda\mu}\right.$$

$$\left. + \frac{C_6^A(q^2)}{m_N^2}q_\lambda q_\mu + \left(\frac{C_3^A(q^2)}{m_N}\gamma^\nu + \frac{C_4^A(q^2)}{m_N^2}p'^\nu\right)(g_{\lambda\mu}g_{\rho\nu} - g_{\lambda\rho}g_{\mu\nu})\,q^\rho\right]u_P(p,s) \tag{5}$$

Under the assumptions that $C_3^A \sim 0$ and $C_4^A \ll C_5^A$ the parity violating asymmetry is proportional to the ratio C_5^A/C_3^V,[12] where C_3^V can be obtained from the electromagnetic N to Δ transition.

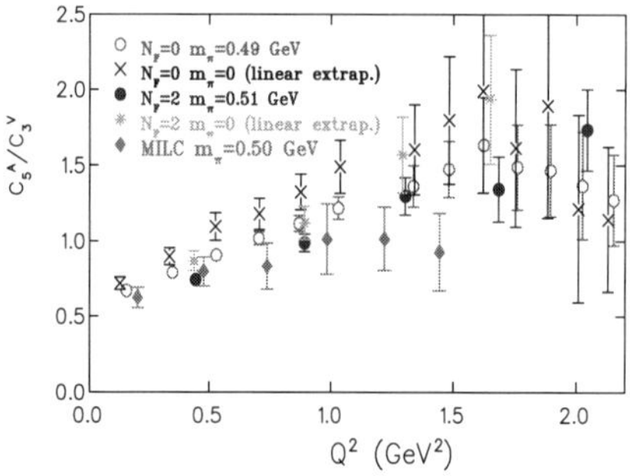

Fig. 3. Parity violating asymmetry.

As shown in Fig. 3, the asymmetry is non-zero when $Q^2 = 0$,[6] increases with Q^2 up to about $Q^2 \sim 1.5$ GeV^2 and shows small unquenching effects for this range of quark masses. Given this weak quark mass dependence, the results can be taken as a prediction for the ratio to be measured by the G0 collaboration .[4]

4.1. *Goldberger-Treiman relations*

Partial conservation of axial current (PCAC), $\partial^\mu A^a_\mu = f_\pi m_\pi^2 \pi^a$, and the axial Ward Identity, $\partial^\mu A^a_\mu = 2m_q P^a$, relate the pion field π^a with the pseudoscalar density: $\pi^a = \left(2m_q P^a / f_\pi m_\pi^2\right)$, where the pion decay constant f_π is determined from the two-point function $< 0|A^a_\mu|\pi^b(p) >= ip_\mu \delta^{ab} f_\pi$. The renormalized quark mass, m_q, is given by $m_q = \frac{m_\pi <0|\tilde{A}^a_0|\pi^a(0)>}{2<0|\tilde{P}^a|\pi^a(0)>}$, where $\tilde{A}^a_0$ and $\tilde{P}^a$ are the renormalized currents. To obtain the πNN and $\pi N\Delta$ form factors we use the decomposition

$$2m_q \langle N(p')|P^3|N(p)\rangle = i\sqrt{\frac{m_N^2}{E_N(\mathbf{p}')E_N(\mathbf{p})}} \frac{f_\pi m_\pi^2 \, G_{\pi NN}(q^2)}{m_\pi^2 - q^2} \, \bar{u}(p')\gamma_5 \frac{\tau^3}{2} u(p)$$

$$2m_q \langle \Delta(p')|P^3|N(p)\rangle = i\sqrt{\frac{2}{3}}\sqrt{\frac{m_\Delta m_N}{E_\Delta(\mathbf{p}')E_N(\mathbf{p})}} \frac{f_\pi m_\pi^2 \, G_{\pi N\Delta}(q^2)}{m_\pi^2 - q^2} \bar{u}^\nu_{\Delta^+}(p') \frac{q_\nu}{2m_N} u_P(p)$$

PCAC relates the axial form factors G_A and G_p with $G_{\pi NN}$ and equivalently C_5^A and C_6^A with $G_{\pi N\Delta}$. These are the well known generalized Goldberger-Treiman relations (GTRs). As mentioned above there are advantages in considering ratios. In Fig. 4 we show two such ratios, namely $G_{\pi N\Delta}/G_{\pi NN}$ and $2C_5^A/G_A$. Both are independent of Q^2 and the quark mass. Fitting to a constant we find $2C_5^A/G_A \sim 1.6 \sim G_{\pi N\Delta}/G_{\pi NN}$, which implies the Goldberger-Treiman relations: $G_{\pi NN}(q^2)\, f_\pi = m_N G_A(q^2)$ and $G_{\pi N\Delta}(q^2)\, f_\pi = 2m_N C_5^A(q^2)$. Assuming pion-pole dominance we can write $G_p(Q^2) = \frac{4m_N^2/m_\pi^2}{1+Q^2/m_\pi^2}\, G_A(Q^2)$ and $C_6^A(Q^2) = \frac{m_N^2/m_\pi^2}{1+Q^2/m_\pi^2}\, C_5^A(Q^2)$. Therefore we have the equality $8C_6^A(Q^2)/G_p(Q^2) = G_{\pi N\Delta}/G_{\pi NN}$. We find that $8C_6^A(Q^2)/G_p(Q^2) \sim 1.7$[13] a few percentage larger than $G_{\pi N\Delta}/G_{\pi NN}$.

In Fig. 5 we present the nucleon and N to Δ axial form factors separately together with fits of G_A and C_5^A to a dipole form, $g_0/(\frac{Q^2}{m_A^2}+1)^2$. Dynamical

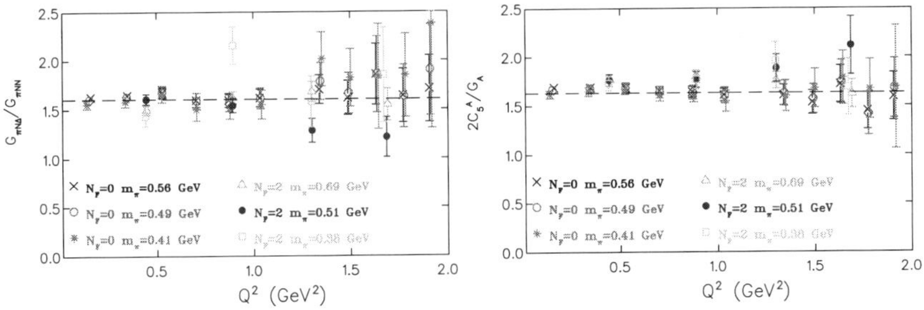

Fig. 4. The ratios $G_{\pi N\Delta}/G_{\pi NN}$ (left) and $2C_5^A/G_A$ (right).

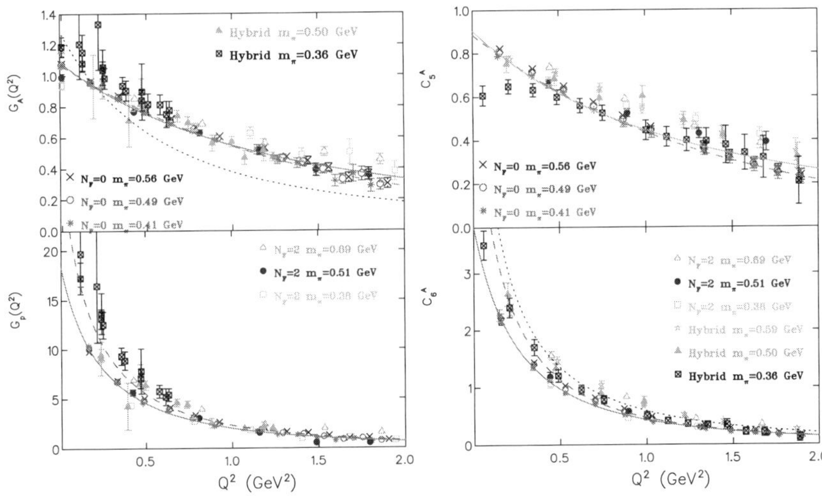

Fig. 5. The nucleon axial form factors G_A (top, left) and G_p (bottom, left) and the N to Δ axial form factors C_5^A (top, right) and C_6^A (bottom, right).

QCD results in the hybrid approach for the smallest pion mass, where we can access low Q^2 values, show large unquenching effects. Having fitted G_A and C_5^A, we can check if pion-pole dominance describes the Q^2-dependence of G_p and C_6^A. The dashed lines correspond to the quenched data and show the behavior for G_p and C_6^A extracted from fits to G_A and C_5^A assuming pion-pole dominance, whereas the dotted line shown for C_5^A is for the hybrid approach, in both cases for the lightest pion mass. As can be seen, they deviate from the lattice results at low Q^2. Instead they are best described by the solid curves, which are obtained by fitting the pole mass.

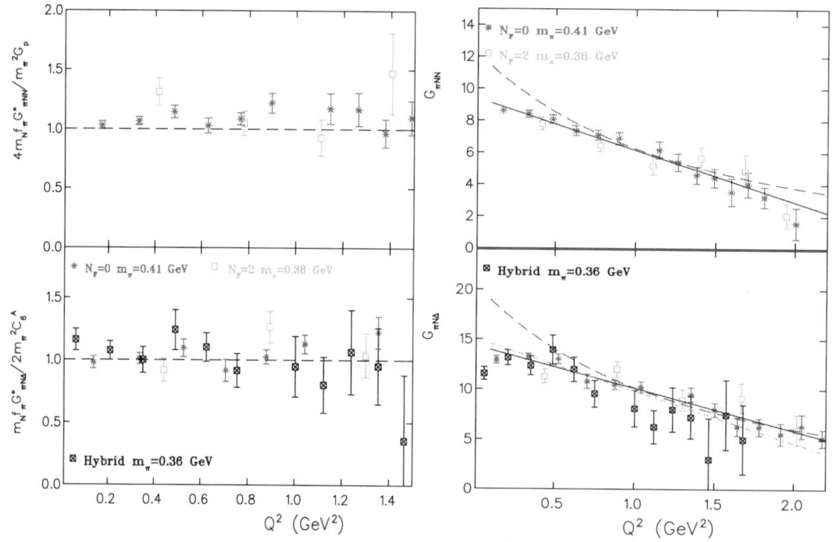

Fig. 6. Left: The ratios R_{NN} (top) and $R_{N\Delta}$ (bottom) and right: $G_{\pi NN}$ (top) and $G_{\pi N\Delta}$ (bottom) for the smallest pion mass in each type of simulation.

In Fig. 6 we show the relations

$$R_{NN} \equiv \frac{4m_N f_\pi G^*_{\pi NN}(Q^2)}{m_\pi^2 G_p(Q^2)}, \qquad R_{N\Delta} \equiv \frac{m_N f_\pi G^*_{\pi N\Delta}(Q^2)}{2m_\pi^2 C_6^A(Q^2)}, \tag{6}$$

where we have defined $G^*_{\pi NN}(Q^2) \equiv G_{\pi NN}(Q^2)/(1 + Q^2/m_\pi^2)$ with a corresponding expression for $G^*_{\pi N\Delta}$. As can be seen these ratios are consistent with unity for all Q^2-values. Finally, in Fig. 6 we show $G_{\pi NN}$ and $G_{\pi N\Delta}$ for the smallest pion mass. The dash lines are obtained from fits of G_A and C_5^A via the GTRs, $G_{\pi NN}(Q^2) = m_N G_A(Q^2)/f_\pi$ and $G_{\pi N\Delta}(Q^2) = 2m_N C_5^A(Q^2)/f_\pi$. As can be seen, there are large deviations at small Q^2. Lattice results at this pion mass give a smaller value in the limit $Q^2 \to 0$ than what is extracted from experimental data namely, $G_{\pi NN}(0) = 13.21(11)$.[14] The solid lines are fits to the form $\left(1 - \Delta \frac{Q^2}{m_\pi^2}\right)$ with a, Δ fit parameters. Using these fits $G_{\pi NN}(0)$ and $G_{\pi NN}(0)$ are extracted (for details see Ref. [13]).

5. Conclusions

Lattice results on the electromagnetic, axial-vector and pseudoscalar form factors for the nucleon and the N to Δ transition are presented in the quenched approximation, for two-flavors of dynamical Wilson fermions and using dynamical staggered sea quarks and domain wall valence quarks (hybrid approach). Results on the quadrupole to dipole ratios EMR and CMR, obtained in the hybrid approach reaching down to a pion mass of 350 MeV and low Q^2-values, are non-zero and of similar magnitude as in experiment. We also find that ratios of form factors, such as $G_{\pi N\Delta}/G_{\pi NN} \sim 1.6$ and $2C_5^A/G_A \sim 1.6$, calculated using Wilson fermions, are in agreement with phenomenology. Our results for the ratio C_5^A/C_3^V as a function of Q^2, can be regarded as a lattice prediction for the parity violating asymmetry to leading order. The deviations from experiment seen for the magnetic dipole N-Δ transition form factor G_m^* and the values of $G_{\pi NN}$ and $G_{\pi N\Delta}$ in the limit $Q^2 \to 0$ need further study. In particular, finite lattice spacing effects, as well as, chiral extrapolation to the physical pion mass must be investigated.

Acknowledgments

I would like to thank my collaborators G. Koutsou, Th. Leontiou, H. Neff, J. W. Negele, W. Schroers and A. Tsapalis for their valuable contributions that made this work possible. This work is supported in part by the EU Integrated Infrastructure Initiative Hadron Physics (I3HP) under contract RII3-CT-2004-506078.

References

1. C. Mertz *et al.*, *Phys. Rev. Lett.* **86**, 2963 (2001).
2. N. F. Sparveris *et al.*, *Phys. Lett. B* **651**, p. 102 (2007).
3. K. Joo *et al.*, *Phys. Rev. Lett.* **88**, p. 122001 (2002).
4. *PAVI 2002* (unpublished, 2002).
5. K. Orginos, D. Toussaint and R. L. Sugar, *Phys. Rev. D* **60**, p. 054503 (1999).
6. C. Alexandrou *et al.*, *Phys. Rev. Lett.* **94**, p. 021601 (2005).
7. O. Orth, T. Lippert and K. Schilling, *Phys. Rev. D* **72**, p. 014503 (2005).
8. C. Urbach *et al.*, *Comput. Phys. Commun.* **174**, p. 87 (2006).
9. C. N. Papanicolas, *Eur. Phys. J.* **A18**, p. 141 (2003).
10. P. Hägler *et al.*, arxiv:0705.4295.
11. S. Adler, *Ann. Phys* **50**, p. 189 (1968).
12. N. C. Mukhopadhyay *et al.*, *Nucl. Phy.* **A633**, p. 481 (1998).
13. C. Alexandrou *et al.*, *Phys. Rev. D (in press),* **arXiv:0706.3011** (2007).
14. H. C. Schröder *et al.*, *Eur. Phys. J.* **C21**, p. 473 (2001).

REAL AND VIRTUAL COMPTON SCATTERING IN PERTURBATIVE QCD

CHUENG-RYONG JI* and RICHARD THOMSON

Department of Physics, North Carolina State University, Raleigh, NC 27695-8202, USA
**E-mail: ji@ncsu.edu*

We present the results of calculations analyzing nucleon Compton scattering to lowest order using perturbative QCD (PQCD) methods. Our recent agreement of theoretical predictions with the results of Brooks and Dixon in 2000 is a significant progress in the PQCD investigation of nucleon real Compton scattering (RCS) process at high momentum transfer. We discuss our extension of RCS calculation to the case of incoming virtual photon and a possible link to the generalized parton distribution (GPD) approach.

Keywords: Perturbative QCD; Nucleon Compton Scattering; Genralized Parton Distributions.

1. Introduction

We are quite excited about the forthcoming 12 GeV upgrade at Jefferson Laboratory (JLab)[1] since it is highly relevant to our hadron structure study at short distances. In particular, nucleon Compton scattering provides an important experimental probe of the structure of both the proton and the neutron. One area where it is possible to perform a theoretical analysis of such experiments is in the high energy limit where the asymptotic perturbative behavior is applicable. The new JLab data[2] from Hall A collaboration at $s = 5 - 11$ GeV^2 and $-t = 2 - 7$ GeV^2 has already updated the old Cornell data[3] at $s = 4.6 - 12.1$ GeV^2 and $-t = 0.7 - 4.3$ GeV^2. Such update revived the interesting discussion on the applicability of perturbative QCD (PQCD) to exclusive processes at high momentum transfer Q. The main question in the applicability issue of PQCD to exclusive processes is "How high is really high in Q?". We believe the forthcoming 12 GeV upgrade at JLab would answer many questions in the discussion of PQCD applicability to exclusive processes and lead us to understand more clearly the short distance hadron structure.

On the theory side of nucleon Compton scattering process, there have been several analyses[4–8] of real Compton scattering (RCS) but none of them agreed to each other, each time with different results. This situation of theory side has been obviously disastrous because it wouldn't make any real sense of comparison with experimental data unless a consistent and identical theoretical prediction is available. Fortunately, we recently found that our own results[9] do agree quite well with the calculation made by Brooks and Dixon.[8] This recent agreement of theoretical predictions is an important progress in the PQCD investigation of nucleon Compton scattering process at high momentum transfer.

While the RCS has been computed a number of times,[4–9] the virtual Compton scattering has been calculated only once previously.[5] Although the virtual calculation is considerably more complex than the real calculation, the extension of RCS to the virtual Compton scattering is very important. As evidenced in the current interest of deeply virtual Compton scattering (DVCS), the generalized parton distribution (GPD) approach is applicable in making predictions for the kinematic region of $Q^2 >> -t$. Since the GPD predictions can be compared with the PQCD results for some overlap region $Q^2 >> -t >> \Lambda^2_{QCD}$ in DVCS process, we believe the applicability of both GPD and PQCD methods can be tested with the analysis of DVCS process in the overlap region.

We thus address both RCS and the virtual Compton scattering where the incoming photon is virtual. In the next section (Section II), we discuss briefly the PQCD method and make remarks on computational methods. In Section III, we present the RCS results in comparison with the new JLab data.[2] In Section IV, we discuss our extension of RCS analysis to the investigation of virtual Compton scattering and remark on our attempt to link our analysis to the GPD approach.

2. PQCD in Light-Front Dynamics

Among the three forms of Hamiltonian dynamics proposed by Dirac in 1949,[10] i.e. instant ($x^0 = 0$), front ($x^+ = x^0 + x^3 = 0$), point ($x_\mu x^\mu = a^2 > 0, x^0 > 0$), the dimension of stability group which leaves the hypersurface of corresponding time invariant is the largest (i.e. seven out of the ten Poincare generators) in the front form which we call the light-front dynamics (LFD). LFD provides a unified framework to analyze various experimental measurements such as generalized parton distributions (GPDs) and single spin asymmetry (SSA) at JLab and DESY (Hermes),[11] B-decays at SLAC (BaBar) and KEK (Belle)[12] as well as quark gluon plasma (QGP) productions at BNL (RHIC) and CERN (ALICE),[13] etc.. Due to the rational energy-momentum dispersion relation, the LFD has distinguished features compared to other forms of Hamiltonian dynamics. In particular, the vacuum fluctuations are suppressed and the kinematic generators are proliferated in LFD. Thus, LFD provides a good framework to formulate the PQCD factorization of exclusive amplitudes.

In PQCD, the helicity amplitude $\mathcal{M}$ for nucleon Compton scattering is given by

$$\mathcal{M}_{hh'}^{\lambda\lambda'}(s,t) = \sum_{d,i} \int [dx][dy]\phi_i(x_1,x_2,x_3)T_i^{(d)}(x,h,\lambda;y,h',\lambda')\phi_i^\star(y_1,y_2,y_3). \tag{1}$$

Here λ (λ') and h (h') are the helicity of the initial (final) photon and nucleon states, respectively; (x_1,x_2,x_3) and (y_1,y_2,y_3) are the momentum fractions carried by the three quarks in the initial and final states, respectively; ϕ_i is the distribution amplitude for the ith spin/flavor state in the nucleon; and $T_i^{(d)}$ is the hard scattering amplitude for a specific Feynman diagram, d. The sum is over all possible diagrams and over the spin/flavor states. In leading-twist approximation, only valence collinear quarks contribute. The integration measure $[dx]$ implies integration over x_1, x_2 and x_3 subject to the constraint $x_1 + x_2 + x_3 = 1$.

To leading order, the hard scattering amplitude is given by the scattering amplitude of the collinear valence quarks with a minimum number of gluon exchanges. For comparison, we use the same notation and classification as in Ref. 6. All the possible Feynman diagrams can be classified according to the arrangement of the gluon lines in the diagram, as shown in Fig. 1. There are 56 ways of attaching the two photons in groups A to F. Therefore altogether there are 6x56 = 336 Feynman diagrams which can contribute to the hard-scattering amplitude.

To simplify the calculations, one can exploit symmetry among the diagrams. Diagrams in groups B, D and F may be obtained from A, C, and E, respectively, by the interchange:

$$\mathcal{E} : x_1 \leftrightarrow x_3, y_1 \leftrightarrow y_3, e_1 \leftrightarrow e_3. \tag{2}$$

By making use of this symmetry, one has only to calculate the diagrams in groups A, C and E. The contributions for groups B, D and F are then accounted for by doubling the results for groups A, C and E. For real Compton scattering, time-reversal symmetry is also available. However, since this symmetry does not apply in the case of virtual Compton scattering, it has

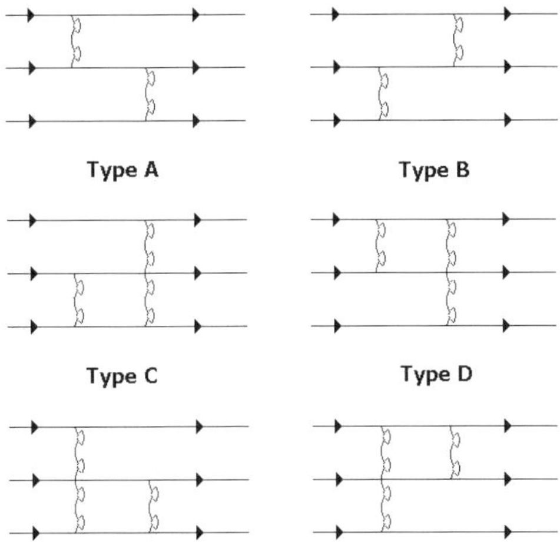

Fig. 1. Classification of diagrams with two gluons attached.

not been used in the calculations (except as a means for checking the RCS results).

In our computation, the generation of the relevant Feynman diagrams has been done using a software package, FeynGen.[14] The corresponding hard scattering amplitudes have also been automatically generated using a software package, FeynComp.[15] Both FeynGen and FeynComp are symbolic software packages developed in Maple at NC State University. In order to calculate the hard scattering amplitudes using the FeynComp package, symbolic definitions must be provided for the incoming/outgoing momenta and for the polarization vectors of the incoming/outgoing photon. We define a quantity $R = 1 + Q^2/s$, which we refer to as the virtulaity parameter. Thus, $R = 1$ refers to the real case ($Q^2 = 0$) and $R = 2$ refers to the deeply virtual case (Q^2 large). Working in the center-of-mass frame, FeynComp has been used to generate the hard scattering amplitudes for all 336 diagrams. It turns out that only 192 of these diagrams have a non-zero amplitude. Exactly one half of these diagrams (96 diagrams) belong to the groups A, C and E. A check was made of the hard scattering expressions by setting $R = 1$. For this case, it was verified that the hard scattering expressions reduce to the expressions given in Table III and Table VI of Ref. 6, with the noted corrections in Ref. 8.

There are several common models for the distribution amplitude ϕ_1 from QCD sum rules. The ones which have been used in our calculations are those of Chernyak, Oglobin and Zhitnitsky (COZ)[16] and King and Sachrajda (KS).[17] In addition we have calculated using the asymptotic distribution amplitude $\phi_{ASY} = 120x_1x_2x_3$.

3. RCS Results

The spin-polarized cross section is given by

$$\frac{d\sigma^{\lambda\lambda'}_{hh'}}{dt} = \frac{1}{16\pi s^2}|\mathcal{M}^{\lambda\lambda'}_{hh'}(s,t)|^2. \tag{3}$$

Plots of cross section and phase against the center-of-mass scattering angle θ were shown in our recent paper [9]. As we have shown there,[9] our results are in excellent agreement with the work of Brooks and Dixon.[8] This agreement is significant since none of the previous results were in agreement.

Fig. 2 shows the unpolarized, real Compton cross section for the proton, comparing our calculations with the new experimental results from JLAB.[2] The figure plots our results for the COZ and KS distribution amplitudes and the experimental results for $s = 6.89$ GeV^2, $s = 8.99$ GeV^2, and $s = 11.00$ GeV^2. All results are scaled, by multiplying by s^6 and then normalized using the Dirac form factor. For the experimental results, we have used

$$Q^4F_1^p(Q^2) = 1.0 \text{ GeV}^4 \tag{4}$$

for Q^2 in the range $7 - 15$ GeV^2, as suggested by Brooks and Dixon.

Some interesting points emerge from the figure. First, it can be seen quite clearly that the experimental data does indicate that the scaled cross section is decreasing as s increases. Thus the data is consistent with the expectation that it will asymptotically approach the PQCD curve. Secondly, the difference between the $s = 11.00$ GeV^2 data and the PQCD curves represents a factor of 2-3, for KS, and 3-4, for COZ. This is a much smaller factor than Brooks and Dixon found, when comparing with the earlier data of Shupe *et al.*[3] We are therefore optimistic that the 12 GeV upgrade at JLab can produce results that are even closer to our PQCD curves, when s is increased still further. A third point that can be deduced from the figure is that the scaling factor s^{-n} for the experimental data corresponds to a value of $n > 6$. Clearly, if this were not the case, we would not see the scaled cross section decreasing as s increases. Since the s^{-6} scaling law is predicted by counting the quarks and photons involved, we can also speculate that there

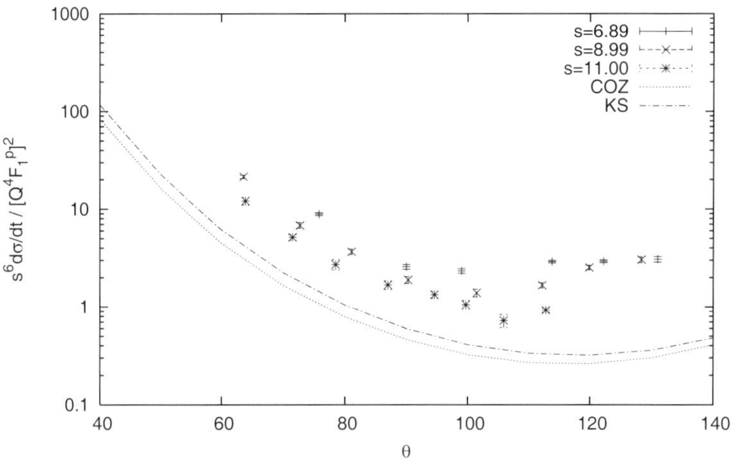

Fig. 2. Scaled unpolarized real Compton cross section: comparing our results with JLAB experiment. Note that the results have been normalized by the scaled form factor, F_1^p. See the text for full explanation.

are additional virtual particles involved in the interaction that cause n to increase. Finally, we note that, at larger angles ($> 110^o$), the experimental data deviates more strongly from the PQCD curve. In this region, the Mandelstam u becomes quite small (about -1 GeV^2), so the applicability of the PQCD curve is questionable.

4. Extension to Virtual Compton Scattering

The virtual calculation is considerably more complex than the real calculation, due to the following reasons. First, there are over twice as many Feynman diagrams that need to be considered. Second, the pole structure occuring in the virtual case is more complex than in the real case. Third, for the virtual case there is also the need to consider longitudinal/temporal polarization of the incoming photon. However, as we discussed in the introduction, the extension of RCS to the virtual Compton scattering is very important. Due to the limited space of this presentation, we do not show here our results of proton virtual Compton scattering process but the detailed plots of cross section and phase against the center-of-mass scattering angle θ have already been presented in our recent paper [9] for different polarizations. Our results differ somewhat from those of Farrar and Zhang.[5] However, there are some good arguments to suggest that our results are more reliable as discussed in Ref. 9.

In an attempt to link our analysis to the GPD approach, we examined the 'same-quark' dominance because of its relationship to the handbag diagram approach to DVCS.[18] For real Compton scattering ($R = 1$), using the COZ distribution amplitude, it was found that, for all forward scattering angles, the amplitude is dominated by diagrams in which the two photons attached to the same quark, as already noted in Ref. 19. As R increases from 1 to 2, the range of angles for which dominance occurs decreases, so that, for DVCS ($R = 2$), 'same-quark' diagrams only dominate for a scattering angle up to 20°. At larger angles, for $R = 2$, the dominance is not observed. The handbag diagram approach is applicable in the case of small

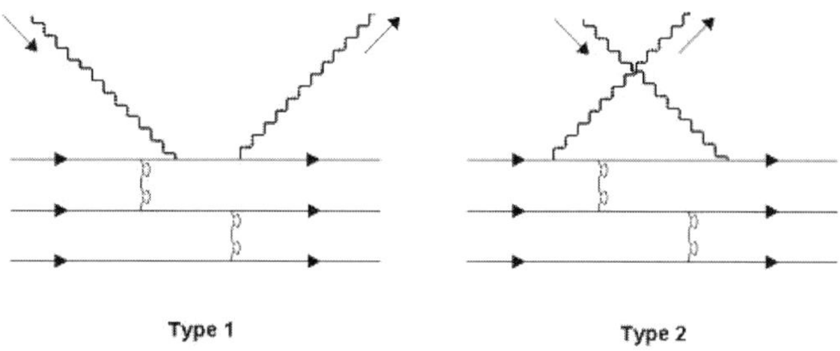

Fig. 3. Diagrams where both photons are attached to the same quark.

momentum transfer and large Q^2, which implies small scattering angle. The condition that the same quark should be struck by both initial and final photons is necessary for the handbag diagrams. However, this is not a sufficient condition for the handbag diagrams. In general, the diagrams in which the same quark is struck by the two photons include some where there is no gluon exchange between the two photons (*type 1* in Fig. 3) and some where there is a gluon exchange between the two photons (*type 2* in Fig. 3). In the handbag diagram, the two photons are attached to the same quark, but there is no gluon exchange in between. This corresponds to the *type 1* diagram in Fig. 3. How do we account for the presence of *type 2* diagrams in our result? We believe that it can be explained by noting that the GPD calculation should be performed in the Light Front gauge, whereas our calculation has been performed in the Feynman gauge. The total amplitude should be gauge invariant, but the contributions from individual diagrams will vary according to the gauge used. We expect that, if our calculations are redone in the Light Front gauge, the amplitude will be dominated by handbag-type diagrams. If this expectation turns out to be correct, it indicates some commonality between the PQCD and GPD approaches. Furthermore, by examining the high energy limit, we expect to see a correlation between the GPD and PQCD approaches. This expected correlation can be used to test the validity of various GPD models proposed in the literature by looking at how closely they reproduce the PQCD predictions. This can also shed some light on the question concerning the energy level at which the PQCD approximation will become valid.

Acknowledgments

We would like to thank the organizers for the arrangement of this stimulating workshop and their hospitality during the workshop. This work was supported in part by a grant from the U.S. Department of Energy under Contract No. DE-FG02-03ER41260.

References

1. L. Cardman, Long Range Plan presented at this workshop.
2. A. Danagoulian, *et al.*, Phys. Rev. Lett. **98**, 152001 (2007).
3. M.A. Shupe, *et al.*, Phys. Rev. D **19**, 1921 (1979).
4. E. Maina and G. R. Farrar, Phys. Lett. B **206**, 120 (1988).
5. G. R. Farrar and H. Zhang, Phys. Rev. D **41**, 3348 (1990); **42**, 2413(E) (1990).
6. A. S. Kronfeld and B. Nižić, Phys. Rev. D **44**, 3445 (1991).
7. M. Vanderhaeghen, P. Guichon, and J. Van de Wiele, presented at workshop on virtual Compton scattering, Clermont-Ferrand, France, 1996.
8. T. Brooks and L. Dixon, Phys. Rev. D **62**, 114021 (2000).
9. R. Thomson, A. Pang and C.-R. Ji, Phys. Rev. D **73**, 054023 (2006); hep-ph/0602164 (2006).
10. P. A. M. Dirac, Rev. Mod. Phys. **21**, 392 (1949).
11. See, e.g., K. Goeke, M. V. Polyakov and M. Vanderhaeghen, Prog. Part. Nucl. Phys. **47**, 401 (2001), and references therein.
12. For a review of B-Physics with emphasis on LFD, see C.-R. Ji and H.-M.Choi, Nucl. Phys. Proc. Suppl. **90**, 93 (2000).
13. For an application of LF wavefunctions to RHIC physics, see B. Hong, C.-R. Ji and D.-P. Min, Phys. Rev. C **73**, 054901 (2006).
14. A. Pang and C.-R. Ji, Computers in Physics, **Vol 9 (No. 6)**, Nov/Dec 1995 p589-593.
15. A. Pang and C.-R. Ji, J. Comp. Phys. **115** 267 (1994).
16. V. Chernyak, A. Ogloblin and I.Zhitnitsky, Sov. J. Nucl. Phys. **48**, 536 (1988);Yad. Fiz. **48**, 841 (1988); Z. Phys. C **42**, 569 (1989).
17. I. King and C. Sachrajda, Nucl. Phys. B **279**, 785 (1987).
18. M. Vanderhaeghen, Eur. Phys. J. A **8**, 455 (2000).
19. A. V. Afanasev, S. J. Brodsky, C. E Carlson, Y.-C Chen, and M. Vanderhaeghen, Phys. Rev. D **72**, 013008 (2005).

DEEPLY VIRTUAL COMPTON SCATTERING AT JEFFERSON LAB

F. SABATIE

CEA Saclay - DAPNIA/SPhN,
91191 Gif-sur-Yvette, FRANCE
E-mail: fsabatie@cea.fr

Seven years of experimental efforts at Jefferson Lab on Deeply Virtual Compton Scattering yielded very significant results such as a strong indication for the handbag dominance (i.e. scaling) even et low energy, the first comparisons to models and a large data sample to constrain their free parameters and even a first model-dependent extraction of the total angular momentum carried by quarks. These proceedings try to summarize the information collected so far without going into details, and in a semi-chronological fashion.

Keywords: DVCS, GPD, nucleon structure

1. Introduction

A small revolution began in the study of the nucleon structure in the last 10 years: the Generalized Parton Distributions (GPD) were invented in the mid-90's, and it did not take long for experimentalists to realize what experiments could be made to access them. Even though a variety of Deep Exclusive Processes should in principle be able to measure GPDs, it rapidly became obvious that Deeply Virtual Compton Scattering (DVCS) is the most promising one, especially in the initial years and/or at low energy.

DVCS is the scattering of a virtual photon on a proton, with the emission of a real photon in the final state. The deep part of this process means that this reaction has to occur in the "deep" regime, i.e. at large Q^2 for fixed x_B. If one adds the condition that the transfer to the proton t is small with respect to the hard scale (Q^2), a factorization theorem states that the leading amplitude derives from a simple handbag diagram, similar to the case of Deeply Inelastic Scattering (DIS) except that it is off-forward. This handbag amplitude can then be directly written as a convolution of a hard scattering kernel (Compton scattering on a quark) with a soft non-perturbative object, namely the nucleon expressed as a function of four GPDs: H, $\widetilde{H}$, E, $\widetilde{E}$. More information about GPDs can be found in these proceedings[1] and in references therein.

To access DVCS, one needs to produce the virtual photon, and this is done at Jefferson Lab using an electron beam. However a complication arises. The electroproduction of a photon is not uniquely achieved via DVCS: the electron lines of this process may also emit the final state photon; this is called the Bethe-Heitler process (BH), and it adds up coherently to the DVCS process in the electroproduction of a real photon. What was first thought as a tremendous issue (since BH is dominant at moderate energy) turned out to be the main reason why one can access DVCS at Jefferson Lab. The amplitudes interfer and this interference can be measured directly using polarisation observables. By either flipping the helicity of the beam or the spin of a target, one can have a direct measurement of the imaginary part of the DVCS amplitude, amplified by the BH. Not only is it now measurable, but the resulting asymmetry is large (of the order 30% in the case of the beam spin asymmetry). Since the first pioneering measurements of this asymmetry by CLAS and HERMES, three experiments took place at Jefferson Lab and another three are on the books. In the rest of this article, I will try and outline the contributions of Jefferson Lab to the study of the DVCS process, from the first non-dedicated analysis to our brand new results. Since this is a review of work which is detailed in other proceedings of this conference, this will be more of a personal overview (or bias!) of these experiments, trying to follow the chronology of events. I decided not to talk about the 12 GeV upgrade since there are two proceedings about the upcoming experiments which are already planned in Hall A and B about DVCS.[2,3]

2. The First DVCS Publications from Hall B/CLAS

After it was pointed out that DVCS indeed was an interesting process to study and even after the very first dedicated proposal to study DVCS was accepted by the Jefferson Lab PAC, it became obvious that multipurpose detectors such as CLAS or HERMES already had the relatively rare (e, p, γ) events on tape. In Hall B, it was soon realized that in order to have a large enough data sample, one should not require detecting all three particles in the final state.

The strategy used to analyse the data was the following: after a clean selection of an electron in CLAS (most of the time the trigger particle), one would look for a proton in the final state and form the (epX) missing mass, looking for the DVCS photon. The remaining problem was however sizeable: the electroproduction of neutral pions could easily mimick DVCS since the π^0 mass is not very far from zero especially considering the moderate CLAS momentum resolution. It was therefore not possible to distinguish between the electroproduction of a photon or a π^0 on an event-by-event basis. However, it was possible to use clean data samples of photons and π^0's separately and constrain the shapes of their missing mass spectra: it was then possible to proceed with a two-gaussian fit in order to extract the photon and π^0 yields separately, as shown in Fig.1-left. The beam spin asymmetry was formed by dividing the difference by the sum of the yields for each beam helicity states. The resulting BSA shown in Fig.1-right was published in 2001.[4] A clean sine wave was observed, which was a first indication that mostly the handbag diagram was contributing to the DVCS amplitude, even at the moderate energy of 4.2 GeV.

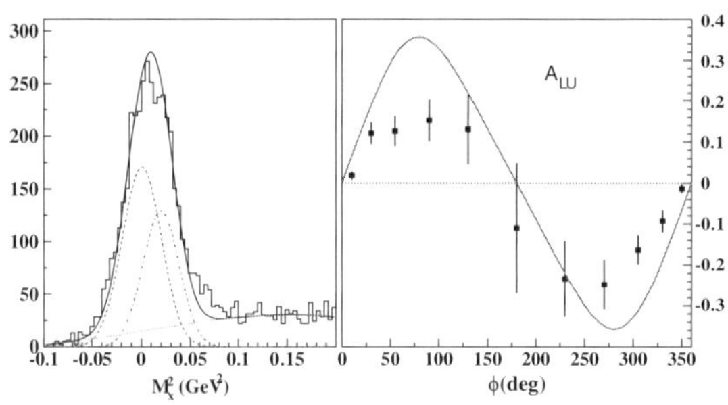

Fig. 1. Left: Missing mass squared for the (epX) system, fitted with its (epγ) contribution (dashed), (epπ^0) contribution (dotted) and a polynomial background. Right: Beam Spin Asymmetry as a function of ϕ as published by the CLAS collaboration. The behavior is mostly a $\sin\phi$, and the basic VGG model[5] overshoots the data by about 30%

An analysis of longitudinally polarized target data showed that even for the target spin asymmetry, a similar sine wave was observed.[6] It is surprising to note that HERMES found a huge $\sin 2\phi$ contribution in the same observable, albeit at a slightly different kinematics.

3. The Hall A Experiments E00-110 and E03-106

The E00-110 experiment in Hall A was proposed as soon as it was realized that DVCS experiments were now feasible at Jefferson Lab. In 2000, the JLab PAC accepted to run this somewhat exploratory experiment, which had a very important purpose: by studying scaling properties of the interference of BH and DVCS, one could check whether or not we were already in the scaling regime. The strongest hint that we could be in such a regime at Q^2 around 2 GeV^2 was that the DIS cross section can be expressed as the imaginary part of the forward Compton amplitude, therefore very similar to DVCS. Deeply Inelastic Scattering is shown to scale down to around 1 GeV^2 or even lower, bearing QCD evolution corrections, so there was hope that DVCS would exhibit the same behavior.

Three years after the first proposal, and after long conversations with Markus Diehl, it was realized that it would be fairly interesting to study DVCS on the neutron. The first idea which comes to mind when one starts studying proton vs. neutron is flavor separation between u and d quarks. However, this was a rather weak point since we were really at a stage where experiments were considered exploratory, and one does not start a flavor separation when you have not even proved that the experiment was feasible! What was really striking in the neutron with respect to the proton is the following: proton DVCS was supposed to be mostly sensitive to the GPD H (around 80% of the signal was expected to be proportional to H, according to models). In contrast, neutron DVCS turned out to be mostly sensitive to the GPD E. Not only was this giving us information about one of the least known of the GPDs (E does not have a forward limit given by the parton distribution functions, like H has), but it gave us a handle on the only really unknown quantity entering Ji's sum rule. It turned out the VGG model used a parametrization of E depending on the total angular momentum carried by the u and d quarks, and the imaginary part of the DVCS amplitude was indeed very dependent on the choice of J_u and J_d in this model. This ended up convincing the PAC to run this neutron DVCS experiment as well, along with E00-110.

The DVCS setup was installed in the summer 2004 and consisted in the usual Hall A cryotarget (15 cm cell). The scattered electron was detected in the left High Resolution Spectrometer of Hall A.[7] With momentum resolution of the order of $\delta p/p \sim 10^{-4}$ and fair angular resolution of about 1 mrad, this gave us a perfect measurement of the leptonic side of the reaction, *i.e.* x_B and Q^2. The acceptance of the spectrometer being rather small (about $60 \cdot 10^{-3} \otimes 120 \cdot 10^{-3}$ sr), it resulted in a small "acceptance" for the virtual photon direction. The strong idea was to center the rest of our experimental apparatus, an electromagnetic calorimeter and an annular array of scintillator blocks, on the central direction of the virtual photon as defined by the spectrometer. The calorimeter was located at 1.1 m from the target and the "shadow" of the virtual photon on the front face of the calorimeter was confined in square of about 6 cm side. The proton detector was slightly closer to the target and covered polar angles 18 to 38°.

The beam time was seperated in the measurement of three different $Q^2 = 1.5, 1.9, 2.3$ GeV^2 points at the same $x_B = 0.36$ using a beam energy of 5.75 GeV. The final analysis did not need the requirement to detect the three-particle final state as our resolution was sufficient to ensure the exclusivity using only the electron and the photon. The π^0 contamination was subtracted bin-per-bin using two-photon π^0 events reconstructed in the calorimeter, renormalized to take into account the full π^0 decay phase space. Once the π^0 contribution was subtracted, the final exclusivity was checked, firstly by the comparison of $(e\gamma X)$ missing mass spectra, bin-per-bin, with the spectra requiring the proton to be detected and also by direct comparison with the Monte-Carlo spectra. The remaining inelastic contribution (production of one or more extra pions) was found to be under 3% of the total yield for all bins.

The plan for this experiment was to perform a scaling check of the imaginary part of the DVCS amplitude we would be sensitive to. In order to access it, we flipped the helicity of the beam and actually measured the difference of cross sections (Fig 2-left). Once a few kinematical coefficients as well as form factors coming from the BH were divided from this cross section, we were left with a quantity expected to correspond to a linear combination of GPDs, which should have exhibited no Q^2 dependence as predicted.[8] Indeed, within the fairly small statistical error, one could see no Q^2 dependence on our observable, as show on Fig. 2-right. Moreover, the difference of cross sections is completely dominated by the $\sin\varphi$ term: together with the Q^2-dependence check, it was the best indication so far that we indeed reached the scaling regime at moderate energy, moderate Q^2 and in the valence quark region.

The neutron experiment E03-106 immediately followed E00-110. The experimental setup was the same except for the target (liquid deuterium instead of liquid hydrogen) and the proton array was upgraded with a veto detector in front in order to discriminate between recoiling protons and neutrons in the final state. The neutron experiment was clearly an even bigger challenge than the proton one: not only did one need to make this proton/neutron separation, but we chose to run at a luminosity four times higher, and the neutron DVCS asymmetry was expected to be of the order 3%, 10 times smaller than for the proton. When we took the data, we soon realized the singles rates in the detector were very high and it was fairly hard to find the neutron signals with enough efficiency considering

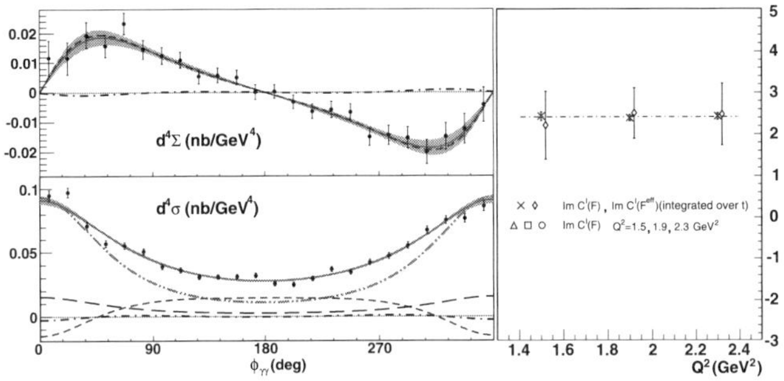

Fig. 2. Left: Difference of cross sections (top) and unpolarized cross section (bottom) for the $ep \to ep\gamma$ reaction as measured by the Hall A DVCS Collaboration.[9] Right: Scaling property of the $\sin\phi$ (crosses) and $\sin 2\phi$ (diamonds) coefficients, showing no Q^2 dependence as expected from the dominance of the handbad amplitude in DVCS.

the high statistical need because of the low asymmetry. However, after a careful calibration of the electromagnetic calorimeter and despite a small smearing effect due to Fermi motion, the resolution on the $(e\gamma X)$ missing mass was found similar to the one measured with the liquid hydrogen target. The solution for the final analysis was to use the deuteron data, subtract the (correctly normalized) proton data which resulted in a mixed neutron + deuteron signal. Due to the good resolution of our apparatus, it was possible to disentangle the neutron and deuteron contributions to the missing mass fitting both contributions at the same time. From the difference of cross sections, the same linear combination of GPDs as for the proton was then extracted for the neutron as a function of t (Fig. 3-left), and compared to some models, and in particular the VGG model, which is parametrized as a function of the total angular momenta carried by the u and d quarks: J_u and J_d. Despite the clear model dependence of such an analysis, it's interesting to see that the neutron data is very sensitive to the GPD E and especially to J_d in this model. HERMES used transversely polarized proton DVCS data to do the same kind of analysis and Fig. 3-right shows the result. The lattice QCD calculations of J_u and J_d are found in good agreement with this model-dependent extraction from DVCS data.

4. The E1-DVCS Hall B/CLAS Experiment

The Hall B E1-DVCS experiment ran in the spring of 2005 for about half its allocated beamtime. Its purpose was the measurement of the beam spin asymmetry over as large a kinematical domain as possible, as permitted using a 5.75 GeV beam in conjonction with the CLAS spectrometer, up-

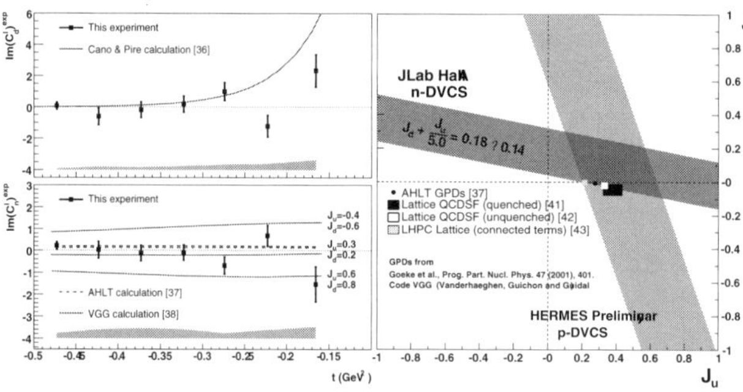

Fig. 3. Left: $\sin \phi$ coefficient of the cross section difference for deuteron (up) and neutron (down) along with GPD-based models, which are in good agreement. Right: Model-dependent constraint to J_u and J_d from the neutron data, which is particularily sensitive to the GPD E which enters Ji's sum rule. More information in the submitted article.[10,11]

graded with a forward electromagnetic calorimeter, protected from Moeller electrons by a supraconducting solenoidal magnet producing a 4.5 T magnetic field. The E1-DVCS analysis was based on the three final-state particle detection associated with various cuts to ensure that the events were in the DVCS kinematics within the expected resolution, hence insuring a close-to-perfect exclusivity. The π^0 contribution was subtracted using the measured two-photon π^0 data as well as a Monte-Carlo evaluation of the ratio of one-photon π^0 to two-photon π^0 yields. The data was then binned in (Q^2, x_b, t) (Fig 4-left) and the beam spin asymmetry was extracted for each individual bin (Fig 4-right). The data is compatible with the Hall A data in the only similar bin common to both experiments. Overall, it still shows a 30% discrepancy with the VGG GPD model at low $|t|$, which gets lower as $|t|$ grows larger. In contrast, a Regge-inspired model from J.M. Laget reproduces the data fairly well at low $|t|$ in the central x_B-Q^2 region, but shows a large discrepancy at large t.[12]

The remaining half of E1-DVCS will run in the summer of 2008, along with the very first dedicated polarized target DVCS run, for about the same period of time. Using the CLAS polarized NH_3 target along with the same forward electromagnetic calorimeter E1-DVCS used, it will measure the target beam asymmetry over a large kinematical domain. This asymmetry has more information about $\widetilde{H}$ than the unpolarized target one, and is therefore mandatory in order to extract information about this particular GPD. Once the full polarized and unpolarized target data sets will be obtained, they will provide a very accurate constraint to GPD models with data spanning an unprecedented kinematical domain.

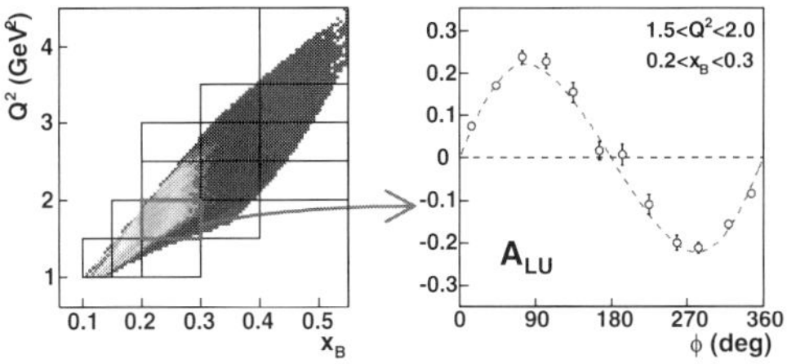

Fig. 4. Left: Q^2 vs. x_B phasespace covered by the E1-DVCS experiment in CLAS,[13] along with a preliminary binning of data. Right: Preliminary example of the t-integrated result for the beam spin asymmetry in the highlighted bin.

5. Summary and Conclusion

Deeply Virtual Compton Scattering studies at Jefferson Lab have been very fruitful for the last 7 years. Results tend to confirm that the handbag diagram is the main contribution to DVCS, at least in the spin observables. The very first, albeit model (and approximation)-dependent extraction of GPDs have been performed and are in reasonable agreement with basic models (at the 30% level). The neutron experiment in Hall A provided the first model-dependent experimental constraint to the total angular moment carried by u and d quarks. Data from Hall B, especially with the upcoming 2008 runs will provide a huge database to contrain models. New experiments are coming up with 6 and 12 GeV beam as well.[2,3] The future of GPDs and their study through DVCS at Jefferson Lab has only started and promise to be the cornerstone of the nucleon structure studies in the next 10 years.

References

1. D. Mueller, *in these proceedings.*
2. J. Roche, *in these proceedings.*
3. L. Elouadriri, *in these proceedings.*
4. S. Stepanyan *et al.*, *Phys. Rev. Lett.* **87**, p. 182002 (2001).
5. M. Vanderhaeghen, M. Guidal, P. Guichon and L. Mossé, *Computer Codes for DVCS and BH Calculations, Private Communications* (2006).
6. S. Chen *et al.*, *Phys. Rev. Lett.* **97**, p. 072002 (2006).
7. J. Alcorn *et al.*, *Nucl. Instrum. Meth.* **A522**, 294 (2004).
8. D. M. A.V. Belitsky and A. Kirchner, *Nucl. Phys.* **B629**, 323 (2002).
9. C. M. Camacho *et al.*, *Phys. Rev. Lett.* **97**, p. 262002 (2006).
10. M. Mazouz, *in these proceedings.*
11. M. Mazouz *et al.*, *arXiv:0709.0450 [nucl-ex], to be published in PRL* (2007).
12. J. M. Laget, *arXiv:0708.1250 [hep-ph].*
13. F. X. Girod, *in these proceedings.*

DVCS AT HERMES: RECENT RESULTS

F. ELLINGHAUS*

University of Colorado,
Department of Physics,
Boulder, Colorado 80309-0390, USA
E-mail: Frank.Ellinghaus@desy.de

for the HERMES Collaboration

Hard exclusive reactions are the tool to learn about generalized parton distributions, which provide a more complete parametrization of the nucleon than the ordinary parton distribution functions. Recent measurements by the HERMES collaboration of the exclusive production of photons, i.e., Deeply-Virtual Compton Scattering, are summarized and compared to model calculations, focusing on the measurements and model comparisons relevant to the extraction of quark orbital angular momentum and on the measurements on heavy nuclei.

Keywords: DVCS, exclusive reactions, GPD, angular momentum, proton

1. Introduction

Similar to the case of inclusive and semi-inclusive DIS, where the nucleon structure is described using **P**arton **D**istribution **F**unctions (PDFs), hard exclusive reactions can be expressed in terms of **G**eneralized **P**arton **D**istributions (GPDs).[1–3] The PDFs and elastic nucleon **F**orm **F**actors (FFs) are included in the GPDs as the limiting cases and moments of GPDs, respectively.[2] While FFs derived in elastic scattering describe the transverse location of partons inside the nucleon and PDFs describe their longitudinal momentum distribution, GPDs are able to provide information on both at the same time. Thus exclusive reactions are able to give a certain 3–dimensional picture of the nucleon structure.[4–6] In particular, GPDs offer for the first time a possibility to determine the total angular momentum carried by the quarks in the nucleon.[2]

Below recent HERMES measurements on the hard exclusive electropro-

*This work is supported in part by the US Department of Energy.

duction of real photons (**D**eeply–**V**irtual **C**ompton **S**cattering, DVCS) are summarized and compared to model calculations. The data has been taken with polarized and unpolarized gas targets using the HERMES spectrometer[7] at the HERA electron/positron–proton collider at DESY, which offers longitudinally polarized 27.6 GeV electron and positron beams.

2. GPD H via beam-charge and beam-spin asymmetries

DVCS amplitudes can be measured through the interference between the DVCS and **B**ethe–**H**eitler (BH) processes, in which the photon is radiated from a parton in the former and from the electron in the latter process. Both processes have an identical final state, i.e., they are indistinguishable, and thus give rise to an interference term I. The photon production cross section depends on the Bjorken scaling variable x_B, the squared virtual–photon four–momentum $-Q^2$, the squared four-momentum transfer t or the reduced four-momentum transfer $t' = (t - t_{min})$ to the target, and the azimuthal angle ϕ defined as the angle between the lepton scattering plane and the (virtual and real) photon production plane. For an unpolarized proton target, and at leading twist, the interference term is given by[8]

$$I \propto -C\,[\,a\,\cos\phi\,\mathrm{Re}\mathcal{M}^{1,1} - b\,P_l\,\sin\phi\,\mathrm{Im}\mathcal{M}^{1,1}\,], \tag{1}$$

where the lepton beam has longitudinal polarization P_l and charge $C = \pm 1$, and a and b are functions of the ratio of the longitudinal to transverse virtual–photon flux. The real (imaginary) part of the DVCS amplitude $M^{1,1}$ can be accessed by measuring the $\cos\phi$ ($\sin\phi$) dependence of a cross section asymmetry with respect to the charge (spin) of the lepton beam. At HERMES kinematics, the DVCS amplitude $M^{1,1}$ gives access to the GPD H. Details and full equations are given in Ref. [9].

The event selection at HERMES requires events with exactly one photon and one charged track, identified as the scattered lepton, with $Q^2 > 1$ GeV^2. For data taken prior to 2006 the recoiling proton is not detected and exclusive events are identified by the fact that the missing mass M_x of the reaction $ep \rightarrow e\gamma X$ corresponds to the proton mass. Due to the finite energy resolution the exclusive sample is defined as $-1.5 < M_x < 1.7$ GeV. A recoil detector used during the recent data taking should reduce the underlying background from presently about 15% to less than 1%.[10]

The beam–spin asymmetry (BSA) and the beam–charge asymmetry (BCA) as a function of ϕ are calculated as

$$A_{LU}(\phi) = \frac{1}{<|P_l|>}\frac{\overrightarrow{N}(\phi) - \overleftarrow{N}(\phi)}{\overrightarrow{N}(\phi) + \overleftarrow{N}(\phi)}, \qquad A_C(\phi) = \frac{N^+(\phi) - N^-(\phi)}{N^+(\phi) + N^-(\phi)}, \tag{2}$$

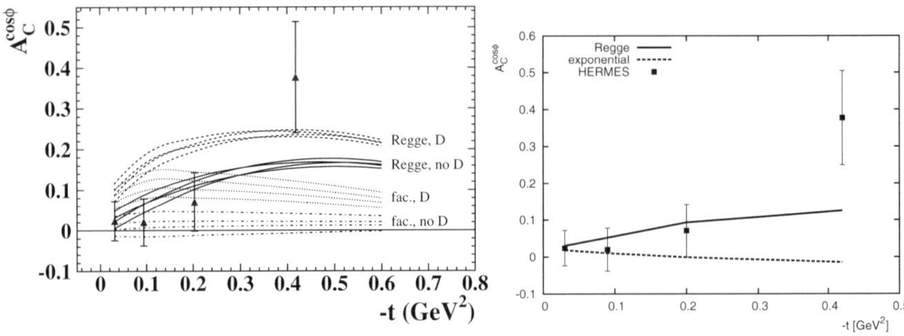

Fig. 1. The $\cos\phi$ amplitude of the beam–charge asymmetry[12] on hydrogen as a function of $-t$. The error bars show the statistical and systematic uncertainties added in quadrature. The calculations in the left panel are based on a double-distribution GPD model using a factorized (fac.) or a Regge–inspired (Regge) t–dependence with (D) or without (no D) a D-term contribution. The parameters b_v and b_s are each set to either unity or infinity. Using the resulting 16 sets of model parameters, the calculated asymmetries fall into four main groups. The variations within these groups are due to the different settings for the b parameters. The calculations in the right panel are based on a dual-parametrization GPD model.[18]

with the normalized yields $\overrightarrow{N}$ ($\overleftarrow{N}$) or N^+ (N^-) using a beam with positive (negative) helicity or a positron (electron) beam, respectively. The BSA (BCA) on the proton as a function of ϕ has been extracted at HERMES[11] ([12]), whereby the predominant $\sin\phi$ ($\cos\phi$) dependence expected from Eqn. 1 has been observed. The $\cos\phi$ amplitudes of the BCA on hydrogen as a function of $-t$ derived from a fit to the BCA in each $-t$ bin[12] are shown in Fig.1, and the recent preliminary BSA result on the kinematic dependences of the $\sin\phi$ amplitudes is shown in Fig.2.

The model calculations shown in the left panel (left column) of Fig.1 (Fig.2) are based on a double-distribution GPD model described in Refs. [13,14]. Since these data are exclusively sensitive to the GPD H as mentioned above, the theoretical calculations shown were derived by only varying the model parameters for the GPD H in the underlying code[15] in order to calculate the asymmetries at the average kinematics of every bin. The model parameters differ with respect to including or neglecting the so–called D–term[16] in the GPD model and whether the t–dependence of the GPD H is calculated in either the simplest ansatz where the t–dependence factorizes from the t–independent part or in the Regge–motivated ansatz. In addition, the so-called skewness parameters b_v and b_s in the profile function[17] have been set to either unity or infinity, the latter value corresponds to a skew-

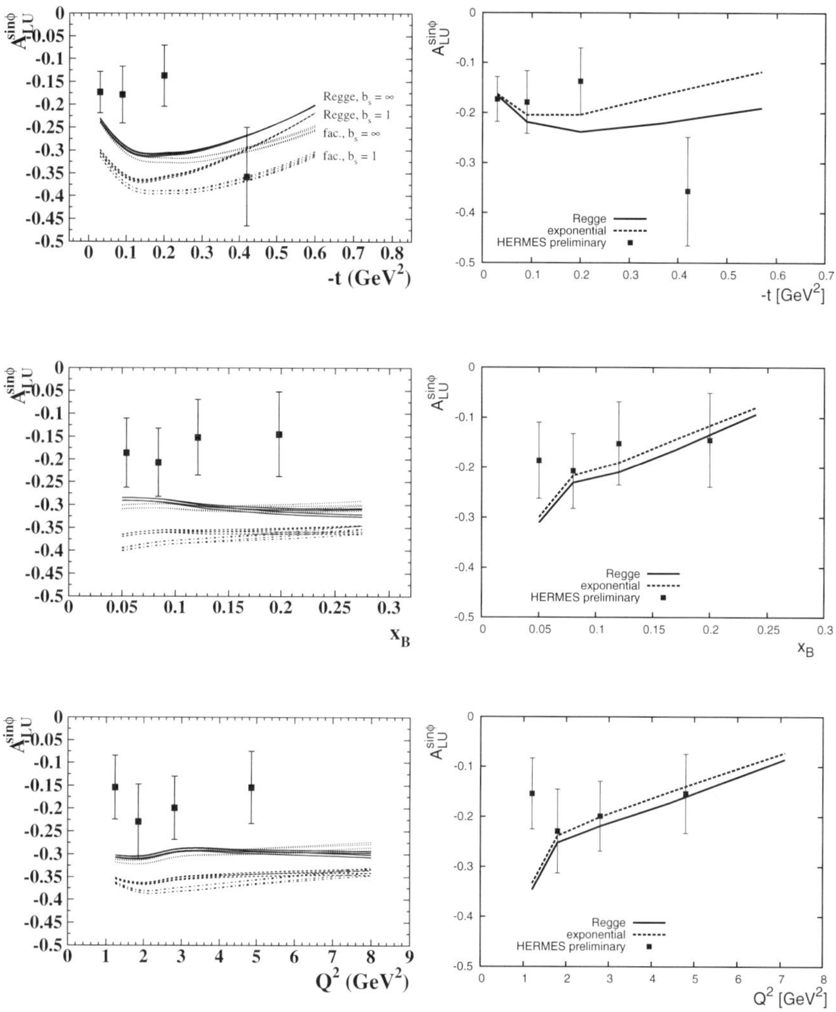

Fig. 2. The $\sin\phi$ amplitude of the beam–spin asymmetry on hydrogen from the 1996-2000 data as a function of $-t$, x_B and Q^2. The error bars show the statistical and systematic uncertainties added in quadrature. The calculations in the left panel are based on a double-distribution GPD model using a factorized (fac.) or a Regge–inspired (Regge) t–dependence with or without a D-term contribution. The parameters b_v and b_s are each set to either unity or infinity. Using the resulting 16 sets of model parameters, the calculated asymmetries fall into four main groups, whereby the BSA appears to be insensitive to the D-Term and to the value of b_v. The calculations in the right column are based on a dual-parametrization GPD model.[18]

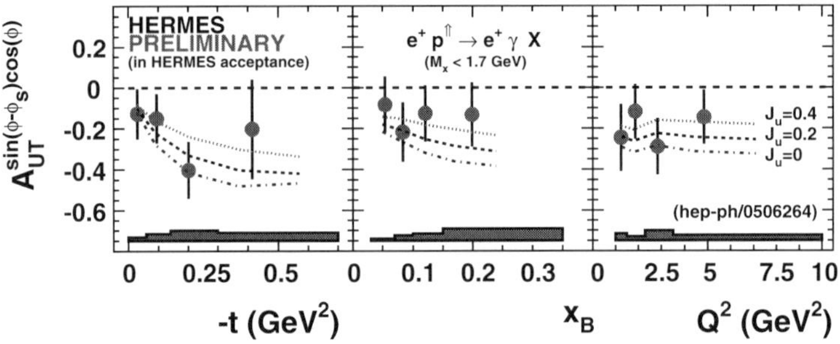

Fig. 3. The $\sin(\phi - \phi_S)\cos\phi$ amplitude of the transverse target-spin asymmetry as a function of $-t$, x_B and Q^2 in comparison to theoretical predictions from Ref. [20].

ness independent ansatz for the GPD. The BCA data appears to disfavor the four parameters sets with the Regge-inspired t-dependence and the D–term contribution as well as the factorized one with the D-term which gives the largest asymmetry ($b_v = 1$, $b_s = \infty$). For the BSA all models seem to overshoot the absolute size of the $\sin\phi$ amplitude. The calculations based on the dual parametrization GPD model[18] shown in the right panel (right column) of Fig.1 (Fig.2) are all in rather good agreement with the data.

3. GPDs *H* and *E* via transverse target-spin asymmetry

Within the next few years HERMES will be able to provide sufficient data to largely constrain the GPD H in the kinematic region of the experiment. It is natural to ask to what extent the GPD E can be accessed, which is the other important GPD necessary in order to determine J_q, the total orbital angular momentum of quarks in the nucleon.[2] For an unpolarized proton target the contribution from the GPD E is suppressed with respect to H, but this is different for transverse target polarization.[9,19] Using an unpolarized beam (U) and a transversely (T) polarized target, a $\sin(\phi - \phi_S)\cos\phi$ modulation in the DVCS transverse target-spin asymmetry (TTSA) gives access to a combination of the GPDs H and E.[20] Here ϕ_s denotes the azimuthal angle of the target polarization vector with respect to the lepton scattering plane.

The results from HERMES data collected on a transversely polarized hydrogen target are shown in Fig. 3. They agree with the model calculations[20] shown in the same figure, which have been calculated for various values of J_u. Based on u-quark dominance the d-quark total angular momentum has been assumed to be zero. Since it was realized that the model calculations are largely insensitive to all model parameters but J_u and J_d, it

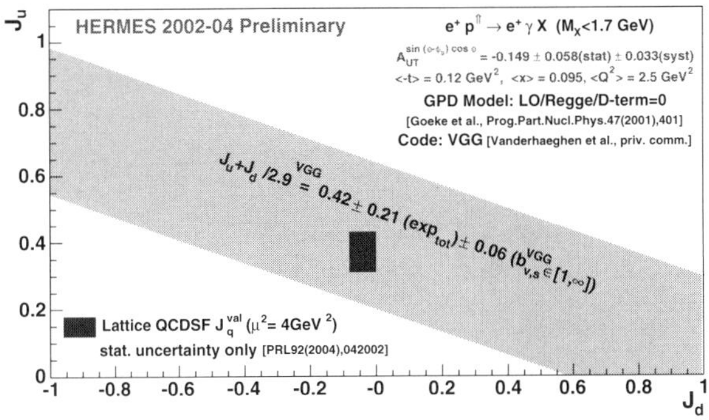

Fig. 4. Model dependent constraint on the quark orbital angular momenta J_u and J_d.

is possible to constrain J_u and J_d even though other model parameters are largely unconstrained.[20] For example, the model calculations in Fig. 3 show little variation if calculated for any of the 16 parameter sets which give very different values for the BCA as shown in Fig.1. Hence the four parameter sets with the Regge–inspired t–dependence and without a contribution of the D–term (solid lines in the left panel of Fig.1), have been chosen in order to determine the parameter space allowed for J_u and J_d. For different values of J_u and J_d these sets have been compared to the DVCS TTSA data shown in Fig. 3, leading to a first model dependent constraint on J_u versus J_d shown in Fig.4.[21] Further improvement can be expected, taking into account that the GPD H and therefore the available theoretical models will be well constrained by the upcoming HERMES data, and that the data shown here is less than half of the data taken on the transversely polarized hydrogen target. Due to the good agreement between the calculations in the dual-parametrization model and the BSA and BCA results as shown above, the next natural step is to also calculate the allowed parameter space for J_u and J_d within this model. This effort is presently ongoing, but calculations for various values of J_u at a fixed $J_d = 0$[18] already indicate that the dual model favors smaller values for J_u than the double-distribution one.

4. DVCS on nuclei

Beam-Spin Asymmetries have also been measured on deuterium and various heavier nuclei (He, N, Ne, Kr, Xe). Coherent processes are dominant for small values of $-t$, while at larger values incoherent scattering on the individual protons and neutrons dominates. Based on MC studies a region

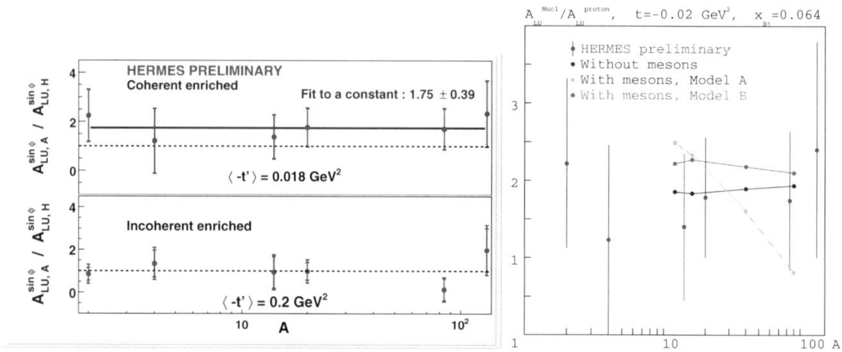

Fig. 5. The $\sin\phi$ amplitude of the BSA on D, He, N, Ne, Kr, Xe divided by the one on the proton in two different kinematic regions dominated by either coherent or incoherent processes (left panel). The 'coherent enriched' result (left panel, upper row) is compared to model calculations[26] in the right panel.

of small $-t'$ ("coherent enriched") has been selected for each nucleus in order to have similar average kinematic values for all targets ($\langle -t' \rangle = 0.018$ GeV^2, $\langle Q^2 \rangle \approx 1.7$ GeV^2, $\langle x_B \rangle \approx 0.065$). According to the MC studies the fraction of coherent processes is approximately 82% for all but the lighter targets (D, He). Similarly, an "incoherent enriched" sample has been selected with an average $-t'$ of 0.2 GeV^2. The ratio of the BSA on the nuclei divided by the one on the proton at the same average kinematics is shown in the left panel of Fig.5. A simple fit to a constant for the coherent enriched sample yields a value above unity by two sigma, while the incoherent enriched sample shows asymmetries very similar to the one on the proton. Note that this is in agreement with earlier HERMES preliminary results[22] where a value consistent with unity was found when comparing the BSA on deuterium and neon to the one on the proton in the full t range. The BSAs on nuclei are expected to be similar to the one on the proton in the incoherent enriched sample since scattering on the protons inside the nuclei should dominate due to the fact that the BH process on the neutron is suppressed. The fit value found for the coherent enriched sample is consistent with a very basic prediction of 5/3 for Spin-0 and Spin-1/2 targets, based on the ratio of the involved valence-quark charges squared,[23] as well as with calculations done specifically for neon and krypton.[24] Also a prediction of R= 1-1.1 for Helium[25] is in agreement with the measurement. The result of a recent calculation[26] is shown in the right panel of Fig.5, whereby one of the three models is disfavored by the data.

5. Summary

The HERMES DVCS data on beam-charge and beam-spin asymmetries is already able to distinguish between some GPD models. Based on a certain model, the DVCS measurements on transversely polarized hydrogen lead to a first model dependent constraint for the total angular momentum of quarks in the nucleon. This method, together with increased statistics and improved models should allow for a constraint with reasonable statistical and theoretical uncertainties in the future.

References

1. D. Müller et al., Fortschr. Phys. **42** (1994) 101.
2. X. Ji, Phys. Rev. Lett. **78** (1997) 610, Phys. Rev. **D55** (1997) 7114.
3. A.V. Radyushkin, Phys. Lett. **B380** (1996) 417, Phys. Rev. **D56** (1997) 5524.
4. M. Burkardt, Phys. Rev. **D62** (2000) 071503; Erratum–ibid. **D66** (2002) 119903.
5. A.V. Belitsky and D. Müller, Nucl. Phys. **A711** (2002) 118.
6. J.P. Ralston and B. Pire, Phys. Rev. **D66** (2002) 111501.
7. HERMES Coll., K. Ackerstaff et al., Nucl. Instr. and Meth. **A417** (1998) 230.
8. M. Diehl et al., Phys. Lett. **B411** (1997) 193.
9. A.V. Belitsky, D. Müller and A. Kirchner, Nucl. Phys. **B629** (2002) 323.
10. R. Perez-Benito, these proceedings.
11. HERMES Coll., A. Airapetian et al., Phys. Rev. Lett. **87** (2001) 182001.
12. HERMES Coll., A. Airapetian et al., Phys. Rev. **D75** (2007) 011103.
13. M. Vanderhaeghen, P.A.M. Guichon and M. Guidal, Phys. Rev. **D60** (1999) 094017.
14. K. Goeke, M.V. Polyakov and M. Vanderhaeghen, Prog. Part. Nucl. Phys. **47** (2001) 401.
15. M. Vanderhaeghen, P.A.M. Guichon and M. Guidal, Computer code for the calculation of DVCS and BH processes, Private Communication, 2001.
16. M.V. Polyakov and C. Weiss, Phys. Rev. **D60** (1999) 114017.
17. I.V. Musatov and A.V. Radyushkin, Phys. Rev. **D61** (2000) 074027.
18. V. Guzey and T. Teckentrup, Phys. Rev. **D74** (2006) 054027.
19. M. Diehl, Phys. Rept. **388** (2003) 41.
20. F. Ellinghaus, W.-D. Nowak, A.V. Vinnikov and Z. Ye, Eur. Phys. J. **C46** (2006) 729.
21. HERMES Coll., Z. Ye, Proceedings of 14th International Workshop on Deep Inelastic Scattering (DIS 2006), Tsukuba, Japan, hep-ex/0606061.
22. HERMES Coll., F. Ellinghaus et al., Proceedings of the 15th International Spin Physics Symposium (Spin 2002), Upton, New York, AIP Conf. Proc. **675** (2002) 303, hep-ex/0212019.
23. A. Kirchner and D. Müller, Eur. Phys. J. **C32** (2003) 347.
24. V. Guzey and M. Strikman, Phys. Rev. **C68** (2003) 015204.
25. S. Liuti and S.K. Taneja, Phys. Rev. **C72** (2005) 032201.
26. V. Guzey and M. Siddikov, J. Phys. **G32** (2006) 251.

DEEPLY VIRTUAL COMPTON SCATTERING WITH CLAS

F.X. GIROD

Hall-B, Jefferson Laboratory
Newport News, VA 23606, USA
E-mail: fxgirod@jlab.org

The beam spin asymmetries of the reaction $\vec{e}p \to ep\gamma$ in the Bjorken regime were measured over a wide kinematical domain using the CLAS detector and a new lead-tungstate calorimeter. Through the interference of the Bethe-Heitler process with Deeply Virtual Compton Scattering, those asymmetries provide constraints for the nucleon Generalized Parton Distributions models. The observed shapes are in agreement with twist-2 dominance predictions.

Keywords: CLAS; DVCS; GPD.

1. Introduction

The introduction of parton Wigner distributions in the nucleon[1] provided us with a rigorous landscape of the different approaches to the hadron structure, linking together all possible hadronic observables. There is unfortunately as of today no known experimental process to access them. Projecting out light-cone energy and transverse momenta, a Fourier transform leads to Generalized Parton Distributions (GPDs). Those encompass Form Factors and Parton Distribution Functions, as well as a wealth of new physical informations. For instance, the Energy-Momentum tensor of partons in the nucleon can be parameterized in terms of GPDs, providing distributions of masses, angular momenta, or forces inside the nucleon.[2]

In the Bjorken limit, the amplitude for Deeply Virtual Compton Scattering (DVCS) $\gamma^* p \to \gamma p$ factorizes (as illustrated in figure 1), and the scattering occurs at the quark level. This is considered to be the cleanest process to access the GPDs. First observations[3] are in agreement with twist-2 dominance in the GPDs formalism. A first dedicated experiment[4] demonstrates that scaling is achieved at moderate values of Q^2, as expected both from theoretical considerations and the successfull description of DIS down to $Q^2 \approx 1$ GeV^2.

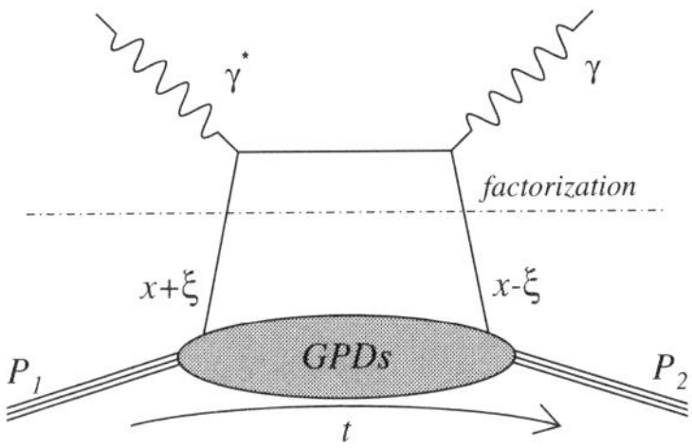

Fig. 1. Handbag diagram for DVCS. t is the squared 4-momentum transfer to the proton, x is the average longitudinal momentum fraction, in terms of $(p+p')/2$, of the active quark in the initial and final states, and $2\xi \approx x_B/(2-x_B)$ parameterizes the difference.

In the reaction $\vec{e}p \rightarrow ep\gamma$, the beam-spin asymmetry of the reaction results from the interference between DVCS and Bethe-Heitler processes, for which the outgoing photon was emitted by the electron. This asymmetry [a] :

$$\mathrm{A} = \frac{\mathrm{d}^4\overrightarrow{\sigma} - \mathrm{d}^4\overleftarrow{\sigma}}{\mathrm{d}^4\overrightarrow{\sigma} + \mathrm{d}^4\overleftarrow{\sigma}} \overset{\text{twist-2}}{\approx} \frac{\alpha \sin\phi}{1+\beta\cos\phi}$$

has been measured as a function of Q^2, x_B, the momentum transfer to the proton t, and the angle[b] ϕ between the leptonic and hadronic planes. Harmonic decompositions[5] have been proposed for the cross-sections. The cross-section difference $\mathrm{d}^4\overrightarrow{\sigma} - \mathrm{d}^4\overleftarrow{\sigma}$ is proportional to the imaginary part of the DVCS amplitude, and sensitive to a specific combination of GPDs H, $\tilde{H}$ and E at $x = \pm\xi$.

2. Experimental context

Using JLab continuous electron beam at energy 5.8 GeV, with average 80% polarisation and intensity 25 nA on a 2.5 cm-long liquid hydrogen target, we collected 45 fb^{-1}. A new lead-tungstate calorimeter was added to CLAS to detect photons between 4.5 and 15°. Avalanche photodiodes collect the light out of the 424 tapered crystals 16 cm long, with average section 2.1 cm^2, each monitored with laser light. Resolutions achieved in energy are better than 5% at 1 GeV, in angle 3 to 4 mrad, and in time less than 1 ns. A superconducting solenoidal magnet actively shields the background of Møller electrons at low angles and energies.

[a]The arrows correspond to the beam helicity.
[b]We use the Trento convention.

3. Analysis and results

We selected events with one and only one electron (trigger), proton and photon. The electron identification uses mainly deposited energy in CLAS calorimeters compared to momentum measured by drift chambers, and Čerenkov counters. The protons were separated from positive pions through time-of-flight and track length compared to momentum. Photon detection threshold was set at 150 MeV.

The mere selection of the 3-particle final state results in clear exclusive peaks in the momentum and energy balance spectra. In addition, we required the missing transverse momentum less than 90 MeV/c, the detected photon within 1.2° from the predicted one, and the angle between the $\gamma^* p$ and γp planes smaller than 1.5°. The neutral pion electroproduction $ep \to ep\pi^0$ can still contaminate our $ep \to ep\gamma X$ sample, in the case were the π^0 decay was asymmetric with one of the photon below the energy threshold, or outside the calorimeters' acceptance. This contamination was estimated in a model-independent way from simulation and subtracted in each elementary bin and for each helicity state.

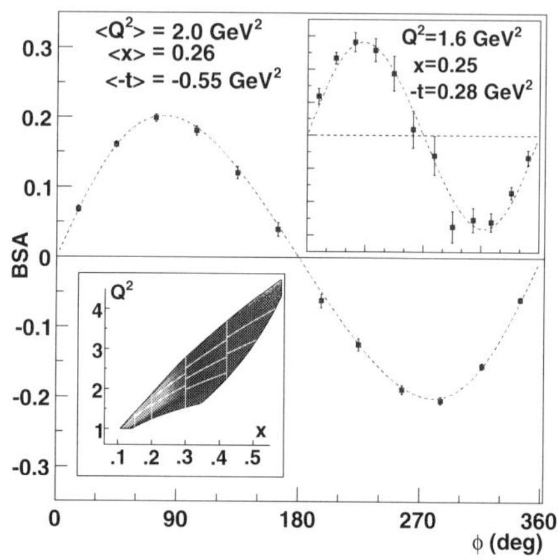

Fig. 2. Integrated result for the asymmetry. Bottom-left insert : (x_B, Q^2) kinematical coverage. Top-right insert : one elementary bin.

The integrated asymmetry is displayed in figure 2, together with the kinematical coverage in the (x_B, Q^2) plane, and one single (x_B, Q^2, t) elementary bin fitted with expected shape. The parameter α estimates the asymmetry at $\phi = \pi/2$ and is shown in (black) circles as a function of $-t$ in the (x_B, Q^2) plane in figure 3. Also appearing are results from

CLAS first observation in a (red) square, and Hall-A in (green) triangles. A GPD semi-phenomenological model[6] appearing in (blue) solid (twist-2) and dashed-dotted (twist-3[c]) lines in this plot reproduces reasonably the data. A Regge-inspired calculation[7] in dotted lines proves fair agreement in some bins. A possible duality between those two descriptions, in terms of Regge-exchanges and GPD formalism, remains to be investigated.

The succesful operation of CLAS and its new calorimeter allowed measurement of the most extensive set of DVCS data, providing strong constraints for GPDs model and future extraction.

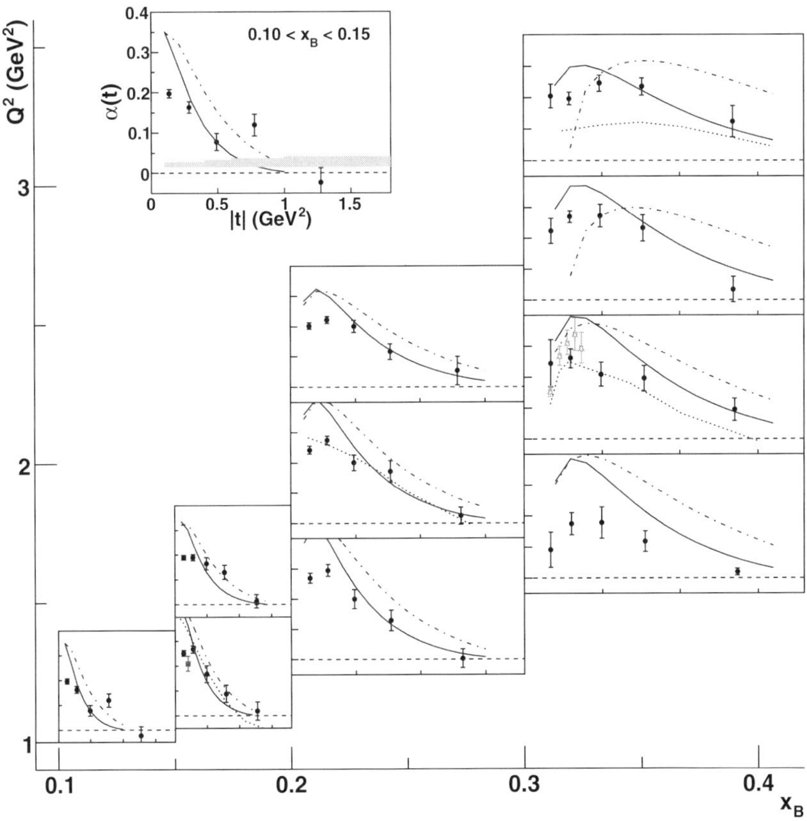

Fig. 3. BSA at $\phi = \pi/2$ estimated by the α parameter, as a function of $-t$ (see text for details).

[c]Kinematical twist-3 only, in the Wandzura-Wilczek approximation.

References

1. X. Ji, *Phys. Rev. Lett.* **91**, 062001 (2003).
2. K. Goeke, J. Grabis, J. Ossmann, M.V. Polyakov, P. Schweitzer, A. Silva, D. Urbano, *Phys. Rev.* **D75**, 094021 (2007).
3. C. Adloff *et al* (H1), *Phys. Lett.*B517 47 (2001); A. Airapetian *et al* (HERMES), *Phys. Rev. Lett.* **87**, 182001 (2001); S. Stepanyan *et al* (CLAS), *Phys. Rev. Lett.* **87**, 182002 (2001); S. Chekanov *et al* (ZEUS), *Phys. Lett.* **573**, 46 (2003); A. Aktas *et al* (H1), *Eur. Phys. J.* **C44** 1 (2005); S. Chen *et al* (CLAS), *Phys. Rev. Lett.* **97**, 072002 (2006); A. Airapetian *et al* (HERMES), *Phys. Rev.* **D75** 011103(R) (2006).
4. C. Muñoz Camacho *et al.* (Jefferson Lab Hall-A), *Phys. Rev. Lett.* **97**, 262002 (2006).
5. M. Diehl, T. Gousset, B. Pire, and J.P. Ralston, *Phys. Lett.* **B411** 193 (1997); A.V. Belitsky, D. Müller, and A. Kirchner, *Nucl. Phys.* **B629** 323 (2002).
6. M. Guidal, M.V. Polyakov, A.V. Radyushkin, and M. Vanderhaeghen, *Phys. Rev.* **D72** 054013 (2005).
7. J.M. Laget, *JLAB-THY-07-702* (arXiv:0708.1250v1).

DEEPLY VIRTUAL COMPTON SCATTERING OFF THE NEUTRON AT JLAB HALL A

M. MAZOUZ

For the Jefferson Lab Hall A and the DVCS Collaborations

Département de Physique, Faculté des Sciences de Monastir, Monastir, 5000, Tunisia
E-mail: mazouz@jlab.org

Deeply Virtual Compton Scattering off the neutron has been investigated in the Hall A at Jefferson Laboratory. This experiment provides, through the measurement of the helicity-dependent $D(\vec{e}, e'\gamma)X$ cross section at Q^2=1.9 GeV^2 and x_B=0.36, a linear combination of generalized parton distributions particularly sensitive to E, the least constrained GPD.

1. Introduction

Generalized parton distributions (GPDs) describe the quark and gluon structure of the nucleon.[1] They provide correlation information between different nucleon states and unify form factors, parton distributions and the angular momentum of partons.[2] Deeply virtual Compton scattering (DVCS) is the simplest hard exclusive process involving GPDs. In the Bjorken regime, where Q^2 is much larger than the invariant momentum transfer t and the quark confinement scale, the factorization theorem separates the reaction amplitude into a known hard scattering kernel and a non-perturbative part described by GPDs.[3] This is depicted in Fig. 1 with the so-called handbag diagram. To leading order in $1/Q$, the DVCS amplitude gives access to GPDs H, $\widetilde{H}$, E and $\widetilde{E}$. A dedicated DVCS experiment on the proton (p-DVCS) at Jefferson Laboratory (JLab) Hall A shows evidence for factorization at Q^2 as low as 2 GeV^2, opening the path to systematic measurements of the GPDs.[4] While H and $\widetilde{H}$ are constrained by parton distributions at the forward limit ($t = \xi = 0$), E and $\widetilde{E}$ do not have similar constraints. Measurements sensitive to E are then of great interest since this particular GPD enters on equal footing with H in the Ji sum rule[2] which relates to the total angular momentum carried by quarks in the nu-

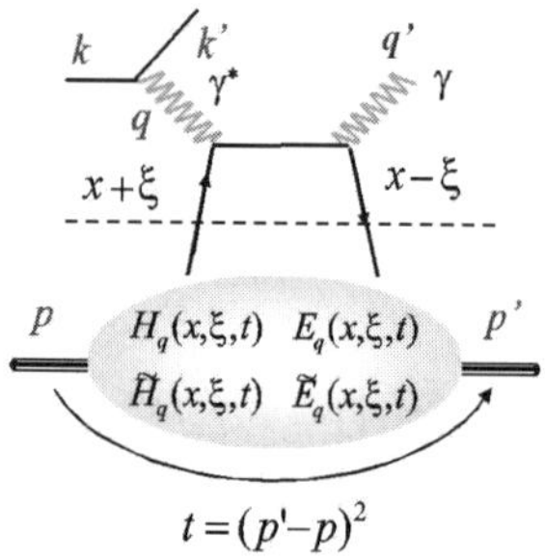

Fig. 1. Handbag diagram for DVCS. A quark with a momentum fraction $x+\xi$ absorbs a virtual photon and then falls back to the nucleon ground state with a momentum fraction $x-\xi$ by radiating a real photon.

cleon. An experiment investigating DVCS on the neutron (n-DVCS),[5] via the helicity-dependent $D(\vec{e}, e'\gamma)X$ cross section measured in the valence quark region ($x_B = 0.36$) at Q^2=1.9 GeV^2, has taken place in the Hall A at JLab. The next section explains why this experiment is mostly sensitive to the GPD E. The experimental setup is then briefly described, and finally the results and their implication on the parametrization of E are discussed.

2. Physics case

Experimentally, the DVCS process is accompanied by the Bethe-Heitler (BH) process where a real photon is emitted by the incoming or outgoing electron. Taking advantage of the interference between the BH and DVCS amplitudes, GPDs can be accessed from the helicity-dependent cross sections. In the differential phase space element $d^5\Phi = dQ^2 dx_{\rm B} dt d\phi_e d\phi_{\gamma\gamma^*}$, where ϕ_e is the azimuthal angle of the scattered electron and $\phi_{\gamma\gamma^*}$ is the angle between the leptonic and hadronic planes, the difference between longitudinally polarized beam cross sections is given by[6]

$$\frac{1}{2}\left[\frac{d^5\sigma^+}{d^5\Phi} - \frac{d^5\sigma^-}{d^5\Phi}\right] \simeq 2\,\mathcal{T}^{BH}\Im m(\mathcal{T}^{DVCS}) = \Gamma^{\Im}\,\Im\mathrm{m}\left[\mathcal{C}^I\right]^{exp}\sin(\phi_{\gamma\gamma^*}) \quad (1)$$

where $\Gamma^{\Im}$ is a kinematical factor and $\Im\mathrm{m}\left[\mathcal{C}^I\right]^{exp}$ is the quantity to be extract from the data. In this equation, only twist-2 contributions are considered as suggested by the p-DVCS experiment results[4] at Q^2=1.9 GeV^2. However, the *exp* superscript is used to indicate eventual higher twist contributions. $\Im\mathrm{m}\left[\mathcal{C}^I\right]$ is a linear combination of GPDs evaluated at $x = \pm\xi$

$$\mathcal{C}^I = F_1(t)\,\mathcal{H}(\xi,t) + \xi(F_1(t)+F_2(t))\,\widetilde{\mathcal{H}}(\xi,t) - \frac{t}{4M^2}F_2(t)\,\mathcal{E}(\xi,t) \quad (2)$$

where F_1 and F_2 are respectively the Dirac and Pauli form factors. $\mathcal{H}$, $\mathcal{E}$ and $\widetilde{\mathcal{H}}$ are twist-2 Compton form factors (CFFs) with for example

$$\Im\mathrm{m}\left[\mathcal{E}\right] = \pi \sum_q e_q^2 \left[E_q(\xi,\xi,t) - E_q(-\xi,\xi,t)\right] , \tag{3}$$

where e_q is the quark charge in unit of the elementary charge. While Eq. 2 for p-DVCS is dominated by $\mathcal{H}$ and $\widetilde{\mathcal{H}}$, it becomes essentially sensitive to $\mathcal{E}$ in the neutron case. That is a consequence of the small value of F_1 and the compensation between u and d polarized parton distribution leading, within a GPD model (VGG),[7] to a small value of $\Im\mathrm{m}\widetilde{\mathcal{H}}$.

3. The n-DVCS experiment

A 5.75 GeV longitudinally polarized electron beam was incident on a 15 cm liquid D_2 target. The beam polarization was about 76% and the beam current of $\sim$4 μA resulted in a luminosity of 4.10^{37} cm^{-2}s^{-1}. Scattered electrons of 2.94 GeV/c were detected in the left High Resolution Spectrometer (HRS-L) defining accurately the virtual photon at fixed Q^2 and x_B. Photons were detected in an electromagnetic calorimeter (11 x 12 array of 3 x 3 x 18.6 cm^3 PbF_2 crystals) centered around the virtual photon.
The selection of DVCS events is ensured by applying a cut on the missing mass $M_X^2 = (q + p - q')^2$ at the pion production threshold. The resulting events correspond, within the impulse approximation, to the sum of the coherent DVCS on the deuteron (d-DVCS) and incoherent p-DVCS and n-DVCS contributions. After the subtraction of p-DVCS events, obtained from the Fermi-broadened $H(\vec{e}, e'\gamma)p$ data, the d-DVCS and n-DVCS contributions are simultaneously extracted in each t-bin from a global analysis taking advantage of their $\Delta M_X^2 = t/2$ kinematical separation.[8]

4. Results

Figure 2 displays the values of $\Im\mathrm{m}\left[\mathcal{C}_n^I\right]^{exp}$ as a function of t.[9] The systematical errors come essentially from the t-dependent uncertainties on the relative calibration between D_2 and H_2 data and estimates of the DVCS-like π^0 channel contamination. The results are compared to three different calculations of the VGG model where E is parametrized by the unknown contribution of the u (J_u) and d (J_d) quarks to the nucleon angular momentum. This comparison indicates that the present data provide constraints on the parametrization of the GPD E that can be expressed, within this model, in terms of a constraint on the quark angular momenta.[9]

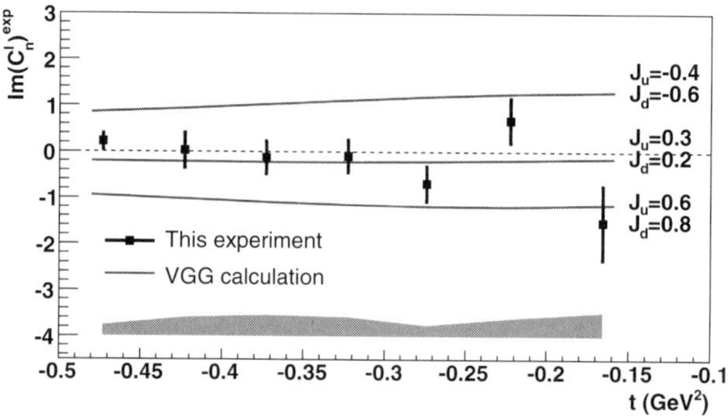

Fig. 2. The t-dependence of $\Im\mathrm{m}\left[\mathcal{C}_n^I\right]^{exp}$. Error bars show statistical uncertainties; systematical uncertainties are indicated by the shaded bands.

5. Conclusion

The present experiment provides a measurement of a linear combinear combination of GPDs particularly sensitive to E, the least constrained GPD. In addition to proton experiments, the n-DVCS experiments access other GPD combinations with a different flavor sensitivity. They appear consequently as a mandatory complement to p-DVCS experiments. These features should open the area of systematic n-DVCS measurements at JLab 6 GeV, and 12 GeV in the near future.

References

1. D. Mueller, D. Robaschik, B. Geyer, F. M. Dittes, and J. Horejsi, Fortschr. Phys. **42**, 101 (1994); A.V. Radyushkin, Phys. Rev. **D 56**, 5524 (1997).
2. X.D. Ji, Phys. Rev. Lett. **78**, 610 (1997).
3. M. Diehl, T. Gousset, B. Pire and J. P. Ralston, Phys. Lett. **B 411**, 193 (1997); X.D. Ji and J. Osborne, Phys. Rev. **D 58**, 094018 (1998); J.C. Collins, A. Freund, Phys. Rev. **D 59**, 074009 (1999).
4. C. Muñoz Camacho et al., Phys. Rev. Lett. **97**, 262002 (2006).
5. P. Y. Bertin, C. Hyde-Wright, F. Sabatié, E. Voutier et al., Jefferson Lab Proposal **E03-106** (2003).
6. A.V. Belitsky, D. Mueller, and A. Kirchner, Nucl. Phys. **B 629**, 323 (2002).
7. M. Vanderhaeghen, P.A.M. Guichon and M. Guidal, Phys. Rev. **D 60**, 094017 (1999); K. Goeke, M.V. Polyakov and M. Vanderhaeghen, Prog. Part. Nucl. Phys. **47**, 401 (2001).
8. M. Mazouz, Thèse de Doctorat, Université Joseph Fourier, Grenoble (France), 2006; M. Mazouz, Nucl. Phys. **A 782**, 41c (2007).
9. M. Mazouz et al., *Submitted to* Phys. Rev. Lett.

THE FUTURE DVCS EXPERIMENTS IN HALL A AT JLAB

J. ROCHE
For the Jefferson Lab Hall A and the DVCS collaborations
Ohio University,
Athens, OH 45701, USA
E-mail: rochej@ohio.edu

Two DVCS experiments off the proton have been recently approved to run in Hall A at JLab to measure electro-production of photon and π^0 absolute cross-sections in the deep inelastic regime. They will exploit the successful experimental technique as well as build on the physics outcomes of the previous generation of DVCS experiments in Hall A. The new experiments will run both before and after the CEBAF upgrade.

Experiment E00-110 on the proton[1,2] measured photon and π^0 electro-production cross sections in the deep inelastic regime using a simple apparatus in Hall A at JLab. A slightly upgraded apparatus was also used to measure the cross sections of photon electro-production off the neutron in the deep inelastic regime during experiment E03-106.[3] Both beam helicity dependent and independent cross sections were measured. From the angular dependance of these cross sections, Compton Form Factors (CFF, related to the Generalized Parton Distribution) are used to paramterize both the imaginary and real part of the process amplitude.[4] Because the Q^2 dependence of the cross-sections measured by E00-110 followed expectations based on the dominance of the leading twist term of the amplitudes ("handbag" approximation[2]), GPDs linear combinations were extracted from those data. The two upcoming DVCS experiments E07-007[5] and E12-06-114[6] are based on the same experimental technique.

1. Critical experimental assessment of the previous generation of DVCS experiments in Hall A at JLab

In the Hall A DVCS experiments, polarized electrons impinged on a liquid target, the scattered electron was detected in one the high resolution spectrometer of Hall A while the emitted photon was detected in a compact electromagnetic calorimeter. The calorimeter was located at small angles with respect to the beam direction (8 deg for the closest block) and its face was 1 m away from the target. Analog ring samplers (ARS[7]) digitalized the calorimeter signals allowing an offline rejection of pile up events. Overall this simple and compact assembly resulted in a good control of the solid angle and ultimately a measurement of the cross-sections with a 5% systematic precision. Looking back at the performances of the apparatus, the main successes of the technique were: (1) The ARS system is efficient at identifying the photon signal at high luminosity. The maximum luminosity achieved off the deuterium target during E03-106 reached $4{\times}10^{37}$ $\mathrm{cm^{-2}s^{-1}}$ per nucleon. This maximum luminosity will be used to design the new experiments. (2) The measurement of $H(e,e'\gamma)X$ missing mass is sufficient to sign the exclusive channel with 3% systematic precision. The original design of the experiment included the use of a plastic scintillator array surrounding the calorimeter to detect the recoil proton in addition to the scattered electron and emitted photon. The analysis of the missing mass spectra[1] proved that the scintillator system is unnecessary. The main short falls of the experimental apparatus as installed in these pilot experiments were: (1) At low quadri-momentum transfer Q^2, the $ep \to ep\pi^0$, main competing channel to $ep \to ep\gamma$, statistics were poor. π^0 are detected by measuring the presence of two decay photons in the calorimeter. The pion energy decreases as Q^2 decreases down to a point where its decay products would not trigger the DAQ. In the next experiments generation the calorimeter trigger threshold will be lowered as well as the DAQ bandwidth and live-time will be improved. (2) The resolution of the calorimeter degraded during the data taking. The accumulation of radiation dose on the calorimeter blocks degrades their transmission properties and ultimately the resolution of the calorimeter. In order to perform longer experiments with a higher luminosity, the blocks will be optically cured in situ during the data taking.[8] Finally, one should note that the size of the calorimeter will be increased by 40%. In addition to the trigger modification, this will increase the range in t over which the CFF can be extracted from 0.20 GeV^2 to 1 GeV^2 providing the first study of the tranverse profile of the nucleon as a function of quark light-cone momentum fraction.

2. E07-007: Complete separation of deeply virtual photon and π^0 electroproduction observables of unpolarized protons

One of the important results of E00-110 is the unpolarized cross-section. The cross section of electro-production of photon is the interference of two amplitudes: the Bethe-Heitler process (BH: either electron radiates the photon) and the Deeply Virtual Compton Scattering process (DVCS: the nucleon radiates the photon). The BH amplitude is fully calculable in QED if one knows the proton form factors. E00-110 showed that for the measured kinematics, the BH process doesn't dominate the total cross-section, that is: the DVCS process is not negligible compared to the BH process. In experiment E07-007, one uses the beam energy dependence of the BH and the DVCS amplitudes to isolate the BH·DVCS interference term from the pure $DVCS^2$ contribution. This is a necessary step in order to extract meaningful information about the GPDs from the unpolarized cross-sections as well as plan future DVCS experiments with the upgraded CEBAF beam. E07-007 was approved for 544h of beam time to measure the cross-sections at fixed x_{Bj}=0.36 for Q^2=2.3, 1.9 and 1.5 GeV^2 with three beam energies ranging from 3.5 to 6 GeV. In addition, E07-007 will also measure the 5 response functions of the exclusive deep virtual π^0 channel and in particular separate $d\sigma_L$ from $d\sigma_T$ with a conventional Rosenbluth separation. This separation, as a function of Q^2, is essential for the testing of the factorization of $d\sigma_L$ in the deeply $ep \to ep\pi^0$ reaction.

3. E12-06-114: Measurements of the electron-helicity dependent cross sections of the DVCS with CEBAF at 12 GeV

For E00-110, the Compton Form Factor were extracted as a function of t between -0.15 and -0.35 GeV^2, x_{Bj}=0.36 and Q^2 = 2.3, 1.9 and 1.5 GeV^2. With the upgraded CEBAF beam reaching up to 12 GeV, the kinematic range (x_{Bj}, Q^2) accessible is widely increased as shown in figure 1. One notes that: (1) with a limited number of setting, E12-06-114 will cover most of the kinematical domain available with a 11 GeV beam, (2) at fixed x_{Bj}, the lever arm as function of Q^2 will be of a factor of 2. This will provide a stringent test of the factorization theorem, and quantify the contribution of higher twist terms. (3) the CFF will be measured as a function of x_{Bj} from 0.36 to 0.6. Data only exist so far at x_{Bj} = 0.36. E12-06-114 was approved to run with beam energies ranging from 6.6 to 11 GeV. The statistic accumulated in each (x_{Bj}, Q^2, t) bin will be comparable to the one of E00-110. Both $ep \to ep\gamma$ and $ep \to ep\pi^0$ will be measured.

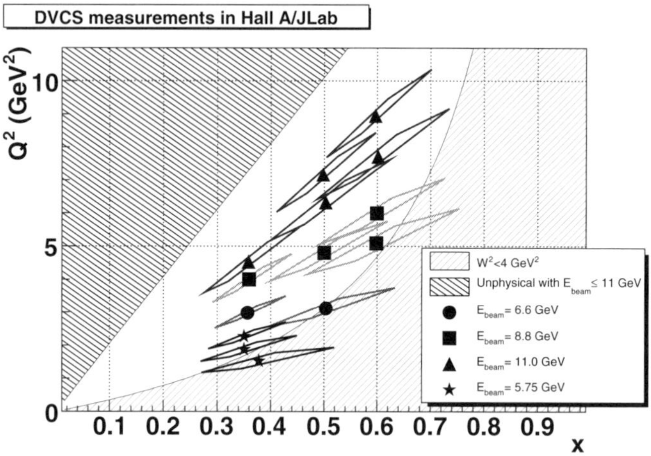

Fig. 1. Kinematic coverage of the DVCS/Hall A experiments. The boundary of the unphysical region corresponds to the maximum possible Q^2 at given x_{Bj} for a 11 GeV beam. The diamond shapes surrounding each symbols trace the approximate acceptance of the detectors at each settings. The data at E_{beam}=5.75 GeV were already measured during E00-110 and will be repeated in E07-007 with different beam energies; the others settings will be measured by E12-06-114.

4. Conclusions

Two new DVCS experiments in Hall A at JLab have been approved for beam time by the JLab PAC. They are based on the same experimental concept than the successful first generation of experiments and use hydrogen as target. In addition to this future program, the Hall A/DVCS collaboration is also preparing proposals with a deuterium target as a follow up of E03-106.[3] Those proposals will be submitted momentarly.

References

1. C. Muñoz Camacho et al., Phys. Rev. Lett. **97**, 262002 (2006)
2. F. Sabatié, these proceedings.
3. M. Mazouz et al., arXiv:0709.0450, submitted to Phys. Rev. Lett and these proceedings.
4. A.V. Belitsky, D. Mueller and A. Kirchner, Nucl. Phys. **B 629**, 323 (2002).
5. P.Y. Bertin, C.E. Hyde, C. Muñoz Camacho, J. Roche et al., http://www.jlab.org/exp_prog/proposals/07/PR-07-007.pdf
6. C.E. Hyde, B. Michel, C. Muñoz Camacho, J. Roche et al., http://www.jlab.org/exp_prog/proposals/06/PR12-06-114.pdf
7. F. Feinstein, Nucl. Inst. Meth. **A504**, 258 (2003).
8. P. Achenbach et al., Nucl. Inst. Meth. **A504**, 258 (2003).

DEEPLY VIRTUAL COMPTON SCATTERING WITH CLAS12

L. ELOUADRHIRI

Jefferson Lab
Newport News, VA 23606, USA
E-mail: latifa@jlab.org

An overview is given about the capabilities provided by the JLab 12 GeV Upgrade to measure deeply virtual exclusive processes with high statistics and covering a large kinematics range in the parameters that are needed to allow reconstruction of a spatial image of the nucleon's quark structure. The measurements planned with CLAS12 will cross section asymmetries with polarized beams and with longitudinally and transversely polarized proton targets in the constrained kinematics $x = \pm\xi$. In addition, unpolarized DVCS cross sections, and doubly polarized beam target asymmetries will be measured as well. In this talk only the beam and target asymmetries will be discussed.

Keywords: GPDs, DVCS

1. Introduction

The concept of GPDs[1–3] has led to completely new methods of "spatial imaging" of the nucleon. The mapping of the nucleon GPDs, and a detailed understanding of the spatial quark distribution of the nucleon, have been widely recognized are a key objectives of nuclear physics of the next decade, and is a key justification for the JLab energy upgrade to 12 GeV. GPDs also allow to quantify how the orbital motion of quarks in the nucleon contributes to the nucleon spin – a question of crucial importance for our understanding of the "mechanics" underlying nucleon structure. This requires a comprehensive program, combining results of measurements of a variety of processes in electron–nucleon scattering with structural information obtained from theoretical studies, as well as with expected results from future lattice QCD simulations.

It is well recognized[4–6] that exclusive processes can be used to probe the GPDs and construct 2-dimensional and 3-dimensional images of the quark content of the nucleon. Deeply virtual Compton scattering and deeply

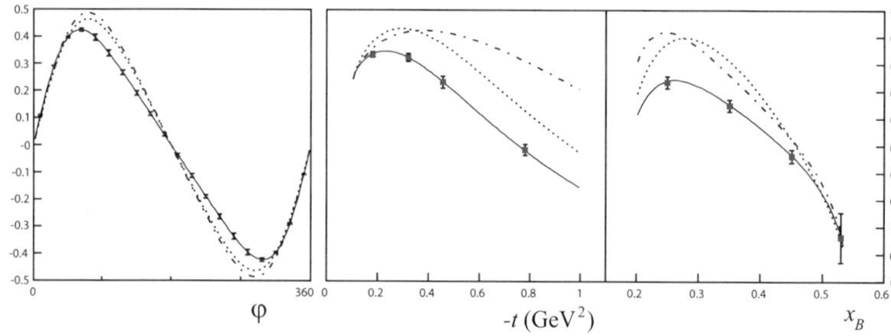

Fig. 1. The beam spin asymmetry showing the DVCS-BH interference for 11 GeV beam energy from the VGG model[14] with uncertainties projected for 11 GeV.[13] Many other bins will be measured simultaneously.

virtual meson production are identified as the processes most suitable to map out the twist-2 vector GPDs H, E and the axial GPDs $\tilde{H}$, $\tilde{E}$ in x, ξ, t, where x is the momentum fraction of the struck quark, ξ the longitudinal momentum transfer to the quark, and t the momentum transfer to the nucleon. Having access to a 3-dimensional image of the nucleon (two dimensions in transverse space, one dimension in longitudinal momentum) opens up completely new insights into the complex structure of the nucleon. For example, the nucleon matrix element of the energy-momentum tensor contains 3 form factors that encode information on the angular momentum distribution $J^q(t)$ of the quarks with flavor q in transverse space, their mass-energy distribution $M_2^q(t)$, and their pressure and force distribution $d_1^q(t)$. These form factors also appear as moments of the vector GPDs,[7] thus offering prospects of accessing these quantities through detailed mapping of GPDs. The quark angular momentum in the nucleon is given by

$$J^q(t) = \int_{-1}^{+1} dx x [H^q(x,\xi,t) + E^q(x,\xi,t)] \ ,$$

and the mass-energy and pressure distribution

$$M_2^q(t) + 4/5 d_1^q(t)\xi^2 = \int_{-1}^{+1} dx x H^q(x,\xi,t) \ .$$

The mass-energy and force-pressure distribution of the quarks are given by the second moment of GPD H, and their relative contribution is controlled by ξ. The separation of $M_2^q(t)$ and $d_1^q(t)$ requires measurement of these moments in a large range of ξ.

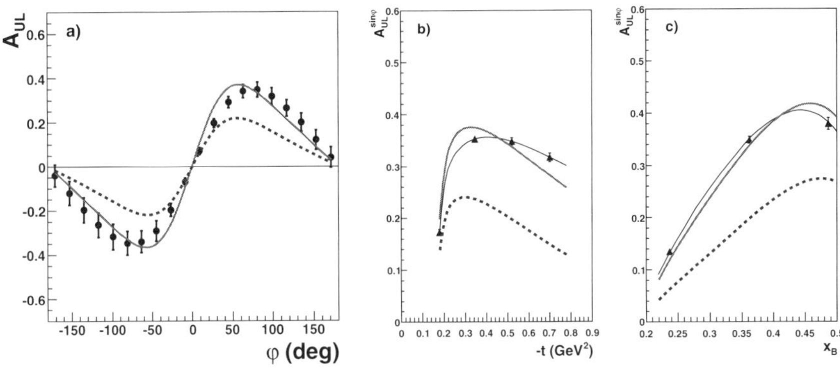

Fig. 2. The longitudinal target spin asymmetry beam spin asymmetry showing the DVCS-BH interference for 11 GeV beam energy from the VGG model[14] with uncertainties projected for 11 GeV.[13] Other bins will be measured simultaneously

2. GPDs and DVCS

DVCS has been shown[8–11] to be the cleanest process to access GPDs at the kinematics accessible today. It is also a relatively rare process and requires high luminosities for the required high statistics measurements. The beam helicity-dependent cross section asymmetry is given in leading twist as

$$A_{LU} \approx \sin\phi[F_1(t)H + \xi(F_1(t) + F_2(t))\tilde{H}]d\phi \ ,$$

where F_1 and F_2 are the Dirac and Pauli form factors, ϕ is the azimuthal angle between the electron scattering plane and the hadronic plane. The kinematically suppressed term with GPD E is omitted. For not too large ξ the asymmetry is mostly sensitive to the GPD $H(x = \xi, \xi, t)$.

The asymmmetry with a longitudinally polarized target is given by

$$A_{UL} \approx \sin\phi[F_1(t)\tilde{H} + \xi(F_1(t) + F_2(t)H] \ .$$

The combination of A_{LU} and A_{UL} allows a separation of GPD $H(x = \xi, \xi, t)$ and $\tilde{H}(x = \xi, \xi, t)$.

Using a transversely polarized target the asymmetry

$$A_{UT} \approx \sin\phi t/4M^2[F_2(t)H - F_1(t)E]$$

can be measured, which depends in leading order on GPD E and is highly sensitive to orbital angular momentum contributions of quarks.

Clearly, determining moments of GPDs for different t will require measurement in a large range of x, in particular at large x. The reconstruction of the transverse spatial quark distribution requires measurement in a large range in t, and the separation of the $d_1^q(t)$ and $M_2^q(t)$ form factors requires a large span in ξ.

3. Upgrade of CLAS to CLAS12

To meet the requirements of high statistics measurements of relatively rare exclusive processes such as DVCS at high photon virtuality Q^2, large t and ξ, the CLAS detector will be upgraded and modified to CLAS12.[16] The main new features of CLAS12 over the current CLAS detector include a high operational luminosity of $10^{35}\text{cm}^{-2}\text{sec}^{-1}$, an order of magnitude increase over CLAS.[15] Improved particle identification will be achieved with additional threshold gas Cerenkov counter, improved timing resolution of the forward time-of-flight system, and a finer granularity electromagnetic preshower calorimeter that, in conjunction with th existing CLAS calorimeter will provide much improved γ/π^0 separation for momenta up to 10 GeV. In addition, a new central detector will be built that uses a high-field solenoid magnet for particle tracking and allows the operation of dynamically polarized solid state targets. With these upgrades CLAS12 will be the workhorse for exclusive electroproduction experiments in the deep inelastic kinematics.

4. Projected Results at 12 GeV with CLAS12

The 12 GeV upgrade offers much improved capabilities to access GPDs. Figure 1 shows the expected statistical precision of the beam DVCS asymmetry for some sample kinematics. At the expected luminosity of $10^{35}\text{cm}^{-2}\text{sec}^{-1}$ and for a run time of 2000 hours, high statistics measurements in a very large kinematics range are possible.

Using a dynamically polarized NH_3 target we can also measure the longitudinal target spin asymmetry A_{UL} with high precision. The projected results are shown in Fig. 2. The statistical accuracy of this measurement will be less than for the A_{LU} asymmetry due to the large dilution factor in the target material, but it will still be a very significant measurement.

Polarizing the target transverse to the beam direction will access a different combination of GPDs, and provide different sensitivity for the y- and x-components of the target polarization. The expected accuracy for one of the polarization projections is shown in Fig. 3. Here the target is assumed to be a frozen HD-Ice target, which has different characteristics from the NH_3 target.

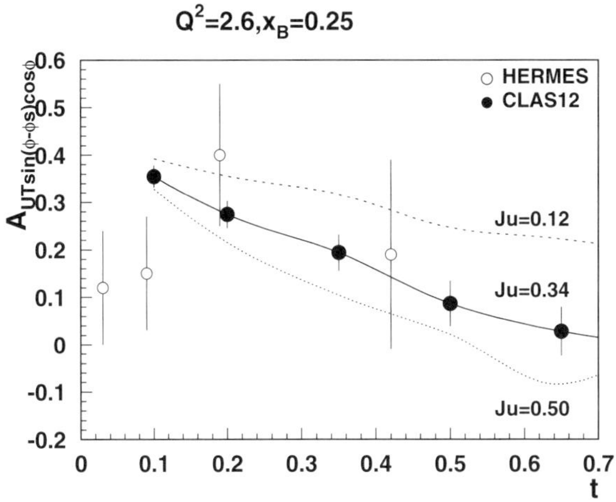

Fig. 3. Projected transverse target asymmetry A_{UT} for DVCS production off protons at 11 GeV beam energy. The curves represent different assumptions on the u-quark contributions to $J(t)$.

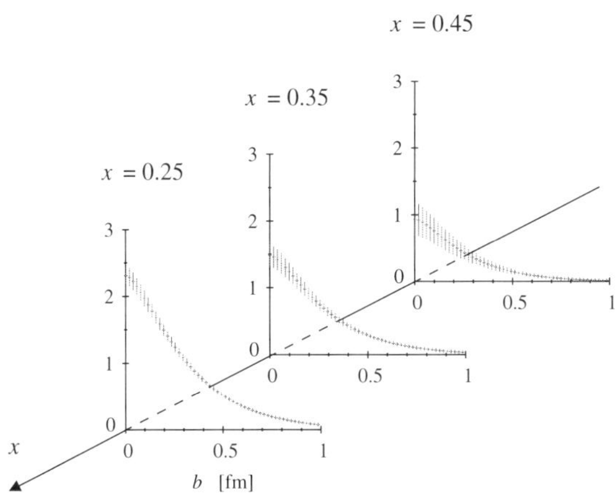

Fig. 4. The u-quark distribution in transverse space as extracted from projected DVCS data with CLAS12.

A measurement of all 3 asymmetries will allow a separate determination of GPDs H, $\tilde{H}$ and E at the above specified kinematics. Through a Fourier transformation the t-dependence of GPD H can be used to determine the

$u-$quark distribution in transverse impact parameter space. Figure 4 shows projected results for such a transformation assuming a model parameterization for the kinematical dependences of GPD H. Knowledge of GPD E will be particularly interesting as it is directly related to the orbital angular momentum distribution of quarks in transverse space.

Acknowledgment

We thank the members of the CLAS collaboration who contributed to the development of the exciting physics program for the JLab upgrade to 12 GeV, and the `CLAS12` detector. Much of the material in this report is taken from the `CLAS12` Technical Design Report Version 3, October 2007.[16]

This work was supported in part by the U.S. Department of Energy and the National Science Foundation, the French Commisariat á l'Energie Atomique, the Italian Instituto Nazionale di Fisica Nucleare, the Korea Research Foundation, and a research grant of the Russian Federation. The Jefferson Science Associates, LLC, operates Jefferson Lab under contract DE-AC05-060R23177.

References

1. D. Müller, D. Robaschik, B. Geyer, F.M. Dittes, J. Horejsi, Fortschr. Phys.**42**, 101 (1994), hep-ph/9812448.
2. X. Ji, Phys. Rev. **D55**, 7114 (1997).
3. A. Radyushkin, Phys. Lett.**B380**, 417 (1996).
4. M. Burkardt, Int. J. Mod. Phys. **A18**, 173 (2003).
5. A. Belitsky, D. Mueller, A. Kirchner, Nucl. Phys. B**629**, 323 (2002).
6. A.V. Belitsky, X. Ji, F. Yuan, Phys. Rev. D**69**, 074014 (2004).
7. K. Goeke *et al.*, Phys. Rev. C**75**, 094021 (2007).
8. S. Stepanyan *et al.*(CLAS collaboration), Phys. Rev. Lett.**87**, 182002 (2001).
9. S. Chen *et al.*(CLAS collaboration), Phys. Rev. Lett.**97**, 072002 (2006).
10. F.X. Girod *et al.*(CLAS collaboration), arXiv:0711.4805, subm. to Phys. Rev. Lett..
11. C. Munoz-Camacho *et al.*, Phys. Rev. Lett.**97**, 262002 (2006).
12. R. De Masi, M. Garcon, B. Zhao, *et al.* (CLAS collaboration), arXiv:0711.4736, submitted to Phys. Rev. Lett.
13. JLAB experiment E12-06-119, F. Sabatie *et al.*
14. M. Vanderhaeghen, P. Guichon, M. Guidal, Phys. Rev. D**60**, 094017 (1999).
15. B. Mecking et al., Nucl. Inst. Meth. A**503**,513 (2003).
16. The complete Technical Design Report document may be obtained from the authors.

QUARK HELICITY FLIP AND THE TRANSVERSE SPIN DEPENDENCE OF INCLUSIVE DIS

A. AFANASEV[1,2], M. STRIKMAN[3], C. WEISS[2]

[1]*Department of Physics, Hampton University, Hampton, VA 23668, USA*
[2]*Theory Center, Jefferson Lab, Newport News, VA 23606, USA*
[3]*Dept. of Physics, Pennsylvania State University, University Park, PA 16802, USA*

Inclusive DIS with unpolarized beam exhibits a subtle dependence on the transverse target spin, arising from the interference of one–photon and two–photon exchange amplitudes in the cross section. We argue that this observable probes mainly the quark helicity–flip amplitudes induced by the non-perturbative vacuum structure of QCD (spontaneous chiral symmetry breaking). This conjecture is based on (a) the absence of significant Sudakov suppression of the helicity–flip process if soft gluon emission in the quark subprocess is limited by the chiral symmetry–breaking scale $\mu^2_{\text{chiral}} \gg \Lambda^2_{\text{QCD}}$; (b) the expectation that the quark helicity–conserving twist–3 contribution is small. The normal target spin asymmetry is estimated to be of the order 10^{-4} in the kinematics of the planned Jefferson Lab Hall A experiment.

A fundamental property of QCD is that the quark helicity is conserved in perturbative interactions, because of chiral invariance in the light–quark sector. It is known, however, that non-perturbative interactions at distances of the order $1/\mu_{\text{chiral}} \sim 0.3\,\text{fm}$ cause the dynamical breaking of chiral symmetry and lead to the appearance of non-zero quark helicity–flip amplitudes. This effect essentially determines the structure and interaction of hadrons at large distances. An interesting question is what the presence of quark helicity–flip amplitudes implies for hard electromagnetic processes (invariant momentum transfer $Q^2 \gg 1\,\text{GeV}^2$). Such processes couple directly to the quark degrees of freedom and in certain cases can be described by factorization of the cross section/amplitude into "hard" and "soft" contributions. Quark helicity flip has been discussed *e.g.* in relation to the high–Q^2 behavior of the proton form factor ratio QF_2/F_1 and wide–angle Compton scattering.[1,2] We have recently pointed out[3] that quark helicity flip in hard processes can be probed in the transverse target spin dependence of inclusive deep–inelastic scattering (DIS) with unpolarized beam,

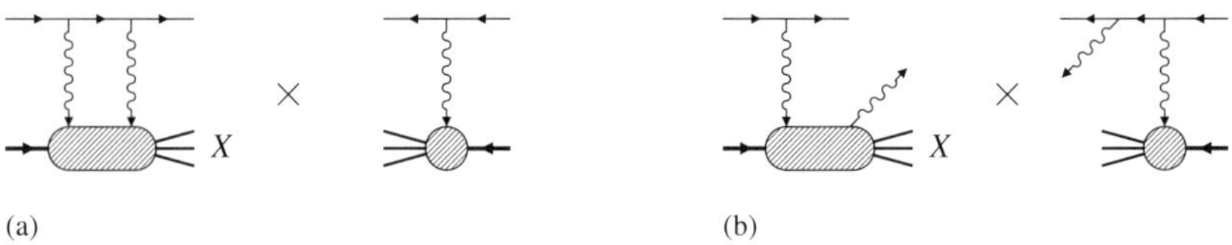

Fig. 1. QED processes contributing to the transverse target spin dependence of the inclusive eN cross section at $O(\alpha^3)$. (a) Interference of one–photon and two–photon exchange. (b) Real photon emission (bremsstrahlung).

$eN \to e'X$. In these proceedings we summarize the arguments relating this observable to quark helicity flip and discuss our estimate of the magnitude of the expected asymmetry (for details, see Ref. 3).

The differential cross section of inclusive eN scattering with unpolarized beam depends on the target spin $\boldsymbol{S}$ as

$$d\sigma = d\sigma_U + \frac{(\boldsymbol{S}, \boldsymbol{l} \times \boldsymbol{l}')}{|\boldsymbol{l} \times \boldsymbol{l}'|} d\sigma_N \tag{1}$$

($\boldsymbol{l}, \boldsymbol{l}'$ are the initial/final electron momenta in the lab frame), and the normal spin asymmetry is defined as

$$A_N \equiv \frac{d\sigma_N}{d\sigma_U}. \tag{2}$$

The spin–dependent term is zero in the one–photon exchange approximation to DIS, by the combination of P and T invariance and the hermiticity of the electromagnetic current operator (Christ–Lee theorem). A non-zero spin dependence appears only in higher orders of the QED coupling constant, from the interference of one–photon and absorptive two–photon exchange amplitudes in the cross section (Fig. 1a), as well as from real photon emission (bremsstrahlung, Fig. 1b). It is important that the two contributions to the spin–dependent cross section are individually free of QED infrared (IR) divergences and thus can be considered as distinct physical effects; this follows from the general factorization properties and the spin–independence of IR photon exchange in QED.[3] The two–photon exchange amplitude is also free of QED collinear divergences (vanishing photon virtuality at non-zero momentum); this is a general consequence of electromagnetic gauge invariance.[3,4] The spin–dependent two–photon exchange cross section, Fig. 1a, thus represents a well–defined observable in QED, which can be used as a new probe of hadron structure.

In DIS kinematics we expect the spin–dependent two–photon exchange cross section to arise predominantly from the amplitudes in which the

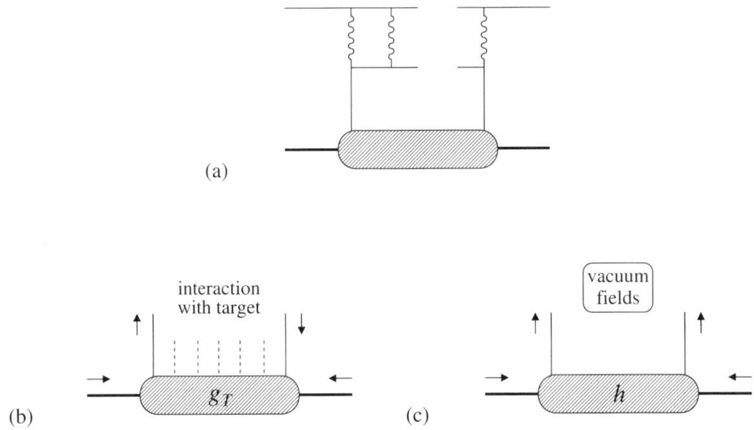

Fig. 2. Transverse spin dependence of the DIS cross section in QCD. (a) Dominance of two–photon exchange with the same quark. (b) Quark helicity–conserving process involving interactions with the target remnants. (c) Quark helicity flip due to interaction with non–perturbative vacuum fields.

two–photon exchange couples to a single quark, namely the same quark which is hit in the interfering one–photon exchange amplitude, see Fig. 2a. This expectation is based on the absence of "anomalous" (IR or collinear–enhanced) contributions in the two–photon exchange amplitude, and on general considerations of scattering from a hadronic system with momentum transfer $Q^2 \gg R^{-2}_{\text{hadron}}$. However, one can easily see that no transverse spin dependence can arise in the parton model approximation in which the electron scatters from an on–shell massless quark, as the transverse spin–dependent cross section for scattering from an on–shell quark requires quark helicity flip and is explicitly proportional to the quark mass. A spin dependence of the hadronic cross section thus can come only from the two mechanisms indicated in Figs. 2b and c.

In the process of Fig. 2b the quark helicity is conserved in the electron–quark subprocess, and the helicity flip between the interfering hadronic amplitudes happens at the level of the quark distribution in the nucleon. This process requires non-zero virtuality (off–shellness) of the active quark, and, at the same time, explicit interaction of the active quark with the spectator system. The two effects are linked by electromagnetic gauge invariance; only the sum of the two maintains transversality of the amplitude and avoids the appearance of unphysical collinear divergences. (An attempt to calculate the transverse spin asymmetry including finite virtuality of the active quark but neglecting explicit interactions has produced a divergent result.[5]) In the process of Fig. 2c the quark helicity is flipped in the electron–quark subprocess. This process requires interaction of the active quark with the non-perturbative vacuum fields which cause the spontaneous breaking of chiral symmetry. It does not require interactions with the spectator system and exists already for on–shell quarks. Perturbative QCD interactions do not "mix" the contributions of Figs. 2b and c. An interesting question is whether in typical DIS kinematics one of the two mechanisms dominates or both make comparable contributions to the transverse spin dependence.

Following Ref. 3, we suggest here that the quark helicity–flip contribution, Fig. 2c, may be sizable and could well be the dominant mechanism in the transverse target spin dependence. This perhaps somewhat surprising assertion rests on the following two observations.

(1) *Insignificant Sudakov suppression of helicity–flip process.* The helicity–flip process in QCD requires propagation through quark configurations with virtualities $\lesssim \mu^2_{\text{chiral}}$, which experience significant interactions with the chiral symmetry breaking vacuum fields. This leads to Sudakov suppression of the cross section compared to the case of unrestricted virtualities. A similar suppression takes place in the usual DIS cross section (where the restriction to small virtualities results from the condition of producing real particles in the final state); however, there it is compensated by real gluon emissions. In the case of the transverse spin asymmetry this compensation is likely to be incomplete, and some residual Sudakov suppression of the interference cross section should remain. Numerical estimates based on the on–shell Sudakov formfactor of QCD show that this suppression should be marginal for $Q^2 \sim$ few GeV^2, if the IR cutoff governing soft gluon emission in the Sudakov formfactor is of the order of $\mu^2_{\text{chiral}} \gg \Lambda^2_{\text{QCD}}$. While we cannot presently prove that this magnitude of the IR cutoff is dictated by chiral symmetry breaking in QCD, it certainly appears natural in the light of dynamical models such as the instanton vacuum, which suppose that gluons of wavelength $< \mu^{-1}_{\text{chiral}}$ are "contained" in the non-perturbative field configurations which break chiral symmetry.

(2) *Non-partonic character of helicity–conserving process.* The quark helicity–conserving process, Fig. 2b, explicitly involves non-zero virtuality of the initial quark and its interaction with the spectator system, and is thus of essentially "non-partonic" character. The calculation of this process within the collinear QCD expansion starts from the "handbag diagram" with the twist–3 quark helicity–conserving distribution $g_T(x)$, which also appears in the calculation of the spin structure function $g_1 + g_2$ measured

with polarized beam and transversely polarized target. However, retaining only the contribution from the "handbag diagram" would not be a consistent approximation for the spin–dependent two–photon exchange cross section, as this would violate electromagnetic gauge invariance and lead to the appearance of QED collinear divergences. The interaction of the active quark with the gluon field in the target needs to be included in order to restore electromagnetic gauge invariance. An interesting question is whether this "restoration" of gauge invariance will altogether eliminate the contribution from the dynamical twist–2 operators originally present in g_T, or whether some part of it survives in the final result. In the former case the helicity–conserving contribution to the asymmetry would be governed by the same mechanism as the twist–3 ("non–Wandzura–Wilczek") part of g_2, which has been shown to be very small compared to g_1 by the SLAC[6] and Jefferson Lab Hall A[7] experiments. In the latter case, one could estimate the relative order–of–magnitude of the contributions from Figs. 2b and c by comparing[3]

$$\frac{\langle \boldsymbol{k}_T^2 \rangle}{M} \, g_{T,f}(x) \quad \longleftrightarrow \quad M_q \, h_f(x), \tag{3}$$

where on the left–hand side $\langle \boldsymbol{k}_T^2 \rangle$ is the typical quark intrinsic transverse momentum, M the nucleon mass, and $g_{T,f}(x)$ is given in terms of the twist–2 quark helicity distribution, $g_f(x)$, by the Wandzura–Wilczek relation, and on the right–hand side $M_q \approx 0.3$–$0.4\,\mathrm{GeV}$ is a typical constituent quark mass, determining the strength of the helicity–flip amplitude for low–virtuality quarks, and h_f the quark transversity distribution ($f = u, d$ denotes the quark flavor). The numerical estimate of Fig. 3 shows that even under these assumptions the quark helicity–conserving process is unlikely to dominate over the quark helicity–flip one.

To summarize, we have argued that a sizable, perhaps dominant, contribution to the transverse spin–dependent cross section comes from the quark helicity–flip process governed by the quark transversity distribution in the nucleon. To validate this conjecture, a more detailed investigation of the effects of electromagnetic gauge invariance on the quark helicity–conserving contribution is needed. We note that this problem bears some similarity to that of gauge invariance in the QCD light–cone expansion of deeply–virtual Compton scattering (DVCS), where it was found that the twist–2 contribution to the amplitude alone is not transverse, and that transversality is restored by including certain "kinematical" twist–3 contributions.[8–10] If the smallness of the helicity–conserving contribution of Figs. 2b could indeed be confirmed by explicit calculation, the normal spin asymmetry in

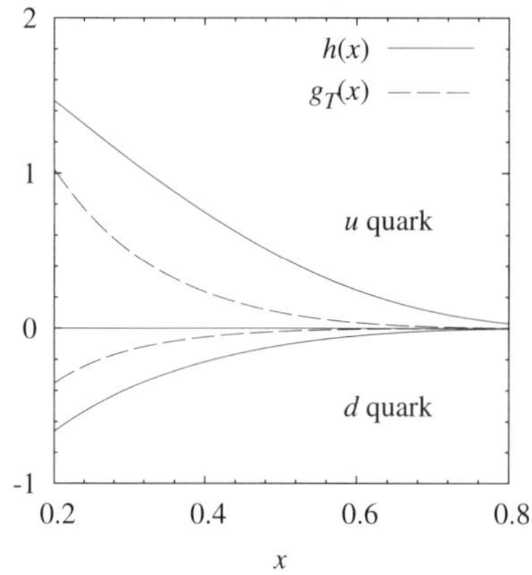

Fig. 3. Comparison of the proton's twist–3 helicity–conserving quark distribution $g_T(x)$ [calculated in terms of the twist–2 quark helicity distribution $g(x)$ using the Wandzura–Wilczek relation] with the twist–2 helicity–flip (transversity) distribution $h(x)$ [estimated assuming that $h(x) = g(x)$, *i.e.*, transversity = helicity distribution]. Shown are the results for both u and d quarks.

inclusive DIS, Eq. (2), would be a very interesting observable for testing the mechanism of non-perturbative quark helicity flip in QCD.

A numerical estimate of the normal spin asymmetry in DIS kinematics was made in Ref. 3 using a light–front constituent quark model, in which DIS is described as elastic scattering from pointlike constituent quarks, and quark helicity–flip amplitudes are generated by the constituent quark mass. In order to have a self–consistent picture, this model was treated in the "composite nucleon" approximation, where we suppose that the quark transverse momenta, which are of the order of the inverse nucleon size, are parametrically small compared to the constituent quark mass,

$$\langle \boldsymbol{k}_T^2 \rangle \quad \sim \quad R_N^{-2} \quad \ll \quad M_q^2. \tag{4}$$

Physically, this corresponds to a picture of the nucleon as an assembly of weakly interacting, massive quarks. In this picture the quark helicity–flip contribution to the transverse spin asymmetry (which is proportional to M_q) dominates over the quark helicity–conserving one (proportional to $\langle \boldsymbol{k}_T^2 \rangle$) and can be calculated in a relativistic impulse approximation with on–shell quarks, which manifestly preserves electromagnetic gauge invariance and is free of QED collinear divergences. We emphasize that Eq. (4) is a theoretical idealization made in order to enable an internally consistent calculation of the two–photon exchange cross section, not necessarily

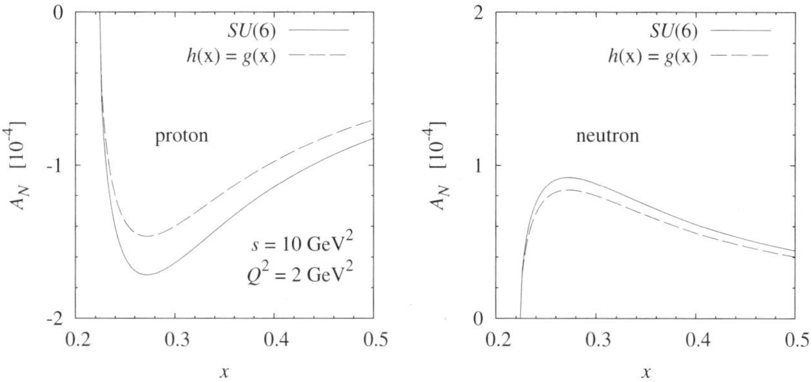

Fig. 4. The normal target spin asymmetry A_N obtained from the light–front constituent quark model with the the composite nucleon approximation, Eq. (4), with different assumptions about the spin–flavor wave function: (a) $SU(6)$ symmetry, (b) transversity = helicity distributions, $h(x) = g(x)$. The kinematics corresponds approximately to the planned Jefferson Lab Hall A experiment ($s = 2E_{\text{beam}}M + M^2$ is the squared electron–nucleon center–of–mass energy).

a reflection of actual nucleon structure. Figure 4 shows the results for the normal target spin asymmetry, Eq. (2), obtained in this picture with two different assumptions about the nucleon spin–flavor wave function: (a) $SU(6)$ symmetry, (b) quark transversity = helicity distributions, $h(x) = g(x)$. One sees that the asymmetry is of the order of 10^{-4}, with different sign for proton and neutron. The dependence on the kinematic variables is investigated further in Ref. 3.

The normal spin asymmetry in electron–neutron DIS is to be measured in a dedicated Jefferson Lab Hall A experiment[11] with a polarized ^{3}He target ($E_{\text{beam}} = 6\,\text{GeV}, x = 0.1 - 0.45, Q^2 = 1 - 3.5\,\text{GeV}^2$), with a projected absolute statistical error of $\delta A_N = 2$–4×10^{-4} in each of the four Q^2 bins. This measurement will improve the sensitivity of the only previous measurement at SLAC[12] by two orders of magnitude (in the SLAC experiment the asymmetry was found to be compatible with zero at the level of $\sim 3.5\%$). The statistical error of the planned JLab measurement is of the same order as the value of the asymmetry predicted by the composite nucleon approximation Eq. (4), see Fig. 4. However, as already stated, this approximation was made for technical reasons and can provide only a rough estimate of the expected asymmetry. More realistic dynamical model calculations of this observable are certainly needed. Work on this problem is in progress.

Acknowledgments

Supported in part by DOE, and by NSF under grants PHY–0114343 and PHY–0301841. Notice: Authored by Jefferson Science Associates, LLC under U.S. DOE Contract No. DE-AC05-06OR23177.

References

1. G. A. Miller and M. R. Frank, *Phys. Rev.* **C65**, 065205 (2002).
2. G. A. Miller, *Phys. Rev.* **C69**, 052201 (2004).
3. A. Afanasev, M. Strikman and C. Weiss, arXiv:0709.0901 [hep-ph].
4. A. V. Afanasev and N. P. Merenkov, *Phys. Lett.* **B599**, 48 (2004).
5. A. Metz, M. Schlegel and K. Goeke, *Phys. Lett.* **B643**, 319 (2006).
6. P. L. Anthony *et al.*, *Phys. Lett.* **B553**, 18 (2003).
7. X. Zheng *et al.*, *Phys. Rev.* **C70**, 065207 (2004).
8. A. V. Radyushkin and C. Weiss, *Phys. Lett.* **B493**, 332 (2000).
9. A. V. Radyushkin and C. Weiss, *Phys. Rev.* **D63**, 114012 (2001).
10. M. Penttinen, M. V. Polyakov, A. G. Shuvaev and M. Strikman, *Phys. Lett.* **B491**, 96 (2000).
11. X. Jiang *et al.*, Target normal single–spin asymmetry in inclusive DIS $n(e, e')$ with a polarized ^{3}He target, Jefferson Lab Hall A Experiment E–07–013.
12. S. Rock *et al.*, *Phys. Rev. Lett.* **24**, 748 (1970).

DEEPLY VIRTUAL PSEUDOSCALAR MESON PRODUCTION

VALERY KUBAROVSKY[1,2] and PAUL STOLER[1]
[1]*Rensselaer Polytechnic Institute,* [2]*Jefferson Lab*

Introduction

Deeply virtual exclusive reactions offer a unique opportunity to study the structure of the nucleon at the parton level as one varies both the size of the probe i.e. the photon virtuality Q^2 and the momentum transfer to the nucleon t. Such processes can reveal much more information about the structure of the nucleon than either inclusive electroproduction (Q^2 only) or elastic form factors ($t = -Q^2$).The characterization of deeply virtual exclusive reactions in terms of their common nucleon structure is one of the major objectives of the Jefferson Lab 12 GeV upgrade.

There have been two commonly used theoretical tools which relate exclusive reactions to the structure of the nucleon. At lower Q^2 where the probe is on the order of the size of hadrons, and the interactions are strong, Regge phenomena has proved effective. At high Q^2 the probe interacts perturbatively with individual quarks, and in the limit $Q^2 \to \infty$ and $-t/Q^2 \to 0$ the QCD factorization theorem[1] unifies all exclusive reactions in terms of their common nucleon structure encoded by generalized parton distributions (GPDs). In this article we consider deeply virtual meson production (DVMP), specifically the reaction $\gamma^* p \to pM$, where M is a meson $(\pi, \eta, \rho, \omega, \phi, etc)$. In the GPD approach, which is schematically shown in Fig. 1, the ingredients involve a hard interaction between a virtual photon and quark which produces a meson whose internal structure is given by the distribution amplitude $\Phi(z)$, and the remaining nucleon whose structure is represented by GPDs. A caveat is that the proof for factorization applies only to the case when the virtual photon has longitudinal polarization. In that case, in the limit $Q^2 \to \infty$ the cross section scales as $\sigma_L \sim 1/Q^6$, and the ratio as $\sigma_T/\sigma_L \sim 1/Q^2$.

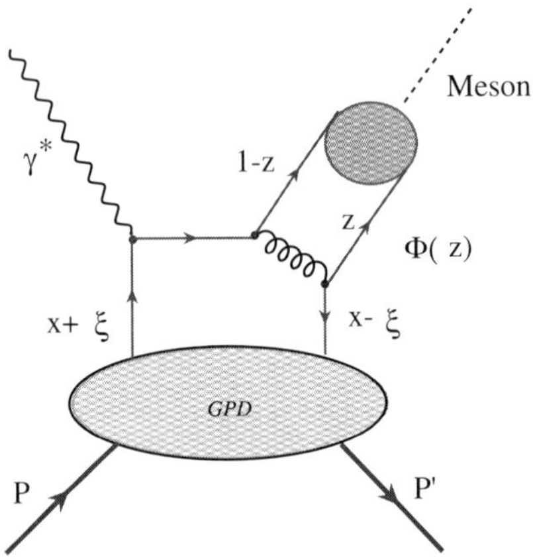

Fig. 1. Schematic illustration of the GPD approach to meson electroproduction.

While most theoretical work on the GPD approach has focused on the high Q^2 - low $|t|$ kinematic region, exclusive production of photons and mesons at large $|t|$ can also be described in terms the nucleon GPDs. Theory also predicts σ_L and σ_T in the high $-t$ low Q^2 region. For example for the pseudoscalar meson electroproduction we have:

$$\frac{d\sigma}{dt} \propto R(t) \int dz \Phi(z) f(z, s, Q^2, t)$$

$$R(t) \propto (e_u R(t)^u + e_d R(t)^d) \qquad R(t) = \int \frac{dx}{x} e^{\alpha t((1-x)/2x} \left[\Delta q(x) - \Delta \bar{q}(x)\right]$$

where $\Phi(z)$'s are the meson distribution amplitudes, $f(z, s, Q^2, t)$'s are parton level amplitudes and $R(t)$'s describe new form factors which are the $1/x$ moments of the GPDs. Δq and $\Delta \bar{q}$ are polarized parton and antiparton distribution for u, d and s quarks. The constant in the exponent α is approximately 1 GeV^{-2}. The Fourier transforms with respect to $\vec{\Delta}_\perp$, describes the correlation of the transverse spatial distribution of quark impact vs x_B in the proton.

Deeply virtual Compton scattering (DVCS) is the cleanest way of accessing GPDs. However, DVCS does not access the helicity dependent GPDs and it is difficult to perform a flavor separation. In the case of pseudoscalar meson production the amplitude involves the axial vector-type GPDs $\tilde{H}$ and $\tilde{E}$. These GPDs are closely related to the distributions of quark spin in the proton, and $\tilde{H}$ reduces to the polarized quark/antiquark densities in the limit of zero momentum transfer. Vector and pseudoscalar meson production allows one to separate flavor and isolate the helicity-dependent GPDs. This is summarized in Table 1.

Table 1. GPD and quark flavor selectivity of pseudoscalar and scalar meson electroproduction.

$\tilde{H}$, $\tilde{E}$	π^+	$\Delta u - \Delta d$
	π^0	$2\Delta u + \Delta d$
	η	$2\Delta u - \Delta d$
H, E	ρ^+	$u - d$
	ρ^0	$2u + d$
	ω	$2u - d$

The extraction of GPDs from electroproduction data is a challenging problem. A detailed understanding of the reaction mechanism is essential before one can compare with theoretical calculations. It is not yet clear at what values of Q^2 the application of GPDs to meson electroproduction becomes valid. However, detailed measurements of observables may test model-independent features of the reaction mechanism, such as t-slopes, flavor ratios, and generally by studying the variation of observables over a wide range of Q^2 and t. Even though current experiments are limited in Q^2 and t, it has been argued[2] that *precocious factorization* ratios of cross sections as a function of x_B could be valid at relatively lower Q^2, than for cross sections themselves. For example, the ratio of cross sections for π^0 and η electroproduction from a proton is related to the helicity structure of the quark flavors as

$$\pi^0/\eta = \frac{1}{2}\left[\frac{2}{3}\Delta u + \frac{1}{3}\Delta d\right]^2 / \frac{1}{6}\left[\frac{2}{3}\Delta u - \frac{1}{3}\Delta d + \frac{1}{3}\Delta s\right]^2$$

The results of a calculation of this ratio as a function of x_B is shown in Fig. 2.

Recent CLAS Measurements of π^0 and η Production

Cross section data for DVCS and DVMP have recently been obtained at Jefferson Lab with the CLAS spectrometer, up to a $Q^2 \sim 5$ GeV^2. This has been made possible by constructing a high quality electromagnetic calorimeter consisting of 424 lead-tungsten glass crystals covering an angular range 4.5° to 15°, which was positioned into the existing CLAS large acceptance detector. The pions and etas are identified through their 2γ decays. A photograph of the new detector and the 2γ invariant mass distribution is shown

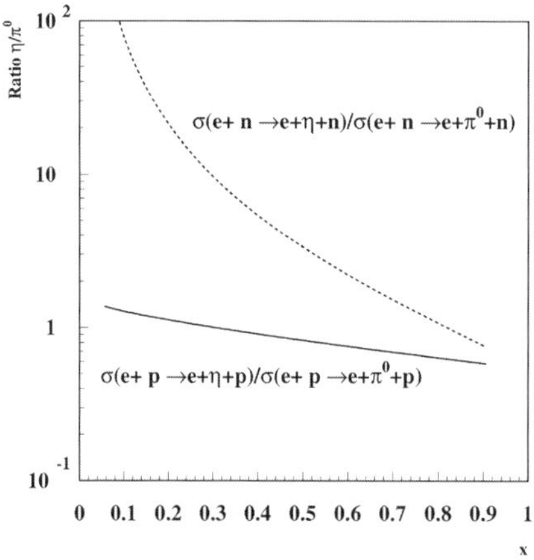

Fig. 2. Predictions of the ratio of cross sections for π^0 and η electroproduction from protons and neutrons[2] utilizing the concept of *precocious factorization*

in Fig. 3. One can see that the pions and etas are clearly observed, even before all final data selection cuts are performed. The kinematic coverage in the variables Q^2, x_B, t and W is shown in Fig. 4.

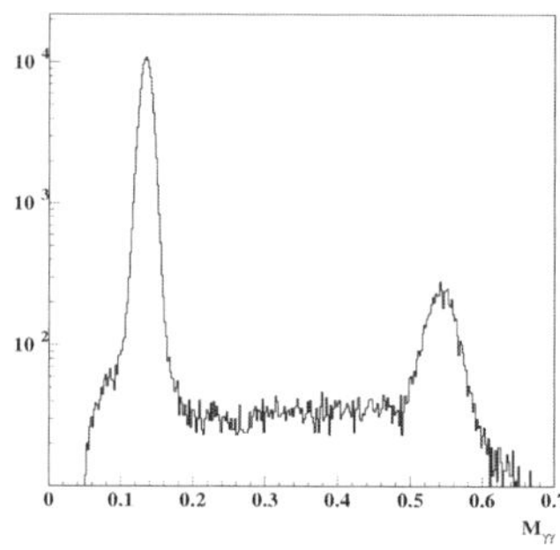

Fig. 3. Left: photograph of the new CLAS lead-tungsten electromagnetic calorimeter. Right: 2γ invariant mass spectrum in which the π^0's and η's are clearly observed (note log scale).

The virtual photon cross section can be written in well known notation as

$$\frac{d\sigma}{d\Omega_\pi} = \sigma_T + \epsilon\sigma_L + \epsilon\sigma_{TT} cos2\phi + \sqrt{2\epsilon(1+\epsilon)/2}\sigma_{LT} cos\phi_\pi + h\sqrt{\epsilon(\epsilon-1)}\sigma'_{LT} sin\phi \quad (1)$$

where h is the electron beam polarization.

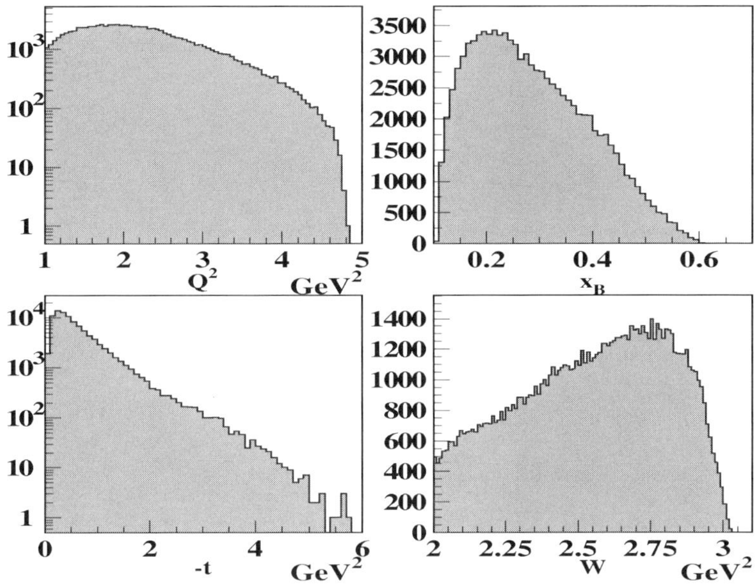

Fig. 4. The kinematic coverage in Q^2, t, x_B and W for neutral pions of the CLAS DVMP experiment.

The large acceptance of CLAS enabled the data to be grouped into intervals in Q^2, t x_B and ϕ. For unpolarized electrons ($h = 0$), separation of the ϕ dependence in moments of a *constant*, $cos\phi$, and $cos2\phi$ allows us to obtain $\sigma_T + \epsilon\sigma_L$, σ_{TT} and σ_{LT}. An example of a ϕ distribution for $t = 0.3$ GeV^2 integrated over Q^2 is shown in Fig. 5. The cross sections are in arbitrary units and radiative correction have not been applied.

The separated structure functions versus t for $Q^2 = 2.3$ GeV^2 is shown in Fig 6. It is observed that all the structure functions have significant non-zero values, and σ_{LT} is comparable to $\sigma_T + \epsilon\sigma_{LT}$ which implies at this relatively low values of Q^2 there are significant contribution of transverse amplitudes, so that factorization cannot be applied. However, one may analyze these data in terms of a hadron based models such as Regge phenomenology.[3] Figure 6 shows the result of such a calculation, which qualitatively follows the sign of the separated structure functions, but not always the shape.

The Fourier transformation of GPD gives information about the impact parameter dependence of parton distributions. The Fourier transformations are given by

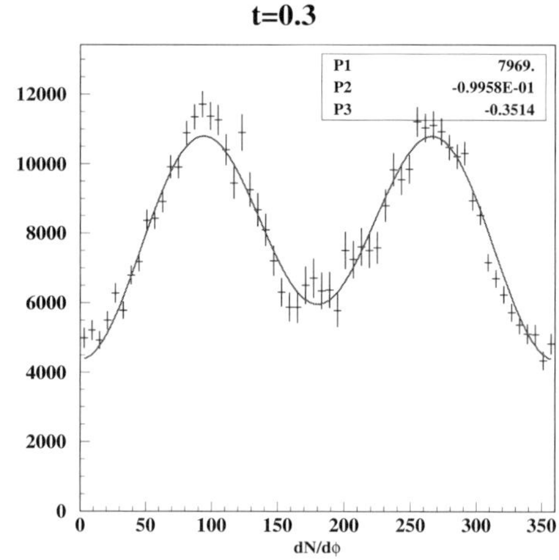

Fig. 5. The angular distribution $d\sigma/dt$ for $t = 0.3$ GeV^2 integrated over $Q^2 > 1$ GeV^2 and $W > 2$ GeV.

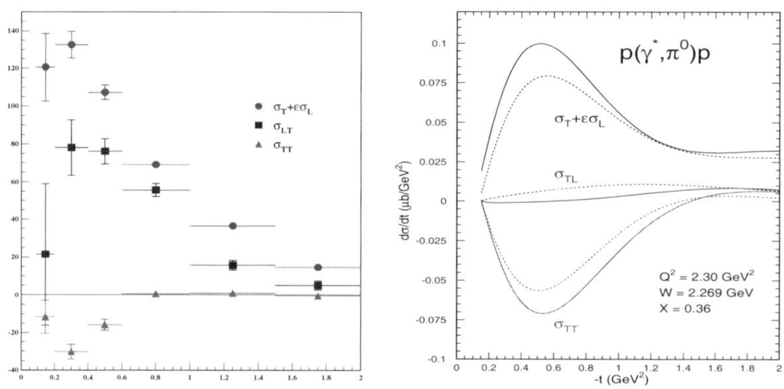

Fig. 6. Left: the separated structure functions $\sigma_T + \epsilon\sigma_L$, σ_{TT} and σ_{LT} as a function of $-t$ at $Q^2 = 2.3$ GeV^2 obtained with the CLAS spectrometer (arbitrary units). Right: the results of a Regge model calculation.[3]

$$F(x,\vec{b}) \propto \int d^2\Delta_\perp e^{i\Delta_\perp b_\perp} \tilde{H}(x,0,\Delta_\perp^2) \tag{2}$$

Due to the significant contribution of transverse amplitudes, at the current kinematics we do not have access to GPDs. However we can apply a Fourier transformation to the cross sections to get impact parameter information. Slope parameters B have been extracted by fitting the t distributions using the parametrization $d\sigma/dt \propto exp\left(B(x_B,Q^2)t\right)$. The result

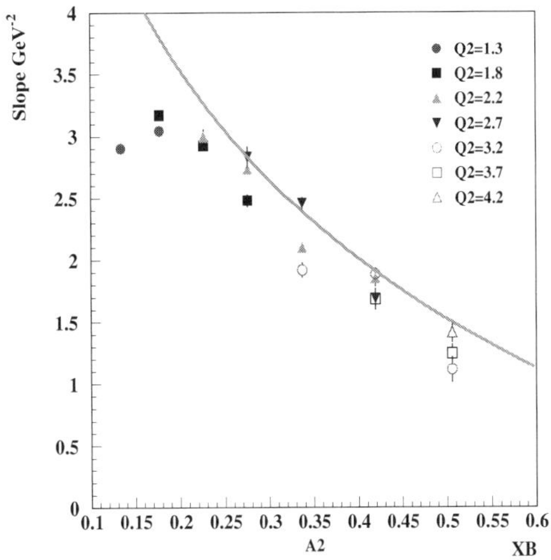

Fig. 7. The experimental slope parameters B obtained from fits to the data for various values of Q^2 and x_B with the function $d\sigma/dt \propto exp\left(B(x_B, Q^2)t\right)$. The solid curve is the Regge inspired parametrization $B(x_B) = 2\alpha ln(1/x_B)$ with $\alpha = 1.1$.

is shown in in Fig. 7. Note that B doesn't appear to significantly depend on Q^2.

In a Regge inspired GPD model, the x_B dependence of the slope parameter is given by $B(x_B) = 2\alpha ln(1/x_B)$, with $\alpha \sim 1$. The curve in Fig. 7 is a plot of this parametrization for B. Remarkably, this curve appears to accurately account for the data with no further parameters or normalization applied.

A fit to the CLAS data finds $\alpha = 0.9$. The resulting distribution is shown in Fig. 8. Note that the impact parameter distribution extracted from the data is broadest at lowest x_B, and becomes narrower at increasing x_B, with with $\langle b_\perp \rangle$ varying between 0.4 fm near $x_B = 0.1$, to 0.2 fm near $x_B = 0.5$. A further narrowing is expected at even higher x_B, approaching the expected zero in the limit $x_B \to 1$.

Beam spin asymmetry

The beam spin asymmetry (BSA) is defined by

$$A = \frac{\overrightarrow{\sigma} - \overleftarrow{\sigma}}{\overrightarrow{\sigma} + \overleftarrow{\sigma}} \sim \alpha sin\phi.$$

From eq. 1 the beam spin asymmetry directly yields the $L - T$ interference structure function σ'_{LT}. Any observation of a non-zero BSA would be indicative of an L-T interference. If σ_L dominates, then σ_{LT}, σ_{TT}, and σ'_{LT}

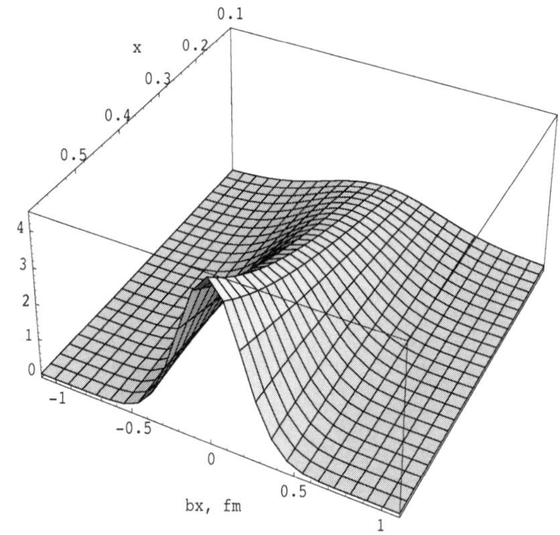

Fig. 8. The impact parameter distribution b_x as a function of x_B. Note that the impact parameter is broadest at lowest x_B, and becomes narrower at increasing x_B.

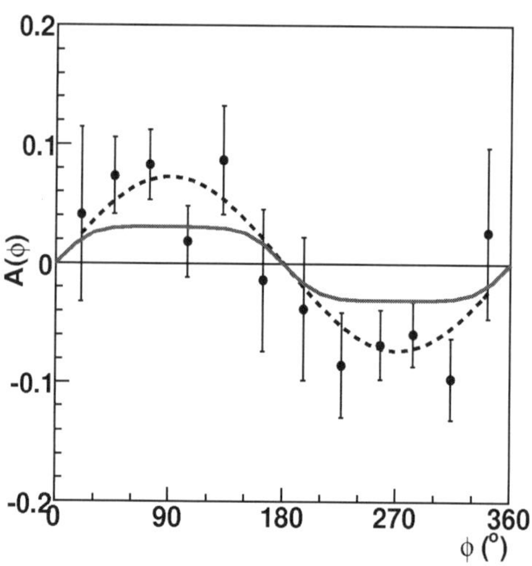

Fig. 9. The angular distribution of the BSA for π^0 at $Q^2 = 2$ GeV^2, $t = -0.3$ GeV^2, and $x_B = 0.25$. The dashed curve is a fit to the function $A = \alpha sin\phi$, and the solid curve is the result of a Regge model[3] calculation.

should be small. An example of a ϕ distribution of the BSA for π^0 and η production at a particular bin is shown in Fig. 9.

Sizable beam-spin asymmetries for exclusive π^0 and η mesons electroproduction have been measured above the resonance region for the first time. These non-zero asymmetries imply that both transverse and longitudinal amplitudes participate in the process. However, the results of a Regge model calculation qualitatively describe the experimental data.

Conclusion

Cross sections and asymmetries for the π^0 and η exclusive electroproduction in a very wide kinematic range of Q^2, t and x_B have been measured and initial analyses already are showing remarkable results. These data will help us to better understand the transition from soft to hard mechanisms. Initial results show that both transverse and longitudinal amplitudes participate in the exclusive processes at currently accessible kinematics. The π^0/η cross section ratio will check the hypothesis of precocious scaling.

We view the work presented here as leading into the program of the Jefferson Lab 12 GeV upgrade. The increased energy and luminosity will allow us to make the analysis presented here at much higher Q^2 and x_B as well as perform Rosenbluth L/T separations. In parallel, we pose the following theoretical questions. What does the t-slope $B(Q^2, x_B)$ tell us? What can we learn from the Q^2 evolution of the cross sections? Can measurement of σ_L, σ_T,σ_{LT}, σ_{TT} and $R \equiv \sigma_L/\sigma_T$ constrain GPDs within the approximations and corrections which have to be made due to non-asymptotic kinematics? How big are the corrections?

References

1. J. C. Collins, L. Frankfurt and M. Strikman, Phys. Rev. D **56**, 2982 (1997) [arXiv:hep-ph/9611433].
2. M. I. Eides, L. L. Frankfurt and M. I. Strikman, Phys. Rev. D **59**, 114025 (1999).
3. J-M. Laget, private communication.

EXCLUSIVE ρ^0 ELECTROPRODUCTION ON THE PROTON: GPDS OR NOT GPDS?

M. GUIDAL*
Institut de Physique Nucléaire d'Orsay
91405 Orsay, FRANCE
E-mail: guidal@ipno.in2p3.fr

S. MORROW
Institut de Physique Nucléaire d'Orsay
91405 Orsay, FRANCE
& CEA-Saclay, Service de Physique Nucléaire
91191 Gif- sur-Yvette,Cedex, FRANCE

We discuss the interpretation of the $ep \to ep\rho^0$ process in terms of, on the one hand, Generalized Parton Distributions and, on the other hand, an effective hadronic model based on Regge theory.

Keywords: Nucleon structure; Generalized Parton Distributions; ρ^0 electroproduction

The $\gamma^{(*)}p \to p\rho^0$ process is, due to its relatively high cross section over a wide range of energy (from threshold at $W \approx 1.75$ GeV up to $W \approx 200$ GeV, where W is the center of mass energy of the (γ, p) system), one of the exclusive processes on the proton the most studied experimentally and, consequently, also theoretically.

In photoproduction (and above the nucleon resonance region), the reaction is understood to proceed, at low W, through the exchange of mesons (σ, f_2,...) in the so-called t-channel (under the form of Regge trajectories) and, at high W, through the exchange of the Pomeron trajectory, which carries the quantum numbers of the vacuum (see Fig. 1). These diagrams are based on *hadronic* degrees of freedom and the complex physics of QCD ("Quantum Chromo-Dynamics") is absorbed into effective meson-nucleon (and electromagnetic meson) coupling constants (plus potentially additionnal hadronic form factors).

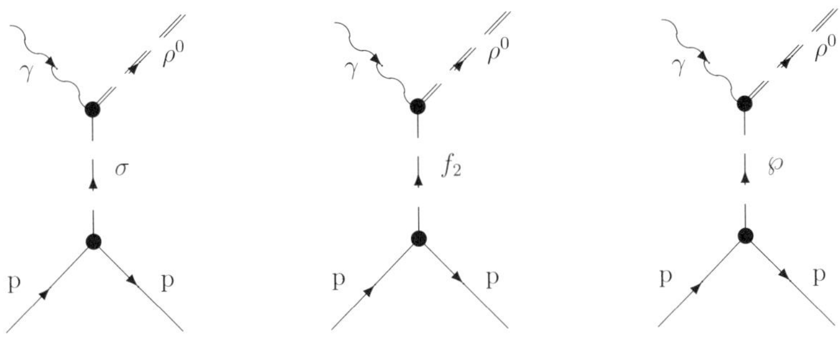

Fig. 1. The dominant t-channel meson exchange diagrams for the reaction $\gamma p \to p\rho^0$.

Going to electroproduction, i.e. increasing Q^2, the virtuality of the initial photon, allows to probe shorter and shorter distances, and be sensitive to *partonic* (quarks and gluons) degrees of freedom. Then, at sufficiently high Q^2, the process should be understable in terms of the so-called "handbag" diagrams illustrated on Fig. 2, where the incoming virtual photon scatters directly off a quark, and whose interaction can be calculable in the framework of perturbative QCD. The complex non-perturbative QCD structure of the nucleon is then represented by (quark and gluon) Generalized Parton Distributions (GPDs), which were shown to factorize with the QCD perturbative process in the handbag process for incoming longitudinal photons in this case.[1]

The GPDs might actually provide, to this day, the most complete information accessible on the structure of the nucleon as they describe the (correlated) spatial and momentum distributions of the quarks in the nucleon (including the polarization aspects), its quark-antiquark content, a way to access the orbital momentum of the quarks, etc... We refer the reader to Refs. [2–4], for example, for very detailed and quasi-exhaustive reviews on the GPD formalism and the definition of some of the variables that will be employed in the following.

The question then arises as to which Q^2, one or another (hadronic or partonic) description is valid and applicable. A first very simple image that one can form is that the *quark* handbag diagram (Fig. 2 left) might be associated to the meson Regge exchange mechanism (Fig. 1 left and center) while the *gluon* handbag diagram (Fig. 2 right) might correspond to the exchange of the quantum numbers of the vacuum (as gluons form an isosinglet state) and thus to the Pomeron exchange mechanism. Indeed, "gluonic" GPDs, which obviously reflect the glue content of the nucleon,

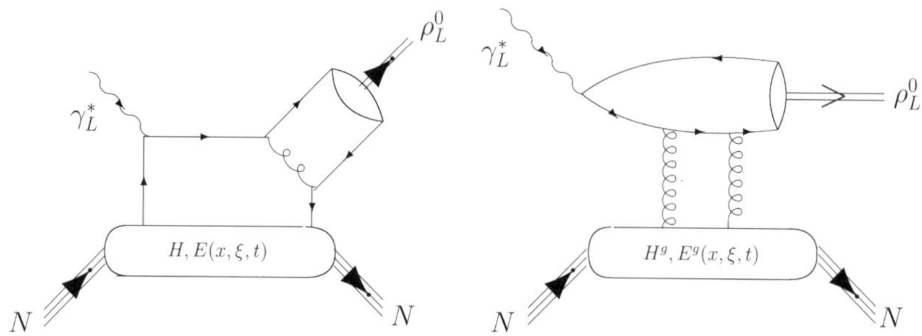

Fig. 2. The "handbag" diagrams for (longitudinal) vector meson leptoproduction. "Quark" GPDs are at play on the left and "gluon" GPDs on the right.

are expected to contribute at high W, i.e. the low x_B "sea" domain, while the "quark" GPDs should be predominant at low W, i.e. the large x_B "valence" domain, where x, the momentum fraction of the quarks or gluons with respect to the nucleon can be related to W by $x_B \approx \xi = \frac{Q^2}{W^2+Q^2-m_N^2}$.

This correspondence is not fully exact in the sense that the *quark* handbag diagram can also carry the quantum numbers of the vacuum (when sea quarks are exchanged) and thus contribute to a Pomeron-type mechanism. Also, GPDs are not only t-channel mechanisms (which are just the $|x| < \xi$ part of the GPDs), there is also a part associated to standard "diagonal" Parton Distribution Functions (PDFs), which is not contained in t-channel Regge exchanges.

Experimentally, in the high W domain, data for the exclusive electroproduction of the ρ^0 meson have been obtained at several Q^2 values by the H1 and ZEUS (at HERA), NMC (at CERN) and E665 (at Fermilab) experiments. It was shown by Frankfurt et al.[5] that these high W data can be well interpreted in terms of the 2-gluon exchange mechanism of Fig. 2 right. However, in order to achieve this, as the Q^2 domain where the data exist is relatively low, corrections have to be incorporated to the pure leading twist calculation of the handbag diagrams. The straightforward leading twist calculation leads, at low Q^2 values, to an overestimation of the data by an important factor (up to 2 at Q^2=10 GeV^2 and higher at smaller Q^2). Also, the leading twist calculation predicts that $\frac{d\sigma_L}{dt}$ should behave as $\frac{1}{Q^6}$, at fixed x_B and a flatter Q^2 dependence is observed in the data.

The origin of the overestimation of the data lies in the presence of the gluon exchange and the associated strong coupling to quarks α_s in the handbag diagrams of Fig. 2. Indeed, α_s is "running" and, in particular, is rising as Q^2, or more generally, the scale associated to the quark-gluon vertex (to which Q^2 is proportional), is decreasing. Also, the gluon exchange is associated to a propagator of the form $\frac{1}{zQ^2}$, where z is the momentum fraction carried by the quark(s) interacting with the gluon. As z runs between 0 and 1, at low z, this can lead to strong enhancements of the cross section.

The handbag diagrams deal mainly with longitudinal degrees of freedom $(x, \xi, ...)$. The idea brought by Frankfurt et al. to remedy the "overshooting" of the leading twist calculation was to also take into account the *transverse* momentum ("$k_\perp$") dependence in the calculations of the handbag processes. This approach, also called the "modified parturbative approach", was originally initiated in Ref. [6]. The consequence is that there will always be a "minimum" momentum scale (the average transverse momentum) and therefore singularities of the type $z \to 0$ will be much attenuated. With such prescription, it was found a remarkable agreement for the x_B (mainly driven by the x dependence of the gluon distributions) and the Q^2 (asymptotic $\frac{1}{Q^6}$ modulated by the "$k_\perp$" correction factor) dependences of the $ep \to ep\rho^0$ cross section (in addition to the absolut normalization).

We have just discussed here the high W domain where, in summary, the ρ^0 electroproduction data looks interpretable in terms of the 2-gluon exchange mechanism of Fig. 2 right (modulo "$k_\perp$" corrections). Now, in the low W region, is, similarly, the quark handbag diagram of Fig. 2 left (modulo "$k_\perp$" corrections) also at play ? There are very few existing data for the $ep \to ep\rho^0$ process in this low W kinematical regime : early data (1976) with a 7.2 GeV electron beam at DESY[7] and with a 11.5 GeV electron beam at Cornell[8] (1981) and more recently, with a 470 GeV muon beam at Fermilab,[9] a 27 GeV positron beam at HERMES,[10] and a 4.2 GeV electron beam at JLab[11] with the CLAS detector. Currently, the experiment E99-105[13] which recently took data with a 5.75 GeV electron beam with the CLAS detector is in the final stage of data analysis and we will have here a quick snapshot to some PRELIMINARY results.

The amplitude for the "quark" handbag diagram of Fig. 2 left (with "$k_\perp$" corrections) has been calculated, for instance, in Ref. [14]. The main input to this calculation is the parametrization of the GPDs. As a first approximation, only $H(x,\xi,t)$ is considered in the following. The (x,ξ) dependence of H is parametrized in terms of the Double Distributions ansatz[15,16] and the (x,t) correlation is based on Regge theory[17] :

$$GPD^q(x,\xi,t) = \int d\alpha d\beta \delta(x-\beta-\xi\alpha) DD(\alpha,\beta,t) \tag{1}$$

with :

$$DD(\alpha,\beta,t) = h(\beta,\alpha) q(\beta) \beta^{-\alpha'(1-\beta)t} \tag{2}$$

and

$$h(\beta,\alpha) = \frac{\Gamma(2b+2)}{2^{2b+1}\Gamma^2(2b+1)} \frac{\left[(1-\mid\beta\mid)^2-\alpha^2\right]^b}{(1-\mid\beta\mid)^{2b+1}} \tag{3}$$

the only free parameters being α' which is fitted, and strongly constrained, to reproduce the nucleon form factors[17] and b which governs the (x,ξ) dependence but has, in effect, relatively little influence on the cross section calculation.

As a second approximation, in the following, the two (quark and gluon) handbags of Fig. 2 are summed at the *cross section* level (note that, in Ref. [18], this sum is treated at the *amplitude* level ; it has been checked that, except in the intermediate W region where interference is maximal, the two -independent- calculations are in remarkable agreement).

Fig. 3 shows the total *longitudinal* cross section for the exclusive ρ^0 electroproduction on the proton as a function of W, over a wide range, for $Q^2 \approx 2.35$ GeV^2, with the current world's data. The cross sections exhibit clearly two different behaviors as a function of W : starting from low W, the cross section decreases with W. Then, at W around ≈ 10 GeV, the W slope changes and the cross section slowly raises. The handbag calculation (sum of the quark and gluon processes), just described above, is shown by the dashed curve and gives a decent description of the high and intermediate (down to ≈ 6 GeV) W region. This result was already observed by the HERMES collaboration.[10] In particular, at high W, the rise of the cross section is due to the gluon and sea contributions.

Now, focusing on lower W values, the GPD calculation clearly misses the data. This discrepancy can reach an order of magnitude at the lowest

W values. The trend of the GPD calculations is to decrease as W decreases whereas the data increase. The trend of the GPD calculation is readily understandable : GPDs are approximatively proportional to the forward quark densities $q(x)$ (the relation is not so direct since, among other aspects, the quark densities are, through the Double Distributions, convoluted with a meson distribution amplitude but, still, the main trends remain). And, as x increases (W decreases), they tend to 0 ($q(x) \approx (1-x)^3$ for x close to 1). There might be a slight local increase around $x \approx 0.3$, due to the valence contribution (which is slightly apparent on Fig. 3) but it can never explain an increase of an order of magnitude, especially at low W.

Fig. 3 also shows (dotted curve) the results of the calculation[19] based on the "hadronic" t-channel diagrams of Fig. 1 which are "reggeized". This is the co-called JML model and it reproduces fairly well the two general behaviors with W just mentionned. Here, the drop of the cross section at low W is due to the t-channel σ and f_2 meson exchange diagrams (the intercept $\alpha(0)$ of the f_2 trajectory is ≈ 0.5 and therefore the cross sections decrease with energy as $\frac{1}{s^{0.5}}$). Then, the (slow) rise of the cross section which is observed after W $\approx$ 10 GeV is attributed to the Pomeron trajectory which has an intercept $\alpha(0) \approx 1+\epsilon$.

Coming back to the GPD calculation, the conclusion that one can draw is then two-fold :

- Either the handbag mechanism formalism is not at all the dominant mechanism in the low W (valence) region and higher twists or (so far) incontrolable non-perturbative effects submerge the handbag mechanism. Now, it remains to explain why the handbag mechanism works in the high/intermediate W (low x) domain and, quite abruptly, no more in the valence region. Higher twist can certainly depend on energy but such a strong variation with W is certainly puzzling.
- Or, an alternative explanation, based on and supported by the success of the handbag mechanism at high and intermediate W values, is that the GPD formalism is indeed at work in the valence region but then, a significant and fundamental contribution, besides Double Distributions, in our parametrization of the GPDs is missing.

What could be such a missing contribution in the low "W" regime? In the framework of the JML model, as we stated, the strong rise of the cross section as W decreases, is due to the t-channel σ and f_2 meson exchange processes. It could then be tempting to associate the potentially missing

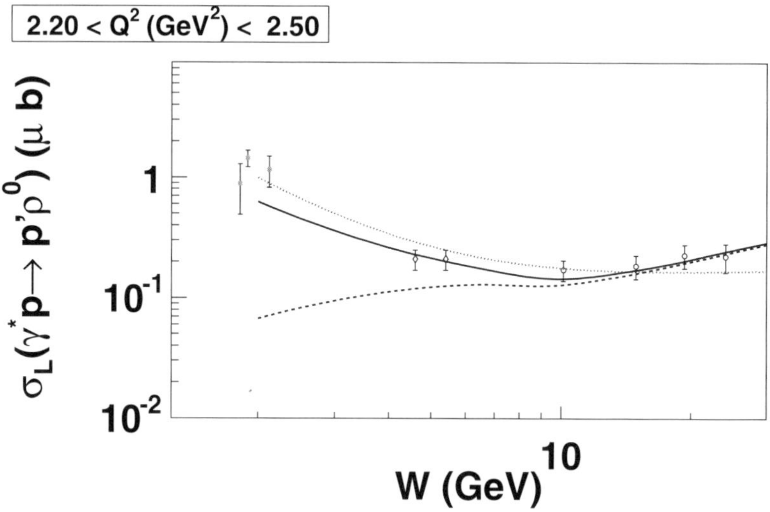

Fig. 3. W dependence of the $\gamma_L p \to p\rho^0$ at $Q^2 \approx 2.35$ GeV^2. Dashed curve : the "standard" handbag calculation (quark and handbag diagrams added incoherently) based on the Double Distribution ansatz for the quark GPDs. The thin solid curve is the result of the calculation including a ("renormalized") D-term-like contribution. The dotted curve is the result of the Regge JML calculation.

piece in the GPDs to t-channel meson exchanges. The best example of such contribution in GPDs is the so-called D-term which was originally introduced in Ref. [20] where it was shown that it was required to introduce such a term in the most general parametrisation of GPDs, in addition to Double Distributions, in order to satisfy the polynomiality rule. We recall that the D-term is non-zero only in the $-\xi < x < \xi$ domain, it is odd in x and it is usually parametrized in terms of Gegenbauer polynomials of argument $\frac{x}{\xi}$. The $-\xi < x < \xi$ region corresponds to the $q\bar{q}$ component of the GPDs and therefore the D-term can be thought of as representing the exchange of mesonic degrees of freedom in the t-channel. We see that a structure that exists only in the $-\xi < x < \xi$ domain would naturally provide a contribution that decreases with W, since $W \approx \frac{1}{\xi}$: in other words, as W increases, ξ decreases and therefore the support of such structure decreases and, as a consequence, its contribution diminishes (up to be null in the extreme $\xi = 0$ limit).

The D-term is associated to a Lorentz *scalar* structure in the GPD definition and thus accounts only for *scalar* (isosinglet) meson exchange (as the σ meson). Other meson contributions (for instance, tensor as the f_2 meson) are not included in the D-term. However, there is, in principle, in the GPD formalism, no reason to restrict these $q\bar{q}$ contributions to the scalar component as it is imposed to the D-term and there is freedom to add more general $q\bar{q}$ contributions, which can be an argument, in the following, to readjust its normalization.. As such contribution "exist" only in the $-\xi < x < \xi$ region (i.e. they vanish at $\xi = 0$), they are not sensitive to the relation linking GPDs to quark densities at ξ=0. Furthermore, depending on their parity in x, these contributions might or not contribute to the sum rules linking the GPDs to form factors (FFs) -for instance, the D-term is odd in x and doesn't contribute to the form factor sum rule-. One sees that, in a very general way, there can be contributions to the GPDs which completely escape any normalization or constraint.

The solid curve shows the results of a calculation with such a D-term which has been renormalized in order to fit the data. It is of course not very satisfying to add such an unconstrained, in terms of normalization, term to the GPDs parametrization. The purpose here is simply to illustrate that such a term provides just the right W dependence and is physically motivated in the sense that it is meant to parametrize $q\bar{q}$ contributions, which are definitely part of the GPD concept. Now, the question of course remains if one can justify such a strong contribution (the normalization of the D-term such as given in Ref. 20] barely changes the dashed curve on Fig. 3).

The CLAS experiment E99-105, which is in the final stage of analysis, will soon bring more than 25 new (Q^2,x_B) data points in this low W domain, along with differential cross sections. For instance, as a snapshot, Fig. 4 shows the PRELIMINARY distribution dN/dt of events for the $\gamma_L p \to p\rho^0$ process. As the analysis is still in progress, the normalization is in arbitrary units but the distributions are, for the most part, corrected for the acceptance of CLAS. One readily sees the large (Q^2,x_B) domain spanned by these data. What is also obvious is that the slope of these t distributions vary with the kinematics and, in particular, increases with decreasing x. In the framework of GPDs, this is a feature which is expected as the t dependence of the GPDs can be related,[21–23] via a Fourier transform, to the transverse spatial distribution of the partons in the nucleon. A general "3-dimensional" image of the nucleon is then that "low x" sea quarks (which can be associated to the "pion cloud") sit at the periphery of the nucleon (corresponding to large impact parameters and therefore large t slopes) while "large x" valence quarks sit in the core of the nucleon (corresponding to low impact parameters and therefore low t slopes). The evolution with x of the t slopes of Fig. 4 is in agreement with this image.

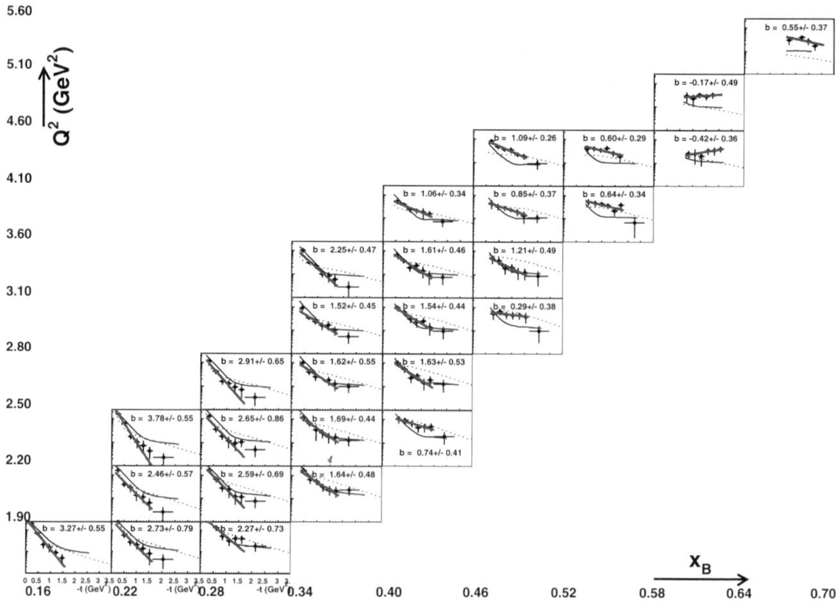

Fig. 4. PRELIMINARY distribution dN/dt of events for the $\gamma_L p \to p\rho^0$ process from the CLAS E99-105 experiment. Units are arbitrary on the y axis. All x and y axis have the same scales in each plot. The thick solid line is an exponential fit Ae^{bt} to the data yielding the b slope parameters. The thin solid line is the result of a GPD calculation including a ("renormalized") D-term-like contribution. The dotted line is the result of the Regge JML calculation.

Finally, if a D-term like contribution is responsible for most of the cross section in this low W domain, it remains the "technical" problem to reconcile a varying t slope (with energy) with a D-term as the polynomiality rule imposes that the t dependence of the D-term be factorized (independent of x and ξ). We propose here a simple ansatz which generalizes the D-term :

$$GPD'(x,\xi,t) = \xi \int d\alpha d\beta \delta(x - \beta - \xi\alpha) DD'(\alpha,\beta,t) \tag{4}$$

with

$$DD'(\alpha,\beta,t) = \alpha h(0,\alpha) \frac{b' \mid t \mid}{\mid \beta \mid^{b't+1}} \tag{5}$$

which has the nice feature to recover the D-term in the forward limit since :

$$\lim_{|t|\to 0} \frac{b' \mid t \mid}{\mid \beta \mid^{b't+1}} = \delta(\beta) \tag{6}$$

The α factor allows to respect the polynomiality rule.

The thin solid line on Fig. 4 shows the result of the calculation with such an ansatz and illustrates the variation of the slope with x (the t slope is, additionally, also non-linear in this calculation). The dashed line is the JML calculation (which uses "saturating" Regge trajectories) and the thick solid line is a simple exponential fit to the data, yielding the b slope parameters which are on the figure.

Let us note the almost flat t slopes at large x. Such flat t slopes are basically impossible to reproduce with the standard double distributions contributions to the H and E GPDs. Indeed, their t-dependence is strongly constrained by the sum rule linking the GPDs to the FFs. If one is indeed sensitive to GPDs in this low W regime, the very flat t slopes observed can only arise from GPD contributions not constrained by the FF sum rule.

Let us finally mention that, at large x_B, the minimum value of t kinematically accessible is large (i.e. ≈ 1.6 GeV^{-2} at $< x_B >= 0.67$). A possible explanation of the disagreement between the "base" (i.e. without additional D-term-like t-channel meson exchange contribution) GPD model, could simply be that, at such large t values, higher twists contributions can be extremely important.

Only additionnal and precise data for the $\gamma_L p \to p\rho^0$ process will allow to distinguish between the two hypothesis raised above : GPDs or not GPDs in this process?

With an important new set of data soon to be released, the E99-105 experiment will already allow to study in a more detailed manner the mechanims at play in this reaction. In the longer term, the upgrade of JLab to 12 GeV, will permit a precise mapping of the Q^2, x_B and t dependences over a broad phase space. This should bring, along with the measurement of new observables, such as the transversely polarized target asymmetry, a more definite answer as to the role or not of GPDs in the valence region for exclusive ρ^0, and more generally, meson production.

Acknowledgment

The authors thank M. Vanderhaeghen, M. Diehl, D. Mueller, B. Pire and M. Polyakov for useful discussions.

References

1. J.C. Collins, L. Frankfurt and M. Strikman, Phys. Rev. D **56**, 2982 (1997).
2. K. Goeke, M. V. Polyakov and M. Vanderhaeghen, Prog. Part. Nucl. Phys. **47**, 401 (2001).
3. M. Diehl, Phys. Rept. **388**, 41 (2003).
4. A.V. Belitsky, A.V. Radyushkin, Phys. Rept. **418**, 1 (2005).
5. L. Frankfurt, W. Koepf and M. Strikman, Phys. Rev. D **54**, 3194 (1996).
6. J. Botts and G. Sterman, Nucl. Phys. B **325**, 62 (1989).
7. P. Joos *et al.*, Nucl. Phys. B **113**, 53 (1976).
8. D.G. Cassel *et al.*, Phys. Rev. D **24**, 2787 (1981).
9. M. R. Adams *et al.*, Z. Phys. C **74**, (1997) 237.
10. A. Airapetian *et al.*, Eur. Phys. J. C **17**, 389 (2000).
11. C. Hadjidakis *et al.*, Phys. Lett. B **605**, 256 (2005).
12. L. A. Ahrens *et al.*, Phys. Rev. Lett. **42**, 208 (1979).
13. JLab approved experiment E-99-105 : "Deeply Virtual Electroproduction of Vector Mesons", spokespersons : M. Garçon, M. Guidal and E. Smith.
14. M. Vanderhaeghen, P.A.M. Guichon, M. Guidal, Phys. Rev. Lett. **80**, 5064 (1998), Phys. Rev. D **60**, 094017 (1999).
15. D. Muller, D. Robaschik, B. Geyer, F. M. Dittes and J. Horejsi, Fortsch. Phys. **42**, 101 (1994).
16. A.V. Radyushkin, Phys. Rev. D **59**, 014030 (1999); Phys. Lett. B **449**, 81 (1999);
17. M. Guidal, M. V. Polyakov, A. V. Radyushkin and M. Vanderhaeghen, Phys. Rev. D **72**, 054013 (2005).
18. S. V. Goloskokov and P. Kroll, Eur. Phys. J. C **42**, (2005) 281 ; Eur. Phys. J. C **50**, (2007) 829.
19. J.-M. Laget, Phys. Lett. B **489**, (2000) 313 ; F. Cano and J.-M. Laget, Phys. Rev. D **65**, (2002) 074022.
20. M. Polyakov and C. Weiss, Phys. Rev. D **60**, 114017 (1999).
21. M. Burkardt, Phys. Rev. D **62**, 071503 (2000) [Erratum-ibid. D **66**, 119903 (2002)] ; Int. J. Mod. Phys. A **18**, 173 (2003).
22. M. Diehl, Eur. Phys. J. C **25**, 223 (2002) [Erratum-ibid. C **31**, 277 (2003)].
23. J. P. Ralston and B. Pire, Phys. Rev. D **66**, 111501 (2002).

ρ^0 TRANSVERSE TARGET SPIN ASYMMETRY AT HERMES

A. AIRAPETIAN
on behalf of the HERMES Collaboration
Randall Laboratory, University of Michigan,
Ann Arbor, Michigan 48109-1040, USA
E-mail: avetik@umich.edu

Preliminary measurements of the Transverse Target Spin Asymmetry, $A_{UT}(\phi)$, for hard exclusive electro-production of ρ^0 mesons at HERMES are presented. Information about the poorly known GPD E can be obtained through measurement of the transverse target spin asymmetry $A_{UT}^{\rho L}(\phi)$ in the longitudinal cross section $\sigma_L(\gamma_L^* p \to \rho_L^0 p\prime)$. Using the self-analyzing character of ρ^0 meson decay, both $A_{UT}^{\rho L}(\phi)$ and $A_{UT}^{\rho T}(\phi)$ were extracted from HERMES data. The TTSA $A_{UT}^{\rho L}(\phi)$ was compared with GPD model predictions.

Keywords: meson; spin; asymmetry; electro-production.

1. Introduction

The recently developed theoretical framework of Generalized Parton Distributions (GPDs) is believed to describe the three-dimensional structure of the nucleon.[1] Knowledge about GPDs will yield valuable information about the total angular momentum carried by partons in the nucleon.[2] Hard exclusive processes can be used to investigate GPDs. In particular it was shown[3] that the Transverse Target Spin Asymmetry (TTSA) A_{UT} in the production of longitudinally polarized ρ_0 mesons (ρ_L^0) on protons is sensitive to the GPDs H and E. A direct connection between the partons' total orbital angular momentum and the A_{UT} asymmetry in ρ_L^0 production was established[4] , allowing predictions based on the VGG code[5] with different ansätze for the parametrization of GPDs.[3]

2. Hard Exclusive ρ^0 Meson Production

The HERMES experiment and spectrometer are described in detail elsewhere.[6] Exclusive ρ^0 events, produced by scattering of 27.6 GeV electrons (positrons) on a transversely polarized hydrogen target, were identified by detecting the scattered lepton in coincidence with two oppositely charged hadrons (h^+h^-), assuming they are pions ($\pi^+\pi^-$). A mass window on the invariant mass of $0.6 < M_{\pi\pi} < 1.0$ GeV was used to identify resonant ρ^0 production, because of the near 100% branching ratio of $\rho^0 \to \pi^+\pi^-$. In addition, the missing energy ΔE was required to be between -2.0 and 0.6 GeV assuring the exclusivity of the process, and assuming the undetected recoiling target to be a proton. Due to the limited mass resolution in the missing mass spectrum, the exclusive reaction of interest $ep \to p\rho^0 \to p\pi^+\pi^-$ was contaminated by semi-inclusive pion-pair production. The magnitude of this contamination was estimated using a PYTHIA Monte Carlo simulation, and the asymmetries were corrected for this approximately 12% background. After correction of the non-exclusive background, the $\pi^+\pi^-$ invariant mass spectrum shows only a few percent of non-resonant background, which was neglected. From the exclusive sample, the TTSA was determined according to

$$A^l_{UT}(\phi, \phi_s) = \frac{1}{P_T} \frac{d\sigma(\phi,\phi_s) - d\sigma(\phi,\phi_s+\pi)}{d\sigma(\phi,\phi_s) + d\sigma(\phi,\phi_s+\pi)} , \qquad (1)$$

with production kinematics and angles defined in Fig. 1,

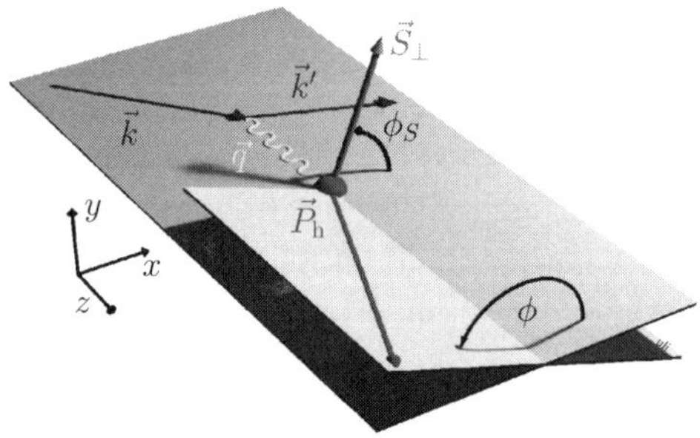

Fig. 1. Definition of azimuthal angles ϕ and ϕ_s .

where the target polarization P_T is defined with respect to the lepton beam direction, and ϕ and ϕ_s are the azimuthal angles of the ρ^0 meson and the target spin direction about the virtual photon. Note that there is a difference in the definition of the TTSA of Ref. [3] $\mathcal{A}_{GPV}$, and the one used here, A_{UT}, given by

$$A_{UT} = -\frac{\pi}{2}\mathcal{A}_{GPV} . \qquad (2)$$

The asymmetry moment $A_{UT}^{sin(\phi-\phi_s)}$ is determined using an unbinned maximum likelihood fit.

3. Separation of Two Helicity States, $\rho^0_{\rm L}$ and $\rho^0_{\rm T}$

Using the fact that the ρ^0 meson decay is self-analyzing and the decay angular distribution has a characteristic dependence on the ρ^0 helicity (polar angle of π^+, θ_π, in the ρ^0 rest frame), the ρ^0 production cross section can be factorized in kinematic and angular terms.[7] If the cross section is integrated over the azimuthal decay angle of π^+ in the ρ^0 rest frame ,ϕ_π, the interference terms between different helicities of ρ^0 production are canceled and the angular distribution $W(\cos\theta_\pi, \phi, \phi_s)$ can be written as:

$$W(\cos\theta_\pi, \phi, \phi_s) \propto$$
$$\Big[\cos^2(\theta_\pi)\hat{W}_{UU}^{\rho_L}\Big(1 + P_T A_{UT}^{\rho_L}(\phi, \phi_s) + A_{UU}^{\rho_L}(\phi)\Big) + \qquad (3)$$
$$\frac{1}{2}\sin^2(\theta_\pi)\hat{W}_{UU}^{\rho_T}\Big(1 + P_T A_{UT}^{\rho_T}(\phi, \phi_s) + A_{UU}^{\rho_T}(\phi)\Big)\Big] ,$$

where $A_{UT}^{\rho_L}(\phi, \phi_s)$ and $A_{UT}^{\rho_T}(\phi, \phi_s)$ are the TTSA moments for longitudinal and transverse ρ^0 production, respectively, which are extracted simultaneously using the unbinned maximum likelihood method. The unpolarized angular distributions $\hat{W}_{UU}$ and the asymmetries A_{UU} of ρ^0_L and ρ^0_T are defined by unpolarized spin density matrix elements[8] already measured at HERMES .

4. Results

Transverse target spin asymmetries for longitudinal $(A_{UT}^{\rho_L}(\phi))$ and transverse $(A_{UT}^{\rho_T}(\phi))$ ρ^0 production have been extracted, as shown in Fig. 2. The rightmost column in Fig. 2 reveals that there might be a different t' dependence for ρ^0_L and ρ^0_T.

Since currently model predictions exist only for ρ^0_L, which are produced by longitudinal photons, only the TTSA $A_{UT}^{\rho_L}(\phi)$ can be compared with GPD model predictions. This is shown in Fig. 3, where the present data favor positive values for J^u, the total angular momentum for u-quarks[4] , in agreement with a HERMES measurement[9] based on the DVCS process.

Acknowledgment

I highly acknowledge W. Lorenzon and W.- D. Nowak for a crytical reading of this manuscript. This research is supported by the U.S. National Science Foundation, under grant PHY-055423.

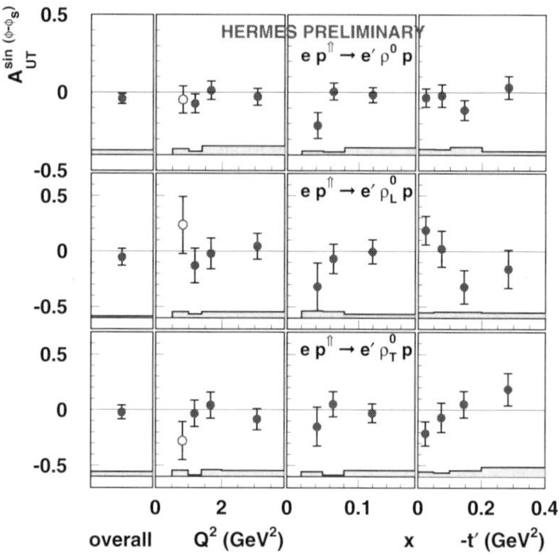

Fig. 2. The overall value, Q^2, x_B, and t' dependences of the $A_{UT}^{sin(\phi-\phi_s)}$ moment of the TTSA of exclusive ρ^0, ρ_L^0 and ρ_T^0 production.

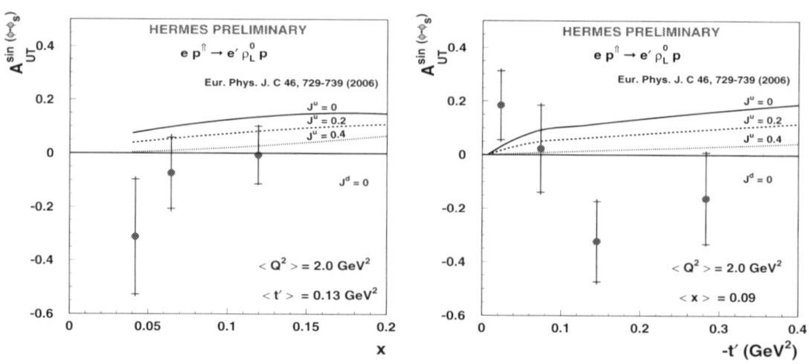

Fig. 3. x_B and t' dependences of $A_{UT}^{sin(\phi-\phi_s)}$ moment of TTSA of ρ_L^0 compared to model calculations.

References

1. D. Muller at al. Fortsch. Phys. **42** (1994) 101;
   A.V. Radyushkin, Phys. Rev. D**56** (1997) 5524.
2. X. Ji, J. Phys. G **24** (1998) 1191.
3. K. Goeke et al., Prog. Part. Nucl. Phys. **41** (2001) 401.
4. F. Ellinghaus et al., Eur. Phys. J C**46** (2006) 729.

5. M. Vanderhaegen et al., Code for the calculation of DVCS and BH process, private communication (2001).
6. HERMES Collaboration, K. Ackerstaff et al., Nucl. Inst. and Meth. A**417** (1998) 230.
7. M. Diehl , S. Sapeta, Eur. Phys. J. C**41** (2005) 515.
8. K. Schilling, G. Wolf, Nucl. Phys. B**61** (1973) 381.
9. W.-D. Nowak, AIP Conf. Proc. **915** (2006) 603.

ELECTROPRODUCTION OF $\phi(1020)$ MESONS

J. P. SANTORO

Physics Department, Catholic University of America,
Washington, D.C. 20064, USA
E-mail: santoro@jlab.org

E. S. SMITH

Physics Division, Jefferson Lab,
Newport News, Virginia 23606, USA
E-mail: elton@jlab.com

FOR THE CLAS COLLABORATION

Electroproduction of exclusive ϕ vector mesons has been studied with the CLAS detector at an average Q^2 of 2.21 GeV^2. The scaling exponent for the total cross section as $1/(Q^2 + M_\phi^2)^n$ was determined to be $n = 1.97 \pm 0.84$, consistent with the Vector Dominance Model prediction of n=2. The data are consistent with the assumption of s-channel helicity conservation (SCHC). Under this assumption we determine the ratio of longitudinal to transverse cross sections to be $R = 0.86 \pm 0.24$.

Keywords: vector mesons; phi meson production; electron scattering.

1. Introduction

This article presents data for exclusive ϕ vector meson production off the proton with a 5.754 GeV electron beam of the CEBAF accelerator and the CLAS detector at Jefferson Lab [1]. One of the leading motivations for the present analysis is the sparse amount of existing ϕ electroproduction data [2–11]. The current data set covers the kinematical regime between $1.6 \leq Q^2 \leq 3.8$ GeV^2, $0.0 \leq t' \leq 3.6$ GeV^2, and $2.0 \leq W \leq 3.0$ GeV. We will present cross sections as a function of the virtuality of the photon Q^2, the momentum transfer $-t$, the azimuthal angle Φ between the electron and hadron scattering planes, as well as angular decay distributions in the rest frame of the ϕ-meson.

A simple phenomenological picture of vector meson production can be formulated in the lab frame. In this frame, the incident virtual photon

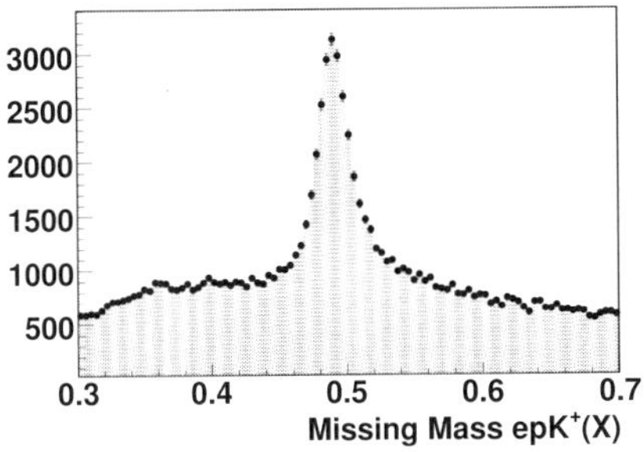

Fig. 1. Plot of $ep \to epK^+X$ missing mass.

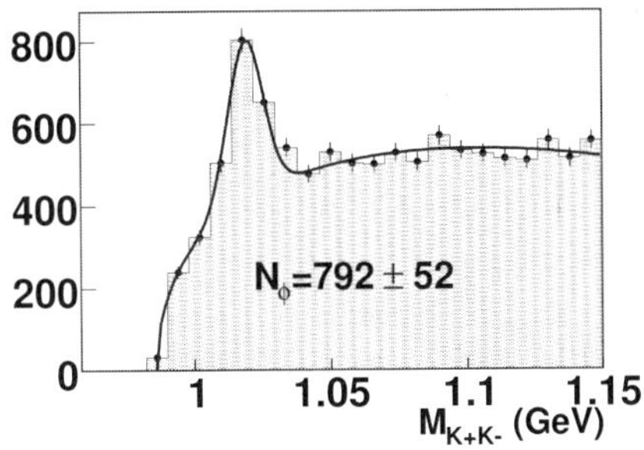

Fig. 2. K^+K^- invariant mass with a fit to the ϕ peak above background.

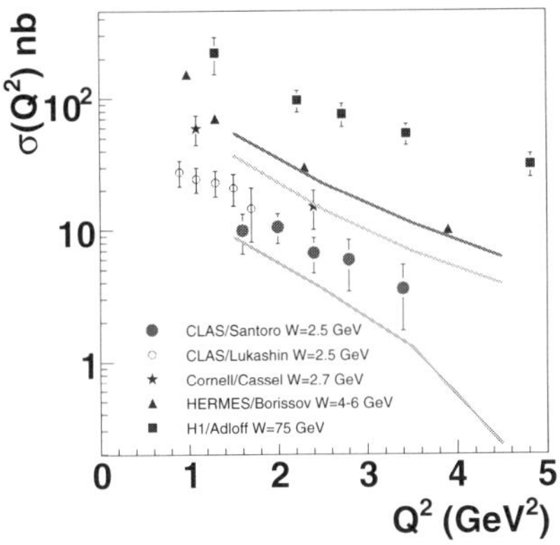

Fig. 3. Cross section as a function of Q^2 for our data (full circles), previous data from JLab (open circles), Cornell data (stars), HERMES (triangles), and HERA (squares). The curves show the JML model at W=2.9, 2.45 and 2.1 GeV (top to bottom).

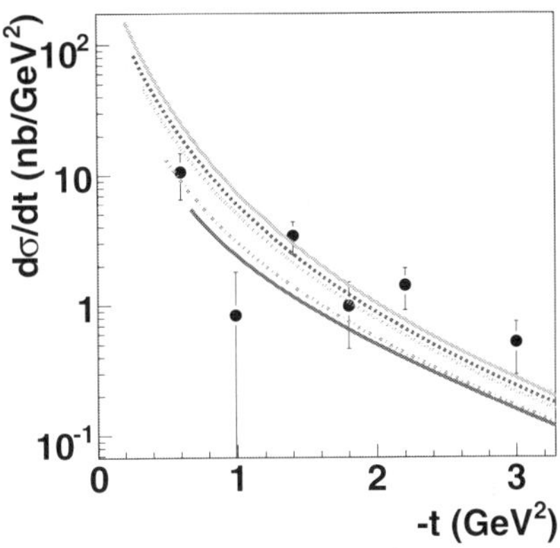

Fig. 4. $d\sigma/dt$ vs $-t$ for the entire Q^2 range and the JML predictions for W=2.5 GeV at five values of Q^2= 1.6, 2.1, 2.6, 3.8 and 5 GeV^2, top to bottom.

fluctuates into a virtual $q\bar{q}$ dipole pair, which elastically scatters off the target. The ϕ-meson production probes the gluonic exchanges between the $s\bar{s}$ dipole and the target nucleon. The data are compared to the JML model [12], which uses effective partonic and hadronic degrees of freedom and models Pomeron exchange by a 2 gluon exchange mechanism.

2. Experiment

The experiment was conducted with the CEBAF Large Acceptance Spectrometer (CLAS) located in Hall B of Thomas Jefferson National Accelerator Facility. The data were collected with a 5.754 GeV electron beam incident on a 5 cm-long liquid hydrogen target [13]. Once the electron, proton and K^+ tracks were identified, the missing mass was computed and is plotted in Fig. 1, which shows a clear peak at the K^- mass. The event sample was obtained using cuts on the missing mass, and includes all physical processes that contribute to the $epK^+(K^-)$ final state as well as accidental background from misidentified pions. Fig. 2 shows the K^+K^- invariant mass and a fit is used to obtain the ϕ yield of $N_\phi = 792 \pm 52$, the mean $M_\phi = 1.0194 \pm 0.0005$ GeV, and a width of $\sigma = 6.5 \pm 0.6$ MeV. To determine the yield in specific kinematic bins, similar distributions were fit by constraining the values of the mean and width.

The production kinematics is defined by Q^2, $-t$, Φ and W. The decay distribution requires two additional variables, $\cos\theta_H$ and ψ. Using a VDM model, a six-dimensional acceptance table was used to correct for a combination of geometrical acceptance and detector efficiencies. Radiative corrections were also applied.

3. Results

The total cross section was obtained as a function of Q^2 by integrating over all other variables (Fig. 3). The data were fit to the form $1/(Q^2 + M_\phi^2)^n$ to determine the scaling behavior. We determined the parameter $n = 1.97 \pm 0.84$. The measured exponent spans the range expected for the dependence on Q^2 due to VDM (n=2) to hard scattering (n=3 at fixed momentum transfer). The measurements of $\sigma(Q^2)$ from the present analysis are shown along with other data on ϕ electroproduction [7,3–5,11] in the figure. The data sets span the range from threshold at W=2 GeV up to HERA energies, but show similar behavior as a function of Q^2.

The differential cross section in $-t$ (integrated over Q^2) is compared to the JML model in Fig. 4. The data tend to have a shallower slope than the calculations, but there is general agreement.

We have extracted values of σ_{TT} and σ_{LT} from the cross section dependence on the angle Φ (Fig. 5). These terms are consistent with zero and the assumption that SCHC is valid for ϕ production. [14]

The ratio $R = \sigma_L/\sigma_T$ has been determined from two projections of the angular decay distribution of the K^+ in the ϕ-meson rest frame and under

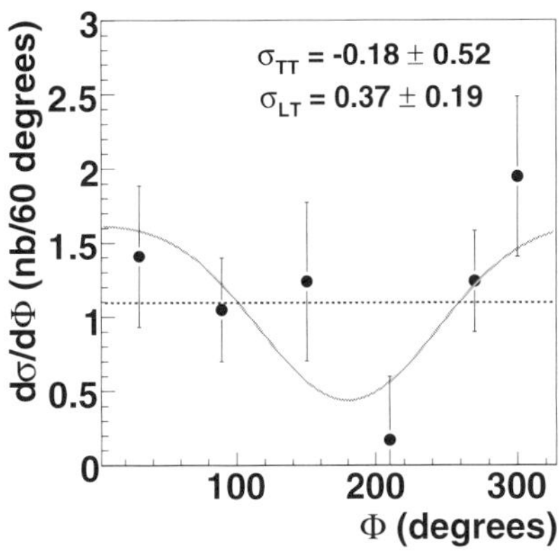

Fig. 5. $d\sigma/d\Phi$ vs Φ. The curve shows a fit to extract σ_{TT} and σ_{LT}, and the line shows a fit to a uniform distribution as expected from SCHC.

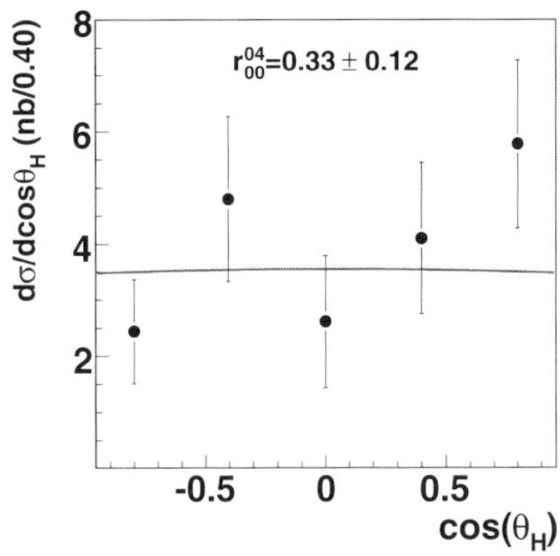

Fig. 6. Polar angular decay distribution of the K^+ integrated over all Q^2 values plus a fit to extract r_{00}^{04}.

the hypothesis of SCHC. The measurement of r_{00}^{04} gives $R = 1.05 \pm 0.38$ (Fig. 6) and the measurement of r_{1-1}^{1} gives a value of $R = 0.72 \pm 0.3$, the weighted average being $R = 0.85 \pm 0.24$.

Acknowledgments

Authored by Jefferson Science Associates, LLC under U.S. DOE Contract No. DE-AC05-06OR23177.

References

1. B. Mecking *et al.*, *Nuclear Instruments and Methods in Physics Research A* **503**, 513 (2003).
2. R. Dixon *et al.*, *Phys. Rev. Letters* **39**, p. 516 (1977).
3. R. Dixon *et al.*, *Phys. Rev. D* **19**, p. 3185 (1979).
4. D. Cassel *et al.*, *Phys. Rev. D* **24**, p. 2787 (1981).
5. A. Borissov, Meson 2000 workshop, in *Cross Sections of Vector Meson Production at intermediate virtual photon energies*, (Cracow, Poland, 2000).
6. A. Airapetian, *Eur. Phys. Journal C.* **17**, 389 (2003).
7. K. Lukashin *et al.*, *Phys. Rev. C* **64**, p. 059901(E) (2001).
8. C. Adloff *et al.*, *Zeitschrift fur Physik C* **75**, 607 (1997).
9. C. Adloff *et al.*, *Phys. Letters B* **483**, 360 (2000).
10. J. Breitwig *et al.*, *European Physics Journal C.* **14**, 213 (2000).
11. S. Chekanov *et al.*, *Nuclear Physics B* **718**, 3 (2005).

12. J. Laget, *Phys. Letters B.* **489**, 313 (2000).
13. J. Santoro *et al.* to be submitted to Phys. Rev. C.
14. K. Schilling, P. Seyboth and G. Wolf, *Nucl. Phys.* **B15**, 397 (1970).

GENERALIZED PARTON DISTRIBUTIONS FROM HADRONIC OBSERVABLES

S. AHMAD, H. HONKANEN, S. LIUTI

University of Virginia - Physics Department
382, McCormick Rd., Charlottesville, Virginia 22904 - USA

S.K. TANEJA

Ecole Polytechnique,
CPHT, F91128 Palaiseau Cedex, France

Following a previous detailed study of unpolarized generalized parton distribution functions in the non-singlet sector, and at zero values of the skewness variable, ζ, we propose a physically motivated parametrization that is valid at $\zeta \neq 0$. Our method makes use of information from the nucleon form factor data, from deep inelastuc scattering parton distribution functions, and from lattice results on the Mellin moments of generalized parton distributions. It provides, therefore, a step towards a model independent extraction of generalized distributions from the data, alternative to the mathematical ansatz of double distributions. Comparisons with recent experimental data on the proton are shown.

1. Introduction

Generalized Parton Distributions (GPDs) parametrize the soft contributions in a variety of hard exclusive processes, from Deeply Virtual Compton Scattering (DVCS) to hard exclusive meson production (see[1,2] for reviews). The feasibility of DVCS-type experiments using a deep inelastic probe with virtuality, Q^2, while detecting a momentum transfer, t, between the initial and final proton, allows one to address a vast, previously inaccessible phenomenology. In particular, one can access parton densities in impact parameter space,[3] and envisage extracting the orbital angular momentum of partons in both nucleons and nuclei.[4]

At present, a central issue is the definition of a quantitative, reliable approach beyond the construction of GPDs from specific models and/or particular limiting cases, that can incorporate new incoming experimental data in a variety of ranges of Q^2, and of the longitudinal (along the lightcone) and transverse components of the four-momentum transfer between the incoming and outgoing protons, ζ and $\Delta_\perp$, respectively (see also discussion in[5]). The matching between measured quantities and Perturbative QCD (PQCD) based predictions for DVCS should proceed, owing to s pecific factorization theorems, similarly to the extraction of Parton Distribution Functions (PDFs) from deep inelastic scattering. A few important caveats are however present since GPDs describe *amplitudes* and are therefore more elusive observables. The comparison between experiment and the formulation of parametrizations necessarily encompasses other strategies using additional constraints, since a direct comparison with the data cannot be performed. Experiments providing sufficiently accurate data to constrain the shape of GPDs have just begun.[6,7]

In Refs.[8,9] we proposed a strategy using a combination of experimental data on nucleon form factors, PDFs, and lattice calculations of Mellin moments with $n \geq 1$. The latter, parametrized in terms of Generalized Form Factors (GFFs), were calculated by both the QCDSF[10] and LHPC[11] collaborations for both the unpolarized and polarized cases up to $n = 3$. At $n \geq 2$, due to the polynomiality property,[1,2] the Mellin moments become dependent on the skewness, ζ.

In this contribution we report on the approach used in [9]; we show our results for both proton and neutron DVCS, and we finally present a preliminay result on the angular momentum sum rule in the deuteron.

2. Generalized Parton Distributions from Lattice Moments

In this contribution we concentrate on the unpolarized scattering GPDs, H, and E, from the vector (γ_μ) and tensor ($\sigma_{\mu\nu}$) interactions, respectively. We adopt the following set of kinematical variables: (ζ, X, t), where $\zeta = Q^2/2(Pq)$ is the longitudinal momentum transfer between the initial and final protons ($\zeta \approx x_{Bj}$ in the asymptotic limit, with Bjorken $x_{Bj} = Q^2/2M\nu$), $X = (kq)/(Pq)$ is the momentum fraction relative to the initial proton carried by the struck parton, $t = -\Delta^2$, is the four-momentum transfer squared. X is not directly observable, it appears in the amplitude as an integration variable.[1,2] The need to deal with a more complicated phase space, in addition to the fact that DVCS interferes coherently with the Bethe-Heitler (BH) process, are in essence the reasons why it is more challenging to extract GPDs from experiment, wherefore guidance from phenomenologically motivated parametrizations becomes important.

We first present a parametrization of H and E in the flavor Non Singlet

(NS) sector, valid in the $X > \zeta$ region, obtained by extending our previous zero skewness form,[8] through proper kinematical shifts:

$$H(X, \zeta, t) = G_{M_X}^{\lambda}(X, \zeta, t)\, R(X, \zeta, t) \tag{1}$$

(a similar form is obtained for $E(X, \zeta, t)$), where $R(X, \zeta, t)$ is a Regge motivated term describing the low X and t behaviors, while $G_{M_X}^{\lambda}(X, \zeta, t)$, was obtained within a spectator model.

In order to model the $X < \zeta$ region, we observe that the higher moments of GPDs give ζ-dependent constraints, in addition to the ones from the nucleon form factors. The $n = 1, 2, 3$ moments of the NS combinations: $H^{u-d} = H^u - H^d$, and $E^{u-d} = E^u - E^d$ are available from lattice QCD.[10,11] They can be written in terms of the isovector components as:

$$H_n^{u-d} \equiv \int dX X^{n-1}(H^u - H^d) = \frac{\tau (H_M^V)_n + (H_E^V)_n}{1+\tau} \tag{2}$$

$$E_n^{u-d} \equiv \int dX X^{n-1}(E^u - E^d) = \frac{(E_M^V)_n - (E_E^V)_n}{1+\tau}, \tag{3}$$

where the l.h.s. quantities are obtained from the lattice moments calculations, whereas $(H_{M(E)}^V)_n$ and $(E_{M(E)}^V)_n$ are amenable to chiral extrapolations. We used lattice calculations for the unpolarized GFFs obtained by the QCDSF collaboration using two flavors of $\mathcal{O}(a)$-improved dynamical fermions for several values of t in the interval $0 \lesssim t \lesssim 5$ GeV^2, and covering a range of pion mass values, $m_\pi \gtrsim 500\,\mathrm{MeV}^2$. Similarly to previous evaluations[11] the GFFs for both H and E, display a dipole type behavior for all three n values, the value of the dipole mass increasing with n. We performed an extrapolation by extending to the $n = 2, 3$ moments a simple ansatz proposed in [12] for the nucleon form factors that: *i)* uses the connection between the dipole mass and the nucleons radius; *ii)* introduces a modification of the non analytic terms in the standard chiral extrapolation that suppresses the contribution of chiral loops at large m_π. Despite its simplicity, the ansatz seems to reproduce both the lattice results trend at large m_π while satisfying the main physical criteria *i)* and *ii)*. Our results for the dipole mass at $n = 2$ are shown in Fig. 1.

3. Reconstruction from Bernstein Polynomials

Similarly to the PDFs case,[13] with a finite number of moments in hand, one can use reconstruction methods attaining weighted averages of the GPDs, around average ranges of X. The weights are provided by the complete set

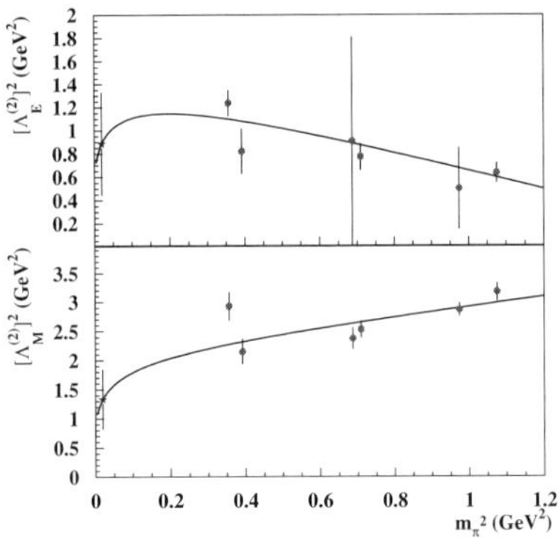

Fig. 1. The dipole masses squared for $n = 2$, for the isovector magnetic (lower panel) and electric (upper panel) contributions obtained by performing fits to the lattice results of [10]. The value at the physical pion mass obtained from our fit is also shown (star). (adapted from [9])

of Bernstein polynomials. The Bernstein polynomials are ideal for reproducing the deep inelastic structure functions in that they are zero at the endpoints, they are normalized to one, and they are peaked in different regions within the interval $x_{Bj} \in [0,1]$. Because of the latter property the Bernstein polynomials allow one to emphasize the behavior of the structure function at given specific regions of x_{Bj}, while suppressing the others. It was found that $n \geq 8$ moments were necessary to give a fully quantitative description of the behavior of $F_2(x_{Bj}, Q^2)$. At present, only $n = 3$ Mellin moments are available from lattice QCD calculations, therefore one cannot reach a similar level of accuracy as in the inclusive case.

In Fig.2 we show H^{u-d} reconstructed using the available lattice moments. We performed the procedure in the $X < \zeta$ region only using:

$$\overline{H}_{k,n}(\zeta, t) = \int_0^\zeta H(X, \zeta, t) b_{k,n}(X, \zeta) dX \quad k = 0, ...n, \tag{4}$$

where: $b_{k,n}(X,\zeta) = X^k (\zeta - X)^{n-k} / \int_0^\zeta X^k (\zeta - X)^{n-k} dX$, and we used subtracted moments, defined as:

$$(H_n)_{X<\zeta} = H_n - \int_\zeta^1 H^I(X, \zeta, t) X^n dX, \tag{5}$$

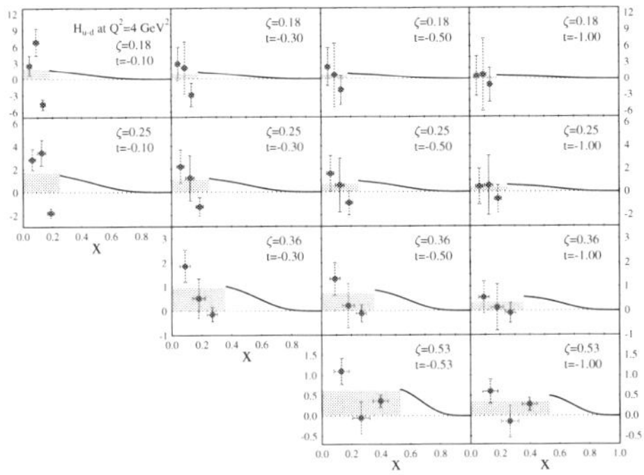

Fig. 2. Comparison of H^{u-d} for different values of $\zeta = 0.18, 0.25, 0.36, 0.53$, and $-t \equiv t_{min} = 0.035, 0.073, 0.18, 0.53$ GeV^2 calculated using the procedure described in the text (adapted from [9]).

where H_n are the Mellin moments, and $H^I(X,\zeta,t)$ was obtained from Eq.(1). For $n = 2$, $k = 0, 1, 2$, the reconstruction procedure yields:[9]

$$\overline{H}_{02}(\zeta X_{02}) = \frac{1}{\zeta^3}\left\{3A_{10}^{\zeta}\,\zeta^2 - 6A_{20}^{\zeta}\,\zeta + 3\left[A_{30}^{\zeta} + \left(-\frac{2\zeta}{2-\zeta}\right)^2 A_{32}\right]\right\},$$

$$\overline{H}_{12}(\zeta X_{12}) = \frac{1}{\zeta^3}\left\{6A_{20}^{\zeta}\,\zeta - 6\left[A_{30}^{\zeta} + \left(-\frac{2\zeta}{2-\zeta}\right)^2 A_{32}\right]\right\},$$

$$\overline{H}_{22}(\zeta X_{22}) = \frac{1}{\zeta^3}\left\{3A_{30} + \left(-\frac{2\zeta}{2-\zeta}\right)^2 A_{32}\right\}, \qquad (6)$$

where $X_{01} = 0.25$, $X_{02} = 0.5$, $X_{03} = 0.75$, and $A_{10}, A_{20}, A_{30}, A_{32}$ are the GFFs from Ref. [10].

4. Comparison with Experiment

The unpolarized Compton form factors were recently extracted from DVCS on proton and deuteron targets at $Q^2 \approx 2$ GeV^2, $\zeta = x_{Bj} = 0.36$, for several values of t in the range $0.15 \leq t \leq 0.35$ GeV^2.[6,7] In Fig.3 and we show the results of our parametrization at $\zeta = X$ for the imaginary part of the Bethe-Heitler BH-DVCS interference term at leading order, for proton and neutron, respectively. The proton results are compared with recent data from Jefferson Lab,[6] at $x_{Bj} = \zeta = 0.36$, and $Q^2 \approx 2$ GeV^2, while the neutron ones are presented at the kinematics of the forthcoming analysis from Ref.[7] .

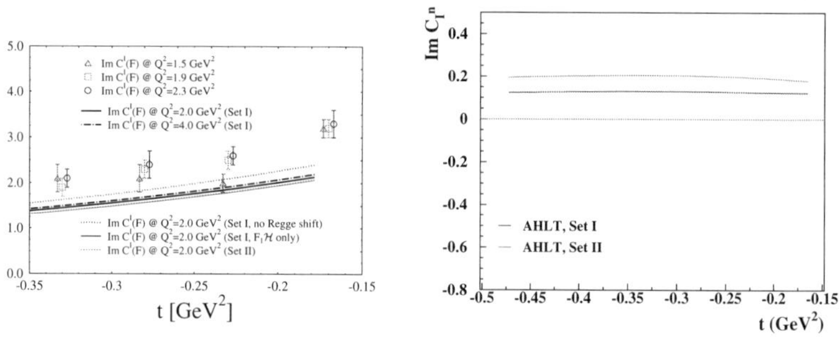

Fig. 3. The imaginary part of the Bethe-Heitler BH-DVCS interference term at leading order, $C^I(F)$, (see [1]). (upper panel) Proton target. Experimental data from Ref.[6]. The full curve shows our prediction, including the theoretical error, evolved to $Q^2 = 2$ GeV^2. The dot-dashed curve, obtained at $Q^2 = 4$ GeV^2, shows that the effect of evolution is relatively small. The other curves represent variations of the parametrization obtained respectively by disregarding the contribution from E, by disregarding the kinematical shift in the Regge term described in the text, and by using E from Set II of Ref.[8]. (lower panel) Neutron target. The two curves show our results using variations of the parametrization for E using Set I and Set II of Ref.[8]. The curves are in agreement with the experimental data from Ref.[7]. (adapted from Ref.[9]).

5. Angular Momentum Sum Rule

We present results derived for the angular momentum for a spin 1 system in terms of second moments of it's generalized parton distributions (a full derivation will be shown in [14]).

The second moment of the parton distributions is relevant to the spin structure of the nucleon. To see this, we first write down the angular momentum operator in QCD as the sum of quark and gluon contributions,[15]

$$\vec{J}_{\rm QCD} = \vec{J}_q + \vec{J}_g \ , \tag{7}$$

where

$$\begin{aligned} \vec{J}_q &= \int d^3x \ \vec{x} \times \vec{T}_q \\ &= \int d^3x \left[\psi^\dagger \frac{\vec{\Sigma}}{2} \psi + \psi^\dagger \vec{x} \times (-i\vec{D}) \psi \right] , \\ \vec{J}_g &= \int d^3x \ \vec{x} \times (\vec{E} \times \vec{B}) \ . \end{aligned} \tag{8}$$

Here $\vec{T}_q$ and $\vec{E} \times \vec{B}$ are the quark and gluon momentum densities, respectively. $\vec{\Sigma}$ is the Dirac spin-matrix and $\vec{D} = \vec{\partial} + ig\vec{A}$ is the covariant derivative. By an analogy with the magnetic moment, one can get the separate quark and gluon contributions to the nucleon spin if the form factors of the momentum density, or equivalently the energy-momentum tensor of QCD, are known at zero momentum transfer. Using Lorentz covariance and other symmetry principles, one can write down six form-factors separately for quark and gluon parts of the energy-momentum tensor,

$$
\begin{aligned}
\langle p'|\theta^{\mu\nu}|p\rangle = & -\frac{1}{2}\left[P^\mu P^\nu\right](\epsilon'^*\epsilon)G_{1,2}(t) - \frac{1}{4}\left[P^\mu P^\nu\right]\frac{(\epsilon P)(\epsilon'^* P)}{M^2}G_{2,2}(t) \\
& -\frac{1}{2}\left[\Delta^\mu\Delta^\nu - g^{\mu\nu}\Delta^2\right](\epsilon'^*\epsilon)G_{3,2}(t) \\
& -\frac{1}{4}\left[\Delta^\mu\Delta^\nu - g^{\mu\nu}\Delta^2\right]\frac{(\epsilon P)(\epsilon'^* P)}{M^2}G_{4,2}(t) \\
& +\frac{1}{4}\left[(\epsilon'^{*\mu}(\epsilon P) + \epsilon^\mu(\epsilon'^* P))P^\nu + \mu \leftrightarrow \nu\right]G_{5,2}(t) \\
& +\left[(\epsilon'^{*\mu}(\epsilon P) - \epsilon^\mu(\epsilon'^* P))\Delta^\nu + \mu \leftrightarrow \nu\right. \\
& \left. + 2g^{\mu\nu}(\epsilon P)(\epsilon'^* P) - (\epsilon'^{*\mu}\epsilon^\nu + \epsilon'^{*\nu}\epsilon^\mu)\Delta^2\right]G_{6,2}(t)
\end{aligned} \tag{9}
$$

where$P^\mu = p^\mu + p^{\mu\prime}$, $\Delta^\mu = p^{\mu\prime} - p^\mu$, and ϵ is the deuteron polarization. Taking $\mu = 0$, in the Breit frame, the matrix element of $\vec{J}_q$, gives

$$
\langle p'|\int d^3x(\vec{x} \times \vec{T}^{0i}_{q,g})_z|p\rangle = G_{5,2}(0)\int d^3x\ p^0 \tag{10}
$$

$$
G_{5,2}(0) = 2J^q_z \tag{11}
$$

Therefore, the generalized form factor, $G_{5,2}(0)$, of the energy momentum tensor of spin 1 system provides the fraction of the spin is carried by quarks and gluons.

6. Conclusions

In conclusion, we provided a fully quantitative parametrization of the GPDs in the non-singlet sector, valid in the region of Jefferson Lab experiments using both proton and deuteron targets.[6,7] We also presented a new result for the angular momentum sum rule in the deuteron. Differently from model calculations, and for the first time to our knowledge, our parametrization makes use of experimental data in combination with lattice results. Given the paucity of direct experimental measurements of GPDs, and the uncertainties related to the partonic interpretation of GPDs in the ERBL region, our goal is to provide more stringent, model independent predictions that will be useful both for model builders, in order to understand the dynamics of GPDs, and for the planning of future hard exclusive scattering experiments.

Acknowledgments

We thank Ph. Haegler, P. Kroll and G. Schierholz for useful comments. We are also grateful to J. Zanotti for providing us with the recent lattice calculations from the QCDSF collaboration. This work is supported by the U.S. Department of Energy grant no. DE-FG02-01ER41200 and NSF grant no.0426971.

References

1. M. Diehl, Phys. Rept. **388**, 41 (2003).
2. A. V. Belitsky and A. V. Radyushkin, Phys. Rept. **418**, 1 (2005)
3. M. Burkardt, Int. J. Mod. Phys. A **18**, 173 (2003); *ibid* Phys. Rev. D **62**, 071503 (2000) [Erratum-ibid. D **66**, 119903 (2002)].
4. X. D. Ji, Phys. Rev. Lett. **78**, 610 (1997) [arXiv:hep-ph/9603249].
5. K. Kumericki, D. Muller and K. Passek-Kumericki, arXiv:0710.5649 [hep-ph].
6. C. Munoz Camacho *et al.* [Jefferson Lab Hall A Collaboration], Phys. Rev. Lett. **97**, 262002 (2006)
7. M. Mazouz *et al.* [Jefferson Lab Hall A Collaboration], arXiv:0709.0450 [nucl-ex].
8. S. Ahmad, H. Honkanen, S. Liuti and S. K. Taneja, Phys. Rev. D **75**, 094003 (2007)
9. S. Ahmad, H. Honkanen, S. Liuti and S. K. Taneja, arXiv:0708.0268 [hep-ph].
10. M. Gockeler *et al.* [QCDSF Collaboration], Phys. Rev. D **73**, 014513 (2006), *and references therein.*
11. R. G. Edwards *et al.* [LHPC Collaboration], PoS **LAT2005**, 056 (2006) [arXiv:hep-lat/0509185], *and references therein.*
12. J. D. Ashley, D. B. Leinweber, A. W. Thomas and R. D. Young, Eur. Phys. J. A **19**, 9 (2004).
13. F.J. Yndurain, Phys. Lett. **74B** (1977) 68.
14. S.K. Taneja and S. Liuti, *in preparation.*
15. X. Ji, Phys. Rev. Lett. 78 (1997) 610.

IMAGING THE PROTON VIA HARD EXCLUSIVE PRODUCTION IN DIFFRACTIVE *pp* SCATTERING

C.E. HYDE[1,2], L. FRANKFURT[3], M. STRIKMAN[4], C. WEISS[5]

[1] *Old Dominion University, Norfolk VA 23529, USA*

[2] *Université Blaise Pascal - Clermont II, 63177 Aubière, France*

[3] *School of Physics and Astronomy, Tel Aviv University, Tel Aviv, Israel*

[4] *Dept. of Physics, Pennsylvania State University, University Park, PA 16802, USA*

[5] *Theory Center, Jefferson Lab, Newport News, VA 23606, USA*

We discuss the prospects for probing Generalized Parton Distributions (GPDs) via exclusive production of a high–mass system (H = heavy quarkonium, di-photon, di-jet, Higgs boson) in diffractive pp scattering, $pp \to p + H + p$. In such processes the interplay of hard and soft interactions gives rise to a diffraction pattern in the final–state proton transverse momenta, which is sensitive to the transverse spatial distribution of partons in the colliding protons. We comment on the plans for diffractive pp measurements at RHIC and LHC. Such studies could complement future measurements of GPDs in hard exclusive ep scattering (JLab, COMPASS, EIC).

It is well–known that the parton densities measured in inclusive deep–inelastic ep scattering describe also the partonic initial state in pp collisions with hard processes (Drell–Yan pair production, high–mass jets). This remarkable universality is a consequence of the QCD factorization of both processes. Recently, the study of the partonic structure of the nucleon has been extended to the Generalized Parton Distributions (GPDs), which describe the transverse spatial distribution of partons and are measured in hard exclusive processes in ep scattering. An interesting question is whether the GPDs, too, can be probed in pp collisions with hard processes. In Ref. 1 we have demonstrated that certain information about the GPDs can be gained from processes in which a high–mass system (H = heavy quarkonium, di-photon, di-jet, Higgs boson) is produced in exclusive diffractive pp scattering,

$$pp \quad \to \quad p + (\text{gap}) + H + (\text{gap}) + p. \tag{1}$$

In such processes the interplay of hard and soft interactions gives rise to a diffraction pattern in the final–state proton transverse momenta, which

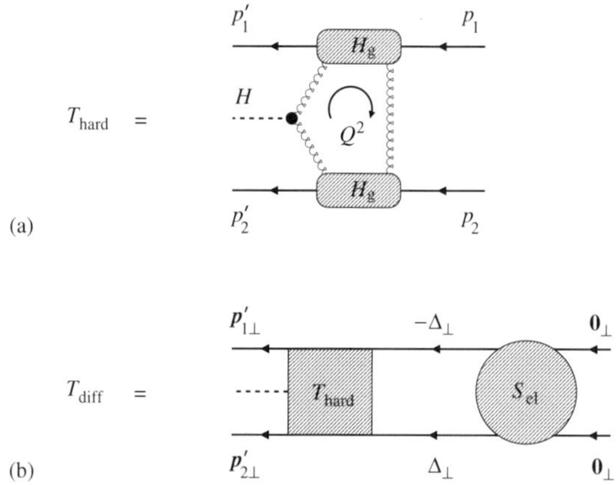

Fig. 1. Interplay of hard and soft interactions in exclusive hard diffraction $pp \to p + H + p$, Eq. (1). (a) Hard scattering process. (b) Modification by soft spectator interactions in the independent interaction approximation (see Ref. 1 for details).

is sensitive to the transverse spatial distribution of partons in the colliding protons. The realization of this program presents a major challenge to theory and experiment. In these proceedings we summarize the role of GPDs in hard exclusive diffraction and comment on the plans for diffractive measurements at RHIC and LHC.

In hard diffractive processes, Eq. (1), the high–mass system H is produced in a hard scattering process involving exchange of two gluons between the protons (see Fig. 1a). The requirement of absence of QCD radiation (imposed experimentally by the selection of exclusive events) forces this process to be highly localized in space–time and ensures that the gluons have virtuality of the order of a scale Q^2 in the range $\Lambda^2_{\rm QCD} \ll Q^2 \ll M_H^2 \ll s$. In analogy with deeply–virtual electromagnetic Compton scattering, the gluon–proton amplitudes in this process can be computed in terms of the gluon GPDs of the protons. Following Ref. 1, the amplitude of the hard scattering process can be represented as

$$T_{\rm hard}(\boldsymbol{p}_{1\perp}', \boldsymbol{p}_{2\perp}'; \boldsymbol{p}_{1\perp}, \boldsymbol{p}_{2\perp}; \xi_1, \xi_2) = \kappa\, H_g(x_1, \xi_1, t_1)\, H_g(x_2, \xi_2, t_2), \quad (2)$$

where H_g is the gluon GPD at the scale Q^2, $\xi_{1,2} = (M_H/\sqrt{s})e^{\pm y_H}$ the longitudinal momentum losses of the protons (y_H is the rapidity of H), $x_{1,2} \approx \xi_{1,2}$, and $t_1 \approx -(\boldsymbol{p}_{1\perp}' - \boldsymbol{p}_{1\perp})^2$ *etc.* the invariant momentum transfers to the protons. The function $\kappa(s, \xi_1, \xi_2)$ contains the $gg \to H$ coupling, which for production of a system with quantum numbers other than 0^+ depends also on the transverse momenta.

The amplitude of the hard scattering process Eq. (2) is modified by soft spectator interactions, which are likewise constrained by the requirement that no additional particles be produced. This effect is the essence of the idea of "rapidity gap survival" in diffraction. It was shown in Ref. 1 that in the approximation where hard and soft interactions are independent the modification can be computed in a model–independent way in terms of the pp elastic scattering amplitude. The amplitude for the diffractive process Eq. (1) is then given by (see Fig. 1b; we suppress all arguments except the proton transverse momenta)

$$T_{\text{diff}}(\boldsymbol{p}'_{1\perp}, \boldsymbol{p}'_{2\perp}) = \int \frac{d^2\Delta_\perp}{(2\pi)^2}\, T_{\text{hard}}\,(\boldsymbol{p}'_{1\perp}, \boldsymbol{p}'_{2\perp}; \boldsymbol{\Delta}_\perp, -\boldsymbol{\Delta}_\perp)\; S_{\text{el}}(s, \boldsymbol{\Delta}_\perp), \quad (3)$$

where S_{el} is the S–matrix of pp elastic scattering, which can be expressed in terms of the elastic scattering amplitude, $T_{\text{el}}(s,t)$, *viz.* its profile function in impact parameter representation, $\Gamma(s, \boldsymbol{b})$, as

$$S_{\text{el}}(s, \boldsymbol{\Delta}_\perp) = (2\pi)^2\, \delta^{(2)}(\boldsymbol{\Delta}_\perp) \;+\; (4\pi i/s)\, T_{\text{el}}(s, t = -\boldsymbol{\Delta}_\perp^2) \quad (4)$$

$$= \int d^2b\, e^{-i\boldsymbol{\Delta}_\perp\cdot\boldsymbol{b}}\, [1 - \Gamma(s, \boldsymbol{b})]\,. \quad (5)$$

The diffractive cross section obtained from Eq. (3) leads to a simple "geometric" expression for the rapidity gap survival probability, which was studied in detail in Ref. 1.

The transverse momentum dependence of the diffractive cross section described by Eq. (3) is essentially an interference phenomenon. The diffractive amplitude is the sum of the original hard amplitude ["1" in Eq. (5)] and the amplitude modified by elastic scattering ["Γ" in Eq. (5)]. To illustrate the resulting diffraction pattern in the $\boldsymbol{p}'_{1\perp}$ and $\boldsymbol{p}'_{2\perp}$ dependence, we evaluate Eq. (3) with simple analytic parametrizations of the ingredients (see Ref. 1 for details). The profile function of the pp elastic amplitude has been determined from fits to the pp elastic and total cross section data. We parametrize it as $\Gamma(s, \boldsymbol{b}) = \exp\left\{-\boldsymbol{b}^2/\,[2B(s)]\right\}$, where $B = 21.8\,\text{GeV}^{-2}$ at $\sqrt{s} = 14\,\text{TeV}$ from extrapolation of fits to the present data. Note that this function approaches the black–disk limit, $\Gamma \to 1$, at $b = 0$; for a detailed discussion of the role of this effect in diffraction see Ref. 1. The t–slope of the gluon GPD has been extracted from measurements of exclusive J/ψ photo/electroproduction;[2,3] see Ref. 4 for an overview. We parametrize the t–dependence as $H_g(x, \xi, t) \propto \exp(B_g t/2)$, with $B_g = 3.24\,\text{GeV}^{-2}$ in the x– and Q^2 range relevant to production of a system with $M_H \approx 100\,\text{GeV}$ at the LHC. Figure 2 shows $|T_{\text{diff}}|^2$ as a function of $\boldsymbol{p}'_{\perp 2}$, for fixed values of $\boldsymbol{p}'_{1\perp}$. The pattern can be understood as the wave generated by T_{hard} diffracting off the "hole" formed by the soft interaction profile, $1 - \Gamma(s, \boldsymbol{b})$. By measuring this dependence one can perform detailed tests of the diffractive reaction mechanism.

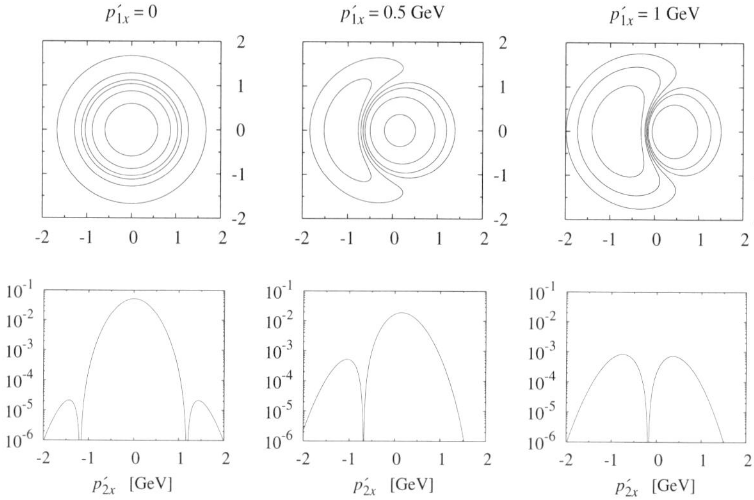

Fig. 2. Transverse momentum dependence of the cross section for exclusive double–gap diffraction, Eq. (1). The plots show $|T_{\rm diff}|^2$ (with $\kappa = 1$) as a function of $\boldsymbol{p}'_{2\perp}$, for three values of $\boldsymbol{p}'_{1\perp} = (p'_{1x}, 0)$. The top plots are 2–dimensional log–scale contour plots in $\boldsymbol{p}'_{2\perp}$; the bottom plots show the distributions for $\boldsymbol{p}'_{2\perp} = (p'_{2x}, 0)$.

The diffraction pattern of Fig. 2 is sensitive to the t–dependence of the gluon GPD. While the slope of the latter is to some degree known from measurements of exclusive J/ψ photo/electroproduction at Fermilab[2] and HERA,[3] more precise data on the shape of the t–dependence would be useful in order to make more accurate predictions. For the valence quark GPDs, such information will come from measurements of hard exclusive processes in ep scattering (deeply–virtual Compton scattering, meson production) with JLab 12 GeV ($0.1 < x < 0.7$) and COMPASS ($0.01 \leq x < 0.1$), extending previous exploratory measurements at JLab 6 GeV and HERMES.[5] The proposed Electron Ion Collider (EIC) would open a new frontier for the study of hard exclusive processes in the range $0.001 < x < 0.1$, with the capability of fully differential measurements in x, Q^2, and t.[5] Depending upon the x–range, different exclusive reaction channels ($ep \to ep\gamma$, $ep \to ep\rho$, $\gamma^* p \to p\, J/\psi$ and others) have different relative sensitivity to the gluon and quark GPDs, allowing one to disentangle the different GPDs. An interesting question is whether measurements of diffractive pp scattering could themselves contribute to our knowledge of the t–dependence of the GPDs, and thus to the program of "quark/gluon imaging" of the proton.

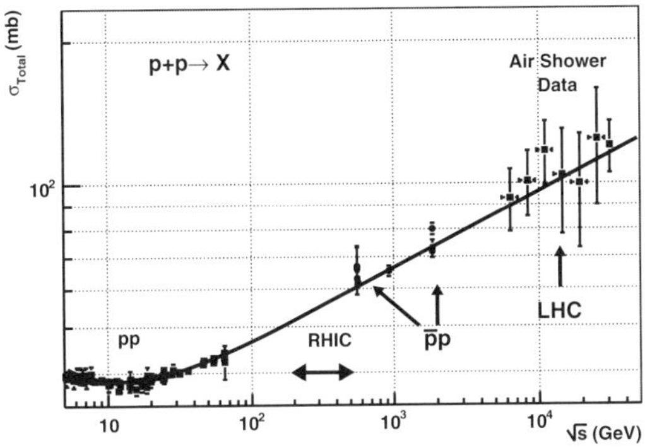

Fig. 3. Summary of high energy pp total cross section data, including Tevatron and SPS $p\bar{p}$ data (PDG compilation[6]). The curve is a phenomenological fit based on the parametrization of Ref. 7. The kinematics of the planned LHC and RHIC measurements are also shown.

The formalism described here also requires precision knowledge of the pp elastic scattering amplitude. Figure 3 summarizes the present status of pp total cross section data.[6] Note that there are no precise pp data above the ISR energy range. The planned measurements at the LHC will greatly enlarge the database. Measurements of polarized pp elastic scattering at RHIC[8] would be very useful for understanding the onset of the black–disk regime at small impact parameters and the role of spin–flip amplitudes.

The derivation of Eq. (3) for the diffractive amplitude is based on the approximation of independence of hard and soft interactions. As we point out in Ref. 1, the fact that at the LHC energy even hard interactions approach the black–disk regime requires qualitative changes of this picture. Preliminary studies show that this effect increases the effective impact parameters in diffraction compared to the independent interaction approximation, thus narrowing the t_1, t_2–distribution. Measurements of the $\boldsymbol{p}'_{1\perp}, \boldsymbol{p}'_{2\perp}$–dependence would in fact be very useful to study this new effect.

Measuring rapidity gap events in a collider requires detectors integrated into the accelerator lattice. The BNL-RHIC pp2pp experiment[8] is currently installing Si tracking stations at 60 m from the STAR Interaction Point (IP). With a dedicated beam tune of $\beta^*(\text{IP}) = 20\,\text{m}$, the experiment will measure elastic scattering at $\sqrt{s} = 200\,\text{GeV}$ for $0.003 < -t < 0.03\,\text{GeV}^2$. Future runs will reach $\sqrt{s} = 500\,\text{GeV}$. The same beam tune will detect central production in STAR, in coincidence with double diffractive (rapidity gap) tagging in the range $0.005 < \xi_{1,2} < 0.05$. A future tracking installation is planned between the DX and D0 magnets. For the full STAR luminosity, this will allow rapidity gap tagging for $0.01 < \xi \leq 0.15$, correlated with $0.5 \geq |\boldsymbol{p}'_\perp| \geq 0\,\text{GeV}/c$. The TOTEM experiment at the LHC[9] has installed forward trackers 200 m on either side of the CMS IP. In a series of special tunes with $\beta^* \leq 1540\,\text{m}$, TOTEM will measure pp elastic scattering for $0.001 \leq -t < 8\,\text{GeV}^2$, as well as σ_{tot} and the Re/Im ratio of the forward scattering amplitude. At full CMS luminosity, TOTEM will tag single and double diffractive events with central production for $0.01 \leq \xi < 0.15$. The FP420 collaboration[10] proposes to install proton trackers at 420 m from the interaction points of CMS and ATLAS. This would extend the tagging range for rapidity gap events down to $\xi \geq 0.001$.

In summary, LHC and RHIC are expected to produce extensive data on double diffractive production over a very wide range of s and ξ, as well as precise pp elastic and total cross section results. To further elaborate how the diffractive data could be interpreted in terms of GPDs, and perhaps be used to gain information about the GPDs, is an outstanding challenge for theory. More generally, the concept of GPDs is emerging as a potentially unifying framework for high–energy ep and pp scattering with hard processes.

Acknowledgments

Work partially supported by the U.S. Department of Energy, the Binational Science Foundation (BSF), and the French CNRS and ANR. Authored by Jefferson Science Associates, LLC under U.S. DOE Contract No. DE-AC05-06OR23177.

References

1. L. Frankfurt, C. E. Hyde, M. Strikman and C. Weiss, *Phys. Rev.* **D75**, 054009 (2007); see also arXiv:0708.3106 [hep-ph].
2. M. Binkley *et al.*, *Phys. Rev. Lett.* **48**, 73 (1982).
3. S. Chekanov *et al.*, *Nucl. Phys.* **B695**, 3 (2004); A. Aktas *et al.*, *Eur. Phys. J.* **C46**, 585 (2006).
4. L. Frankfurt, M. Strikman and C. Weiss, Ann. Rev. Nucl. Part. Sci. **55**, 403 (2005).
5. H. Abramowicz *et al.*, White Paper contributed to APS DNP Joint Town Meetings on Quantum Chromodynamics, Rutgers University, Jan. 12–14, 2007, available at: `http://www.physics.rutgers.edu/np/gpdwp.pdf`
6. W. M. Yao *et al.*, *J. Phys.* **G33**, 1 (2006).
7. A. Donnachie and P. V. Landshoff, *Phys. Lett.* **B296**, 227 (1992).
8. S. Bultmann *et al.*, *Phys. Lett.* **B647**, 98 (2007); and `www.rhic.bnl.gov/pp2pp/`
9. CERN TOTEM Experiment, `http://totem.web.cern.ch/Totem/`
10. FP420 R&D Project, `http://www.fp420.com`

REGGE CONTRIBUTIONS TO EXCLUSIVE ELECTRO-PRODUCTION

A. P. SZCZEPANIAK* and J.T.LONDERGAN

Physics Department and Nuclear Theory Center, Indiana University,
Bloomington, IN 47405, USA
**E-mail: aszczepa@indiana.edu*
www.physics.indiana.edu

We discuss the role of t-channel Regge trajectories in the Bjorken limit of Deeply Virtual Compton Scattering (DVCS). We show that in the limit of high energy and small t, the natural interpretation of DVCS is in terms of photon scattering off the meson cloud of the nucleon rather than in terms of scattering off the partons inside the bare nucleon.

Keywords: Regge exchange, Bjorken limit, DVCS.

1. Introduction

It has been observed that the recent DVCS and meson electro-production data appear to contain significant contributions from t-channel exchanges.[1–4] The observation that t-channel meson exchanges may be relevant in hard pion electro-production processes was previously made in [5], where it was shown that in the Bjorken limit the hand-bag diagram and t-channel pion exchange have the same scaling behavior. Regge exchanges can be built by summing over all contributing t-channel resonances (Regge poles) and multi-particle states (Regge cuts). If indeed all these are relevant in the Bjorken limit, $Q^2, p \cdot q \to \infty$, with $x_B = Q^2/2p \cdot q =$ fixed, they may lead to scaling violations and a different microscopic interpretation of the hard exclusive reaction. In the following we discuss how this occurs in DVCS and point to differences between DVCS and deep inelastic scattering (DIS), where leading Regge exchanges are known to be relevant only at low x_B.

2. t-channel description of DIS and DVCS reactions

At very low x_B DIS structure functions $f(x)$ are known to exhibit Regge behavior,[6]

$$f(x_B \to 0) \to {x_B}^{-\alpha(0)}, \tag{1}$$

where $\alpha(0)$ is the intercept of the leading Regge trajectory, *e.g.* $\alpha(0) \sim 1$ for the Pomeron exchange which dominates isosinglet structure functions, and $\alpha(0) \sim 0.5$ for meson exchanges in the isovector channel. The origin of this behavior is illustrated in Fig. 1.

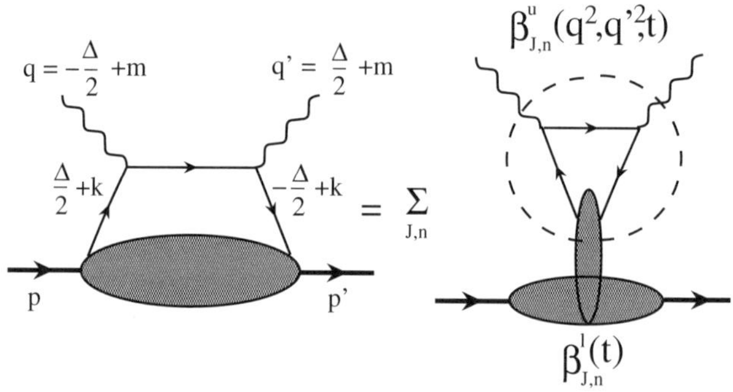

Fig. 1. t-channel meson contribution to the hadronic tensor for exclusive lepto-production. The amplitude is summed over all spins J that can contribute, and depends on the product of the residue functions β at the upper and lower vertices.

The cross-section for the inclusive electro-production process $\gamma^* p \to X$ is given by the discontinuity in $s = (q+p)^2$ of the forward, doubly virtual Compton amplitude, $\gamma^* p \to \gamma^* p$ (for DIS reactions in Fig. 1 one would set $m = 0$, $t = 0$). In the $Q^2 \to \infty$ limit, the dominant contribution originates from the space-time region where the separation between the two photons is along the light-cone and can be described in terms of an exchange of a free quark. This leads to the hand-bag picture shown on the left side of Fig. 1. In the spirit of the traditional approach to high-energy reactions,[7] on the right side we represent the amplitude in terms of a sum of all possible t-channel exchanges.[8] If all external particles are hadrons, in the limit $-t/s \to 0$ the sum over all t-channel exchanges becomes saturated by the leading Regge exchange, *i.e.* the trajectory with the largest intercept $\alpha(t)$. This occurs because Regge amplitudes are of the form,

$$A(s, t = 0) \sim \beta^l(t = 0)\beta^u(t = 0)s^{\alpha(t=0)} \tag{2}$$

where $\beta^l(t), (\beta^u(t))$, are the lower (upper) vertex functions representing coupling of the t-channel region to the external particles. In the case when one external leg represents a virtual photon, the upper vertex also depends on Q^2. It is this dependence that requires special attention. Also it needs to be verified whether Regge-factorization of the t and s dependence given in Eq. (2) holds for large photon virtuality. To address these issues, we write Regge exchange as a sum over exchanges of all contributing spins. The amplitude for exchange of a particle with spin J has the form

$$A_J(s, t = 0, Q^2) \sim \frac{\beta_J^u(t = 0, Q^2)\beta_J^l(t = 0)}{J - \alpha(t = 0)} s^J. \tag{3}$$

The upper residue $\beta_J^u(t = 0, Q^2)$ corresponds to an amplitude $\gamma^* X_J \to \gamma^*$, which describes a coupling of a hadron X_J with spin J to two virtual photons. In the t-channel the two photons probe the quark content of X_J at relative spatial separations of the order of $(1/Q^2)^J$.[9] The spatial wave function for small distances is suppressed due to the angular momentum barrier, which leads to

$$\beta_J^u(t = 0, Q^2) \sim \beta_J^u(t = 0)(Q^2)^{-J}. \tag{4}$$

Since in the Bjorken limit $s \to Q^2(1 - x_B)/x_B$ we find that individual t-channel exchange amplitudes become Q^2 independent, and for small x_B they have the form

$$A_J \sim \beta(t = 0){x_B}^{-J}. \tag{5}$$

After summing over all spins this results in the behavior given in Eq. (1), which is correct for both the real and imaginary parts of the amplitude.[10] In general summing over all possible Regge exchanges leads to

$$f(x_B) \sim {x_B}^{-\alpha(0)} \left[1 + \sum_i x^{\alpha(0)-\alpha_i(0)} \right] \tag{6}$$

where the sum in Eq. (6) extends over daughter trajectories with $\alpha_i(0) < \alpha(0)$. Thus as $x_B \to 0$ the leading behavior of the structure functions is determined by the properties of the leading Regge trajectory. The situation is quite different in the valence region, $x_B \sim \mathcal{O}(1)$. In this region all Regge exchanges contribute, the t-channel picture becomes complicated and the alternative picture of s-channel scattering off the nucleon constituents (partons) becomes more natural.

Now let us consider DVCS reactions. In this case the upper residue can be related to the amplitude $\gamma^* X_J \to \gamma$ which is known to have the same Q^2 dependence regardless of the spin J of the hadron, X_J.[8,9] We thus conclude that in this case the Q^2 dependence of the fixed-J amplitude,

$$A_J(t = \text{ finite}, Q^2) \sim \beta(t = \text{ finite})(Q^2)^\lambda (Q^2/x_B)^J \tag{7}$$

remains J-dependent; here λ is some fixed power related to the Q^2 fall-off of the form factor $\gamma^* X_J \to \gamma$. Upon summing over all spins we find that the DVCS amplitude exhibits the behavior[9]

$$A(t, Q^2, x_B) \sim (Q^2)^\lambda (Q^2/x_B)^{\alpha(t)} \tag{8}$$

i.e. the amplitude acquires an additional, non-scaling dependence on Q^2 proportional to $(Q^2)^{\alpha(t)}$. The $(Q^2)^\lambda$ behavior is identical to the case of a parton model description based on generalized parton distributions, *i.e.* $\lambda = 0$ for transverse DVCS processes or $\lambda = -1/2$ for meson production.[11]

3. *s*-channel description of DVCS

In the previous section we discussed how a non-scaling $(Q^2)^{\alpha(t)}$ dependence of DVCS amplitudes arises from t-channel meson exchanges. The same result can be obtained by considering the hand-bag diagram (the left side panel in Fig. 1) directly and focusing on the contribution from the region of large parton-nucleon invariant mass $s_q \equiv (p - \Delta/2 - k)^2$. As with any other hadronic amplitude, for large $s_q/(-t)$ the parton-nucleon amplitude is expected to behave as[12]

$$A_{\text{parton-nucleon} \to \text{parton-nucleon}}(s, t) \sim \beta(t) s_q^{\alpha(t)}. \tag{9}$$

Nucleons are on their mass-shell and it is the region of small parton virtuality that contributes in the Bjorken-limit (it is the parton between the exchanged photons that carries large virtuality). The DIS process $\gamma^* p \to \gamma^* p$ was studied in this approach in Ref. [12] where it was shown that the large s_q behavior of the parton-nucleon amplitude indeed determines the small-x_B behavior of the structure functions. Furthermore, it was also shown that Eq. (9) leads to a constant contribution to the real part of the $\gamma^* p \to \gamma^* p$ amplitude, which in terms of the complex angular momentum corresponds to an exchange of $\alpha(t) = 0$ singularity. The need for such a fixed-pole contribution has been later demonstrated in Ref. [13]. This is an example that illustrates how Regge exchange contributions to the full amplitude might be different from those that contribute to the imaginary part (*i.e.* structure functions) alone. Thus models based on extrapolation of the small-x_B behavior of structure functions to the properties of GPD's may be too naive. In Ref. [14] we studied the hand-bag diagram with the parton-nucleon amplitude that has the asymptotic behavior given by Eq. (9). We found that indeed, the Regge contribution to the real part of electroproduction amplitudes is quite different in the diagonal $\gamma^* p \to \gamma^* p$ (DIS) and off-diagonal $\gamma^* p \to \gamma p$ (DVCS) cases. For DVCS reactions we have confirmed the appearance of the extra $(Q^2)^{\alpha(t)}$ factor that we originally found in the t-channel based analysis.[8]

4. Summary

Regge effects make sizeable contributions to hard exclusive amplitudes. In particular, in the presence of Regge behavior in the parton-nucleon amplitude we have shown that there is a breakdown of collinear factorization, at least in the particular case of exclusive lepto-production processes such as DVCS and exclusive meson production. This means that the soft amplitudes are not universal, but are process-dependent. Thus the validity of QCD factorization for exclusive lepto-production will require, in addition to a hard scale Q^2, a sizeable value of t. As t increases Regge intercepts become negative and at this point Bjorken scaling is recovered. There remains, however the fixed-pole contributions to the real parts of the amplitudes. In the region of small t, Regge effects will make substantial contributions to DVCS and exclusive meson lepto-production. We have demonstrated that exclusive lepto-production processes should show violation of QCD scaling behavior. These scaling violations will persist regardless of the magnitude of Q^2, in contradiction to the expectations of QCD scaling.

Recently a Hall A Collaboration at Jefferson Laboratory has carried out a test of scaling in DVCS reactions.[15] The data appear to be in very good agreement with the Q^2-independent DVCS amplitude predicted by QCD,[17] however the available Q^2 window is quite small, from 1.5 – 2.5 GeV^2. Within the published experimental errors one cannot rule out a power-like dependence of the amplitude, $A \propto (Q^2)^\alpha$, with α as large as 0.25. Even more surprising, "standard" Regge-exchange models have proved successful in describing a variety of cross sections for exclusive electroproduction reactions,[1–4] in the kinematic range where scaling would be expected based on comparisons with DIS. Our calculations show that the success of the Regge picture is to be expected, and is not accidental.

Acknowledgments

Some of the work here was done in collaboration with F. Llanes-Estrada. The authors would like to thank S. Brodsky, J-M. Laget, W. Melnitchouk, D. Mueller, A. Radyushkin, M. Strikman and C. Weiss for useful comments and discussions. APS was supported by DOE contract DE-FG0287ER40365 and JTL by NSF contract PHY-0555232.

References

1. M.Guidal, *this proceedings*
2. A.Rostomyan, *this proceedings*
3. L. Morand *et al.* [CLAS Collaboration], Eur. Phys. J. A **24**, 445 (2005) [arXiv:hep-ex/0504057].
4. J. M. Laget, arXiv:0708.1250 [hep-ph].
5. L. Mankiewicz, G. Piller and A. Radyushkin, Eur. Phys. J. C **10**, 307 (1999) [arXiv:hep-ph/9812467].
6. For recent analysis of low-x_B DIS data see, for example A. Donnachie and P. V. Landshoff, Phys. Lett. B **595**, 393 (2004) [arXiv:hep-ph/0402081].
7. A.D. Martin and T.D. Spearman, *Elementary Particle Theory*, (North-Holland, Amsterdam 1970).
8. A. P. Szczepaniak and J. T. Londergan, Phys. Lett. B **643**, 17 (2006) [arXiv:hep-ph/0604266].
9. V. L. Chernyak and A. R. Zhitnitsky, Phys. Rept. **112**, 173 (1984).
10. S. J. Brodsky, F. E. Close and J. F. Gunion, Phys. Rev. D **6**, 177 (1972).
11. M. Vanderhaeghen, P. A. M. Guichon and M. Guidal, Phys. Rev. D **60**, 094017 (1999) [arXiv:hep-ph/9905372].
12. S. J. Brodsky, F. E. Close and J. F. Gunion, Phys. Rev. D **8**, 3678 (1973).
13. A. Zee, Phys. Rev. D **5**, 2829 (1972) [Erratum-ibid. D **6**, 938 (1972)].
14. A. P. Szczepaniak, J. T. Londergan and F. J. Llanes-Estrada, arXiv:0707.1239 [hep-ph].
15. C. Munoz Camacho *et al.* [Jefferson Lab Hall A Collaboration], Phys. Rev. Lett. **97**, 262002 (2006) [arXiv:nucl-ex/0607029].
16. M. Diehl, W. Kugler, A. Schafer and C. Weiss, Phys. Rev. D **72**, 034034 (2005) [Erratum-ibid. D **72**, 059902 (2005)] [arXiv:hep-ph/0506171].
17. M. Vanderhaeghen, P. A. M. Guichon and M. Guidal, Phys. Rev. Lett. **80**, 5064 (1998).

HIGH ENERGY BREAK-UP OF FEW-NUCLEON SYSTEMS

MISAK SARGSIAN

Department of Physics, Florida International University, Miami, FL 33199

Abstract We discus recent developments in theory of high energy two-body break-up reactions of few-nucleon systems. The characteristics of these reactions are such that the hard two-body quasielastic subprocess can be clearly separated from the accompanying soft subprocesses. We discuss in details the hard rescattering model (HRM) in which hard photodisintegration develops in two stages. At first, photon knocks-out an energetic quark which rescatters subsequently with a quark of the other nucleon. The latter provides a mechanism of sharing the initial high momentum of the photon by the outgoing two nucleons. Within HRM we discuss hard break-up reactions involving ^2D and ^3He targets. Another development of HRM is the prediction of new helicity selection mechanism for hard two-body reactions, which was apparently confirmed in the recent JLab experiment.

1. Introduction to Nuclear QCD

There is class of high-energy and momentum transfer (semi) exclusive nuclear reactions which suits very well for studies of Nuclear QCD. There are several common features in theoretical approaches for studying these reactions: (a) one can clearly identify a hard sub-process; (b) which can be factorized from the soft nuclear part of the reaction; (c) the subprocess is hard enough for pQCD to be applicable; (d) soft part of the reaction could be expressed through the measurable/extractable quantities such as partonic distribution functions, hadron-hadron scattering amplitudes, form-factors and calculable nuclear wave functions. If all above requirements are achieved then such theoretical approaches may yield (sometimes) parameter free predictions (see e.g. [1]) and practically always will allow us to study the aspects of strong interaction dynamics which otherwise can not be investigated without using nuclear targets.

Several of such reactions which satisfy above criteria are, (i) semiinclusive deep-inelastic nuclear reactions aimed at studies of nuclear modifications of PDFs (EMC effects)(see e.g. [2]); (ii) high momentum transfer elastic scattering off few-nucleon systems that can be used to study the onset of quark degrees of freedom in strongly correlated few nucleon systems,[3] (iii) electrodisintegration of few-nucleon systems in the kinematics dominated by final state interaction that is used to probe the dominance of point like configurations in the hadronic wave function at large Q^2 (color transparency/coherence phenomena);[4–7] (iv) DIS nuclear scattering at $x_{Bj} > 1$ as a framework for investigation of the mechanism of generation of super-fast quarks[2,8] (v) as well as, high-energy break-up of two nucleons in nuclei which can be used to probe the dynamics of strong interaction in two-nucleon system at intermediate to short distances.[1,3,9,10]

2. High Energy Break-Up of Two Nucleons in Nuclei

In this presentation we focus on the last class (v) of the reactions, in which high energy photon produces two energetic nucleons which equally share the initial energy of the photon. These reaction kinematically corresponds to the break-up of 2N system at 90^0 angle in the $\gamma-$ $2N$ center of mass reference frame.

Due to completely symmetric configuration, photon predominantly probes the structure and the dynamics of the exchanged particle in the NN system (Fig.1). High energy of photon in this case provides necessary resolution to probe the QCD content of NN interaction whether it proceeds through the $q\bar{q}$ exchange (Fig.1b), quark interchange (Fig.1c) or gluon exchange (Fig.1d).

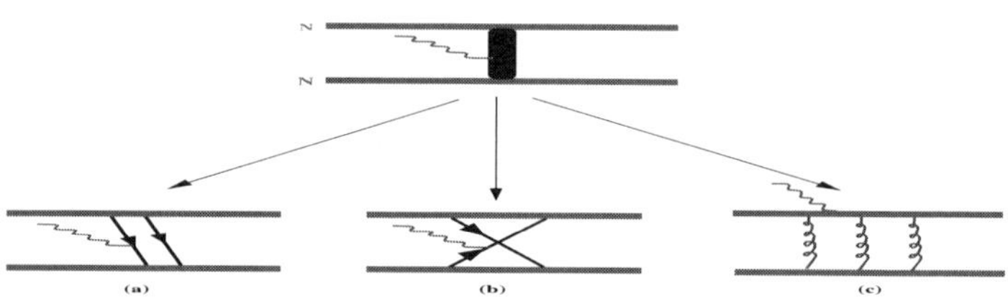

Fig. 1. Possible QCD dynamics of NN interaction.

Effectiveness of these processes in probing QCD aspects of nuclear interaction can be seen (for $\gamma d \rightarrow pn$ reactions) from the following kinematical considerations[3,9,10]) in which:

$$s = (k_\gamma + p_d)^2 = 2M_dE_\gamma + M_d^2; \quad t = (k_\gamma - p_N)^2 = (cos(\theta_{cm}) - 1)\frac{s - M_d^2}{2}. \tag{1}$$

Simple estimate shows that already at $E_\gamma > 2$ GeV the invariant momentum transfer $-t\mid_{90^0} > 4$ GeV^2 and invariant mass of the NN system $M_{NN} = \sqrt{s} > 2$ GeV. These are conditions for which one expects an onset of quark degrees of freedom in the dynamics of strong interaction.[11]

One of the first theoretical predictions for high energy and large CM angle $\gamma d \to pn$ reactions was the prediction of the s dependence of the differential cross section based on the quark counting rule,[3] which yields: $\frac{d\sigma}{dt} \sim s^{-11}$. This prediction was experimentally confirmed already starting at $E_\gamma = 1$ GeV for several set of experiments at SLAC[12,13] and Jefferson Lab.[14–17]

The quark counting predictions are based on the hypothesis that the Fock states with minimal number of partonic constituents dominate in two-body large angle hard collisions.[18] Although successful in describing energy dependences of the number of hard processes, this hypothesis does not allow to make calculation of the absolute values of cross sections. Especially for reactions involving baryons, calculations within perturbative QCD underestimate the measured cross sections by orders of magnitude (see e.g.[19]). This may be an indication that in the accessible range of energies bulk of the interaction is in the domain of nonperturbative QCD.[19,20] However, the main problem is that even if we fully realize the importance of nonperturbative interactions the theoretical methods of calculations in the nonperturbative domain are very restricted.

3. Hard Rescattering Mechanism of Two-Body Break-up Reactions

The underlying assumption in hard rescattering model (HRM)[1] is that high energy photodisintegration of two-nucleon system proceeds through the two stages in which an absorption of photon by a quark of one nucleon is followed by a high-momentum transfer (hard) rescattering with a quark from the second nucleon. The latter rescattering produces a final two nucleon state with large relative momenta. A typical diagram representing such a scenario is presented in Fig.2.

Analyzing the type of diagrams as in Fig.2 allows us to do the following observations:

- the dominant contribution comes from the soft vertices of $d \to NN$ transition, while quark rescattering proceeds trough the hard gluon exchange,
- the $d \to NN$ transition can be evaluated through the conventional deuteron wave functions,

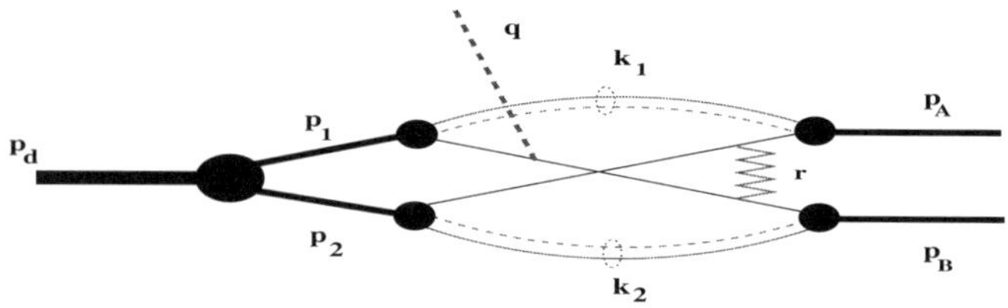

Fig. 2. Typical diagram for hard rescattering mechanism.

- the structure of hard quark interchange interaction in the rescattering part of the reaction is similar to that of hard NN scattering,
- as a result the sum of the multitude of diagrams with incalculable nonperturbative part of the interaction can be expressed through the experimentally measured amplitude of hard NN scattering.

Based on these observations, calculation of the $\gamma + d \rightarrow pn$ amplitude yields[1,21]

$$\langle p_{\lambda_A}, n_{\lambda_B} \mid A \mid \lambda_\gamma, \lambda_D \rangle = \sum_{\lambda_2} \frac{f(\theta_{cm})}{3\sqrt{2s'}} \int \Psi^{\lambda_D,\lambda_\gamma,\lambda_2}(\alpha_c, p_\perp) \frac{d^2 p_\perp}{(2\pi)^2} \times$$
$$\left(\langle p_{\lambda_A}, n_{\lambda_B} | A_{pn}(s, t_n) | p_{\lambda_\gamma}, n_{\lambda_2} \rangle - \langle p_{\lambda_A}, n_{\lambda_B} | A_{pn}(s, u_n) | n_{\lambda_\gamma} p_{\lambda_2} \rangle \right), \tag{2}$$

where A_{pn} is high momentum transfer elastic pn scattering amplitude, $\mid N_\lambda \rangle$ $(N = p, n)$ represents the helicity wave function of nucleon and λ_γ is the helicity of incoming photon. Based on Eq.(2) one obtains the following expression for the differential cross section of $\gamma + d \rightarrow pn$ reaction:

$$\frac{d\sigma^{\gamma d \rightarrow pn}}{dt} = \frac{8\alpha}{9}\pi^4 \frac{1}{s'} C(\frac{\tilde{t}}{s}) \frac{d\sigma^{pn \rightarrow pn}}{dt} \left| \int \Psi_d^{NR}(p_z = 0, p_t) \sqrt{m_n} \frac{d^2 p_t}{(2\pi)^2} \right|^2, \tag{3}$$

where $s' = s - 4m_N^2$ and $\tilde{t} = (p_n - m_n)^2$. The interesting properly of the function C is that $C(\theta_{cm} = 90^0) \approx 1$. Therefore for 90^0 CM scattering HRM prediction is parameter free. Since pn cross section in high momentum transfer behaves like s^{-10}, Eq.(3) predicts same s^{-11} dependence as quark counting rule without requiring an onset of pQCD regime. Also, due to angular dependence of the pn amplitude, HRM predicts an angular distribution being not symmetric around 90^0 CM. These predictions agree reasonably well with the experimental data[14,16](see e.g. Fig.3).

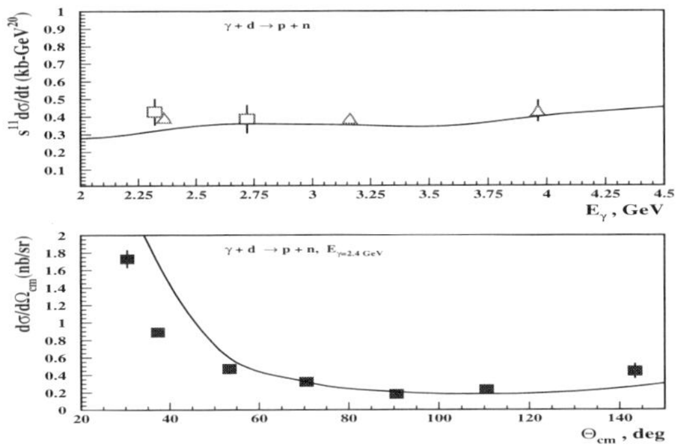

Fig. 3. Energy dependence of the scaled cross section at 90^0 CM scattering (top) and angular dependence of the cross section at $E_\gamma = 2.4$ GeV (bottom).

4. Hard Break-Up of Two Protons from ^3He

With all its success and accuracies yet to be improved HRM is only one of the approaches in describing hard photodisintegration reactions. Other models such as reduced nuclear amplitude (RNA) formalism[24] and quark-gluon string (QGS) model[25] describe many features of hard photodisintegration reaction, with QGS being rather successful in describing lower energy data. However RNA and QGS require an absolute normalization.

Recently it was suggested[22,23] that the break-up of pp pair from ^3He will further advance our understanding of the dynamics of hard photodisintegration of two-nucleon systems and allow further discrimination between above mentioned models.

Within HRM the typical diagram describing two-proton break-up is shown in Fig.4. HRM calculation similar to that of deuteron break-up reaction yields:

$$\frac{d\sigma}{dtd^3p_n} = \left(\frac{14}{15}\right)^2 \frac{16\pi^4\alpha}{S - M^2_{^3He}} (\frac{2c^2}{1+2c^2}) \frac{d\sigma^{pp}}{dt}(s_{pp}, t_n)\frac{S_{34}}{E_n}, \tag{4}$$

where $S_{34} = \sum\limits_{\lambda_1=-\lambda_2,\lambda_3=-\frac{1}{2}}^{\frac{1}{2}} \left| \int \psi^{\frac{1}{2}}_{^3He}(\lambda_1,\lambda_2,\lambda_3) m \frac{d^2p_{2\perp}}{(2\pi)^2}\right|^2$ and $c = \frac{|\phi_{3,4}|}{|\phi_1|}$ with ϕ_i being pp helicity amplitudes. Since the cross section of pp scattering enters in Eq.(4) one of the interesting predictions of HRM is the possibility of observation of energy oscillations at 90^0 CM scattering similar to one observed in elastic pp scattering. Another interesting feature of two-

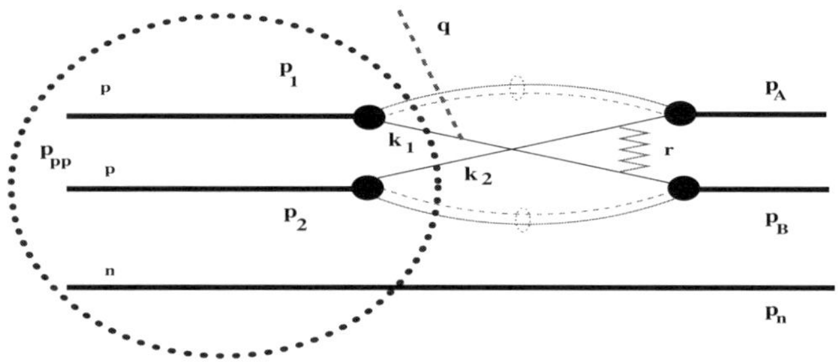

Fig. 4. Typical diagram for hard break-up of pp pair from ^3He.

proton break-up reactions is the fact that at lower energies this reaction is three-step[26] rather than two-step process. HRM in fact predicts that with an increase of energy due to the onset of quark-interchange (rather than meson-exchange) mechanism the two-body processes will dominate the cross section. The pioneering experiment of high energy *pp* break-up reaction[27] was recently performed at Jefferson Lab which may shed new light on many issues of hard rescattering processes.

5. Helicity Selection Mechanism

In addition to the cross section measurements, polarization observables may provide a new insight into the dynamics of hard photodisintegration. Original motivation for polarization measurements in high energy photodisintegration reaction was the expectation that the onset of the pQCD regime in the reaction dynamics will be accompanied by an observation of the helicity conservation in polarized reaction. Both energy and angular distributions of several polarization observables have been measured at JLab.[28,29] Although the energy range covered was rather restricted, it provided an interesting insight into the structure of HRM.

One of the unique features of HRM is that the struck quark carries the helicity of incoming photon. As a result one of the final nucleons will carry the bulk of the polarization of incident photon (see e.g. Eq.(2)). Thus in HRM photon plays as a helicity selector for the final nucleons. This yields a prediction for large asymmetry[27] ($C_{z'}$) for the longitudinal polarization of outgoing nucleons. In Ref.[21] we predicted a sizable asymmetry for $C_{z'}$ even though the existing data[28] (with rather large errors) at that time were indicating on vanishing values of $C_{z'}$. However, recent data,[29] appears to confirm HRM prediction for large values of $C_{z'}$ (see Fig.5). It will be

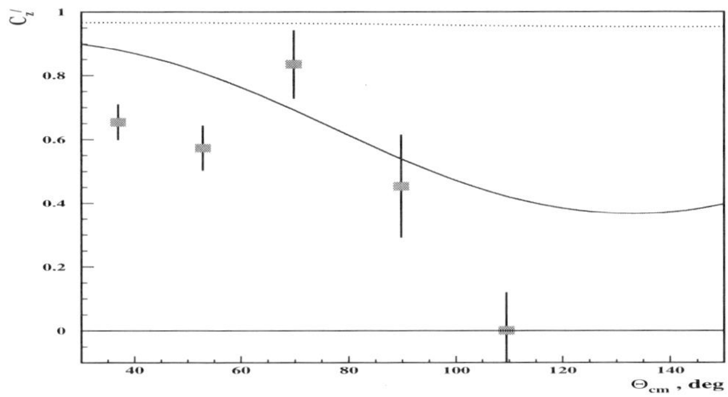

Fig. 5. Angular dependence of $C_{z'}$ for $E_\gamma = 1.9$ GeV.

interesting also to check the other HRM prediction that $C_{z'}$ will continue to approach to unity with an increase of photon energy at 90^0 CM scattering.

It is very interesting that above described helicity selection mechanism of HRM predicts (*an opposite*) vanishing value of $C_{z'}$ for two-body break up of proton pair from ^3He. This follows from the fact that the dominant part of the amplitude which represents two final state nucleons polarized in same direction are proportional to the nuclear ground state wave function with two initial nucleons having same helicities. Due to Pauli principle this part of the amplitude is strongly suppressed for the proton-pair in ^3He target. No such suppression exists for pn break up reactions.

6. Summary and Outlook

There is an accumulating evidence that hard rescattering mechanism explains the underlying dynamics of high energy and large CM angle break-up of a nucleon pair from ^2D and ^3He targets. One of the important features of HRM is that its prediction of unpolarized cross section at 90^0 center of mass photodisintegration of deuteron is parameter free and no further adjustments are required.

HRM predicts that energy dependence of two-proton break-up reaction should resemble that of hard elastic pp cross section.

Another feature of HRM, observed recently, is the prediction of large longitudinal asymmetries due to helicity selection mechanism characteristic to hard rescattering model.

If HRM will prove to be a true mechanism of hard photodisintegration reaction involving two nucleons, it will advance also our understanding of the dynamics of NN interaction at short distances.

A new venue for advancing our understanding of the dynamics of hard break-up reactions could be an extension of these studies to the kinematics in which two excited baryonic states (like Δ-isobars) are produced at large center of mass angles of $\gamma - NN$ system.

References

1. L.L. Frankfurt, G.A. Miller, M. M. Sargsian and M. I. Strikman, Phys. Rev. Lett. **84**, 3045 (2000).
2. M. M. Sargsian *et al.*, J. Phys. G **29**, R1 (2003) [arXiv:nucl-th/0210025].
3. S.J. Brodsky and B.T. Chertok, Phys. Rev. Lett. **37**, 269 (1976).
4. K. Egiyan *et al.*, Nucl. Phys. A **580**, 365 (1994) [arXiv:nucl-th/9401002].
5. L. L. Frankfurt *et al.*, Z. Phys. A **352**, 97 (1995) [arXiv:nucl-th/9501009].
6. L. L. Frankfurt *et al.*, Phys. Rev. C **56**, 1124 (1997)[arXiv:nucl-th/9603018].
7. M. M. Sargsian, Int. J. Mod. Phys. E **10**, 405 (2001) [arXiv:nucl-th/0110053].
8. L. L. Frankfurt and M. I. Strikman, Phys. Rept. **160**, 235 (1988).
9. C. E. Carlson *et al.*, Ann. Rev. Nucl. Part. Sci. **47**, 395 (1997).
10. R. A. Gilman and F. Gross, J. Phys. G **28**, R37 (2002).
11. R. Feynman, *Photon Hadron Interactions*, W.A. Benjamin Inc., 1972.
12. J. Napolitano *et al.*, Phys. Rev. Lett. **61**, 2530 (1988); S.J. Freedman *et al.*, Phys. Rev. **C48**, 1864 (1993).
13. J.E. Belz *et al.*, Phys. Rev. Lett. **74**, 646 (1995).
14. C. Bochna *et al.*, Phys. Rev. Lett. **81**, 4576 (1998).
15. E.C. Schulte *et al.*, Phys. Rev. Lett. **87**, 102302 (2001).
16. E. C. Schulte *et al.*, Phys. Rev. C **66**, 042201 (2002).
17. M. Mirazita *et al.*[CLAS Collaboration], Phys. Rev. C **70**, 014005 (2004)
18. S.J. Brodsky and G.R. Farrar, Phys. Rev. Lett. **31**, 1153; V. Matveev, R.M. Muradyan and A.N. Tavkhelidze, Lett. Nuovo Cimento **7**, 719 (1973).
19. N. Isgur and C.H. Llewellyn Smith, Phys. Rev. Lett. **52**, (1984) 1080.
20. A. Radyushkin, Acta Phys. Pol. **B15**, 403 (1984).
21. M. M. Sargsian, Phys. Lett. B **587**, 41 (2004) [arXiv:nucl-th/0309054].
22. M. M. Sargsian, arXiv:nucl-th/0208027.
23. S. J. Brodsky *et al.*, Phys. Lett. B **578**, 69 (2004) [arXiv:nucl-th/0305068].
24. S.J. Brodsky and J.R. Hiller, Phys. Rev. C **28**, 475 (1983).
25. L. A. Kondratyuk *et al.*, Phys. Rev. C **48**, 2491 (1993); V. Y. Grishina, *et al.* Eur. Phys. J. A **10**, 355 (2001).
26. J.M. Laget, Nucl. Phys. A **497**, 391 (1989).
27. R. Gilman and E. Piasetzky (Spokespersons). JLab Experiment, E-03-101.
28. K. Wijesooriya *et al.* [Jefferson Lab Hall A Collaboration.], Phys. Rev. Lett. **86**, 2975 (2001).
29. X. Jiang *et al.* [Jefferson Lab Hall A Coll.], Phys. Rev. Lett. **98**, 182302 (2007)

PHOTODISINTEGRATION OF THE DEUTERON, AND ^{3}He

R. GILMAN

Department of Physics & Astronomy, Rutgers University,
Piscataway, NJ 08854, USA
and
Jefferson Lab
Newport News, VA 23606, USA
E-mail: rgilman@physics.rutgers.edu
http://www.physics.rutgers.edu/~rgilman

In this article, I review what has been learned about the underlying mechanisms of deuteron photodisintegration, mostly above the resonance region, and the prospects for improving our understanding of high-energy photodisintegration through proton-proton photodisintegration in ^{3}He.

Keywords: Deuteron photodisintegration.

1. Introduction

Determining the quark-hadron transition has been a central goal of nuclear physics for some time. Studies of the deuteron, particularly elastic scattering and photodisintegration, have been a primary source of information on this transition in nuclei.[1] Ideally, as one probes the nucleus with increasing energy and momentum transfer, cross sections and polarization observables undergo a transition in their behavior, akin to a phase transition, that clearly signals the transition from a low energy, low momentum transfer hadronic regime to a high energy, high momentum transfer quark-gluon regime. Even with no clear signal in the data, one expects a transition in the ability to formulate a description in quark vs. in hadronic degrees of freedom.

The quark-hadron transition in certain reactions, such as inclusive (e, e') scattering, is reasonably well understood.[2] There is, for example, the deep inelastic scattering quark regime, and the duality regime, where the hadronic description averages to the quark description.

Observing the transition in exclusive reactions has been more problematic. Reactions such as quasi-free $d,^{3}He(e, e'p)n$ and ed elastic scattering, measured to large momentum transfers, are generally well understood with hadronic theories.[3–8] In inclusive reactions, one needs both large energy - more specifically large center of mass energy W - and large momentum transfer Q^2 to reach the deep inelastic regime. Thus, it makes sense that quasi-free scattering, with $W = m_p$, and elastic scattering, with $W = m_d$ for the deuteron as a whole, but $W^2 < m_p^2$ for the struck nucleon, show little if any indication of quark behavior. Whatever the quark effects are, they are already largely incorporated into the hadronic theory.

To obtain high energy and momentum transfer, we turn to photodisintegration of the deuteron. Above $E_\gamma = 1$ GeV, with $W^2 = s = 2E_\gamma m_d + m_d^2$, W is always above 2.7 GeV – for a proton target, the 1 GeV photon leads to $W > 1.66$ GeV. Also, at $\theta_{c.m.} = 90°$, we have a four momentum transfer $-t > 1$ GeV^2; $-t$ in real photo-reactions corresponds to Q^2 in virtual photon transfer. Thus, the kinematics for photodisintegration above 1 GeV roughly correspond to the kinematics for inclusive deep inelastic scattering.

2. High Energy Deuteron Photodisintegration

Experiments[9–14] at SLAC and Jefferson Lab have shown that the photodisintegration cross section for $E_\gamma > 1$ GeV and $-t > 1.1$ GeV^2 roughly follow the constituent counting prediction,[15] $d\sigma/dt \propto s^{-11}$ at constant center of mass angle, with s the square of the center of mass energy. The behavior is amazingly good;[16] the cross section has been measured to fall about a factor of $\approx$30,000 with an s^{-11} dependence, from 1 to 4 GeV at $\theta_{c.m.} = 90°$. This can be taken as an indication that quarks might be the appropriate underlying degrees of freedom, but it is not conclusive; in simple models hadronic degrees of freedom can lead to roughly the same behavior.

There is no satisfactory hadronic theory nor is there a prospect of a good conventional hadronic theory above 1 GeV photon energy.[1] There are simply too many possible resonance channels with poorly known amplitudes to have a reliable prediction. But there are no clean indications of any resonances in the data, and quark models, which automatically average over the resonances, are possible. Most existing quark models are based on the idea that the incoming photon is absorbed on a pair of quarks being exchanged between the two nucleons in the deuteron. Some models do approximate evaluations of this diagram;[17,18] others evaluate it by relating the photodisintegration to nucleon-nucleon scattering.[19,20] Finally, the quark-gluon string model[21] is based on the idea that the dominance of planar diagrams in QCD leads to the dominance of 3-quark exchange in photodisintegration, and evaluates the reaction using Regge phenomenology.

These models, and the underlying reaction dynamics, can be tested further with polarization observables. Only three high-energy experiments have been done. The linearly polarized photon asymmetry, Σ, was measured at Yerevan,[22] up to 1.6 GeV at $\theta_{c.m.} = 90°$. The data show that above 1 GeV Σ is moderately sized, 0.2 or 0.3, and positive; the highest energy point indicates Σ might be increasing towards 1, but it is also only 1σ above 0.3. While it was previously believed that perturbative QCD leads to hadron helicity conservation,[23] which leads to $\Sigma \rightarrow -1$, it is now understood that both chains of this argument have problems. Orbital angular momentum can lead to interesting spin effects, preventing hadron helicity conservation.[24] The limit of -1 relies on the photo-coupling being isoscalar; $\Sigma \rightarrow 1$ if the photo-coupling is isovector.[25] Thus, the data might be indicating isovector dominance, and a more extensive measurement of the Σ asymmetry, which can easily be done at Jefferson Laboratory Hall B, is highly desirable.

Recoil proton polarizations have been measured in two high-energy experiments in Jefferson Lab Hall A. E89-019[26] found that at $\theta_{c.m.} = 90°$ the induced polarization steadily decreases in magnitude, going from near -1 at 500 MeV to essentially 0 by 1 GeV. This contradicts the expected hadronic behavior, in which resonances lead to large, strongly energy-dependent induced polarizations. The transferred polarizations are moderate in size near 1 GeV, but appear to steadily decrease with energy at higher energies. The ensuing experiment, E00-007,[27] measured a 5-point angular distribution for $E_\gamma \approx 2$ GeV, from 37° to 110°. The polarizations have a smooth dependence with angle. The induced polarization p_y and transferred polarization $C_{x'}$ (transverse in the reaction plane) start out negative and moderately sized, but cross zero near $\theta_{c.m.} = 90°$ and are positive at larger angles. The transferred longitudinal polarization $C_{z'}$ starts out large and positive, decreasing to be zero near $\theta_{c.m.} = 110°$. From the hard rescattering model,[28] if isovector photon coupling dominates, then the NN amplitude ϕ_5 dominates, which drives p_y and $C_{x'}$ to zero – these are each proportional to ϕ_5 multiplied by a sum of other amplitudes. It is interesting that the polarization observables Σ, p_y, and $C_{x'}$ all seem to indicate a dominantly isovector coupling.

3. Low Energy Deuteron Photodisintegration

Hadronic models have been much improved over the last several years, particularly due to the work of Schwamb and Arenhövel.[29] While various problems have been resolved, there remain some prominent disagreements, most notably as the energy increases above 300 MeV, the Δ resonance region, in the induced polarization p_y, which led to much discussion of exotic dibaryons in deuteron photodisintegration back in the 1970s and 1980s.[a] Two recent experiments have probed this region in much more detail.

A Novosibirsk experiment[30] measured tensor polarizations across a wide angle range, $24° - 102°$, for energies from 25 to 440 MeV. Generally, below pion production threshold there is little model dependence to the calculations, and there is an excellent prediction of the data. At energies near the Δ resonance, the most modern calculations of Schamb and Arenhövel improve the description of T_{20} and T_{22}, but hurt the description of T_{21}; it is not clear whether the discrepancy arises from uncertainties in the calculation or missing underlying dynamics.

Experiment 05-103 in JLab Hall A[31] measured the recoil proton polarizations for energies from 280 – 360 MeV and angles from $20° - 110°$. This is the energy region in which p_y starts to grow in magnitude, contradicting calculations; even the modern Schwamb and Arenhövel calculations show that the induced polarization tends to be small near $\theta_{c.m.} \approx 90°$.[b] The aim of the experiment was to provide a systematic set of polarization data to try to help identify the missing reaction dynamics. Some preliminary data, shown in Fig. 1, are available for the polarization transfers; possible instrumental asymmetries make the online analysis of p_y more problematic. For P_z^c – which we have previously referred to as $C_{z'}$ – the most recent calculations are subtly different from earlier ones, but it is not clear which are superior; the predicted polarizations fall faster than, and cross over, the data. For P_x^c – which we have previously referred to as $C_{x'}$ – the most recent calculations are again slightly different from earlier ones, but here it is clear that the older calculations are in superior agreement with the data, as they show similar structures and are only slightly offset. It is not clear why the more recent calculations are slightly worse. All parameters in the calculations are taken from NN and photo-nucleon reactions, and the treatment of these reactions has been refined in the latest calculations.

---

[a] Another still unresolved notable problem from the same period is the induced neutron polarization at low energy.

[b] In the 1970s and early 1980s, this difference led to much consideration of dibaryons in deuteron photodisintegration, leading to much of the existing polarization data; while this explanation was eventually discarded, the missing reaction dynamics has yet to be identified.

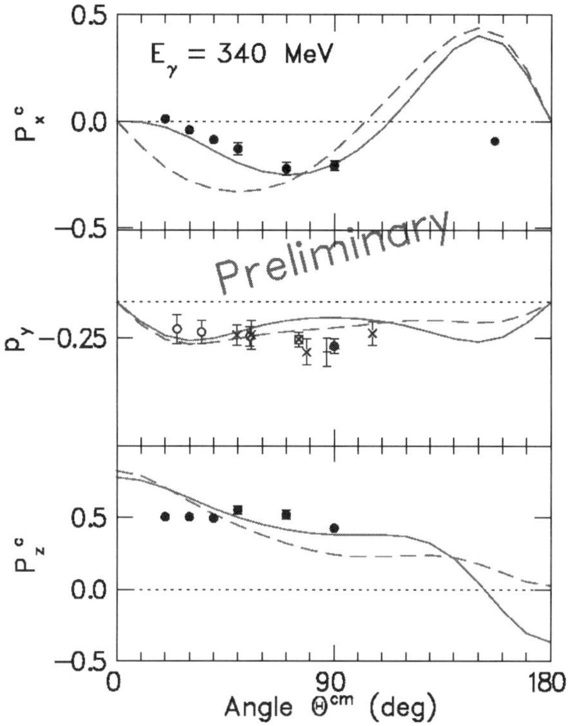

Fig. 1. Preliminary data from E05-103 for recoil transfer polarizations as a function of angle, for a 20 MeV energy bin about $E_\gamma = 340$ MeV. The induced polarization data come from earlier experiments. The solid curve comes from Schwamb and Arenhövel; the dashed curve is a more recent improved calculation from Schwamb.

4. *pp* photodisintegration in ^{3}He

The underlying mechanisms of *pn* or deuteron photodisintegration can be illuminated by comparison with *pp* photodisintegration. Since there is no *pp* bound state, it is most natural to use ^{3}He, as with only one undetected neutron the reaction kinematics can be completely reconstructed and final state interactions are minimized.

At low energies, *pp* disintegration is known to be small, which is understood to result from the two protons being largely in an $s = 0$ state, with no net magnetic moment. At high energies, different ideas about the underlying reaction dynamics lead to a range of predictions for the *pp* disintegration cross section, from much larger than deuteron photodisintegration to much smaller.[32]

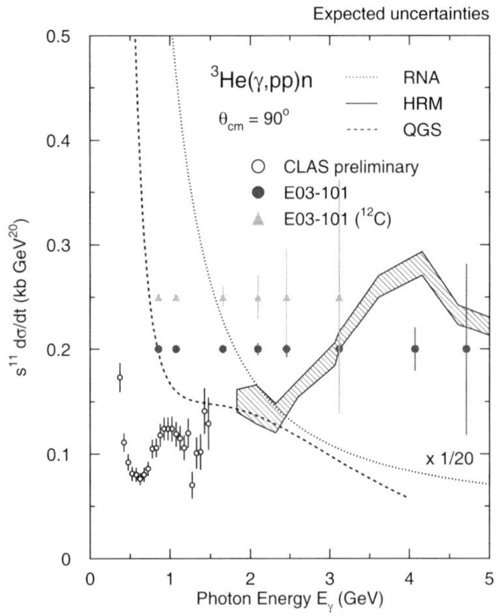

Fig. 2. The current state of high-energy $\gamma^3\text{He} \to pp + n$. CLAS data are preliminary. The energies and statistical uncertainties of Hall A E03-101 are indicated, but the vertical positions of the points are arbitrary. The Reduced Nuclear Amplitudes (RNA) calculation is divided by 20, so it is visible on the same scale. The Quark-Gluon String model (QGS) predicts pp photodisintegration is small compared to pn photodisintegration, while the Hard Rescattering Model (HRM) predicts the reactions are of similar size.

Experiment 03-101[33] ran in Hall A during summer, 2007, shortly after the *Exclusive Reactions* meeting. The data were largely at $\theta_{c.m.} = 90°$, with eight photon energies from 0.8 to 4.8 GeV, as shown in Fig. 2. Calibration data taken with carbon might also allow us to determine pp disintegration in carbon for all but the two highest energies. Online reconstructions show that we have generally clean proton spectra in each of the HRS spectrometers, clean coincidences between the spectrometers with little background, and clear Bremsstrahlung endpoints in the reconstructed photon energy spectra. The statistical uncertainties are generally in the range of a few to 10 % for all energies up to 4 GeV; systematic uncertainties are expected to be about 10%.

One important point in the photodisintegration of ^{3}He is that one can calculate the light cone momentum fraction of the undetected neutron. In general, $\alpha = \frac{E-p_z}{m}$; since α is a conserved quantity, we calculate $\alpha_n = \alpha_\gamma + \alpha_{^3He} - \alpha_{p_1} - \alpha_{p_2} = 0+3-\alpha_{p_1} - \alpha_{p_2}$, using the two protons measured in the HRS spectrometers to determine α_{p_1} and α_{p_2}. The interesting thing is that α_n provides an independent check of whether the pp photodisintegration is long or short range; a short range process leads to a broader, flatter distribution centered near $\alpha_n = 1$. In addition, the asymmetry of this α_n distribution reflects how fast the cross section is falling with s. Thus, pp photodisintegration potentially has more information about the underlying dynamics than does deuteron photodisintegration.

Another interesting observation is that pp elastic scattering has prominent oscillations about the scaling prediction of energy dependence; in pn elastic scattering, oscillations are not so apparent. Thus, models which relate photodisintegration to NN scattering predict oscillations in the photodisintegration, compared to the scaling energy dependence. Final experimental results should be available within a year.

5. Summary

Recent experiments at a few hundred MeV are providing stringent tests of improved modern hadronic calculations. Recent measurements of ^{3}He photodisintegration should give us a much clearer picture of the underlying dynamics of deuteron photodisintegration. A new comprehensive measurement of the Σ asymmetry in Hall B is highly desirable, for its possible impact on the issue of isoscalar vs. isovector photo-coupling.

Acknowledgments and Appendices

The author acknowledges the support of the U.S. National Science Foundation, through grant PHY 06-52713 to Rutgers University, and the U.S. Department of Energy, which supports Jefferson Laboratory through Contract No. DE-AC05-06OR23177 to Jefferson Science Associates, LLC. The U.S. Government retains a non-exclusive, paid-up, irrevocable, world-wide license to publish or reproduce this manuscript for U.S. Government purposes.

The author thanks S. Brodsky, G. Miller, M. Sargsian, M. Schwamb, and M. Strikman for many helpful discussions. The author thanks his colleagues for their work on the experiments described, particularly for the recent experiments the students J. Dumas, J. Glister, B.W. Lee, E. McCullough, I. Pomerantz, G. Ron, and Y. Rousseau, as well as my co-spokespeople D. Higinbotham, X. Jiang, E. Piasetzky, A. Sarty, and S. Strauch.

References

1. R. Gilman and Franz Gross, J. Phys. G **28**, R37 (2002).
2. W. Melnitchouk, R. Ent, and C. Keppel, Phys. Rep. **406**, 127 (2005).
3. R. Gilman, Nucl. Phys. A **737**, 156-164 (2004).
4. L.C. Alexa *et al.*, Phys. Rev. Lett. **82**, 1374 (1999).
5. D. Abbott *et al.*, Phys. Rev. Lett. **84**, 5053 (2000).
6. P.E. Ulmer *et al.*, Phys. Rev. Lett. **89**, 062301 (2002).
7. F. Benmokhtar *et al.*, Phys. Rev. Lett. **94**, 082305 (2005)
8. M. Rvachev *et al.*, Phys. Rev. Lett. **94**, 192302 (2005).
9. J. Napolitano *et al.*, Phys. Rev. Lett. **61**, 2530 (1988); S.J. Freedman *et al.*, Phys. Rev. C **48**, 1864 (1993).
10. J.E. Belz *et al.*, Phys. Rev. Lett. **74**, 646 (1995).
11. C. Bochna *et al.*, Phys. Rev. Lett. **81**, 4576 (1998).
12. E.C. Schulte *et al.*, Phys. Rev. Lett. **87**, 102302 (2001).
13. E.C. Schulte *et al.*, Phys. Rev. C **66**, 042201R (2002).
14. M. Mirazita *et al.*, Phys. Rev. C **70**, 014005 (2004).
15. S.J. Brodsky and G.R. Farrar, Phys. Rev. Lett. **31**, 1153 (1973); V. Matveev *et al.*, Nuovo Cim. Lett. **7**, 719 (1973).
16. P. Rossi *et al.*, Phys. Rev. Lett. **94**, 012301 (2005).
17. S.J. Brodsky and J.R. Hiller, Phys. Rev. C **28**, 475 (1983).
18. A. Radyushkin, private communication.
19. L.L. Frankfurt, G.A. Miller, M.M. Sargsian, and M.I. Strikman, Phys. Rev. Lett. **84**, 3045 (2000); L.L. Frankfurt, G.A. Miller, M.M. Sargsian, and M.I. Strikman, Nucl. Phys. A **663**, 349 (2000).
20. B. Julia-Diaz and T.-S. H. Lee, Mod. Phys. Lett. A **18**, 200 (2003).
21. E. De Sanctis *et al.*, Few Body Syst. Suppl. **6**, 229 (1992); L. A. Kondratyuk *et al.*, Phys. Rev. C **48**, 2491 (1993); V.Yu Grishina *et al.*, Eur. Phys. J. A **10**, 355 (2001).
22. F. Adamian *et al.*, Eur. Phys. J. A **8**, 423 (2000).
23. See S.J. Brodsky and G.P. Lepage, Phys. Rev. D **24**, 2848 (1981).
24. T. Gousset, B. Pire, and J.P. Ralston, Phys. Rev. D **53**, 1202 (1996).
25. V.Yu Grishina *et al.*, Eur. Phys. J. A **18**, 207 (2003); V.Yu Grishina *et al.*, Eur. Phys. J. A **19**, 117 (2004).
26. K. Wijesooriya *et al.*, Phys. Rev. Lett. **86**, 2975 (2001).
27. X. Jiang *et al.*, Phys. Rev. Lett. **98**, 182302 (2007).
28. M.M. Sargsian, Phys. Lett. B **587**, 41 (2004).
29. M. Schwamb and H. Arenhövel, Nucl. Phys. A **690**, 682 (2001),Nucl. Phys. A **696**, 556 (2001),Mod. Phys. Lett. A **18**, 190 (2003);M. Schwamb, habilitation thesis, Johannes Gutenberg-Universität Mainz, 2006.
30. I. Rachek *et al.*, Phys. Rev. Lett. **98**, 182303 (2007).
31. R. Gilman, A. Sarty, S. Strauch *et al.*, Jefferson Lab E05-103; J. Glister, G. Ron, B.W. Lee *et al.*, to be published.
32. S.J. Brodsky *et al.*, Phys. Lett. B **578**, 69 (2003); E. Piasetzky *et al.*, Jefferson Lab proposal E03-101.
33. E. Piasetzky, R. Gilman *et al.* Jefferson Lab E03-101; I. Pomerantz *et al.*, to be published.

A REVIEW OF THE FEW-BODY FORM FACTORS

GERASSIMOS G. PETRATOS

Physics Department, Kent State University,
Kent, Ohio 44242, USA
**E-mail: gpetrato@kent.edu*

A review of measurements of the elastic form factors of the deuteron and the isotopes of helium is presented. The existing data are compared to selected theoretical calculations. Recent JLab measurements, currently in the analysis phase, of the helium form factors at large momentum transfers are described. Plans for future JLab measurements are also summarized.

Measurements of the elastic form factors of the few-body nuclear systems are of crucial importance in understanding their electromagnetic structure.[1,2] They offer unique opportunities to study the short-range nucleon-nucleon interaction, meson-exchange currents (MEC), isobar configurations (IC) and three-body force contributions, and effects from possible quark-cluster wave function admixtures. Large momentum transfer measurements can test "nuclear chromodynamics" predictions based on quark dimensional scaling and perturbative Quantum Chromodynamics (pQCD).[3]

The conventional theoretical approach of elastic electron scattering from few-body systems is based on the impulse approximation (IA), where the electron interacts with one of their nucleons. The few-body form factors are then convolutions of the nuclear wave function with the form factors of the constituent nucleons. It has long been understood and overwhelmingly supported by the available data that these form factors are sensitive to the presence of MEC and IC that augment the IA picture.[1] At large momentum transfers, Q, the effects of relativity cannot be ignored and either IA corrections or fully relativistic approaches are necessary.[4,5]

It is widely recognized that at distances much less than the nucleon size, the underlying quark substructure of the nucleons cannot be ignored. This has lead to the formulation of so called hybrid quark models[6] that treat few-body nuclei as quark clusters when the internucleon separations become smaller than ~1 fm. At sufficiently "large" Q^2, the few-body form factors are expected to be described in terms of only quarks and gluons within the

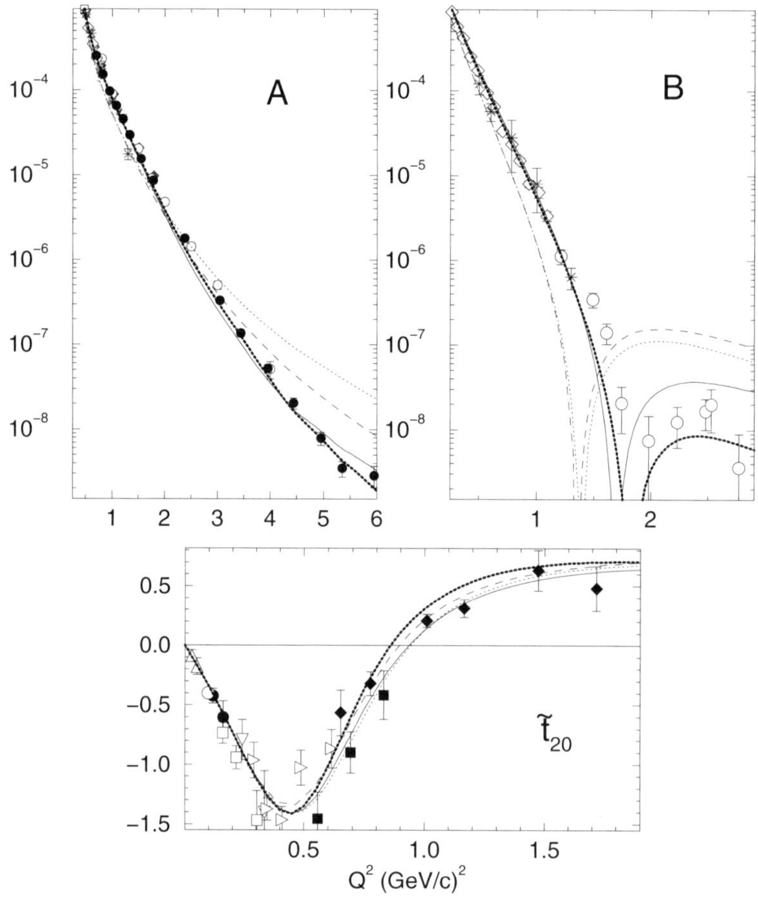

Fig. 1. Experimental data on the deuteron $A(Q^2)$, $B(Q^2)$ and $\tilde{t}_{20}$ compared with selected theoretical calculations based on Propagator Dynamics (from Ref. 2, see text).

framework of pQCD. The first attempt for such a description was based on the dimensional-scaling quark model (DSQM),[7] where the underlying dynamical mechanism during elastic scattering is the hard rescattering of the constituent quarks via exchange of hard gluons. The Q^2 dependence of this process is then predicted by simply counting the number n of gluon propagators. For the deuteron case (n=5), the "electric" elastic structure function, $A(Q^2)$, should follow the power law: $\sqrt{A(Q^2)} \sim (Q^2)^{-5}$, an asymptotic behavior which was subsequently explicitely substantiated in the pQCD framework.[8]

Measurements of the deuteron from factors have been performed over the last 50 years at several laboratories.[2,4] The existing data for $A(Q^2)$ (es-

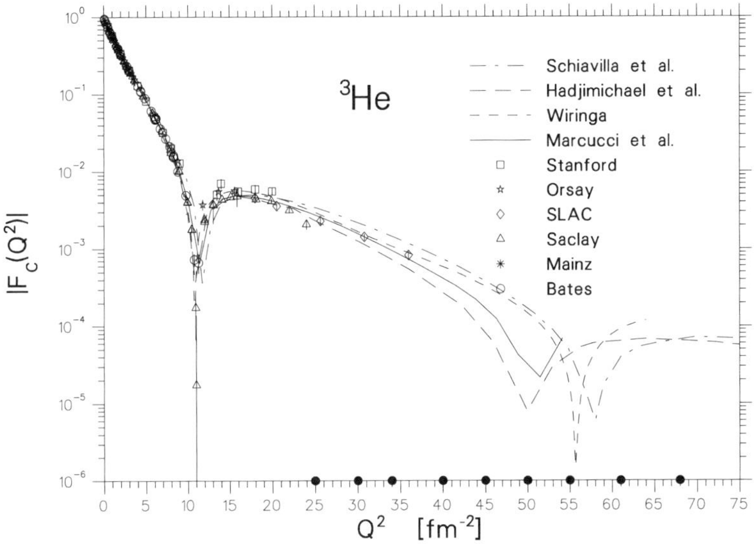

Fig. 2. ^{3}He F_C data compared to four different calculations based on the IA with inclusion of MEC (for details see Ref. 14). The solid black circles indicate the Q^2 values of new JLab measurements under analysis.[14]

sentially a linear combination of the squares of the charge and quadrupole form factors, $F_C^2(Q^2)$ and $F_Q^2(Q^2)$), the "magnetic" elastic structure function $B(Q^2)$ (proportional to the square of the magnetic form factor, $F_M^2(Q^2)$), and the reduced tensor tensor polarization $\tilde{t}_{20}(Q^2)$ (function of the ratio F_C/F_Q) are shown, along with selected relativistic calculations. The calculations are based on Propagator Dynamics using three-dimensional reductions of the Bethe-Salpeter equation [sometimes referred to as the relativistic impulse approximation (RIA)]. The data for $A(Q^2)$ indicate a smooth fall-off with no apparent diffractive structure. The data for $B(Q^2)$ indicate the presence of a diffraction minimum in the vicinity of $Q^2 = 1.8$ (GeV/c)^2. The $\tilde{t}_{20}(Q^2)$ data together with the $A(Q^2)$ and $B(Q^2)$ data show that the charge form factor exhibits a diffractive structure with a minimum at $Q^2 = 0.7$ (GeV/c)^2 and that the quadrupole form factor falls off exponentially in the measured Q^2 region. A complete compilation of all available data and an extended overview of the theory are given in Ref. 4.

In Fig. 1, the solid, dotted and dashed curves represent the RIA calculations of Van Orden, Devine and Gross (VDG),[9] Hummel and Tjon (HT),[10] and Phillips, Wallace and Devine (PWD),[11] respectively. The VDG curve is based on the Gross quasipotential equation with a one-boson-exchange interaction and assumes that the electron interacts with an off-mass-shell

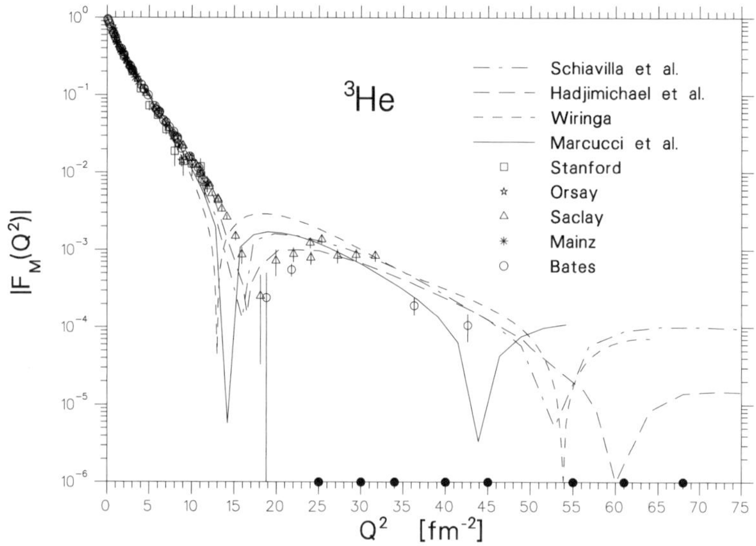

Fig. 3. ^{3}He F_M data compared to four different calculations based on the IA with inclusion of MEC (for details see Ref. 14). The solid black circles indicate the Q^2 values of new JLab measurements under analysis.[14]

nucleon or a nucleon that is on-mass-shell right before or after the interaction. The HT curve is based on a one-boson-exchange quasipotential approximation of the Bethe-Salpeter equation where the two nucleons are treated symmetrically by putting them equally off their mass-shell with zero relative energy. The PWD curve is based on an also one-boson-exchange interaction but with a single-time equation that constraints the relative time to be zero. All three calculations are augmented by the $\rho\pi\gamma$ MEC contribution. The thick dotted curve is an alternate calculation by the VDG group with a different nucleon form factor choice in the off-shell modification of the single nucleon current required to insure current conservation. Although the difference in the models is indicative of the size of theoretical uncertainties and ambiguities, it appears that the relativistic impulse approximation can, as in the case of the alternate VDG model, reproduce all deuteron form factor data fairly well. It should be noted that non-relativistic IA calculations or relativistic calculatations based on Hamiltonian Dynamics[12] (instant/point form or light front) fail to simultaneously describe all available deuteron form factor data.

The study of the (stable) helium isotope form factors has also been an integral part of all electron scattering laboratories over the last 50 years.[13] The conventional theoretical calculations for them are based on the IA.

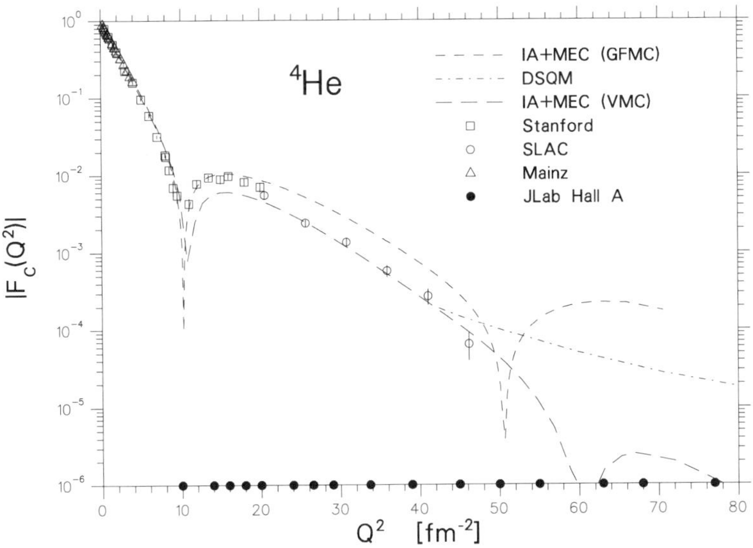

Fig. 4. ^{4}He F_C data compared to two IA calculations with inclusion of MEC using Green's Function Monte Carlo (GFMC) and Variational Monte Carlo (VMC) techniques (for details see Ref. 14). Shown is also the asymptotic DSQM prediction.[7] The solid black circles indicate the Q^2 values of new JLab measurements under analysis.[14]

The problem of solving for the nuclear ground states relies on using the Faddeev-Yakubosky equations, correlated hyperspherical harmonic functions or Green's function and variational Monte Carlo methods. The existing data on the charge and magnetic form factors, F_C and F_M, of ^{3}He are shown in Figs. 2 and 3, and on the charge form factor of ^{4}He in Fig. 4. All three form factors exhibit a diffractive future in qualitative agreement with theoretical calculations based on the IA, augmented by MEC (shown in Figs. 2-4), whose inclusion, as for the deuteron case, brings the theory in better agreement with the data. To obtain a very good agreement with the data, the theory must by augmented by phenomenological contributions from multi-quark clusters.[6]

A recently completed JLab experiment measured all three helium form factors at larger momentum transfers. The experiment, E04-018,[14] was carried out in the fall 2006 and spring and summer 2007 in the Lab's Hall A Facility. Elastic electron scattering off ^{3}He was measured at forward and backward electron scattering angles to extract its F_C and F_M form factors using the Rosenbluth separation method. For ^{4}He, only forward elastic electron scattering was measured to extract directly its F_C form factor. The experiment used high intensity electron beams with energy in the range between 0.7 and 4.4 GeV, and a high density cryogenic helium and hydrogen (for calibration) target systems. Scattered electrons and recoil nuclei were detected in coincidence in the Electron and Hadron High Resolution Spectrometers (HRS), respectively. The identification of electron-nucleus coincidences was accomplished by double-arm time-of-flight measurements between the electron and recoil detector electronic trigger signals. Data have been obtained in the Q^2 range from 25 to 65 fm^{-2} for ^{3}He, and from 10 to 75 fm^{-2} for ^{4}He, as indicated by the solid black circles in Figs 2, 3 and 4. The data analysis is in progress and preliminary results are expected within a year.

Finally, it should be noted that the high luminosities accecible at JLab offer the possibility of extending the deuteron $B(Q^2)$ measurements up to Q^2 = 5-6 $(\text{GeV/c})^2$ by detecting backward electrons in a custom made spectrometer and forward recoiling deuterons in one of the Hall A HRS's.[15] Also, the 12 GeV energy upgrade of JLab will offer the possibility for even larger Q^2 measurements of elastic electron scattering of few-body systems[16] and in particular to extend the deuteron $A(Q^2)$ measurements up to $Q^2 = 8$-9 $(\text{GeV/c})^2$ to test the recent relativistic calculations and the long-standing predictions of quark dimensional scaling and perturbative QCD.

References

1. J. Carlson and R. Schiavilla, *Rev. Mod. Phys.* **70**, 743 (1998).
2. M. Garcon and J. W. Van Orden, *Adv. Nucl. Phys.* **26**, 293 (2001).
3. C. E. Carlson, J. R. Hiller and R. J. Holt, *Annu. Rev. Nucl. Part. Sci.* **47**, 395 (1997).
4. R. Gilman and F. Gross, *J. Phys. G* **28**, R37 (2002).
5. F. Gross, A. Staedler and M. Pena, *Phys. Rev. C* **69**, 034007 (2004).
6. H. Dijk and B. L. G. Bakker, *Nucl. Phys. A* **494**, 438 (1989); W.-H. Ma and L. .S. Kisslinger, *Nucl. Phys. A* **531**, 493 (1991); and references therein.
7. S. J. Brodsky and G. R. Farrar, *Phys. Rev. Lett.* **31**, 1153 (1973); V. A. Matveev, R. M. Muradyan and A. N. Tavkhelidze, *Lett. Nuovo Cimento* **7**, 719 (1973); B. T. Chertok, *Phys. Rev. Lett.* **41**, 1155 (1978).
8. S. J. Brodsky, C-R. Ji and G. P. Lepage, *Phys. Rev. Lett.* **51**, 83 (1983).
9. J. W. Van Orden, N. Devine and F. Gross, *Phys. Rev. Lett.* **75**, 4369 (1995).
10. E. Hummel and J. A. Tjon, *Phys. Rev. C* **42**, 423 (1990).
11. D. R. Phillips, S. J. Wallace and N. K. Devine, *Phys. Rev. C* **58**, 2261 (1998).
12. B. D. Keister and W. N. Polyzou, *Adv. Nucl. Phys.* **20**, 225 (1991); J. Carbonell *et al.*, *Phys. Rep.* **300**, 215 (1998).
13. I. Sick, *Prog. Part. Nucl. Phys.* **47**, 245 (2001).
14. G. G. Petratos *et al.*, *JLab Proposal PR-04-018* (2004).
15. G. G. Petratos *et al.*, *JLab Proposal PR-03-112* (2003).
16. *The Science Driving the 12 Gev Upgrade of CEBAF*, JLab Report (2001).

NUCLEON FORM FACTOR MEASUREMENTS AND INTERPRETATION

C.F. PERDRISAT

the College of William and Mary, Williamsburg, VA 23187
E-mail: perdrisa@jlab.org

The data base for the form factors of the nucleon obtained from elastic *ep* scattering is discussed, as well as some the recent developments in their calculation.

I will first introduce the subject of measurements of the two form factors (FF) of the nucleon, G_E and G_M, or alternatively F_1 and F_2. A short description of the two methods used to obtain them, cross section and double polarization measurements follows, illustrated by displays of the complete data base for proton and neutron. The now well known "discrepancy", namely the systematic and significant difference between results from cross section, and from polarization experiments, will be discussed next, together with radiative corrections. Following will be a succinct description of recent progress in calculating the nucleon FFs. I will conclude with an outline of forthcoming experiments dedicated to the characterization of the nucleon FFs. Recent reviews of the subject can be found in [1–4].

Electromagnetic FFs are observables within the context of the Born approximation. They contain structure information on the many-body system of quarks and gluon which is the nucleon. However, the FFs extracted from experiment are relativistic invariants only to the extend that the probe is the single virtual photon exchanged between electron and nucleon. Higher order terms, regarded as radiative corrections, destroy this invariance.

In Born approximation the transition matrix is the product of the electron current, $j_\mu =< e'|\gamma_\mu|e >$ and of the nucleon current $J_\mu =< p'|\Gamma_\mu|p >$, where Γ_μ, describes the structure of the nucleon, and is of the form $\Gamma_\mu = \gamma_\mu F_1(Q^2) + \frac{i\sigma_{\mu\nu}}{2M} q^\nu F_2(Q^2)$. F_1 and F_2 are the Dirac and Pauli FFs; they contain the information on the spatial extent of the charge and magnetic distributions inside the nucleon. Of equally common usage are the

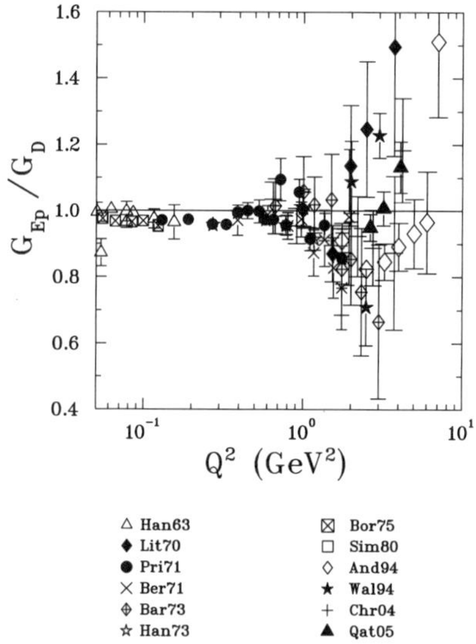

Fig. 1. *Complete data base for G_{Ep}/G_D, from Rosenbluth experiments.*

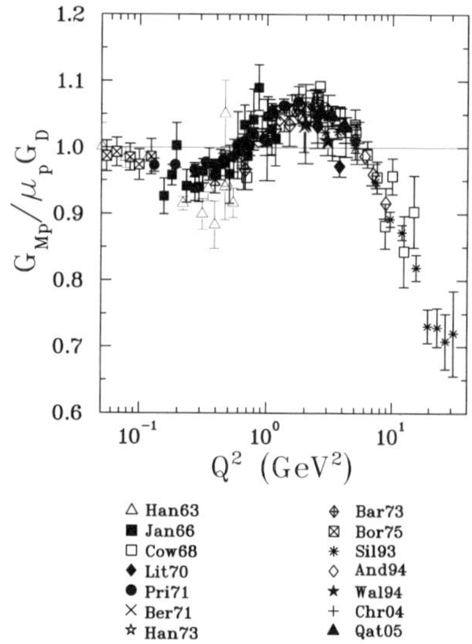

Fig. 2. Same as Fig. 1, for $G_{Mp}/\mu_p G_D$.

electric- and magnetic Sachs FFs:

$$G_E(Q^2) = F_1(Q^2) - \tau F_2(Q^2) \text{ and } G_M(Q^2) = F_1(Q^2) + F_2(Q^2).$$

The Sachs FFs appear in the Rosenbluth cross section squared:

$$\left(\frac{d\sigma}{d\Omega}\right)_{ep} = \left(\frac{d\sigma}{d\Omega}\right)_{Mott} \left(G_E^2 + \frac{\tau}{\epsilon} G_M^2\right),$$

with $\epsilon = [1 + 2(1+\tau)\tan^2 \frac{\theta_e}{2}]^{-1}$ and $\tau = Q^2/4M^2$, but as $G_E \times G_M$ in double polarization observables; the polarization of the recoiling nucleon emerging from the scattering of 100% longitudinally polarized electrons, has the two in-the-reaction plane components:

$$P_t = -2\sqrt{\tau(1+\tau)} G_E G_M \tan\frac{\theta_e}{2}/I_0 \text{ and}$$
$$P_\ell = \frac{(E_e + E_{e'})}{M}\sqrt{\tau(1+\tau)} G_M^2 \tan^2\frac{\theta_e}{2}/I_0,$$

where E_e, E_e' and θ_e are the incident energies, and scattered electron energy and angle, respectively, and $I_o = G_E^2 + \frac{\epsilon}{\tau} G_M^2$. The ratio of G_E and G_M is

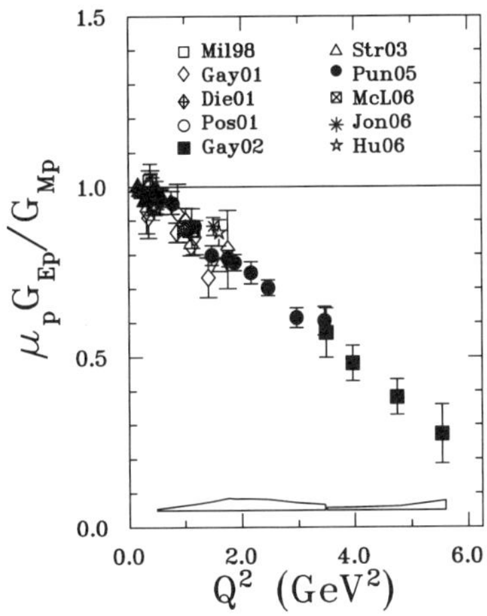

Fig. 3. *$\mu_p G_{Ep}/G_{Mp}$ from recoil polarization experiments.*

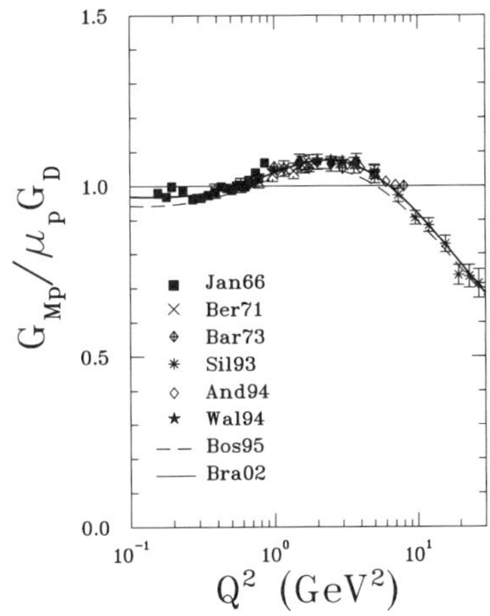

Fig. 4. *$G_{Mp}/\mu_p G_D$ after reanalyze in Ref. [9].*

then related to the ratio of the two polarization components:

$$\frac{G_E}{G_M} = -\frac{P_t}{P_\ell}\frac{(E_e + E_{e'})}{2M}\tan\left(\frac{\theta_e}{2}\right).$$

The results from all Rosenbluth measurements published so far, divided by the dipole form factor $G_D = (1 + Q^2/0.71)^{-2}$, are in Figs. 1 and 2. The unacceptable scatter of data seen in Fig. 1 existed in 1989 when the first proposal to use recoil polarization to obtain G_{Ep}/G_{Mp} with much greater sensitivity and smaller systematics, was presented.[5] In stark contrast is the internal consistency of the ratio data obtained in the first JLab recoil polarization experiment seen in Fig. 3.[6–8] Assuming that in fact Rosenbluth data above Q^2 = 1-2 GeV^2 should be ignored, the G_{Mp} data become even more internally consistent, as shown in the analysis of Ref. [9] shown in Fig. 4.

So what causes the different results, from cross section and from polarization? Part of the answer is that radiative corrections at large Q^2 are both large and strongly ϵ dependent. This is demonstrated in Fig. 5, where the Rosenbluth data of Andivahis *et al* [10] are shown as the reduced cross section:

$$\sigma_R = \frac{(1+\tau)}{\tau\epsilon}\left(\frac{d\sigma}{d\Omega}\right)_{ep} / \left(\frac{d\sigma}{d\Omega}\right)_{Mott} = \frac{\epsilon}{\tau}G_{Ep}^2 + G_{Mp}^2,$$

versus ϵ, without (bottom) and with (top) radiative correction. The striking

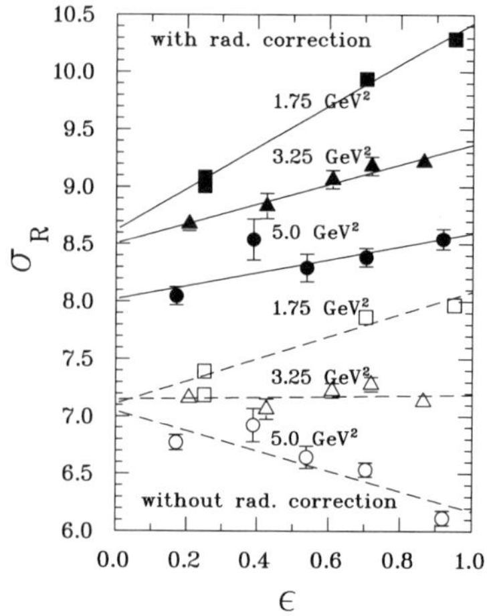

Fig. 5. *Reduced cross sections without and with radiative corrections .*[10] *Q^2 in GeV^2*

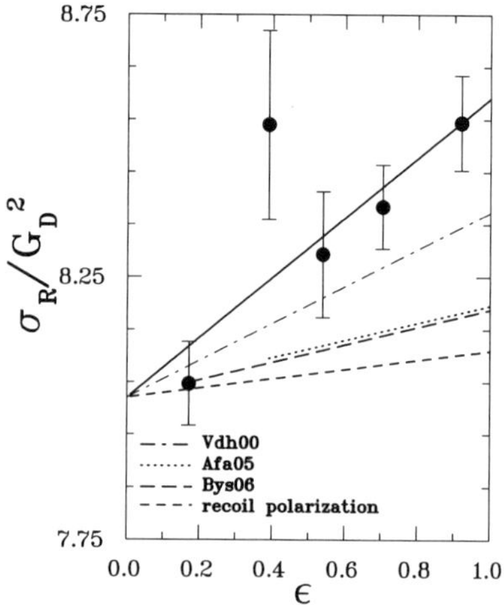

Fig. 6. *Several estimates of radiative corrections: [10,14,15,17]; 5 GeV^2 data from [10].*

feature seen in Fig. 5 is the change of slope resulting from the radiative correction at $Q^2 = 5$ GeV^2; the slope is G_{Ep}^2/τ. The accuracy of the radiative correction has to be of order 1% at $\epsilon = 1$ for this or larger Q^2. In Fig. 6 I show the radiative correction applied in [10], which is based on the Mo and Tsai[11] approach, and modified by Walker *et al*[12] (solid line). Also shown is the result of a calculation based on the Maximon and Tjon approach,[13] as obtained from a code developed by Vanderhaeghen[14] (dot-dashed line). The difference is about 3% at $\epsilon = 1$, i.e. half of the discrepancy between the data of [10] and the JLab polarization data (short dashed line). Also shown, is the result of Bystritskiy *et al* [15] (thick dashed line); even though it does not appear that the same energy cuts have been applied in this calculation as in [10], the point is made that radiative correction calculations are not stable at the required level of a few % .

The JLab results have led to a reexamination of the two-photon contribution, which is included in standard radiative corrections only in the limit of one of the photon energies being small. Recent calculations of the contribution of two-photon exchange with both photons approximately sharing the momentum transfer, first by Guichon *et al*,[16] then Afanasev *et al*,[17] and Blunden *et al*,[18] have shown that it probably contributes a correction to the standard radiative correction of several % at ϵ =1. As an example the calculation in [17] is shown as the dotted line in Fig. 6.

The final answer to the question of the origin of the discrepancy will have to come from experiments. There are several experiments designed to provide answers to the question. For example:

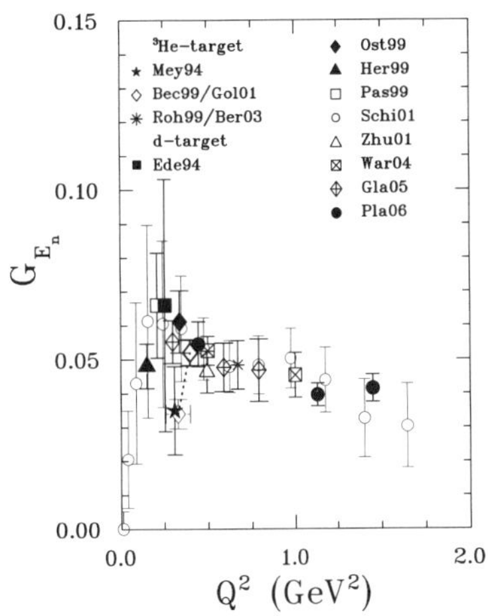

Fig. 7. *Data base for G_{En}.*

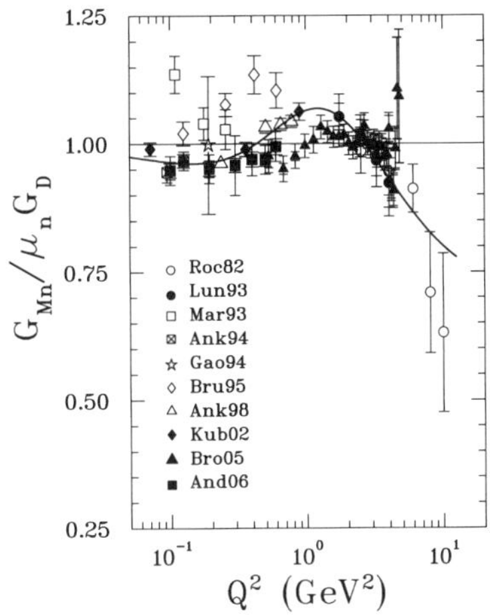

Fig. 8. *Data base for G_{Mn}.*

1) ϵ-dependence of the G_E/G_M ratio at constant Q^2, obtained from recoil polarization: E04-019 in Hall C, Fall 2007.[19]

2) cross section difference in e^+ and e^- proton scattering; in Hall B.[20]

3) non-linearity of Rosenbluth plot; E05-017, data taking completed.[21]

All three determine the real part of the 2-photon amplitude.

4) Also, the imaginary part of the two-photon amplitude can be obtained from measurements of the induced, out-of-plane polarization component; such information is likely to be a by-product of E04-019,[19] and single spin target asymmetry on the neutron: E05-015[22] with a $^3He\uparrow$-target.

5) Single spin beam asymmetry measurement at Bates[23] and Mainz ,[24] are likewise related to the imaginary part of the two-photon amplitude but with asymmetries at the level of parts per million.

The corresponding data for the neutron have improved too; all recent experiments measuring G_{En} have used the double polarization technique; the current data base is shown in Fig. 7. New data up to 3.4 GeV^2 from Hall C are currently being analyzed.[25] The complete data base for G_{Mn} is shown in Fig. 8; included in this figure are older results probably affected by systematics (open symbols), as well as the preliminary Hall B data of Brooks et al.[26]

As shown in Fig. 9, the long standing pQCD prediction from Brodsky

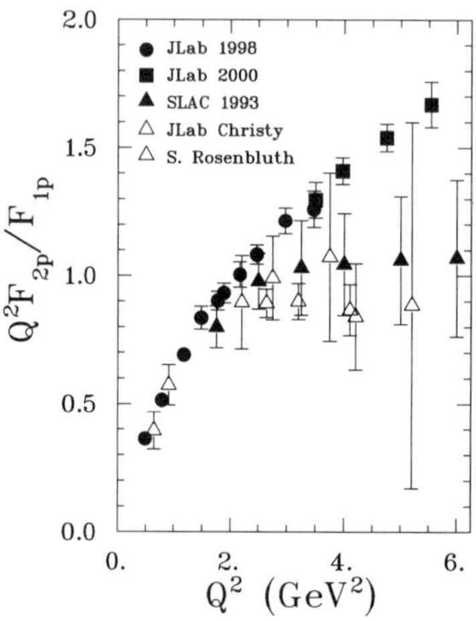

Fig. 9. *F_2/F_1 and perturbative QCD.*

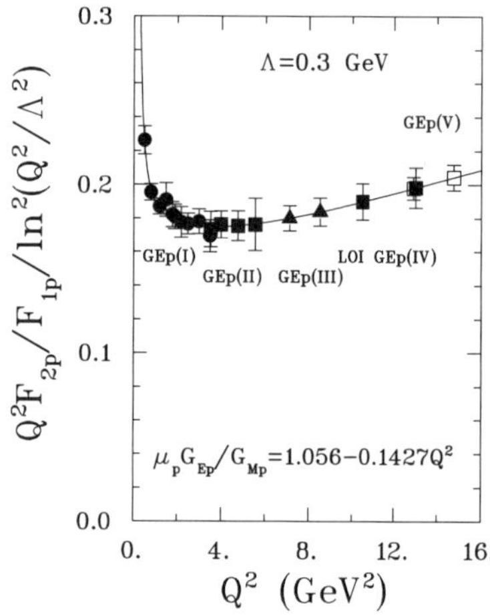

Fig. 10. *Logarithmic scaling according to Belitzky et al.*

and Farrar[27] is not followed by the proton FFs F_1 and F_2; the ratio F_2/F_1 exhibits instead a $\sim 1/Q$ behavior. In Fig. 10 it is shown that scaling may occur at rather low Q^2 for the ratio $Q^2 F_1 F_2 / \ln^2(Q^2/\Lambda^2)$; Belitsky et al[28] have argued that this behavior would be a signature of the contribution of quark orbital angular momentum.

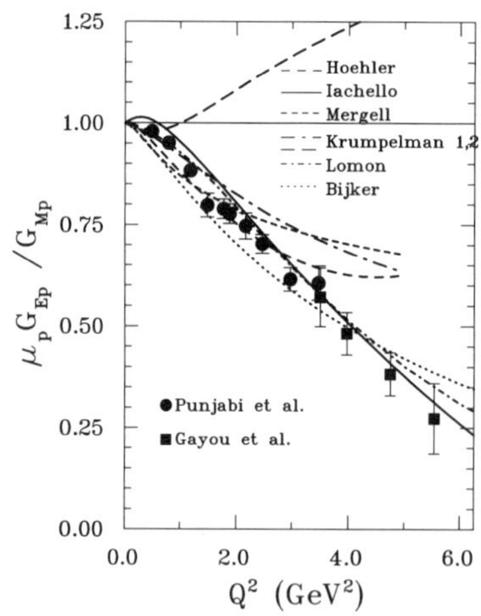

Fig. 11. *VMD models of the FF ratio; data from Refs.[7,8].*

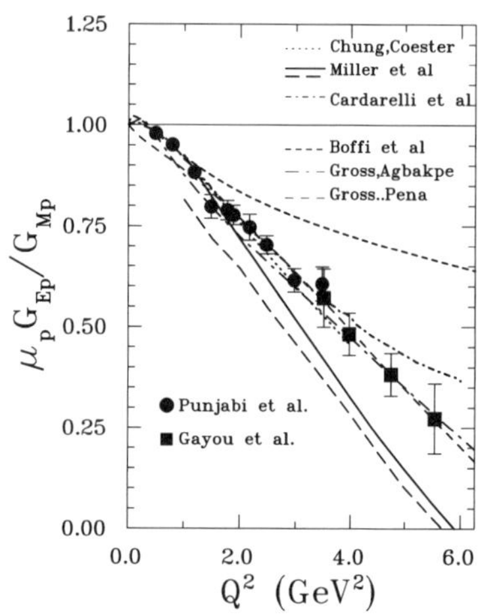

Fig. 12. *CQM models of the FF ratio; data from Refs.[7,8].*

The earliest attempt at understanding proton FF data were based on the Vector Dominance Model, with the work of Iachello,[29] the Hoehler group[30] and Gari and Krumpelmann.[31] More recent work from the Bonn and COSY group includes;[32,33] independently, Lomon has systematically refined the Gari and Krumpelman approach;[34] the results of these analyzes are in Fig. 11.

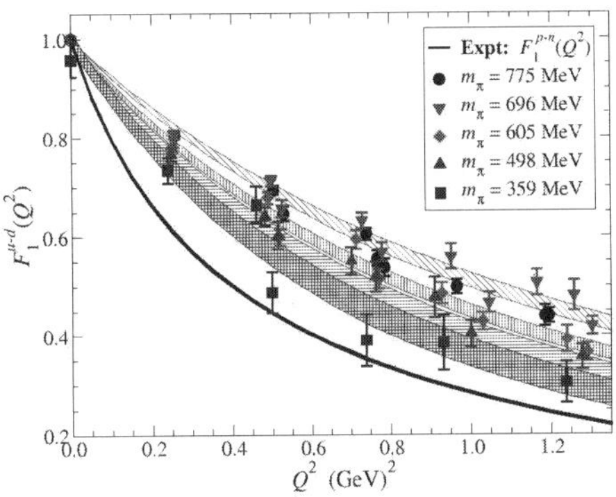

Fig. 13. *Isovector FF in lattice QCD.*[42]

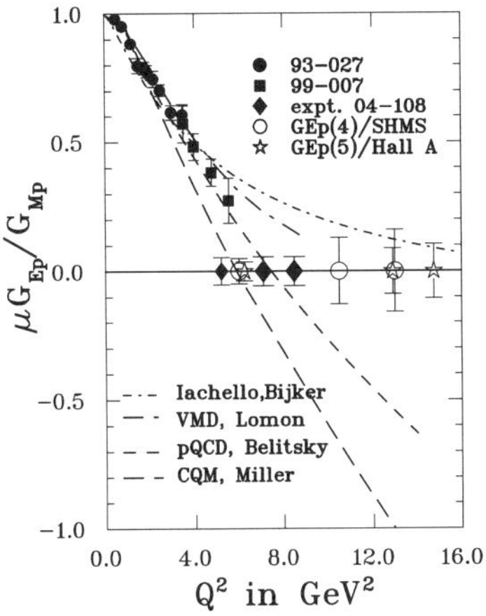

Fig. 14. *Error bars of planned measurements of G_{Ep}/G_{Mp}.*

In Fig. 12 a few of the typical relativistic constituent quark model (CQM) calculations are shown. The model was originally proposed by Isgur and Karl,[35] and the three forms of dynamics possible in the relativistic version of the model, had been already discussed by Dirac:[36] instant, point-like and light-front. An early example of the light-front form is due to Chung and Coester,[37] as are calculations from Miller and collaborators.[38] A more recent example of the wave-front form is from Cardarelli.[39] An example of point-form calculation from Boffi *et al* is also shown in the figure. Also shown are the covariant spectator predictions of Gross and Agbakpe,[40] and Gross, Ramalho and Peňa.[41]

The nucleon FF can also be calculated from first principle, solving the QCD Lagrangian on the lattice. Predictions are still limited by computer power to pion mass m_π larger than natural. The isovector FF $F_1^V = F_1^p - F_1^n$ can be obtained without the need to include disconnected $q\bar{q}$-loops. As seen in Fig. 13, the latest calculations of F_1^V are in fair agreement with experiment.[42]

The measurement of nucleon FFs to the highest possible Q^2 is one of the prime tenets of the JLab 12 GeV upgrade. Several experiments will still be done with the 6 GeV beam; in particular E-04-108, which will measure G_{Ep}/G_{Mp} to about 9 GeV^2. A letter of intent has been submitted,[43] and a proposal approved,[44] to pursue these measurements into the region of Q^2 13 to 15 GeV^2 with 11 GeV; the anticipated statistical error bars of all future G_{Ep}/G_{Mp} measurements are shown in Fig. 14.

To conclude, the experimental characterization of the FFs of the nucleon, and their phenomenological understanding, have changed drastically since the publication of the first ratio measurements in 2000.[6] The rapid decrease of G_{Ep} should not have been a complete surprise: at least three papers[29,38,45] had predicted it prior to this experiment. A full understanding of the implications of the JLab findings is not in yet. Where does the pQCD behavior kick in? are the nucleons spherical in their ground state? Are we "really" seeing the effect of quark orbital angular momentum? If the shape of the FFs is not diffractive, what does GEp=0 or GEp<0 mean? Also to come is a full understanding of two-photon effects, and revision of standard radiative correction calculation codes. Several of the forthcoming experiments will address these questions.

Notice: Authored by Jefferson Science Associates, LLC under U.S. DOE Contract No. DE-AC05-060R23177. The U.S. Government retains a non-exclusive, paid-up, irrevocable, world-wide license to publish or reproduce this manuscript for U.S. Government purpose.

Acknowledgment

I like to thank my colleague V. Punjabi for help in preparing this manuscript; I acknowledge support from a NSF grant, PHY-0456645.

References

1. H.Y. Gao, Int. J. Mod. Phys. E**12**, 1 (2003).
2. C.E. Hyde-Wright and K. de Jager, Ann. Rev. Nucl. Part. Sci **54**, 217 (2004)
3. J. Arrington, C. D. Roberts and J. M. Zanotti, J. Phys. G **34**, S23 (2007)
4. C.F. Perdrisat, V. Punjabi and M. Vanderhaeghen, Journal of Progress in Particle and Nuclear Physics, (2007): http://sciendirect.com/science/journal/01466410.
5. "Electric Form Factor of the Proton by Recoil Polarization", C.F. Perdrisat and V. Punjabi, E-93-027 (1987).
6. M.K. Jones, *et al*, Phys. Rev. Lett. **84**, 1398 (2000).
7. V. Punjabi *et al*, Phys. Rev.C *71*, 055202 (2005, errat. ibid C *71*, 069902 (2005).

8. O. Gayou *et al*, Phys. Rev. Lett. **88**, 092301 (2001); also O. Gayou, William and Mary PhD thesis unpublished (2002).
9. E.J. Brash, A. Kozlov, S. Li and G.M. Huber, Phys. Rev. C **65**, 051001 (2002).
10. L. Andivahis *et al*, Phys. Rev. D **50**, 5491 (1994).
11. L.W. Mo and Y.S. Tsai, Rev. Mod. Phys. **41**, 205 (1969).
12. Walker R.C. *et al*, Phys. Rev. D **49**, 5671 (1994).
13. L.C. Maximon and J.A. Tjon, Phys. Rev. C**62**,054320 (2000).
14. M. Vanderhaeghen *et al*, Phys. Rev. C**6**, 025501 (2000).
15. Yu. M. Bystriskiy, E.A. Kuraev and E. Tomasi-Gustafsson, Phys. Rev. C**75**, 015207 (2007).
16. P.A.M. Guichon,, and M. Vanderhaeghen, Phys. Rev. Lett. **91**, 142303 (2003).
17. A.V. Afanasev, *et al*, Phys. Rev. D**71**, 013008 (2005).
18. P.G. Blunder, W. Melnitchouk and T.A. Tjon, Phys. Rev.Lett. **91**, 142304 (2003), and Phys. Rev. C**72**, 034612 (2005).
19. R. Gilman, L. Pentchev, C.F. Perdrisat and R. Suleiman, Hall C experiment E04-019 (2004).
20. A. Afanasev *et al*, E07-005 (alias E04-116).
21. "A measurement of Two-Photon Exchange in Unpolarized Elastic *ep* Scattering, J. Arrington E-05-017 (2005).
22. T. Averett *et al.* E-05-015 (2005).
23. S.P. Wells *et al*, Phys. Rev. C**63**, 064001 (2001).
24. F.E. Maas *et al*, Phys. Rev. Lett **94**, 082001 (2005).
25. G. Cates, K. McCormick, B. Reitz and B. Wojtsekhowski, JLab experiment E-02-013 (2002).
26. W.K. Brooks and J.D. Lachniet, Nucl. Phys. A**755**, 261 (2005).
27. S. Brodsky and G.R. Farrar, Phys. Rev. D **11**, 1309 (1975).
28. A.V. Belitsky, X. Ji and F. Yuan, Phys. Rev. Lett. **91**. 092003 (2003).
29. F. Iachello, A.D. Jackson and A. Landé, Phys. Lett. B **43**, 191 (1973).
30. G. Hoehler it et all, Nucl. Phys. B [bf 114, 505 (1976).
31. M.F. Gari and W. Kruempelmann, Z. Phys. A **322**, 689 (1985) and Phys. Lett. B **274**, 159 (1992).
32. P. Mergell, U.G. Meissner and D. Drechsel, Nucl. Phys. A**596**, 367 (1996).
33. M.A. Belushkin, H.W. Hammer and U.G. Meissner, Phys. Lett. B**633**, 507 (2006).
34. E.L. Lomon, Phys. Rev. C **64**, 035204 (2001), Phys. Rev. C**66**, 045501 (2002) and arXiv:nucl-th/0609020 (2006).
35. N. Isgur and G. Karl, Phys. Rev. D**18**, 4187 (1978); **19**, 2653 (1979); **20** 1191 (1979); C. Hayne and N. Isgur,*ibid* **25**,1944 (1982).
36. P.A.M. Dirac, Rev. Mod. Phys. **21**, 392 (1949).
37. Chung P. L. and F. Coester, Phys. Rev. D **44**, 229 (1991).
38. M.R. Frank, B.K. Jennings, and G.A. Miller, Phys. Rev. C **54**, 920 (1996); G.A. Miller and M.R. Frank, Phys. Rev. C **65**, 065205 (2002).
39. Cardarelli F. and S. Simula, Phys. Rev. C **62**, 65201 (2000).
40. F. Gross and P. Agbakpe, Phys. Rev. C**73**, 015203 (2006).

41. F. Gross, G. Ramalho and M.T. Peňa, arXiv:nucl-th/0606029 (2006).
42. R.G. Edwards *et al*, (LHPC collaboration), arXiv:hep-lat/0610007 (2006).
43. "G_{Ep}/G_{Mp} with an 11 GeV beam", C.F. Perdrisat, V.Punjabi, M.K. Jones and E.J. Brash. JLab LOI12-06-103 (2006).
44. "Large Acceptance Proton Form Factor Ratio Measurements at 13 and 15 GeV^2 Using Recoil Polarization", L. Pentchev, C.F. Perdrisat, E. Cisbani, V.Punjabi and B. Wojtsekhowski, E12-07-109.
45. G. Holzwarth, Z. Phys. A**356**, 339 (1996).

IMPLICATIONS OF $G_E^p(Q^2)/G_M^p(Q^2)$

S. DUBNIČKA

Institute of Physics, Slovak Academy. of Sciences, Bratislava, Slovak Republic
E-mail: _fyzidubn@savba.sk
www.savba.sk

A. Z. DUBNIČKOVA

Department of Theoretical physics, Faculty of mat., phys. and informatics, Comenius University, Bratislava, Slovak Republic
E-mail: _dubnickova@fmph.uniba.sk

The analysis of the JLab proton polarization data on the ratio $\mu_p G_{Ep}(Q^2)/G_{Mp}(Q^2)$, together with all other existing nucleon form factor data, by the unitary and analytic model of nucleon electromagnetic structure reveals them to be consistent with all known nucleon form factor properties. Moreover, it implicates surprising non-dipole behavior of the proton electric form factor in the space-like region with the zero around the momentum transfer squared value $Q^2 = 13GeV^2$. Consequences of such $G_{Ep}(Q^2)$-behavior on various physical quantities are briefly reviewed.

Keywords: form factor, sum rules, electromagnetic structure of hadrons, electron-proton scattering, annihilation process

1. INTRODUCTION

The unpolarized electron-proton elastic scattering was used to be the most common way of the proton electromagnetic (EM) structure study from the half of 50's of the last century[1] and abundant data on proton electric $G_{Ep}(t)$ and magnetic $G_{Mp}(t)$ form factors (FFs) in the space-like region $-Q^2 = q^2 = t < 0$ appeared (for references see [2] and [3]). They have been obtained from the measured differential cross-section in the laboratory system by the Rosenbluth technique. Their ratio $\mu_p G_{Ep}(t)/G_{Mp}(t)$ roughly equals one (see Fig. 1 crosses), showing the electric and magnetic distributions in the proton to be equal, in this manner FFs having more or less dipole behavior.

Recently in JLab[4–6] measuring transverse and longitudinal components of the recoil proton's polarization (as suggested in Refs. [7]) in the electron scattering plane of the polarization transfer process simultaneously, the very precise and surprising data on the ratio $\mu_p G_{Ep}(t)/G_{Mp}(t)$ have been obtained (see Fig.1 open circles),

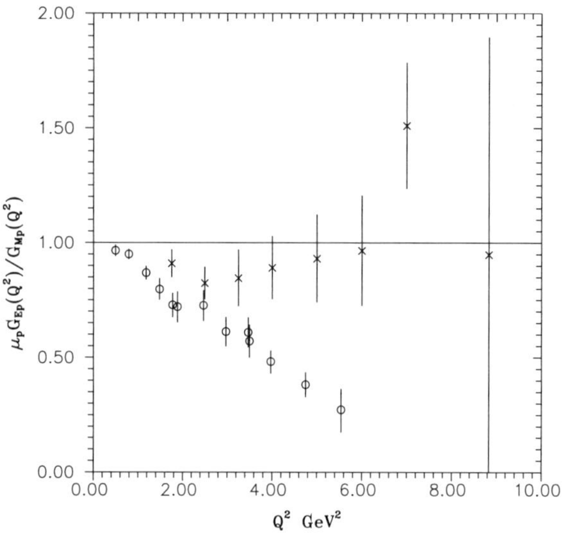

Fig. 1. Rosenbluth and JLab proton polarization data on the ratio $\mu_p G_{Ep}(t)/G_{Mp}(t)$ denoted by crosses and open circles, respectively.

A lot of effort has been devoted[8–15] to the search of a definite solution of the problem. The explanation is likely to be related to solve the problem by inclusion of additional radiative correction terms, corresponding to two-photon exchange approximations, which may reach 40% on the unpolarized cross-section, but are negligible on the polarization ratio. Yet all authors were restricted only to the experimentally measured region of JLab proton polarization data.

In our investigations[16,17] we have taken a phenomenological point of view to investigate the arisen problem and explicitly demonstrate that the proton electric FF $G_{Ep}(t)$, and in no case the magnetic one, is responsible for the discrepancy in the above-mentioned sets of results. Moreover, due to the analyticity of the used unitary and analytic ($U\&A$) model of the nucleon EM structure we have predicted the new surprising non-dipole $G_{Ep}(t)$-behavior in the space-like region with the zero around the $t = -Q^2 = -13$ GeV^2 (see full line in Fig. 2a).

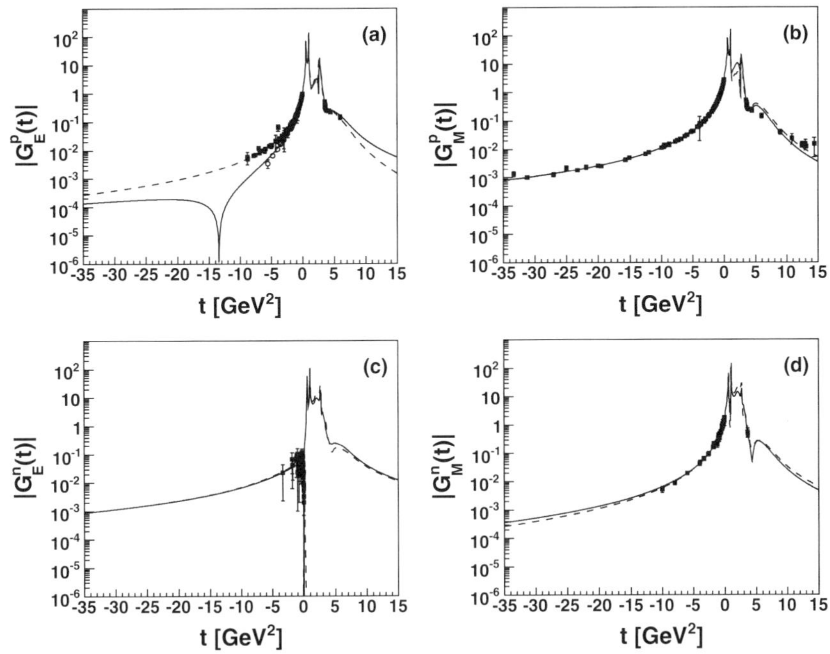

Fig. 2. Results of the analysis of Rosenbluth (dashed lines) and JLab (full lines) data by $U\&A$ model.

2. CONSEQUENCES ON CHARGE DISTRIBUTION WITHIN PROTON

The proton charge distribution (assuming to be spherically symmetric) is an inverse Fourier transform of the proton electric FF

$$\rho_p(r) = \frac{1}{(2\pi)^3} \int e^{-iQr} G_{Ep}(Q^2) d^3Q \tag{1}$$

from where

$$\rho_p(r) = \frac{4\pi}{(2\pi)^3} \int_0^\infty G_{Ep}(Q^2) \frac{\sin(Qr)}{Qr} Q^2 dQ. \tag{2}$$

Substituting for the $G_{Ep}(Q^2)$ under the integral either the dashed line from Fig. 2a corresponding to the dipole $G_{Ep}(t)$-behavior, or non-dipole $G_{Ep}(t)$-behavior with the zero given by full line in Fig. 2a, one gets different charge distributions within the proton given in Fig.3 by dashed and full lines, respectively.

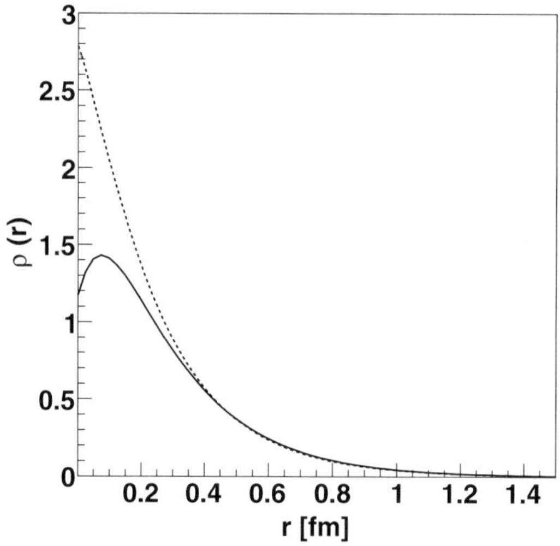

Fig. 3. Charge distribution behavior within the proton, dashed line from dipole and full line from non-dipole $G_{Ep}(t)$-behavior.

They lead to different mean square proton charge radii. That obtained by the dipole $G_{Ep}(t)$-behavior takes the value $\langle r_p^2 \rangle = 0.68\,\mathrm{fm}^2$ and the radius obtained by non-dipole $G_{Ep}(t)$-behavior $\langle r_p^2 \rangle = 0.72\,\mathrm{fm}^2$ takes larger value.

3. POSSIBLE INSIGHT BY DIS

Recently the new nucleon sum rule[18]

$$\begin{aligned} F_{1p}^2(-\mathbf{q}^2) + \frac{\mathbf{q}^2}{4m_p^2} F_{2p}^2(-\mathbf{q}^2) \\ -F_{1n}^2(-\mathbf{q}^2) - \frac{\mathbf{q}^2}{4m_n^2} F_{2n}^2(-\mathbf{q}^2) = \\ 1 - 2\frac{(\mathbf{q}^2)^2}{\pi\alpha^2}\left(\frac{d\sigma^{e^- p \to e^- X}}{d\mathbf{q}^2} - \frac{d\sigma^{e^- n \to e^- X}}{d\mathbf{q}^2}\right), \end{aligned} \tag{3}$$

giving into a relation nucleon electromagnetic form factors with difference of deep inelastic electron-proton and electron-neutron Q^2-dependent differential cross-sections, has been derived. By the calculation of the left-hand side of (3) once taking the dipole and then the non-dipole $G_{Ep}(t)$-behavior of

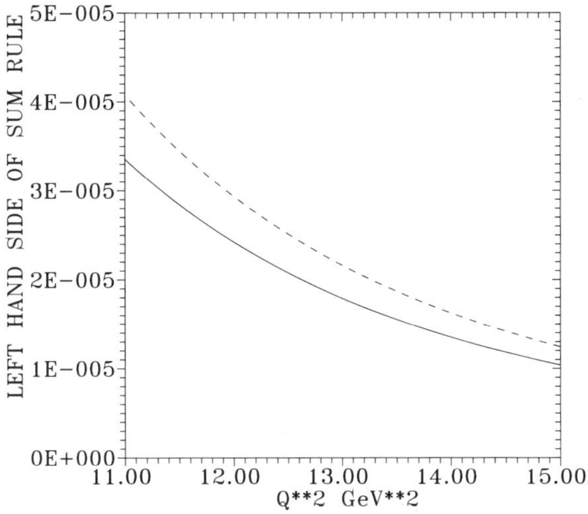

Fig. 4. Difference of DIS Q^2-dependent proton and neutron cross-sections.

the electric proton form factor one gets two different behaviors (see Fig. 4) of the difference of deep inelastic electron-proton and electron-neutron Q^2-dependent differential cross-sections to be in principle verified by DIS experiments.

4. SENSITIVITY OF STRANGE NUCLEON FFs ON DIPOLE AND NON-DIPOLE $G_{Ep}(Q^2)$-BEHAVIOR

The momentum dependence of the nucleon matrix element of the strange-quark vector current $J^s_\mu = \bar{s}\gamma_\mu s$ is contained in the Dirac $F^s_1(t)$ and Pauli $F^s_2(t)$ strange nucleon FFs, or in the strange electric and strange magnetic nucleon FFs

$$G^s_E(t) = F^s_1(t) + \frac{t}{4m^2_N} F^s_2(t), \quad G^s_M(t) = F^s_1(t) + F^s_2(t), \tag{4}$$

which, as a consequence of the isospin zero value of the strange quark, contribute only to the behavior of the iso-scalar nucleon FFs and never to iso-vector ones. So, the nucleon Dirac and Pauli strange FFs $F^s_1(t)$, $F^s_2(t)$ can be found just from the behavior of $F^{I=0}_1(t)$, $F^{I=0}_2(t)$.[19] As one can see from right-hand side of Fig.5, a reasonable value of the strangeness nucleon magnetic moment is predicted $\mu_s = +0.19\,[\mu_N]$.

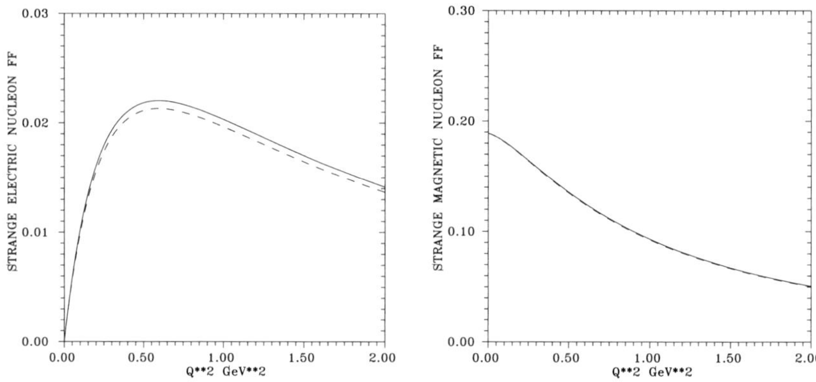

Fig. 5. Theoretical prediction of strange electric and magnetic form factors.

The behavior of strange nucleon FFs doesn't feel too much the difference in dipole and non-dipole behaviors of $G_{Ep}(t)$ in space-like region.

A reasonable description of the recent data[20] on the combination $G_E^s(Q^2) + \eta(Q^2)G_M^s(Q^2)$ for $0.12\,\mathrm{GeV}^2 < Q^2 < 1.0\,\mathrm{GeV}^2$ is achieved (see Fig.6)

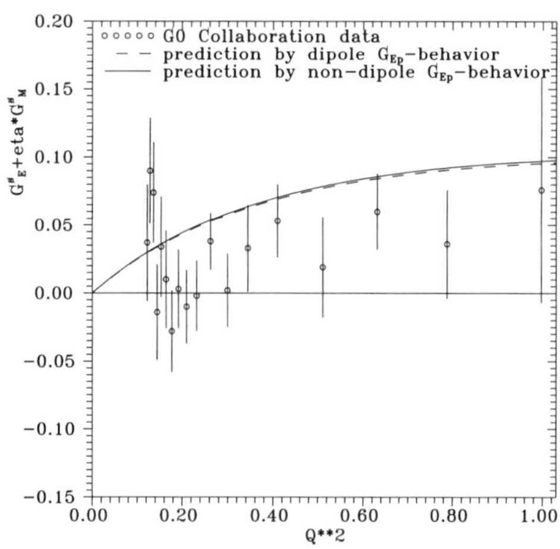

Fig. 6. Prediction for the behavior of the combination of strange nucleon form factors $G_E^s(Q^2) + \eta(Q^2)G_M^s(Q^2)$.

The work was partly supported by Slovak Grant Agency for Sciences VEGA, Grant No. 2/7116/27 (S.D. and A.Z.D.).

References

1. R. Hofstadter, F. Bumiller and M. Yearian, *Rev. Mod. Phys.* **30** 482 (1958).
2. S. Dubnicka, A. Z. Dubnickova and P. Weisenpacher, *J. Phys. G* **29** 405 (2003).
3. J. Friedrich and Th. Walcher, *Eur. Phys. J. A* **17** 607 (2003).
4. M. K. Jones et al, *Phys. Rev. Lett.* **84** 1398 (2000).
5. O. Gayou et al, *Phys. Rev. Lett.* **88** 092301 (2002).
6. V. Punjabi et al, *Phys. Rev.C* **71** 055202 (2005).
7. A. I. Akhiezer and M. P. Rekalo, *Sov. Phys. Dokl.* **13** 572 (1968), *Sov. J. Part. Nucl.* **4** 277 (1974).
8. A. Afanasiev, I. Akushevich and N. Merenkov, *Phys. Rev. D* **64** 113009 (2001).
9. P. A. M. Guichon and M. Vanderhaeghen, *Phys. Rev. Lett.* **91** 142303 (2003).
10. P. G. Blunden, W. Melnitchouk and J. A. Tjon, *Phys. Rev. Lett.* **91** 142304 (2003). Phys. Rev. C **72** 034612 (2005).
11. S. Dubnicka, E. Kuraev, M. Secansky and A. Vinnikov, hep-ph/0507242.
12. A. V. Afanasiev, S. J. Brodsky, C. E. Carlson, Y. C. Chen and M. Vanderhaeghen, Phys. Rev. D **72** 013008 (2005).
13. L. Andivahis et al, *Phys. Rev. D* **50** 5491 (1994).
14. E. Tomasi-Gustafsson and G. I. Gakh, *Phys. Rev. C* **72** 015209 (2005.
15. Yu. M. Bystritskiy, E. A. Kuraev and E. Tomasi-Gustafsson, hep-ph/0603132.
16. S. Dubnicka, A. Z. Dubnickova, *Fizika B* **13** 287 (2004).
17. C. Adamuscin, S. Dubnicka, A. Z. Dubnickova and P. Weisenpacher, *Prog. Part. Nucl. Phys.* **55** 228 (2005).
18. E. Bartos, S. Dubnicka and E. A. Kuraev, *Phys. Rev. D* **74**, 034023-1 (2004).
19. S. Dubnicka and A.Z. Dubnickova, *Eur. Phys. J. A* **32** 399 (2007).
20. D.S. Armstrong et al, *Phys. Rev. Lett.* **95** 092001 (2005).

HIGH Q^2 LARGE ACCEPTANCE G_E^p/G_M^p MEASUREMENTS USING POLARIZATION TRANSFER

L. PENTCHEV* and C.F. PERDRISAT

Department of Physics, The College of William and Mary,
Williamsburg, VA 23187, USA
**E-mail: pentchev@jlab.org*

B. WOJTSEKHOWSKI

Hall A, Jefferson Laboratory, 12000 Jefferson Ave.,
Newport News, VA 23606, USA

A method to measure the proton form factor ratio up to Q^2 of 15 GeV^2 using polarization transfer technique is described. A large detector acceptance needed for such high Q^2 measurements is achieved by means of a single dipole detector system to detect the recoil proton that includes a polarimeter and a hadron calorimeter, and a highly granulated electromagnetic calorimeter for the electron registration. For the proton tracking, Gas Electron Multiplier (GEM) technology is used that can sustain the high fluxes of particles due to direct view from the target. The impact of such high precision measurements on the theoretical description of the proton is briefly discussed.

The measurement of the polarization transfer in elastic *ep* scattering has proved to be a reliable method for extracting the proton form factor ratio because of the low systematic uncertainties it affords. The price one has to pay is the high statistics needed to measure the proton polarization because of the required second scattering. Such experiments became possible only because of the very high current, high beam polarization and continuous beam of the JLab 6 GeV accelerator. This beam energy allows measurements of the form-factors close to Q^2 of 9 GeV^2. The 12 GeV upgrade will increase significantly the kinematical limit, but the actual limit of such measurements will depend on the detector acceptances, and ultimately on the decreasing analyzing power of the secondary scattering.

The first two recoil polarization experiments[1–3] at JLab used the two High Resolution Spectrometers in Hall A and measured the form factor

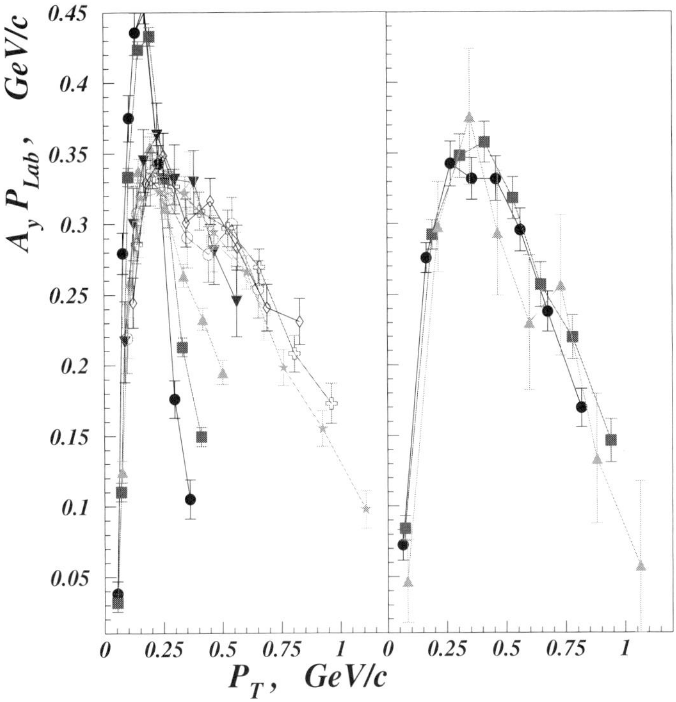

Fig. 1. Analyzing power times proton momentum, $A_y P_{Lab}$, versus secondary particle transverse momentum, P_T. Left, for graphite analyzer and proton momenta: solid circles, boxes, triangles up, triangles down – 0.76, 0.99, 1.27, 1.46 GeV/c; open circles, boxes, dimonds, crosses, and stars – – 1.65, 1.71, 2.07, 2.36, 2.64 GeV/c. Right, for CH_2 analyzer and proton momenta: circles, boxes, and triangles – 2.92, 3.37, 3.81 GeV/c.

ratio up to Q^2 of 5.6 GeV^2. A polarimeter placed in the focal plane of the proton arm measured the polarization components, P_t^{polar} and P_n^{polar}, of the proton perpendicular to its momentum. The form factor ratio is directly proportional to the ratio of the polarization components at the target, P_t and P_ℓ, that were obtained from the measured P_t^{polar} and P_n^{polar}, using a spin transport model of the spectrometer. Due to the increase of the electron to proton solid angle ratio with Q^2, for the second GEP experiment[3] the electron spectrometer was replaced by a lead-glass calorimeter. A similar solution was chosen for the forthcoming experiment E04-108[4] that will extend the measurements close to 9 GeV^2. It will use the HMS spectrometer in Hall C and the specially built lead-glass calorimeter, BigCal, with 1744 lead-glass blocks each with a $4 \times 4\,cm^2$ frontal area.

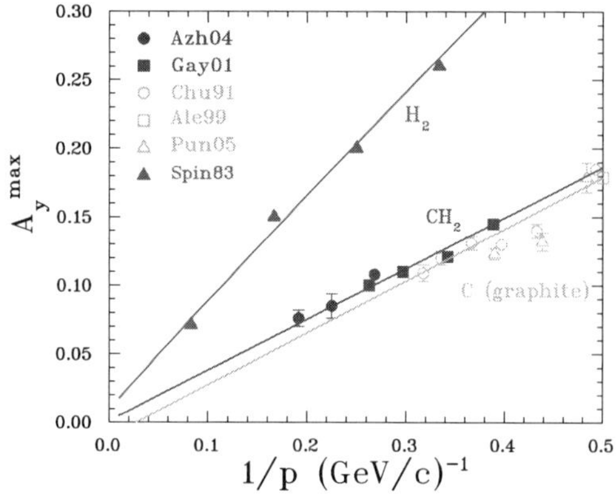

Fig. 2. Maximal value of analyzing power, A_y^{max}, for the $p + C/CH_2 \rightarrow 1charged + X$ reactions (open circles, boxes, and triangles/solid circles, and boxes) from,[2,5,7,9,10] and for the elastic pp reaction (solid triangles)[8] versus the inverse of the proton lab momentum.

The elastic ep cross-section, $d\sigma_{el}/d\Omega_{Lab}$ drops as $\sim E^2/Q^{12}$. The analyzing power, A_y, of the reaction used to measure the proton spin falls as $1/p_p \sim 1/Q^2$,[5,6] where p_p is the proton momentum in the lab frame. This is illustrated at Fig. 1 for the first two GEP experiments.[1–3] In Fig. 2 we compare the maximum value of the analyzing power, A_y^{max}, for C and CH_2 including data up to p_p of 5.3 GeV/c.[5] In the same figure the A_y^{max} value for the elastic pp reaction[8] is plotted, showing a similar $\sim 1/p_p$ behavior up to proton momenta of 12 GeV/c. The overall experimental Figure-of-Merit is proportional to $\sigma_{el}A_y^2 \sim E^2/Q^{16}$. Therefore, for increasing Q^2 in order to get statistical errors similar to the low Q^2 measurements without using excessive beam time, one needs detectors with acceptances an order of magnitude larger than the one of typical spectrometers. The proposed experimental set-up is illustrated at Fig. 3. A simple, non-superconducting magnet is placed close to a long (40 cm) target providing a solid angle > 33 msr for the proton. The magnet is followed by a polarimeter and a hadron calorimeter, HCAL. The polarimeter consists of two CH_2 analyzers with tracking chambers on both sides of each analyzer. The hadron calorimeter, with an energy resolution of about 30%, is used to reduce the trigger rate. The price one has to pay for having a large acceptance are the high background rates especially at the front chambers which are in direct view of the target. However, the invention by F. Sauli of the Gas Electron Multiplier (GEM)[11] has stimulated amazing progress in tracking technology during the last decade, and makes it possible to handle rates

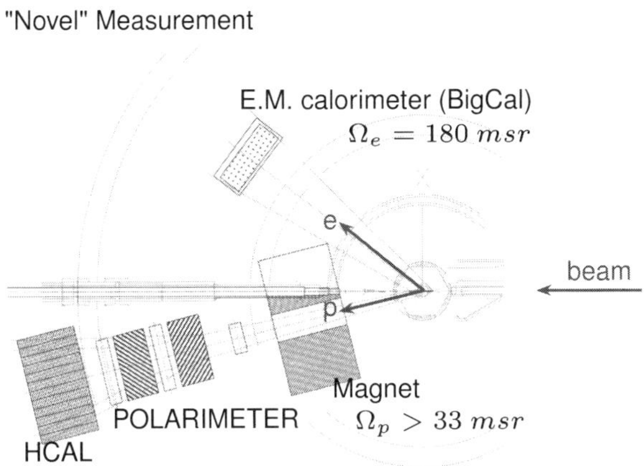

Fig. 3. Proposed experimental set-up.

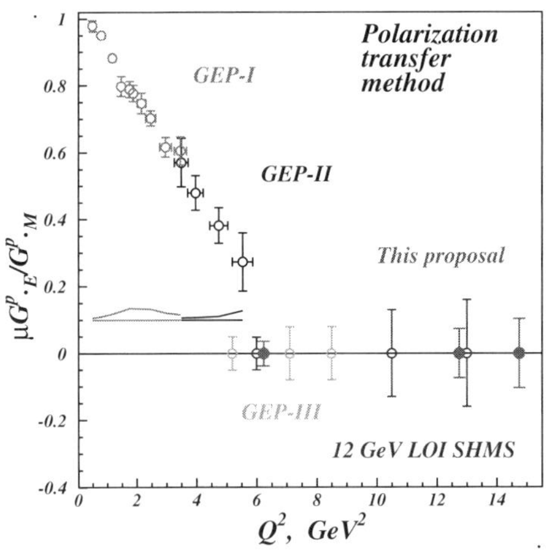

Fig. 4. Errors of the proposed form factor ratio measurements (solid circles), compared to other experiments (open circles).

orders of magnitude higher than needed for this experiment. We propose to use the GEM based trackers described in the articles.[11,12] The GEM is based on gas avalanche multiplication within small holes, etched or drilled in a metalized Kapton foil. Such a technology gives a true two-dimensional localization of the radiation. The avalanche is confined in the hole, free of photon-mediated secondary effects, giving very fast signals. All these properties result in very high rate capabilities of up to 100 MHz per cm^2. The highly segmented electromagnetic calorimeter, BigCal, allows to correlate kinematically the elastic electron angle with the proton angle, making it possible to reconstruct the proton track in the presence of a high hit multiplicity at the front chambers.

The anticipated results of the proposed experiment, together with previous[1–3] and other[4,14] planned experiments at JLab, are shown in Fig 4. The striking feature of the form factor ratio observed in the first experiments is the unexpected fall of the ratio with Q^2. It will be investigated further at much higher Q^2 with a similar precision. The experiment will be decisive for many theories that try to explain the difference in the falling rate with Q^2 of the electric and magnetic form factors. The extension of the measurements to high Q^2 will shed light on the most challenging transition region between non-perturbative and perturbative QCD (pQCD). In particular the experiment may observe the pQCD asymptotic behavior of the form factors, or put a lower limit of the pQCD onset at 15 GeV^2. More discussions about the importance of high Q^2 form factor measurements can be found at,[15] see Fig.14 there.

The proposed experiment[13] was recently approved by JLab PAC32.

References

1. M.K. Jones *et al.*, *Phys. Rev. Lett.* **84**, 1398 (2000).
2. V. Punjabi *et al.*, *Phys. Rev.* **C71**, 055202 (2005), [Erratum-ibid. **C71**, 069902 (2005)].
3. O. Gayou *et al.*, *Phys. Rev. Lett.* **88**, 092301 (2002).
4. C.F. Perdrisat, V. Punjabi, M.K. Jones, and E. Brash, JLab PAC20 proposal 01-109 (2001); JLab PAC26 update 04-108 (2006).
5. L.S. Azhgirei *et al.*, *Nucl. Instr. Meth.* **A538**, (2005) 431.
6. I.M. Sitnik, *Nucl. Instr. Meth.* **A527**, (2004) 278.
7. O. Gayou *et al.*, *Phys. Rev.* **C64**, 038202 (2001).
8. H. Spinka *et al.*, *Nucl. Instr. Meth.* **A211**, 239 (1983).
9. N.E. Cheung *et al.*, *Nucl. Instr. Meth.* **A363** (1995) 561.
10. I.A. Alekseev *et al.*, *Nucl. Instr. Meth.* **A434**, 254 (1999).
11. F. Sauli, *Nucl. Instr. Meth.* **A386** (1997) 531.
12. R. Chechik *et al.*, *Nucl. Instr. Meth.* **A535** (2004) 303.
13. C.F. Perdrisat, L. Pentchev, E. Cisbani, V. Punjabi, and B. Wojtsekhowski, JLab 12 GEV PAC32 proposal 12-07-109 (2007).
14. E. Brash, M.K. Jones, C.F. Perdrisat, and V. Punjabi, JLab 12 GEV PAC30, LOI (2006).
15. C.F. Perdrisat, "Nucleon Form Factor Measurements and Interpretation", Proceedings of this Workshop (2007).

A PRECISE MEASUREMENT OF THE NEUTRON MAGNETIC FORM FACTOR G_M^n IN THE FEW-GEV^2 REGION

G.P. GILFOYLE*, W.K. BROOKS, J.D. LACHNIET, B. QUINN,
M.F. VINEYARD, *et al.*, (the CLAS Collaboration)
** E-mail: ggilfoyl@richmond.edu*
www.richmond.edu

The neutron elastic magnetic form factor G_M^n has been extracted from quasielastic scattering from deuterium in the CEBAF Large Acceptance Spectrometer (CLAS) at Jefferson Lab. The kinematic coverage of the measurement is continuous over a broad range, extending from below 1 (GeV/c)^2 to nearly 5 (GeV/c)^2 in four-momentum transfer squared. High precision was achieved by employing a ratio technique in which many uncertainties cancel, and by a simultaneous in-situ calibration of the neutron detection efficiency, the largest correction to the data. Neutrons were detected using electromagnetic calorimeters and time-of-flight scintillators. Data were taken at two different electron beam energies, allowing up to four semi-independent measurements of G_M^n to be made. The dipole parameterization is found to provide a good description of the data for $Q^2 > 1$ (GeV/c)^2.

Keywords: Form factors; Proceedings; World Scientific Publishing.

1. INTRODUCTION

The elastic form factors of the proton and neutron are fundamental quantities which have been studied for decades. The dominant features of the larger form factors G_M^p, G_E^p, and G_M^n at $Q^2 < 10$ (GeV/c)^2 are described by the dipole form $G_D = (1+Q^2/0.71)^{-2}$ within 10%. Current directions in the field include precise measurements of the neutron electric form factor,[1] extractions of the strange electric and magnetic form factors for the proton,[2] as well as time-like form factors.[3] There has been renewed theoretical interest on several fronts.[4] First, models of the nucleon ground state can often be used to predict several of these quantities, and it has proved to be very difficult to describe all of the modern data simultaneously in a single approach. Second, lattice calculations are now becoming feasible in the few-GeV^2 range, and over the next decade these calculations will become increasingly precise. Finally, since elastic form factors are a limiting case of the generalized parton distributions (GPDs), they can be used to constrain GPD models.[5] High precision and large Q^2 coverage is important.[5] At present the neutron magnetic form factor at large Q^2 is more poorly known than the proton form factors.

2. THE CLAS MEASUREMENT

This measurement[6] makes use of quasielastic scattering on deuterium where final state protons and neutrons are detected. The ratio of ^{2}H(e, e′n) to ^{2}H(e, e′p) in quasi-free kinematics is approximately equal to the ratio of elastic scattering from the free neutron and proton. The ratio is:

$$R_D = \frac{\frac{d\sigma}{d\Omega}[^2\mathrm{H(e,e'n)_{QE}}]}{\frac{d\sigma}{d\Omega}[^2\mathrm{H(e,e'p)_{QE}}]} = a \cdot \frac{\frac{(G_E^n)^2+\tau(G_M^n)^2}{1+\tau} + 2\tau(G_M^n)^2\tan^2(\frac{\theta}{2})}{\frac{(G_E^p)^2+\tau(G_M^p)^2}{1+\tau} + 2\tau(G_M^p)^2\tan^2(\frac{\theta}{2})} \tag{1}$$

Using deuteron models one can accurately compute the correction factor $a(Q^2, \theta_{pq})$ where θ_{pq} is the angle between the scattered nucleon direction and the three-momentum transfer. It is nearly unity for quasielastic kinematics at higher Q^2. The value of G_M^n is then obtained from the measured value of R_D and the experimentally known values of G_E^n, G_M^p, and G_E^p. This method has been used previously.[7] The $(e, e'n)$ and $(e, e'p)$ reactions were measured in CLAS at the same time from the same target. Use of the ratio R_D under these circumstances reduces or eliminates several experimental uncertainties, such as those associated with the luminosity measurement or radiative corrections. The remaining major correction is the neutron detection efficiency.

Neutrons were measured in two CLAS scintillator-based detectors: the forward-angle electromagnetic shower calorimeters and the time-of-flight (TOF) scintillators. The efficiency measurement was performed using tagged neutrons from the ^{1}H(e, e′π^+)X reaction where the mass of the final state M_X was chosen to be that of the neutron. Since the precise value of the detection efficiency can vary with time-dependent and rate-dependent quantities such as photomultiplier tube gain, the detection efficiency was measured *simultaneously* with the primary deuterium measurement. Two separate targets were positioned in the beam at the same time, one for deuterium and the other for hydrogen, separated by less than 5 cm. Plots of the resulting neutron detection efficiencies are shown in Fig. 1. The main plot shows the results for the forward electromagnetic shower calorimeter, while the inset shows the results for the time-of-flight scintillators. Note the agreement between the two different beam energies. The systematic uncertainty

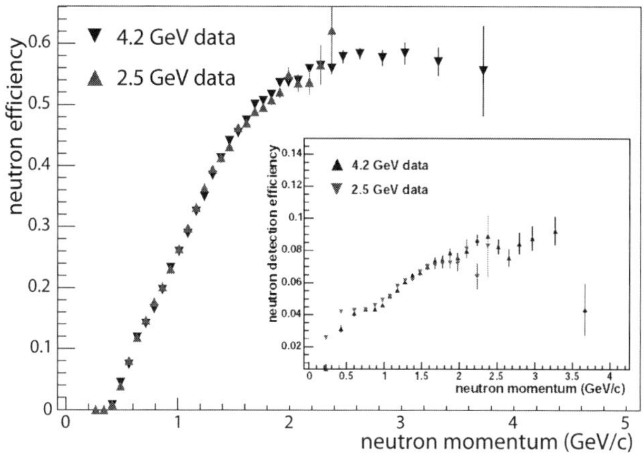

Fig. 1. Detection efficiency versus momentum for neutrons detected in the forward-angle electromagnetic calorimeters and in the TOF system (inset) at two different beam energies. The efficiency has been integrated over all six sectors of the CLAS spectrometer.

in the neutron detection efficiency was estimated by using different parameterizations of the fit to the data and extracting the difference as a function of neutron momentum. Uncertainties in the range 1-2% were obtained. The full inventory of systematic uncertainties is described elsewhere.[6] The total uncertainty is less than 2.5% across the full Q^2 range.

The CLAS extraction of G_M^n consists of multiple overlapping measurements. The time-of-flight scintillators cover the full angular range of the spectrometer, while the forward calorimeters cover a subset of these angles so G_M^n can be obtained from two independent measures of the neutron detection efficiency. In addition, the experiment was carried out with two different beam energies that had overlapping coverage in Q^2, so that the detection of the protons of a given Q^2 took place in two different regions of the drift chambers. As a result, essentially four measurements of G_M^n have been obtained from the CLAS data that could have four independent sets of systematic errors. We have found the four measurements are consistent within the statistical errors, suggesting that the systematic errors are well-controlled and small.

3. RESULTS

The preliminary results for the neutron magnetic form factor are shown in Fig. 2 together with a sample of existing data. The error bars shown are due only to statistical uncertainties. Systematic uncertainties are represented by the red band. The data shown are the weighted averages of the four overlapping individual measurements. A few features are noteworthy. First, the quality and coverage of the data is a substantial improvement over the existing world's data set. Second, the dipole form describes the data over the Q^2 range measured. This feature is at variance at higher Q^2 with parameterizations based on previous data which tend to show a more strongly decreasing trend for $G_M^n/(\mu_n G_D)$ with increasing Q^2.[4]

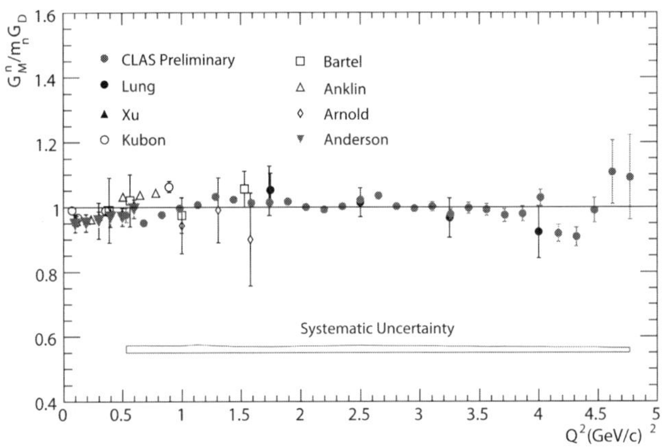

Fig. 2. Preliminary results for $G_M^n/(\mu_n G_D)$ from CLAS are compared with a selection of previous data[4,8]

References

1. R. Madey *et al.*, *Phys. Rev. Lett.* **91**, p. 122002 (2003).
2. K. A. Aniol *et al.*, *Phys. Lett.* **B365**, p. 275 (2006).
3. F. Iachello and Q. Wan, *Phys. Rev. C* **69**, p. 055204 (2004).
4. C. Hyde-Wright and K. deJager, *Ann. Rev. Nucl. Part. Sci.* **54**, p. 217 (2004).
5. M. Diehl, T. Feldmann, R. Jakob and P. Kroll, *Eur.Phys.J.C* **39**, p. 1 (2005).
6. J. D. Lachniet, A high precision measurement of the neutron magnetic form factor using the clas detector, PhD thesis, Carnegie-Mellon University, (Pittsburgh, PA, USA, 2005).
7. G. Kubon *et al.*, *Phys.Lett.* **B524**, 26 (2002).
8. B. Anderson *et al.*, *Phys. Rev. C* **75**, p. 034003 (2007).

MAGNETIC FORM FACTOR OF THE NEUTRON UP TO 8 $(\mathrm{GeV}/c)^2$

B. QUINN

Physics Dept., Carnegie Mellon University,
Pittsburgh, PA, 15213
E-mail:bquinn@cmu.edu

A feasibility study has been made for an experiment to greatly extend the Q^2-range of precision measurements of the neutron's magnetic form factor using the existing JLab 6 GeV beam. The results are promising.

Keywords: neutron, nucleon, form factor

1. Introduction

This work was done in collaboration with Dr. Bogdan Wojtsekhowski of JLab. We have investigated the possibility of using Hall A detectors and the existing JLab beam to extend the kinematic range of high precision measurements of the neutron magnetic form factor.

Nucleon elastic form factors are of fundamental interest because they give access to more precise structure information than can be obtained for other hadrons. As lattice QCD calculations approach the point of being able to make meaningful predictions of hadronic structure, it will be important to have precision measurements against which to test them. Nucleon form factors may be expected to approach perturbative QCD predictions for Q^2 beyond 10 $(\mathrm{GeV}/c)^2$, so the region of several $(\mathrm{GeV}/c)^2$ may be an interesting transition region from a non-perturbative description at lower Q^2. The form factors are also an important touch-stone in the more ambitious program to determine the generalized parton distributions, for which the elastic form factor provides a sum rule.

The most precise measurements of the neutron's magnetic form factor are the soon-to-be-published CLAS results (Fig. 1). These data were analyzed by Jeff Lachniet for his Ph.D. thesis[1] as my student. They cover a Q^2 range to beyond 4.5 $(\mathrm{GeV}/c)^2$, although the errors grow to several percent in the last few points. SLAC measurements[2] extend to higher Q^2, but give far less precision since they are based on single-arm measurements with the proton cross section subtracted from that of the deuteron.

While precision measurements are planned[3] up to Q^2=14 $(GeV/c)^2$ after the JLab 12 GeV upgrade, it is interesting to see how far the knowledge of G_M^n can be extended with the existing 6 GeV beam. The kinematic points we consider are give in Table 1. While variation in scattered electron energy is moderate, there is a significant range of nucleon momenta.

Table 1. Kinematic of points considered

Q^2 $(GeV/c)^2$	E_{beam} (GeV)	θ_e	θ_N	E′ (GeV)	P_N (GeV/c)
3.5	4	37.5°	29.2°	2.1	2.65
4.5	4	49.5°	22.4°	1.6	3.2
6	5	48.1°	19.5°	1.8	4.0
8	6	52.°	15.5°	1.7	5.1

2. Experimental Technique

The technique considered is the 'ratio method', which is also used for the existing[1] and proposed[3] CLAS measurements and is far less susceptible to systematic errors than the proton-subtraction technique.[2,4] The measured quantity, R, is the ratio of the neutron-coincidence to proton-coincidence quasi-elastic (Q.E.) electron scattering cross sections on the deuteron. With almost negligible nuclear corrections, this is the ratio of the elastic scattering cross sections of the neutron and proton. With a very small correction for the neutron electric form factor, this yields the quantity of interest, $(G_M^n)^2$, in terms of the relatively well known elastic scattering cross section of the proton. With the usual kinematic factors, η, τ, and ϵ,

$$R = \frac{\frac{d\sigma}{d\Omega}\Big|_{d(e,e'n)}}{\frac{d\sigma}{d\Omega}\Big|_{d(e,e'p)}} \approx \frac{\frac{d\sigma}{d\Omega}\Big|_{n}}{\frac{d\sigma}{d\Omega}\Big|_{p}} \rightarrow \frac{\eta\frac{\tau/\epsilon}{1+\tau}\sigma_{Mott}(G_M^n)^2}{\frac{d\sigma}{d\Omega}\Big|_{p}}$$

The BigBite spectrometer will be detect the Q.E. scattered electrons while recoiling neutrons and protons will be detected with BigHAND (Hall A Nucleon Detector), a large array of plastic scintillators interspersed with iron converters to initiate hadronic showers. The front veto layer of BigHAND could provide proton/neutron discrimination, but this suffers from hadronic interactions in the heavy shielding upstream of the veto. Addition of a large-aperture dipole at the beginning of the nucleon's path allows protons to be distinguished from neutrons by the deflection from the electron's momentum-transfer vector. For Q.E. scattering, the predicted position is

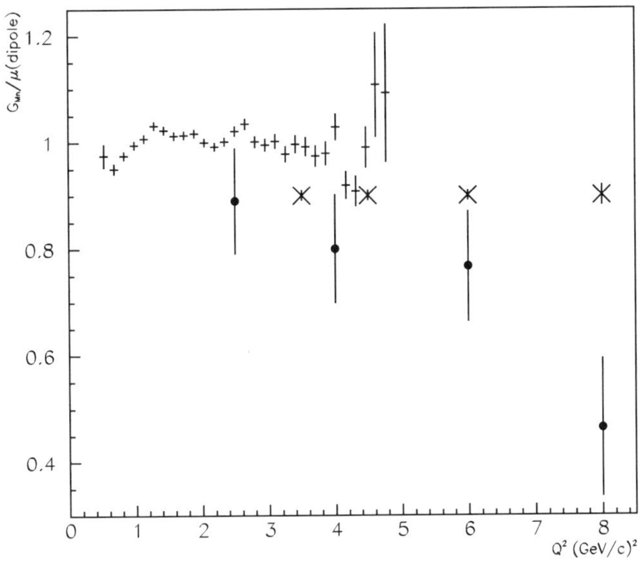

Fig. 1. Existing and projected measurements of G^n_M, as ratio to the scaled-dipole parameterization. Unpublished **PRELIMINARY** measurements of CLAS[1] are horizontal lines with ***statistical errors only***. SLAC measurements[2] are circles (a higher Q^2 point is off-scale). X's (arbitrarily placed at 0.9 to avoid overlap with other points) show kinematics and projected statistical errors for four measurements considered here.

smeared by Fermi motion. Even with Fermi-broadening, the overlap of the neutron- and proton-images can be reduced to the 5% level by a 'kick' of 200 MeV/c which can be achieved with a modest $\int B\, dl$ of 0.66 T·m. The veto layer can be used to accurately correct for the remaining 5% mis-identification.

3. Calibration

The efficiency and acceptance of BigBite cancels in the the ratio, R. This is not true of the BigHAND detection of the neutron and proton. The acceptance can be largely matched by fiducial cuts on the BigBite acceptance to remove events in which the scattered nucleon is likely to be near the edges of BigHAND. It is important to have accurate measurements of proton and neutron efficiencies. Concurrent measurement of the efficiencies with the data is not critical here, however, since the detection efficiency is mostly sensitive to geometry and hadronic interaction probabilities and so does not fluctuate as it might if tracking chambers were involved.

Proton efficiency will be calibrated by using BigBite to detect elastic scattering off a Hydrogen target and determining the efficiency for detecting the coincident protons. Similarly $^1H(\gamma, \pi^+)n$ will be used as a source of coincident neutrons for calibration of neutron efficiency. A 6% radiator will provide photons and BigBite will detect the π^+. Unambiguous identification of the desired calibration reaction is possible when the photon is close enough to the bremsstrahlung end point so the the pion momentum is too high to have been produced by other reactions. Setting the threshold for pion momentum 1.5% above the kinematic limit from two-pion production determines the range of the end-point spectrum which is useful. Modest running time without a radiator allows subtraction to isolate the real-photon contribution. Because the reactions' kinematics are almost identical, the calibration-nucleon momenta closely match that of the Q.E. peak.

4. Rate Estimates and Conclusions

Rates were estimated for $\mathcal{L} = 10^{37}/\text{cm}^2/\text{s}$, consistent with previous running of BigBite. The scaled dipole was used for nucleon cross sections and s^7-scaling (with empirical angular variation) was used for $H(\gamma, \pi^+)$. The combined acceptance of the spectrometers was used, with fiducial cuts for the Q.E. data. Rates of over 100 counts/hour were found, even at the highest Q^2, for all four reactions of interest (p- and n-coincident Q.E. and the calibrations). With less than 500 beam hours, 1% measurements could be made on all reactions at the four kinematic points of Table 1. Error-propagation in calibrations and ratios then predicts about 2% errors on R which in turn gives 1% statistical errors on G_M^n. To reflect the correction for inclusive background overlapping the Q.E. peak, the error on the highest Q^2 point has been assumed to be 2% in Fig. 1.

The projected points in (shown as X's) in Fig. 1 show this is a promising experiment for the near-term extension of the Q^2 coverage of high-precision measurements of G_M^n.

References

1. J. Lachniet, A high precision measurement of the neutron magnetic form factor using the CLAS detector, PhD thesis, Carnegie Mellon University, (PA, USA, 2005), unpublished, (http://www-meg.phys.cmu.edu/ bquinn/jl_thesis.pdf).
2. S. Rock *et al.*, *Phys. Rev. Lett.* **49**, 1139 (1982).
3. JLab experiment E12-07-104
4. A. Lung *et al.*, *Phys. Rev. Lett.* **70**, 718 (1993).

TIMELIKE FORM FACTORS

KAMAL K. SETH

Northwestern University,
Evanston, IL 60208, USA
E-mail: kseth@northwestern.edu

Form factors of nucleons and mesons with timelike momentum transfers are discussed. New experimental results for protons, pions, and kaons at large momentum transfers are presented, and the inadequacy of existing theoretical ideas about these is pointed out.

Keywords: Form factors, pions, kaons, protons

1. Introduction

It is exactly 100 years that the proton (more correctly, the positive nucleus) was discovered, and 60 years since the pion and kaon were identified. Since 1964, when the quark model was first proposed, we have known that baryons and mesons are all made up of quarks and antiquarks. One would therefore expect that by now we could have figured out exactly how quarks fit into the hadrons and lead to their observable properties, mass, size, spin, charges and curents. But Nature is much more devious, and does not allow easy insight into its workings. One of the tools that has been successfully used to gain insight into the structure of hadrons is the measurement of electromagnetic form factors as a function of momentum transfer.

Electromagnetic form factors of a hadron are the most direct link to the structure of the hadron in terms of its constituents. They describe the coupling of a photon with a certain four–momentum to the distribution of charges and currents in the hadron.

The four–momentum transfer Q^2 in the collision of two particles with four-momenta p_1 and p_2 can be positive or space-like (in scattering) or negative or time-like (in annihilation/production).

No understanding of form factors can be considered complete unless it includes an explanation of form factors for both spacelike and timelike momentum transfers, which are just two sides of the same coin.

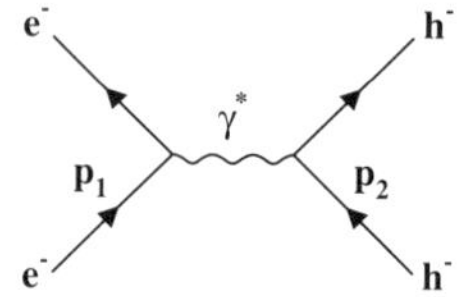

Scattering, Spacelike
positive $Q^2 = t$

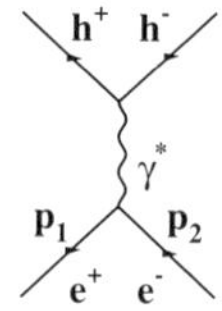

Annihilation, Production
negative $Q^2 = s$

1.1. *Form Factors for Space-like Momentum Transfers*

The extensive nucleon form factor measurements done at SLAC, JLab, and other electron accelerators are made with electron beams elastically scattered from targets of p, d, etc., and are exclusively for spacelike momentum transfers. Measurements of spacelike form factors of mesons at large Q^2 are extremely difficult, if not impossible to make because meson targets do not exist! Measurements by means of either scattering of pion and kaon beams with atomic electrons, or by electroproduction of pions are largely confined to small momentum transfers

1.2. *Form Factors for Time-like Momentum Transfers*

Measurements of form factors for time-like momentum transfers are done at e^+e^- colliders, and they can, in principle, be used to measure form factors of any mesons or baryons. The $p\bar{p}$ annihilations have so far been only done with $\bar{p}$ beams incident on fixed proton targets. These, of course, only lead to proton form factors.

It is important to note that form factors are analytic functions of Q^2. Therefore, the Cauchy theorem alone guarantees that $F(Q^2, \text{timelike}) \xrightarrow{Q^2\to\infty} F(Q^2, \text{spacelike})$.

1.3. *Cross Sections for Time-like Momentum Transfers*

For protons, there are two form factors, Pauli and Dirac Form Factors, or more familiarly, the magnetic $G_M(s)$ and the electric $G_E(s)$ form factors, and the cross section $e^+e^- \to p\bar{p}$ is

$$\sigma_0(s) = \frac{4\pi\alpha^2}{3s}\beta_p\left[|G_M^p(s)|^2 + \frac{\tau}{2}|G_E^p(s)|^2\right], \quad \tau \equiv 4m_p^2/s$$

At large momentum transfers separation between $G_M(s)$ and $G_E(s)$

becomes difficult, and the results which are generally reported assume $G_E(s) = 0$, or $G_E(s) = G_M(s)$.

For pions and kaons, both of which have spin 0, there is no magnetic contribution, and only the electric form factor $F(s)$ exists. In this case the cross section for $e^+e^- \to m^+m^-$ is

$$\sigma_0(s) = \frac{\pi\alpha^2}{3s}\beta_m^3|F_m(s)|^2$$

Jumping the gun a little, let me point out that pQCD counting rules predict that the baryon form factors are proportional to Q^{-4} (or s^{-2}) and the meson form factors are proportional to Q^{-2} (or s^{-1}), so that

$$\sigma_0(s)_{\text{proton}} \propto s^{-5}, \qquad \sigma_0(s)_{\text{meson}} \propto s^{-3}$$

This tells you how rapidly the cross sections fall, and how difficult it becomes to measure any form factors at large momentum transfers.

For example, $\sigma(e^+e^- \to p\bar{p}) \approx 1$ pb at $s = Q^2 = 13.5$ GeV^2. At $s = 25$ GeV^2 one expects to drop down by a factor ~ 20, to ~ 50 fb!

2. Baryon Form Factors

The only baryon form factors that have ever been measured are for nucleons, mainly for the proton. Timelike form factors can also be measured for other baryons, Λ, Σ, etc., but so far no such measurements exist.

2.1. *Spacelike Form Factors of the Proton*

The spacelike magnetic form factors $G_M(Q^2)$ of the proton were measured with precision in the ep scattering experiments at SLAC, all the way up to $Q^2 = 31$ GeV^2.[1] For $Q^2 \geq 15$ GeV^2, their variation follows the pQCD counting rule prediction that $Q^4 G_M(|Q^2|)/\mu_p$ is essentially constant and varies only as α^2(strong).

In the pQCD factorization formalism of Brodsky and Lepage,[2] the form factor can be factorized into the hard scattering amplitude, which can be calculated perturbatively, and hadron distribution amplitudes (DA), which contain all the non-perturbative physics.

The asymptotic distribution amplitude for the proton leads to $G_M^P(|Q^2|) = 0$ for all Q^2, and many different variations of asymmetric DA's have been considered, with and without Sudakov corrections, and with and without transverse momenta. QCD sum–rule predictions, and predictions based on GPD and meson–cloud pictures have also been made. It is not surprising that with an appropriate choice of the parameters, the spacelike form factors of the proton can be fitted by nearly all model calculations.

2.2. *Timelike Form Factors of the Proton*

Prior to the Fermilab (E760/E835) measurements in 1993/2000 [8,9,10] of the timelike form factors of the proton by the reaction $p\bar{p} \to e^+e^-$, the data were sparse, had large errors, and were confined to $|Q^2| < 5,7$ GeV^2. The Fermilab measurements[3,4] obtained $G_M^p(|Q^2|)$ for four $|Q^2|$ between 8.9 and 13.11 GeV^2. As Fig. 1 (left) shows, while $Q^4 G_M^p(|Q^2|)$ was found to vary as α^2(strong), the value of the timelike form factor was found to be twice as large as the spacelike form factor, i.e.,

$$R \equiv G_M^p(|Q^2|)(\text{timelike})/G_M^p(|Q^2|)(\text{spacelike}) \approx 2$$

Prior to the Fermilab measurements there were few theoretical predictions of the timelike form factor of the proton. Following the Fermilab measurements, Hyer[5] reported predictions for timelike form factors within the pQCD formalism including Sudakov suppression. Hyer's predictions, showed large sensitivity to the assumed distribution amplitude, but did not address the question of the experimental ratio $R \approx 2$. Iachello and Wan[6] gave made predictions based on a picture of the bare meson surrounded by a vector meson cloud. As shown in Fig. 1 (right), a typical Hyer prediction gives a $Q^4 G_M^p(|Q^2|)$ which is nearly constant with $|Q^2|$, and the Iachello prediction gives $Q^4 G_M^p(|Q^2|)$ which falls rapidly with $|Q^2|$. Neither fits the data.

In order to explain the ratio $R \approx 2$ Kroll and collaborators[7] proposed the diquark–quark model of the nucleon. While this model has at least two extra parameters, it explains both spacelike and timelike $G_M^p(|Q^2|)$, and $R \approx 2$ very nicely. The three predictions mentioned above are shown in Fig. 1 (right), where the more recent results for $G_M^p(|Q^2|)$ from the e^+e^- annihilation measurments by Cornell[8] at $|Q^2| = 13.5$ GeV^2, by BES[9] at ten values of $|Q^2| = 4 - 9.4$ GeV^2, and by BaBar[10] using ISR from $\Upsilon(4S)$ at $|Q^2| = 3.6 - 20.3$ GeV^2 are also shown. All these measurements are consistent with each other, and confirm $R \approx 2$. BaBar has gone a step beyond, and has also attempted to derive G_E/G_M from their ISR data.

Before I leave the proton, and go on to the pion and kaon, let me point out some possible consequences of the recent JLab measurements[11] of $R(|Q^2|) \equiv \mu_p G_E(|Q^2|)/G_M(|Q^2|)$ for spacelike $|Q^2| < 6$ GeV^2. As is well known, these polarization measurements show that R decreases monotonically as $|Q^2|$ increases. If this trend is extrapolated, one reaches $R \approx 0$ at $|Q^2| \approx 8$ GeV^2, and R becomes negative for larger $|Q^2|$, e.g., $R \equiv -0.8$ at $|Q|^2 = 13.5$ GeV^2. I do not know what zero and negative $G_E(|Q^2|)$ mean,

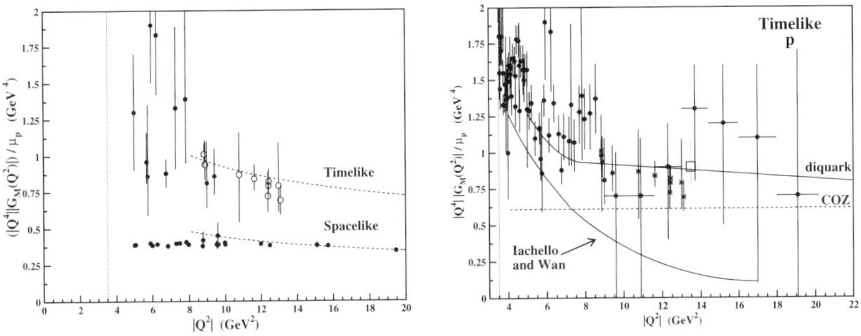

Fig. 1. $Q^4 G_M^p(|Q^2|)/\mu_p$ as a function of $|Q^2|$ for timelike momentum transfers. (Left): Illustrating the nearly factor 2 larger values of timelike $|Q^2|$ as measured in the Fermilab experiments. (Right) A sample of theoretical predictions together with all the present experimental data from Fermilab, CLEO, BES, and BaBar.

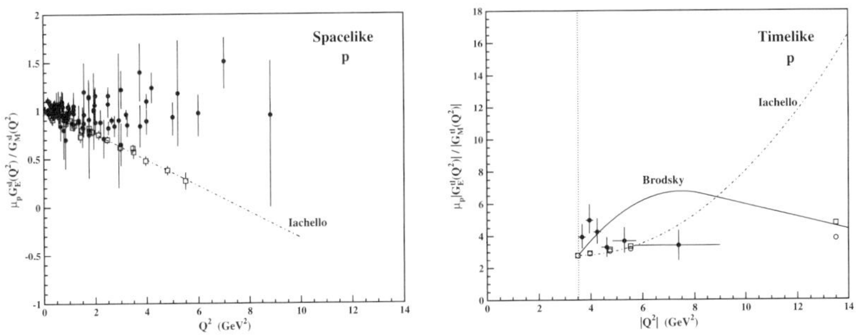

Fig. 2. (Left) $R \equiv \mu_p G_E^p(|Q^2|)/G_M^p(|Q^2|)$ for spacelike momentum transfers, (Right) R for timelike momentum transfers. See text.

but I am tempted to speculate about what these would imply for timeilike $R(|Q^2|)$.

Strangely enough, if we assume that the ratio of the Pauli and Dirac form factors, $F_2(|Q^2|)/F_1(|Q^2|)$, is the same for timelike $|Q^2|$ as it is for spacelike, for $|Q^2| = 13.5$ GeV^2 we obtain almost the same result, $[\mu_p G_E(13.5)/G_M(13.5)]_{\text{timelike}} = 3.9$ and 4.9, whether $R(13.5))_{\text{spacelike}} = 1$ or -0.8.

3. Meson Form Factors

Mesons represent much simpler systems than baryons; two quark systems are expected to be easier to understand than three quark systems. It is because of this that the now-classic debate about when $|Q^2|$ is large enough

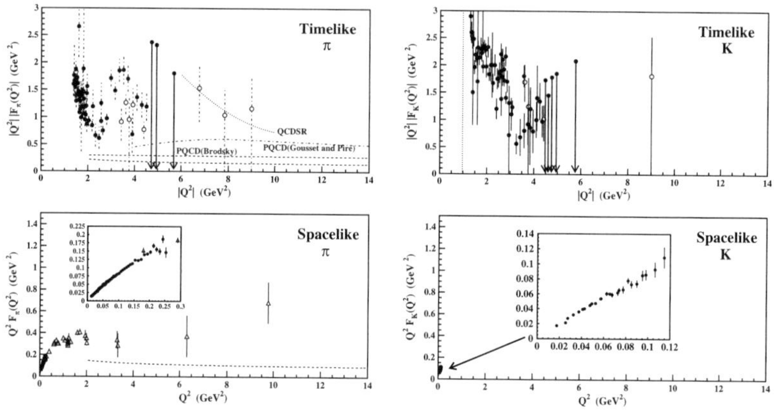

Fig. 3. World data for form factors for spacelike and timelike momentum transfers for pions (left) and kaons (right) before the CLEO measurements.

for the validity of pQCD took place in the 1980s between Brodsky and collaborators on one side and Isgur and Llwellyn Smith on the other side. Unfortunately, the then existing experimental data on pion form factors was extremely poor, especially in the large $|Q^2|$ region which was the subject of the entire debate. All the pion and kaon form factor data available before the recent CLEO measurements are shown in Fig. 3.

3.1. *Pion Form Factors*

Spacelike form factors of pions for $Q^2 > 0.3$ GeV^2 are exclusively determined by electroproduction measurements, $e^-p \to e^-\pi^-p$, $e^-\pi^+n$. The longitudinal part of the pion electroproduction cross section $\sigma_L(t)$ is related to $F_\pi^2(|Q^2|)_{\text{spacelike}}$ via the pion–nucleon coupling constant $g_{\pi NN}(t)$. The latest JLab electroproduction measurements[12] for $|Q^2| = 0.6 - 2.45$ GeV^2 make the longitudinal/transverse separation for the first time, but still suffer from the uncertainties inherent in the t dependence of $g_{\pi NN}(t)$ and the need to extrapolate the cross section to the physical pion pole at $t = m_\pi^2$. The old larger Q^2 measurements from Cornell have the additional problem of very large ($\gtrsim \pm 40\%$) errors. Despite these limitations, the spacelike form factor data have been fitted by many model calculations with suitable choices of parameters. For timelike form factors of the pion, the available data were sparse and generally of poor quality.

3.2. *Kaon Form Factors*

For kaon spacelike form factors, there are no electroproduction measurements so far, and the available data are limited to $Q^2 < 0.12$ GeV^2. The data for timelike form factors of kaons had the same limitations, both in the quality and the range of Q^2 as for pions. There were no direct theoretical predictions, except that in the lowest order, it is expected that for all timelike $|Q^2|$, $F_\pi(Q^2)/F_K(Q^2) = f_\pi^2/f_K^2 = 0.67 \pm 0.01$.

3.3. *The CLEO Measurements of Pion and Kaon Form Factors*

CLEO[8] has recently reported measurements of the pion and kaon form factors for the timelike momentum transfer of $|Q^2| = 13.48$ GeV^2. Precision at the level of ±6% for kaons, and ±13% for pions has been achieved. This unprecedented level of precision for a large $|Q^2|$ provides for the first time data which present a serious challenge to the theorists.

CLEO measurements of π and K form factors presented formidable background problems. The form factor cross sections $\sigma(e^+e^- \to m\overline{m}) \approx 5$–10 pb, while $\sigma(e^+e^- \to e^+e^-,\ \mu^+\mu^-)$ are 10^3 to 10^5 times larger. Therefore, in addition to the standard track and shower quality requirements, very clever use of dE/dx, E_{CC}, and RICH information was done in order to obtain small but background free samples of $\pi^+\pi^-$ and K^+K^- events.

The CLEO measurements were made using 20.7 pb^{-1} of e^+e^- data taken at $\sqrt{s} = 3.671$ GeV, i.e., 15 MeV below the ψ' resonance. The data were originally taken for background studies for the ψ' decays which were being studied. It is ironic that these background studies have provided the world's best measurements of pion and kaon form factors.

These data were analyzed to obtain for $|Q^2| = 13.48$ GeV^2: $|Q^2|F_\pi(|Q^2|) = 1.01 \pm 0.11 \pm 0.07$ GeV^2, $|Q^2|F_K(|Q^2|) = 0.85 \pm 0.05 \pm 0.02$ GeV^2, and $F_\pi(13.48\ \text{GeV}^2)/F_K(13.48\ \text{GeV}^2) = 1.19 \pm 0.07$.

These are the world's first measurements of the form factors of any mesons at this large a momentum transfer, and with precision of this level. They are shown in Fig. 5 along with the old world data, and arbitrarily normalized curves showing the pQCD predicted variation of $|Q^2|F_\pi$ and $|Q^2|F_K$ with α_S.

In the figures for both pions and kaons, in addition to the results of the CLEO measurements points marked J/ψ are shown. These have been obtained by using the relation

$$\frac{\mathcal{B}(J/\psi \to \gamma^* \to m\overline{m})}{J/\psi \to \gamma^* \to e^+e^-} = 2F_m^2(M_{J/\psi}^2) \times \left(\frac{p_m}{M_{J/\psi}}\right)^3$$

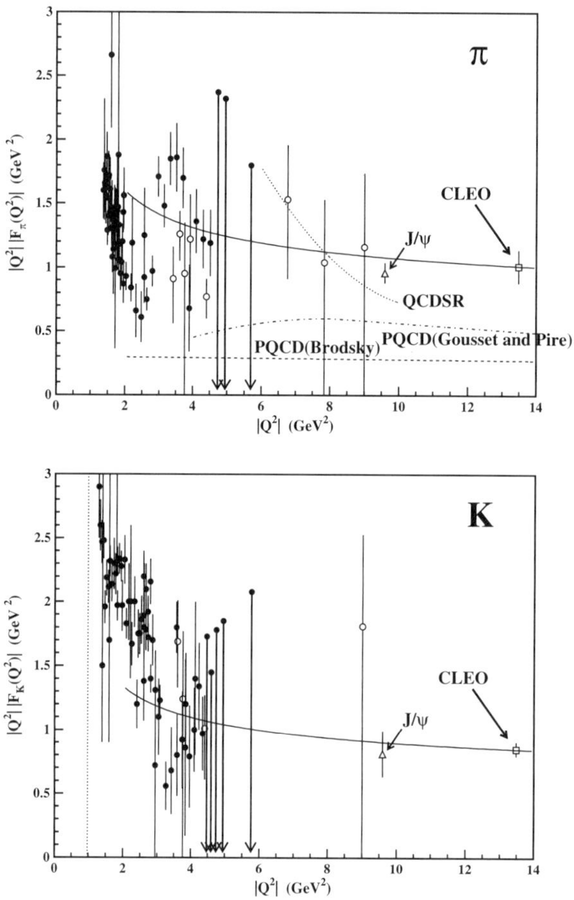

Fig. 4. World data for timelike form factors, including CLEO and J/ψ results for pions (top) and kaons (bottom). A sample of theoretical predictions available for pions is shown in addition to the arbitrarily normalized pQCD prediction of α_S variation (solid lines).

In general $\mathcal{B}(J/\psi \to m\overline{m}) = K|A_\gamma + A_{ggg} + A_{\gamma gg}|^2$. It was noted by Milana et al.[13] that for $\pi^+\pi^-$, A_{ggg} and $A_{\gamma gg}$ are negligably small, so that $\mathcal{B}(J/\psi \to \gamma^* \to \pi^+\pi^-) = \mathcal{B}(J/\psi \to \pi^+\pi^-)$. This was extended by us[14] to K^+K^- by noting that

$$\mathcal{B}(J/\psi \to \gamma^* \to K^+K^-) \approx \mathcal{B}(J/\psi \to K^+K^-) - \mathcal{B}(J/\psi \to K_S K_L)$$

Using literature values for the branching fractions, the results are $|Q^2|F_K(9.6\ \text{GeV}^2) = 0.81 \pm 0.06\ \text{GeV}^2$, $|Q^2|F_\pi(9.6\ \text{GeV}^2) = 1.01 \pm$

0.13 GeV^2, both of which are in good agreement with the CLEO measurements for $|Q^2| = 13.5$ GeV^2. The ratio $F_\pi/F_K = 1.19 \pm 0.07$ and $1.16 \pm 0,27$ for $|Q^2| = 13.5$ GeV^2 and $|Q^2| = 9.6$ GeV^2, respectively, which are both in disagreement with the pQCD expectation that $F_\pi/F_K = f_\pi^2/f_K^2 = 0.67 \pm 0.01$.

To summarize, for the first time we now have precision results for the form factors of charged pions and kaons for large timelike momentum transfers of 9.6 GeV^2 and 13.5 GeV^2. None of the theoretical calculations, which exist only for pions, come even close to the experimental results. In absence of precsion experimental results this could be tolerated. Now there is no excuse. The theorists must now go back to work on new QCD–based models for form factors which are among the most important measures of hadron structure. It is interesting to note that no help is expected in this endeavour from Lattice practitioners who work in Euclidean time. For the experimentalists, the challenge is to extend the precision measurements to as large momentum transfers as possible, and to other mesons and baryons. BES III is ideally placed to meet this challenge.

References

1. A. F. Sill et al., Phys. Rev. **D48**, 29 (1993).
2. G. P. Lepage and S. J. Brodsky, Phys. Rev. **D22**, 2157 (1980).
3. E760 Collaboration, T. A. Armstrong et al., Phys. Rev. Lett. **70**, 121 (1993).
4. E835 Collaboration, M. Ambrogiani et al., Phys. Rev. **D60**, 032002 (1999); E835 Collaboration, M. Andreotti et al., Phys. Lett. **B559**, 20 (2003).
5. T. Hyer, Phys. Rev. **D47**, 3875 (1993).
6. F. Iachello and Q. Wan, Phys. Rev. **C 69**, 055204 (2004).
7. P. Kroll et al., Phys. Lett **B 316**, 546 (1993).
8. CLEO Collaboration, T. Pedlar et al., Phys. Rev. Lett. **95**, 261803 (2005).
9. BES Collaboration, M. Ablikim et al., Phys. Lett. **B630**, 14 (2003).
10. BaBar Collaboration, B. Aubert et al., Phys. Rev. **D73**, 012005 (2006).
11. M. K. Jones et al., Phys. Rev. Lett. **84**, 1398 (2000); O. Gayou et al., Phys. Rev. Lett. **88**, 092301 (2002).
12. J. Volmer et al., Phys. Rev. Lett. **86**, 1713 (2001); V. Tadevosyan et al., nucl-ex/0607007; T. Horn et al., Phys. Rev. Lett. **97**, 192001 (2006).
13. J. Milana, S. Nussinov, and M. G. Olsson, Phys. Rev. Lett. **71**, 2533 (1988).
14. K. K. Seth, Phys. Rev. **D 75**, 017301 (2007).

POLARIZATION PHENOMENA IN $e^+e^- \rightarrow p\bar{p}$ REVISITED

A. Z. DUBNIČKOVA

Department of Theoretical physics, Faculty of math. phys. and informatics, Comenius University, Bratislava, Slovak Republic
E-mail: _dubnickova@fmph.uniba.sk
www.university_uniba.sk

S. DUBNIČKA

Institute of Physics, Slovak Academy. of Sciences, Bratislava, Slovak Republic
E-mail: _fyzidubn@savba.sk

Polarization effects in $e^+e^- \rightarrow N\bar{N}$ processes are reinvestigated. The explicit form of components of the single-spin and double-spin polarization of the created nucleon N in the annihilation of of e^+e^- colliding beams is presented. The sensitivity in a behavior of the single spin/or double spin polarization components of the nucleon is demonstrated graphically by using two recent formulations of the unitary and analytic model of the nucleon electromagnetic structure.

Keywords: form factor, single and double polarization, annihilation process

1. Introduction

Prior to the year 2000 all data on form factors (FFs) $G_{Ep}(t)$ and $G_{Mp}(t)$ in the space-like ($t < 0$) region were obtained by measuring the differential cross-section of elastic scattering on proton in the laboratory frame utilizing the Rosenbluth technique.[1] Both of them have more or less dipole behavior and their ratio in error bars is equal one.

More recently at JLab[2–4] by means of double polarization experiment a great success was achieved measuring simultaneously transverse and longitudinal components of the recoil proton's polarization in the electron scattering plane of the polarization transfer process. The data on the ratio in the region $0.3GeV^2 \leq Q^2 \leq 5.6GeV^2$ are obtained, which clearly demonstrate a remarkable fall of $G_{Ep}(t)$ with increased Q^2 in comparison with $G_{Mp}(t)$ and so, this data are in a rather strong disagreement with the data obtained by Rosenbluth technique.

The latter example revealed an importance of investigation of polarization effects in particle physics.

Here we are concerned with the theoretical reinvestigation of the polarization phenomena in the electron-positron annihilation into proton-antiproton pairs.[5]

The above-mentioned process is interesting that there are noticeable polarization effects even if there are no polarized particles in the initial state. The appearance of polarization effects is due to $G_{Ep}(s)$ and $G_{Mp}(s)$ being complex with non-zero relative phase. On that account there are also nontrivial polarization effects in the scattering of longitudinally polarized electrons on unpolarized target.

This contribution is devoted to the analysis of polarization effects in the process $e^+e^- \to p\bar{p}$ calculated in the framework of the one-photon exchange approximation.

In terms of two electromagnetic (EM) FFs G_{Ep} and G_{Mp} we calculate single spin and double spin polarization of the recoil proton (antiproton) by using the Unitary and Analytic (U&A) model of the nucleon EM structure.[6]

2. Single and double spin polarization observables

In order to calculate the single spin polarization observables (either for proton or for antiproton) for the case of

- unpolarized incoming leptons
- or incoming electron to be longitudinally polarized.

we are in need of the matrix element of the process $e^+e^- \to p\bar{p}$ in the one-photon exchange approximation

$$\begin{aligned} \mathcal{M} &= \frac{e^2}{k^2} j_\mu J_\mu, \\ j_\mu &= \bar{u}(-k_2)\gamma_\mu u(k_1), \\ J_\mu &= \bar{u}(p_1)\big[F_{1p}(s)\gamma_\mu - F_{2p}(s)\frac{\sigma_{\mu\nu}k_\nu}{2m_p}\big]u(-p_2), \end{aligned} \tag{1}$$

where $s = k^2 \geq 4m_p^2$. The c.m. system of the reaction $e^+e^- \to p\bar{p}$ is the most suitable for the analysis of polarization effects.

Then EM currents j_μ and J_μ are conserved $k \cdot j = k \cdot J = 0$ and the matrix element $\mathcal{M}$ is determined by product of spatial components of the currents $\vec{j}$ and $\vec{J}$.

The electromagnetic current $\vec{J}$ can be expressed through two-component spinors φ_1 and φ_2

$$\vec{J} = \sqrt{s}\varphi_1^+ \left[G_{Mp}(s)(\vec{\sigma} - \vec{n}\vec{\sigma} \cdot \vec{n}) + \frac{2m_p}{\sqrt{s}} G_{Ep}(s)\vec{n}\vec{\sigma} \cdot \vec{n} \right] \varphi_2, \qquad (2)$$

where we denote

$$\vec{F} = \sqrt{s}\,[\,G_M(s)(\vec{\sigma} - \vec{n}\,\vec{\sigma}.\vec{n}) + \frac{2M}{\sqrt{s}}\, G_E(s)\vec{n}\,\vec{\sigma}.\vec{n}\,], \qquad (3)$$

$\vec{\sigma}$ are Pauli matrices and $\vec{n}$ is the unit vector along the three momentum $\vec{q}$ of the proton, $\vec{m}$ is the unit vector of incoming electron.

In order to find the corresponding cross-section, one has to calculate $|\mathcal{M}|^2$.

In the case of unpolarized initial leptons the corresponding lepton tensor takes the form

$$j_{ij} = 2s(\delta_{ij} - m_i m_j) \qquad (4)$$

and the vector polarization is

$$\vec{P} = \frac{j_{ij} Tr[F_i F_j^\dagger \vec{\sigma}]}{j_{ij} Tr[F_i F_j^\dagger]}. \qquad (5)$$

Calculating the corresponding trace in the numerator, and similarly in the denominator, one obtains nonzero only P_y component of the vector polarization to be

$$\begin{aligned} P_x &= 0 \\ P_y &= -\frac{\frac{1}{\sqrt{\tau}} Im(G_M^*(t)G_E(t)) \sin 2\vartheta}{\frac{1}{\tau}|G_E|^2 \sin^2\vartheta + |G_M|^2(1+\cos^2\vartheta)} \\ P_z &= 0. \end{aligned} \qquad (6)$$

The y-axis is orthogonal to the scattering plane defined by the unit vectors $\vec{m}$ and $\vec{n}$ along the three-momentum of the electron and the three-momentum of the created proton, respectively.

The contributions of P_x and P_z in proton polarization are different from zero only if the electron is longitudinally polarized, i.e. the lepton tensor takes the following form

$$j_{ij} = 2s(\,\delta_{ij} - m_i m_j + \lambda i \varepsilon_{ijl} m_l\,). \qquad (7)$$

Then the components of the vector polarization $\vec{P}$ of the reaction

$e^+e^- \to p\bar{p}$ are

$$P_x = -\frac{2\sin\theta \cdot Re[G_{Ep}(s)G^*_{Mp}(s)]\tau}{|G_{Ep}(s)|^2\sin^2\theta/\tau + |G_{Mp}(s)|^2(1+\cos^2\theta)} \tag{8}$$
$$P_y = -\frac{\frac{1}{\sqrt{\tau}}Im(G^*_M(t)G_E(t))\sin 2\vartheta}{\frac{1}{\tau}|G_E(t)|^2\sin^2\vartheta + |G_M(t)|^2(1+\cos^2\vartheta)}$$
$$P_z = \frac{2\cos\theta|G_{Mp}(s)|^2}{|G_{Ep}(s)|^2\sin^2\theta/\tau + |G_{Mp}(s)|^2(1+\cos^2\theta)},$$

assuming 100% i.e. $\lambda = 1$ longitudinal polarization of one lepton.

In a similar procedure one can find explicit forms of double spin polarization observables in the $e^+e^- \to p\bar{p}$ process, where we are interested for polarizations of created proton and antiproton simultaneously. The corresponding tensor of the polarization is

$$P_{kl} = \frac{j_{ij}Tr[F_i\sigma_k F_j^\dagger \sigma_l]}{j_{ij}Tr[F_iF_j^\dagger]}, \tag{9}$$

where $k, l = x, y, z$. Calculating the trace in the numerator and similarly the trace in denominator, considering unpolarized incoming leptons one finds

$$\begin{aligned}
P_{xx} &= \frac{|G_M(t)|^2\cos^2\vartheta - \frac{1}{\tau}|G_E(t)|^2\sin^2\vartheta}{\frac{1}{\tau}|G_E(t)|^2\sin^2\vartheta + |G_M(t)|^2(1+\cos^2\vartheta)};\\
P_{yy} &= \frac{|G_M(t)|^2(1+\sin^2\vartheta) - \frac{1}{\tau}|G_E(t)|^2\sin^2\vartheta}{\frac{1}{\tau}|G_E(t)|^2\sin^2\vartheta + |G_M(t)|^2(1+\cos^2\vartheta)};\\
P_{zz} &= \frac{|G_M(t)|^2\sin^2\vartheta - \frac{1}{\tau}|G_E(t)|^2\cos^2\vartheta}{\frac{1}{\tau}|G_E(t)|^2\sin^2\vartheta + |G_M(t)|^2(1+\cos^2\vartheta)};\\
P_{xy} &= P_{yx} = 0;\\
P_{xz} &= P_{zx} = \frac{\frac{1}{\sqrt{\tau}}Re[G^*_M(t)G_E(t)]\sin 2\vartheta}{\frac{1}{\tau}|G_E(t)|^2\sin^2\vartheta + |G_M(t)|^2(1+\cos^2\vartheta)};\\
P_{yz} &= P_{zy} = 0.
\end{aligned} \tag{10}$$

Now, for the contraction, using the expression for lepton tensor with longitudinally polarized electron or positron one obtains for components P_{xx}, P_{yy}, P_{zz}, P_{xy}, P_{yx}, P_{xz}, P_{zx} are the same as in the previous case and only the following components

$$P_{yz} = P_{zy} = -2\frac{\frac{1}{\sqrt{\tau}}Im[G^*_M(t)G_E(t)]\sin\vartheta}{\frac{1}{\tau}|G_E(t)|^2\sin^2\vartheta + |G_M(t)|^2(1+\cos^2\vartheta)} \tag{11}$$

are different.

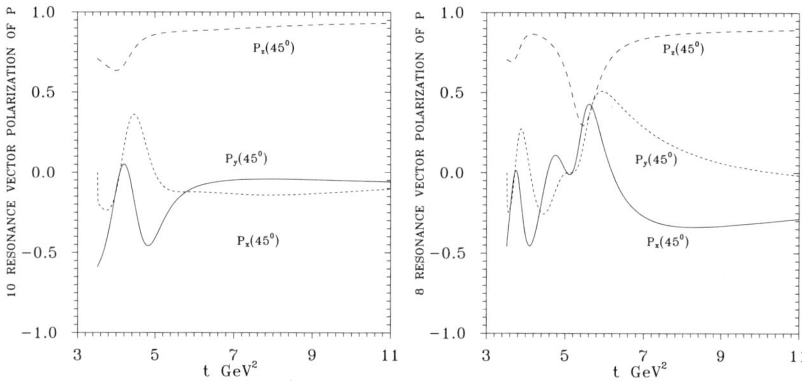

Fig. 1. Prediction of the single polarizations observables by ten-resonance (left-hand) and eight-resonance (right-hand)U&A model.

Every of components P_{kl} characterize a polarization of the proton p at the direction of k, if antiproton $\bar{p}$ is polarized at the direction of l and quantities P_{kl} are calculated for 100% polarization of one of the initial leptons.

Now, exploiting behaviors of proton electric and magnetic FFs in the time-like region as predicted by our ten-resonance U&A model,[6] we find behaviors of single and double spin polarization observables of the $e^+e^- \rightarrow p\bar{p}$ process as they are presented in Figures 1-2. For the comparison there we show also behaviors found by our eight resonance U&A model[5](see Fig.1).

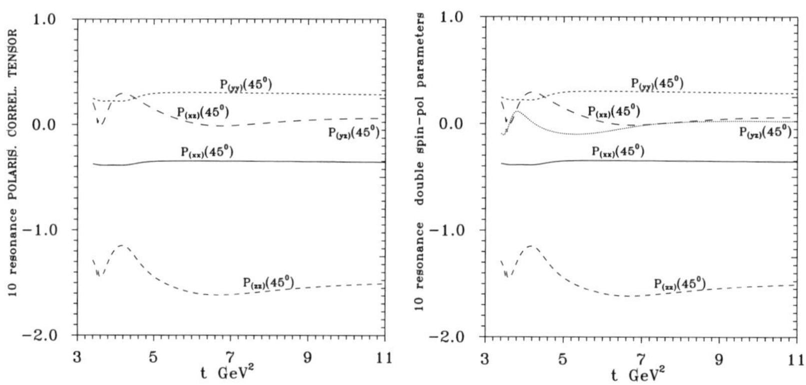

Fig. 2. 10 resonance U&A - double polarization observables for unpolarized and polarized leptons.

3. Summary

We have reinvestigated the single and double polarization observables in the $e^+e^- \rightarrow p\bar{p}$ process by using predictions of the proton EM FFs from eight- and ten-resonance our U&A model of nucleon EM structure, which creates naturally the nonzero imaginary parts of the nucleon FFs. The new analyzes were carried out with completed new experimental information. It was clearly demonstrated that even in the framework of the same U&A model, however, with different number of resonances (8 and 10) the predictions of polarization observables are different and only experimental measurements of single and double spin polarization observables in the $e^+e^- \rightarrow p\bar{p}$ can decide which model is correct.

The work was partly supported by Slovak Grant Agency for Sciences VEGA, Grant No. 2/7116/27 (A.Z.D. and S.D.).

References

1. R. Hofstadter, F. Bumiller, M. R. Yearian, *Rev. Mod. Phys.* **30** 483 (1958).
2. M. K. Jones et al,*Phys. Rev. Lett.* **84** 1398 (2000).
3. O. Gayou et al, *Phys. Rev. Lett.* **88** 092301 (2002).
4. V. Punjabi et al, *Phys. Rev C* **71** 055202 (2005).
5. A. Z. Dubničková, S. Dubnička, M. P. Rekalo,*Nuovo Cim. A* **109** 241 (1996).
6. S. Dubnička, A. Z. Dubničková, P.Weisenpacher, *J. Phys. G* **29** 405 (2003).

LIGHT-CONE SUM RULES FOR FORM FACTORS OF $N\gamma\Delta$ TRANSITION AT $Q^2 = 0$

J. ROHRWILD

Institut für Theoretische Physik, Regensburg University,
Regensburg, D-93040, Germany
E-mail: juergen.rohrwild@physik.uni-r.de

The radiative $\Delta \to \gamma N$ transition is examined at the real photon point $Q^2 = 0$ using the framework of light-cone QCD sum rules. In particular, we determine the sum rules for the transition form factors $G_M(0)$ and R_{EM} up to twist 4.

Keywords: hard exclusive processes, distribution amplitudes, form factors

1. Introduction

In this talk we present a light-cone sum rule (LCSR) calculation for the $N\gamma\Delta$ transition form factors at $Q^2 = 0$. The form factors are by themselves interesting quantities as they contain information on the baryon structure. Especially, it is known that a non-vanishing electromagnetic quadrupole transition form factor $G_E(0)$ indicates a deformation of the nucleon-Δ system [1]. Therefore, the $N\gamma\Delta$ transition has been studied extensively by both experimentalists [2–10] and theorists [11–15].

Our calculation is primarily motivated by the fact that an early SVZ sum rule-based examination of the transition did not yield a definite result [16]. Moreover, a recent light-cone sum rule calculation[17] based on nucleon distribution amplitudes (DAs) [18] shows a notable deviation of the magnetic dipole form factor $G_M(Q^2)$ for 1 $\text{GeV}^2 < Q^2 < 2$ GeV^2 from experiment, see Fig.1, while this approach works perfectly for the nucleon electromagnetic form factors [19–21]. The reason for this discrepancy is not yet understood and the presented analysis is a first step towards a better understanding. Different LCSR analyses of the $N\gamma\Delta$ transition at $Q^2 = 0$ can be found in [22] and, as part of a more general analysis, in [23].

In the following, we will give a short introduction to LCSR calculations with photon DAs using the example of the proton magnetic moment and present the results for the form factors of the $N\gamma\Delta$ transition at $Q^2 = 0$.

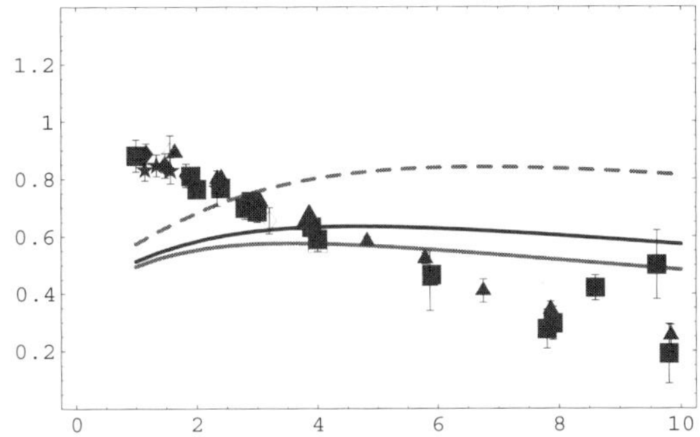

Fig. 1. LCSRs for the $\gamma^* N \to \Delta$ transition form factor $G_M(Q^2)/(3G_{\text{Dipole}})$ vs. Q^2 from [17]. The different lines correspond to different shapes of the nucleon distribution amplitude. For references to the experimental data see [17].

2. Light-cone sum rules for form factors

The starting point for the LCSR approach at $Q^2 = 0$ is a correlation function of the form

$$\Pi(p,q) = i^2 \int d^4x \int d^4y \, e^{ipx+iqy} \langle 0| \mathcal{T}\{\eta'(x)\, j^{\mu}_{\text{em}}(y)\, \overline{\eta}(0)\} |0\rangle \, e^{(\lambda)}_{\mu} \,, \quad (1)$$

which describes the transition of a baryon $B(p+q)$ to a baryon $B'(p)$ by an interaction with the electromagnetic current $j^{\mu}_{\text{em}}(y)$ ($e^{(\lambda)}_{\mu}$ is the 4-polarization vector of the photon). The baryons are created by three-quark currents η' and η, respectively.

For simplicity, we will consider the explicit example of the magnetic moment of protons, which has been considered e.g. in [24]. The interpolating fields η and η' are replaced by the Ioffe-current for the proton

$$\eta_{\text{Ioffe}}(x) = \left(u^a(x)\mathcal{C}\gamma^{\lambda}u^b(x)\right)\gamma_5\gamma_{\lambda}d^c(x)\varepsilon^{abc} \quad (2)$$

and the electromagnetic current can be written in the from

$$j^{\mu}_{\text{em}}(y) = e_d\overline{d}(y)\gamma_{\mu}d(y) + e_u\overline{u}(y)\gamma_{\mu}u(y). \quad (3)$$

The standard strategy of the light-cone sum rule approach involves the calculation of the correlation function (2) in two different regimes:

- on the hadron level the correlator is expressed in terms of form factor (in case of the proton the Dirac and Pauli form factors F_1 and F_2)
- on the level of quarks the correlation function is expanded in terms of photon distribution amplitudes of definite twist.

As the interpolating field η has non-vanishing overlap with states of higher mass, whose quantum numbers coincide with those of the current η, Eq.(2) contains contribution of all these states. After equating both hadronic and quark representation a Borel transformation can be employed to suppress the higher mass states exponentially. As the momenta p and $p+q$ are independent, it is possible to perform a so-called double Borel transformation and introduce two Borel parameters M_1 and M_2, corresponding to p and $p+q$. Introducing a model for the excited states and the continuum one can subtract these contributions from both sides and express the form factors via the photon distribution amplitudes[a], which are known up to twist 4 [26]. Besides the dependence on M_1 and M_2 the form factors are also functions of the continuum subtraction threshold S_0.

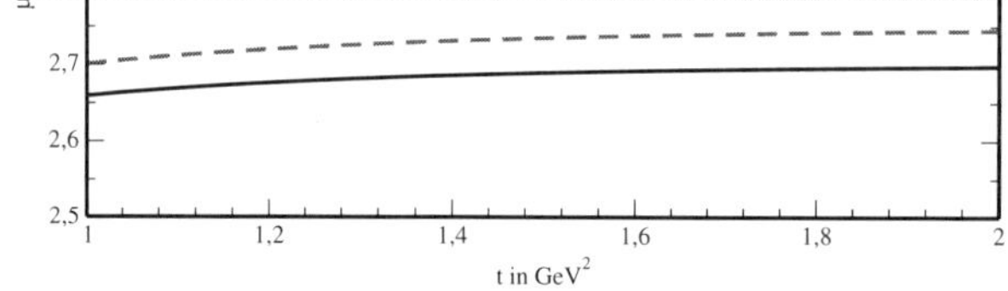

Fig. 2. Magnetic moment of the proton μ_p vs. Borelparameter $t = (M_1^2 + M_2^2)/M_1^2 M_2^2$ for different central values of the photon wave function $\varphi(1/2)$ (Plot from [29]). The solid line is obtained by the choice of an asymptotic shape and a magnetic susceptibility of the quark condensate of $\chi(\mu = 1\text{GeV}^2) = 3.15 \pm 0.3\text{GeV}^{-2}$ [26]. The hatched line represents the experimental value.

[a]The distribution amplitudes are non-perturbative in nature and can be accessed e.g. via QCD sum rule estimates[25,26], LCSRs [27] or lattice calculations.

The process $p \to p\gamma$ is inherently symmetric in the initial and final hadron state. Therefore, it is natural to choose both Borel parameters equal: $M_1^2 = M_2^2$. With this choice the leading-twist photon wave function at the middle point $\varphi(1/2)$, see e.g. [24, 28], is the main non-perturbative input parameter. The result for the magnetic moment of the proton for different values of $\varphi(1/2)$ is shown in Fig.2. The agreement with experiment is very good.

3. Nucleon-Δ-transition

The radiative $\Delta \to N$ decay can also be described by the correlation function (2), the current η' has to be replaced by

$$\eta_\Delta^\mu(x) = \left[\left(u^a(x)\mathcal{C}\gamma^\mu u^b(x)\right) d^c(x) + 2\left(u^a(x)\mathcal{C}\gamma^\mu d^b(x)\right) u^c(x)\right] \varepsilon^{abc} \,. \quad (4)$$

Compared to the standard approach presented in the previous section, several subtleties have to be taken into account. First and foremost, the interpolating field (4) is known to have non-vanishing overlap with states of spin 1/2 and negative parity, these have to be removed "by hand" by a specific choice of the Lorentz basis before the continuum subtraction is performed, see e.g. [29, 30] for two possible choices. Furthermore, it has been argued that a symmetric choice of the two Borel parameters, $M_1^2 = M_2^2$, is not optimal. In order to take into account the mass difference of 300 MeV between the nucleon and the Δ the choice of $M_1^2/M_2^2 = m_\Delta^2/m_N^2$ is advantageous. This strategy was advocated in [25].

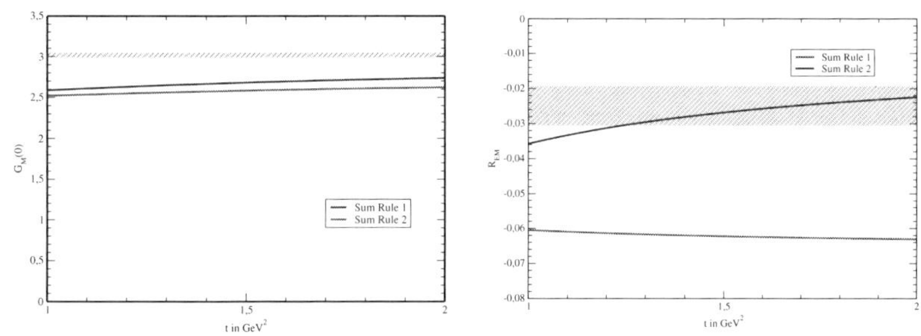

Fig. 3. Light-cone sum rules for the $N\gamma\Delta$ transition form factors (left panel: $G_M(0)$ vs. t ; right panel: R_{EM} vs. t) in each case obtained from two different Lorentz structures. The experimental value is given by the hatched region.

The result for the magnetic dipole form factor $G_M(0)$ is shown in the left panel of Fig.3. The two solid lines correspond to two different sum rules that stem from different Lorentz structures. Within errors, the sum rule prediction [29]

$$G_M(0) = 2.70 \pm 0.27 \quad (5)$$

agrees well with the experimental value of $G_M(0) = 3.03 \pm 0.03$ [2], which is displayed by the hatched region. The right panel in Fig.3 shows the ratio of the electric quadrupole form factor $G_E(0)$ and the magnetic dipole form factor $G_M(0)$, $R_{EM} := G_E(0)/G_M(0)$. The solid lines again correspond to sum rules for different Lorentz structures. Although the upper line seems to agree very well with the value of $R_{EM} = -2.5 \pm 0.4\%$ given by PDG [31], this is actually misleading. Not every structure is suitable for a sum rule and, indeed, the lower line, which yields

$$R_{EM}(0) = -(6.4 \pm 0.8)\% \quad (6)$$

corresponds to the more "trustworthy" sum rule [29]. This is further supported by its marginal dependence on the unphysical Borel parameter $t = (M_1^2 + M_2^2)/M_1^2 M_2^2$. Our result agrees with experiment only within a factor 2, however, it should be pointed out that the smallness of R_{EM} is largely due to cancellations. The quantity is therefore intrinsically difficult to access via a sum rule.

4. Conclusions

Photon distribution amplitudes provide an intriguing device to treat radiative decays within the method of LCSRs and allow the determination of transition from factors at $Q^2 = 0$, which are usually not accessible when using baryon distribution amplitudes. Our result for μ_p is in very good agreement with experiment and lends further support to this method.

$G_M(0)$ is rather close to the corresponding experimental value. While this is desirable, it does not shed much light on the peculiar behavior of LCSR predictions for large values of Q^2 [17], which are almost a factor 2 below data for momentum transfers below 2 GeV^2. However, our result indicates that the effect observed in [30, 17] is probably not due to the choice of the interpolating field. In order to close the gap to aforementioned calculation, it is necessary to expand our approach from the real photon point to virtualities ranging from 0 to -1GeV^2. This requires photon distribution amplitudes for virtual photons, see e.g. [32].

Moreover, it would be desirable to have more precise values for the non-perturbative input parameters appearing in the photon distribution amplitudes, since e.g. the parameters of twist 4 are only known up to 50%.

Acknowledgements

I would like to thank V. Braun and A. Lenz for many enlightening discussions and the organizers of this workshop for their perfect work.

References

1. S. L. Glashow, *Physica* **A96**, 27 (1979).
2. V. V. Frolov *et al.*, *Phys. Rev. Lett.* **82**, 45 (1999).
3. K. Joo *et al.*, *Phys. Rev. Lett.* **88**, p. 122001 (2002).
4. C. Kunz *et al.*, *Phys. Lett.* **B564**, 21 (2003).
5. M. Ungaro *et al.*, *Phys. Rev. Lett.* **97**, p. 112003 (2006).
6. N. F. Sparveris *et al.*, *Phys. Lett.* **B651**, 102 (2007).
7. G. Blanpied *et al.*, *Phys. Rev.* **C64**, p. 025203 (2001).
8. R. Beck *et al.*, *Phys. Rev.* **C61**, p. 035204 (2000).
9. A. M. Sandorfi *et al.*, *Nucl. Phys.* **A629**, 171c (1998).
10. R. Beck *et al.*, *Phys. Rev. Lett.* **78**, 606 (1997).
11. N. Isgur, G. Karl and R. Koniuk, *Phys. Rev.* **D25**, p. 2394 (1982).
12. A. Wirzba and W. Weise, *Phys. Lett.* **B188**, p. 6 (1987).
13. V. Pascalutsa and M. Vanderhaeghen, *Phys. Rev.* **D73**, p. 034003 (2006).
14. C. Alexandrou *et al.*, *Phys. Rev. Lett.* **94**, p. 021601 (2005).
15. V. Pascalutsa, M. Vanderhaeghen and S. N. Yang, *Phys. Rept.* **437**, 125 (2007), (and references therein).
16. B. L. Ioffe and A. V. Smilga, *Nucl. Phys.* **B232**, p. 109 (1984).
17. V. M. Braun, A. Lenz, G. Peters and A. V. Radyushkin, *Phys. Rev.* **D73**, p. 034020 (2006).
18. V. Braun, R. J. Fries, N. Mahnke and E. Stein, *Nucl. Phys.* **B589**, 381 (2000).
19. V. M. Braun, A. Lenz and M. Wittmann, *Phys. Rev.* **D73**, p. 094019 (2006).
20. V. M. Braun, A. Lenz, N. Mahnke and E. Stein, *Phys. Rev.* **D65**, p. 074011 (2002).
21. A. Lenz, M. Wittmann and E. Stein, *Phys. Lett.* **B581**, 199 (2004).
22. T. M. Aliev and M. Savci, *Phys. Rev.* **D60**, p. 114031 (1999).
23. T. M. Aliev and A. Ozpineci, *Nucl. Phys.* **B732**, 291 (2006).
24. V. M. Braun and I. E. Filyanov, *Z. Phys.* **C44**, p. 157 (1989).
25. I. I. Balitsky, V. M. Braun and A. V. Kolesnichenko, *Nucl. Phys.* **B312**, 509 (1989).
26. P. Ball, V. M. Braun and N. Kivel, *Nucl. Phys.* **B649**, 263 (2003).
27. J. Rohrwild, *arXiv:0708.1405* (2007).
28. V. M. Belyaev, V. M. Braun, A. Khodjamirian and R. Ruckl, *Phys. Rev.* **D51**, 6177 (1995).
29. J. Rohrwild, *Phys. Rev.* **D75**, p. 074025 (2007).
30. V. M. Belyaev and A. V. Radyushkin, *Phys. Lett.* **B359**, 194 (1995).
31. W. M. Yao *et al.*, *J. Phys.* **G33**, 1 (2006).
32. R. Yu, J. P. Liu and K. Zhu, *Eur. Phys. J.* **A28**, 329 (2006).

EXCLUSIVE ELECTROPRODUCTION OF π^0 MESONS

A. N. VILLANO for the Jlab E01-002 Collaboration
Physics, Applied Physics and Astronomy, Rensselaer Polytechnic Institute,
Troy, NY 12180, US
E-mail: villaa@jlab.org

A measurement of the π^0 electroproduction cross section via the process $^1H(e,e'p)\pi^0$ has been made in the momentum transfer region 6.3 GeV $< Q^2 <$ 7.8 GeV and for hadronic invariant masses from pion threshold up to 1.4 GeV. The information contained in the differential cross section is used to probe the Δ resonance region where transition form factors such as $G^*_M(Q^2)$ can be extracted. The data will be used to extract the ratio of electric quadrupole to magnetic dipole Δ excitation amplitudes, $\frac{E2}{M1}$.

Keywords: Delta; Rem; Baryon.

1. Overview

Measurements in the first resonant region have a long history including studies using both photoproduction and electroproduction of pions.[14] The accurate parameterization of pion production cross sections can be useful to extract resonance parameters in this region of invariant mass. In particular, the invariant mass region of the current study contains the well known Δ^+ state, whose resonance parameters in the high momentum transfer region might contain important evidence concerning the behavior of Quantum Chromodynamics (QCD).[4,5,13,15]

The Jefferson Laboratory experiment e01-002 produced high quality unpolarized data at the largest momentum transfer yet attained which will add to the study of the first resonance region and therefore the study of QCD. The process $^1H(e,e'p)\pi^0$ was measured in the Jefferson Lab Hall C with a two spectrometer setup. In the following the initial and final lepton 4-momenta are denoted k and k' respectively. The symbol k_π are the pion 4-momentum while the initial and final proton 4-momenta are denoted as

p and p'.

$$k = (E, \mathbf{k}); \quad k' = (E', \mathbf{k}'); \\ p = (E_p, \mathbf{p}); \quad p' = (E'_p, \mathbf{p}'); \quad k_\pi = (E_\pi, \mathbf{k}_\pi) \tag{1}$$

$$q^\mu \equiv k^\mu - k'^\mu \tag{2}$$

$$cos(\theta_e) \equiv \frac{\mathbf{k}' \cdot \mathbf{k}}{|\mathbf{k}'||\mathbf{k}|} \quad ; \quad cos(\theta) \equiv \frac{\mathbf{q} \cdot \mathbf{k}_\pi}{|\mathbf{q}||\mathbf{k}_\pi|} \tag{3}$$

It is also useful to express the electron variables in terms of Lorentz invariants. Since the one photon exchange approximation is used exclusively these invariants will help parameterize the properties of the "virtual" photon. With that parameterization the process $\gamma^* p \rightarrow p'\pi^0$ in the center of mass of the initial photon-proton system can be considered as directly probing the hadronic current. The relevant kinematic variables are:

$$Q^2 \equiv -q^2 = 2EE'(1 - \cos\theta_e) \\ W \equiv \sqrt{(q+p)^2} = \sqrt{m_p^2 + 2q^0 m_p - Q^2} \tag{4}$$

The picture of the photon-hadron process in the center of mass is displayed in Figure 1. When referring to a non-covariant object, an asterisk will denote

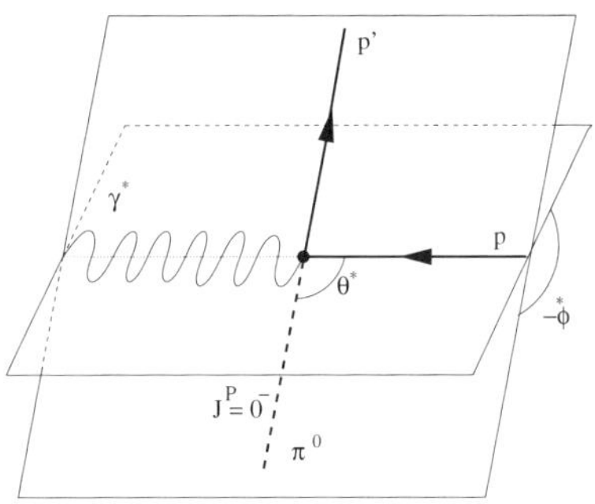

Fig. 1. Center of mass frame neutral pion production.

the center of mass frame and no accent will represent the lab frame. For example, θ is the pion angle in the lab and θ^* is the pion angle in the center of mass.

One can write the differential cross section for the neutral pion production by considering the process as being induced by a virtual photon in the center of mass frame relative to the target proton. Although in the one photon exchange approximation it is a simple matter to work in the center of mass, rigorous justification for the factorization of the electroproduction process will not be given here. It has been shown that the electron current can be parameterized in terms of a virtual photon flux factor.[11] The proper point of view is that each lepton scattering event produces a virtual photon "beam" with a certain density matrix, which need not represent a pure state.[6] Equation 5 shows how the 5-fold differential cross section using the laboratory electron variables and center of mass hadron variables can be written as a flux factor times the center of mass cross section.

$$\frac{d\sigma}{dE' d\Omega_e d\Omega_\pi^*} = \Gamma \frac{d\sigma}{d\Omega_\pi^*} \tag{5}$$

The factor Γ is the virtual photon flux factor and ϵ represents the ratio of longitudinal to transverse photon polarization.

$$\begin{aligned} \Gamma &\equiv \frac{\alpha}{2\pi^2} \frac{E'}{E} \frac{W^2 - m_p^2}{2m_p Q^2} \frac{1}{1-\epsilon} \\ \epsilon &\equiv \left(1 + 2\tan^2\left(\frac{\theta_e}{2}\right) \frac{|\mathbf{q}|^2}{Q^2}\right)^{-1} \end{aligned} \tag{6}$$

Because of the factorization it is useful to think of the process as $\gamma^* p \to p'\pi^0$, for which the amplitude can be written in first order covariant perturbation theory as shown in equation 7, with the appropriate specification of ϵ_μ.

$$\mathcal{M} = e\epsilon_\mu \langle \pi^0, p | J^\mu_{had} | p \rangle \tag{7}$$

Using the photon density matrix to assign interpretation to each portion of the square of the above amplitude allows one to separate the center of mass cross section based on the virtual photon polarization. Equation 8 displays the portions of the center of mass cross section along with their ϕ^* dependencies, which are constrained by parity symmetry and the spacial rotation properties of the photon density matrix components.[11]

$$\begin{aligned} \frac{d\sigma}{d\Omega_\pi^*} = \frac{d\sigma_T}{d\cos\theta^*} + \epsilon \frac{d\sigma_L}{d\cos\theta^*} + \epsilon \frac{d\sigma_{TT}}{d\cos\theta^*} \cos 2\phi^* \\ + \sqrt{2\epsilon(1+\epsilon)} \frac{d\sigma_{LT}}{d\cos\theta^*} \cos\phi^* \end{aligned} \tag{8}$$

The $\cos\theta^*$ dependence is carried in the "response functions" which have been designated with the labels L,T,LT, or TT. These labels stand for longitudinal, transverse, longitudinal-transverse interference, or transverse-transverse interference respectively.

Each of the functions which hold the $\cos\theta^*$ dependence can be parameterized by an infinite but denumerable set of complex coefficients. The set with the most transparent interpretation are the multipole amplitudes in the center of mass. The quantity $\epsilon_\mu \langle \pi^0, p | J^\mu_{had} | p \rangle$ can be expanded with these amplitudes using derivatives of Legendre polynomials to specify the $\cos\theta^*$ dependence.[3] The complex coefficients are denoted $E_{l\pm}$, $M_{l\pm}$ and $S_{l\pm}$. These are electric, magnetic and scalar multipoles respectively, where l specifies the orbital angular momentum of the final state and $\pm$ the spin of the final state nucleon.

2. Physical Motivation

One of the most interesting aspects of hadronic physics is the possible onset of perturbative QCD (pQCD) at high momentum transfers. In this framework quark and gluon degrees of freedom should describe hadronic observables via a well defined perturbative expansion. It is predicted that at very large momentum transfer a well behaved perturbation series can be used for QCD calculations.[10] Perturbative QCD calculations have varying degrees of success for hadronic observables, thus it becomes important to quantify which observables can be described by pQCD and at what momentum scale these descriptions become valid.

The current data has two important indicators relating to the validity of pQCD. The first is related to *helicity conservation*. A massless vector particle (like a gluon or photon) cannot transfer helicity to a massless fermion (like a quark). Given this fact, the helicity of the initial and final state hadrons cannot be changed in the limit that the quarks are massless. Since the high energy limit of QCD is chirally symmetric, this situation should be attained. It turns out that in the rest frame of a produced Δ state in the reaction $ep \rightarrow \Delta \rightarrow e'p'\pi^0$ the $E2$ amplitude will exactly equal the $M1$ amplitude if helicity is conserved[a]. This means that the $\frac{E2}{M1}$ ratio should approach unity in this limit. It is clear from previous studies that this is far from realized in data up to a momentum transfer of $Q^2 \sim 4.0$ GeV, in fact the value is small in magnitude and negative.[9] The second important indicator for relevance of pQCD is a $\frac{1}{Q^3}$ behavior of the helicity conserving amplitude in the $\gamma^* p \rightarrow \Delta$ reaction.[5] This will reveal itself as a fall off for the transition form factor G^*_M consistent with the dipole form (Q^{-4}). A

[a] Assuming pure resonance excitation the amplitudes are both purely imaginary at the resonance position.

fall off of the form factor which is faster than this will indicate that soft processes are still important.

3. Experimental Overview

The experiment was carried out using the High Momentum Spectrometer (HMS) and the Short Orbit Spectrometer (SOS) housed in Jefferson Lab's Hall C. A detailed review of the spectrometer parameters is beyond the scope of this article but has been considered elsewhere [2,16].

The outgoing electrons were detected in the SOS and the outgoing protons were detected in the HMS. The events where there is a single missing pion are selected by requiring the missing mass squared, m_x^2, to correspond to that of the pion at about .019 GeV^2.

$$m_x^2 \equiv (q + p - p')^2 \tag{9}$$

Since the experiment depends on detecting an electron and proton which emerge from the same event, strict timing constraints must be used. The relative trigger times from each of the spectrometers serves as a useful parameter for selecting events from the same vertex. Figure 2 displays the timing variable $t_c^\star$ versus the HMS spectrometer momentum. The variable $t_c^\star$ is corrected so that protons will appear universally at time zero. The timing parameter allows for clear separation of the proton coincidence events and events with other hadron signals in the HMS spectrometer.

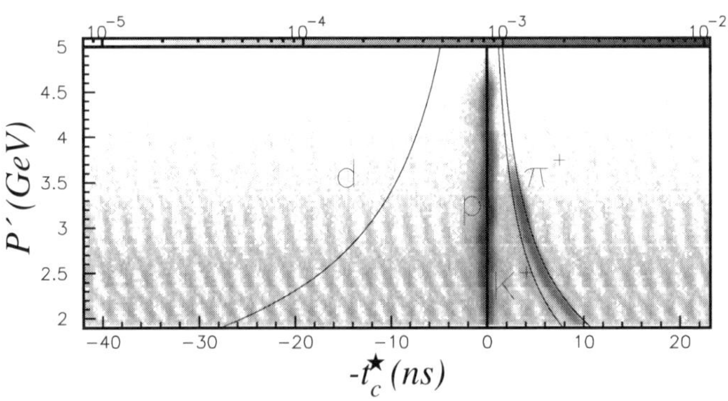

Fig. 2. Timing parameter vs. hadron momentum normalized to make the proton events display vertically.

Setting the beam energy to 5.5 GeV allowed the experiment to access from near threshold through the first resonance region[b] at momentum transfer of $Q^2 \sim 6.4$ GeV^2 and $Q^2 \sim 7.7$ GeV^2.

4. Corrections and Backgrounds

There are several backgrounds that one must deal with in order to extract the signal process from this study. The most notable (and the only ones to be mentioned here) are due to elastic or meson production radiative effects.

Elastic scattering can be accompanied by emission of a photon which may shift the measured invariant energy, W, to a higher value than the usual elastic value of m_p. Though these events have a missing mass of zero, resolution effects can cause them to overlap significantly with the low-mass π^0 peak and therefore must be subtracted. The elastic radiative processes have been studied extensively[8,12] and can be subtracted or excluded with high accuracy.

The emission of photons can also accompany the meson production process. Radiated photons with low energies will be selected with the signal events within the detector systems missing mass resolution. It is therefore inevitable that the photon emission amplitudes play a role in the meson production data. Since no realistic detector system can reject such events, they must be corrected for. Luckily, a deep knowledge of QED allows one to treat these processes and they have been calculated to high accuracy.[1] Thus the corrections are applied to the present data with little introduced error.

5. Differential Cross Sections

Partial wave expansions are beyond the scope of this work. However, to the extent that W is low and the $\Delta(1232)$ contribution dominates, the present series expansion will only include s and p waves. The $\cos\theta^*$ dependent cross sections of equation 8 can then be written explicitly and fit to experimental data. The dependence that one obtains by including only the lowest two angular momentum quanta is quoted in several references.[7] It is then possible to write the s and p wave expansion in terms of three unknown functions which depend on W and Q^2 and are well defined functions of $\cos\theta^*$ but not functions of ϕ^*.

$$\frac{d\sigma}{d\Omega^*_\pi} = A(\cos\theta^*) + \epsilon B(\cos\theta^*)\cos 2\phi^* + \sqrt{2\epsilon(1+\epsilon)}C(\cos\theta^*)\cos\phi^* \quad (10)$$

[b]The experiment actually has data up to approximately W=1.8 GeV which is being used for η production and ${}^1H(e, e'\pi^+)n$ studies as well.

The σ_L and σ_T contributions get lumped into one representative parameter, A, since the experiment at hand does not vary ϵ and therefore cannot separate these contributions. Using the truncated partial wave expansion one can then write the explicit angular dependence.

$$\begin{aligned} A(\cos\theta^*) &\equiv A_0 + A_1\cos\theta^* + A_2\cos^2\theta^* \\ B(\cos\theta^*) &\equiv B_0\sin^2\theta^* \\ C(\cos\theta^*) &\equiv (C_0 + C_1\cos\theta^*)\sin\theta^* \end{aligned} \tag{11}$$

The parameters A_i, B_i, and C_i are now only functions of the electron variables W and Q^2, and not functions of the hadronic center of mass angles. A simple way to proceed in characterizing the extracted cross sections is to fit the angular distributions in each W and Q^2 bin independently. This is the point of view advocated in this work, while more general procedures can be used when the data is coupled with phenomenological models.[3]

Figure 3 shows the differential pion production cross section in the center of mass frame for a 40 MeV energy bin centered about W=1.232 GeV. The curves displayed on the figure are those which result from a fit to the differential cross section using the parameters introduced in equation 11. The fits can be used to obtain a measurement of the total cross section

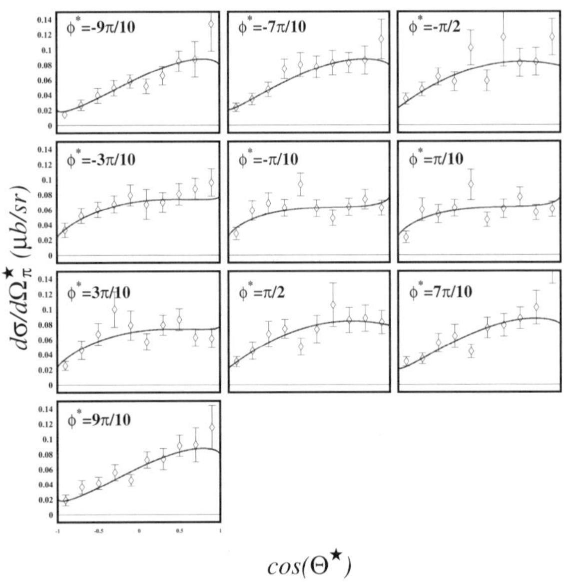

Fig. 3. Preliminary fit to differential cross sections at W=1.232 GeV and Q^2=6.36 GeV^2.

in each W bin. Figure 4 shows the total π^0 center of mass virtual photon differential cross sections for each Q^2 point obtained by the experiment. The variation of the total cross section at the point W=1.232 GeV between

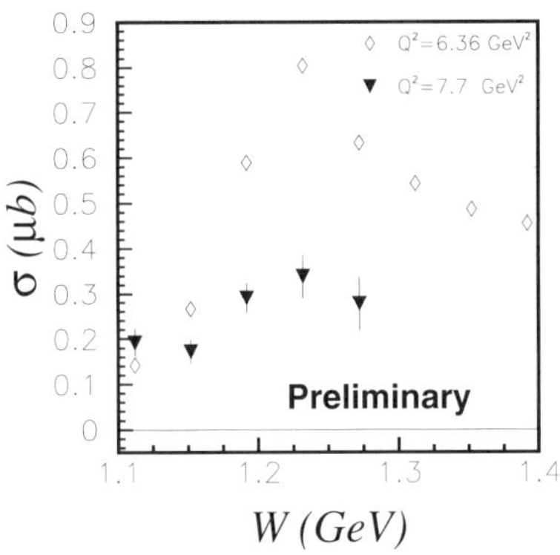

Fig. 4. Preliminary total virtual photon π^0 center of mass cross sections.

the two Q^2 points is an indication of the variation of G^*_M. This parameter is to be extracted from this data in the near future.

The ratio of electric quadrupole to magnetic dipole excitations can be extracted in a simple M_{1+} dominance model. Here only terms which multiply M_{1+} are kept when relating the A_i, B_i and C_i parameters back to the multipole amplitudes. Figure 5 shows the result of the M_{1+} dominated fit for the $Q^2 \sim 6.36$ GeV data. The parameter $\frac{Re(M_{1+}E^*_{1+})}{|M_{1+}|^2}$ is extracted for each invariant energy point. This quantity, taken at the resonance position, is the $\frac{E2}{M1}$ ratio for the Δ in the approximation that the Δ dominates the cross section and that M_{1+} is pure imaginary and dominates all other multipoles. Note that as one evolves to either side of the delta resonance peak position the assumption of M_{1+} dominance becomes less valid. This measure of $\frac{E2}{M1}$ yields a value of 7±4%. The fact that this measurement indicates a positive ratio is significant. However, it is expected that non-resonant backgrounds are important at these larger Q^2, so that fits constrained by larger amounts of world data without imposing M_{1+} dominance will give a more accurate determination. Previous measurements indicate that the ratio $\frac{E2}{M1}$ is small in magnitude and *negative*. The magnitude has thus been consistent with a quark-model-type spin flip transition (dominantly magnetic). Perturbative QCD, however, predicts the ratio to be unity. Insights can be gained by studying the interpolating behavior between the photo-

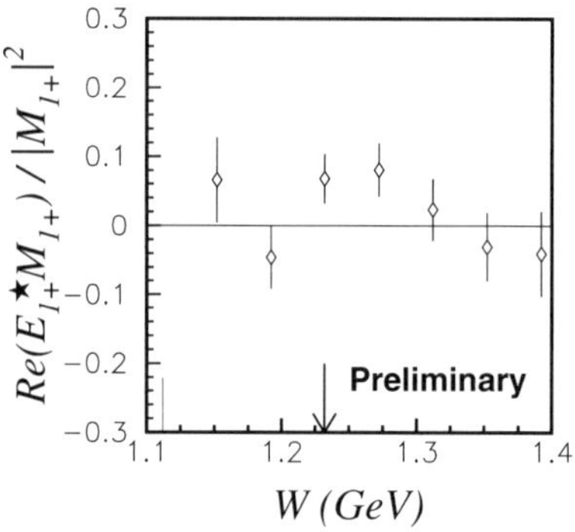

Fig. 5. Preliminary extracted $\frac{Re(E_{1+}^* M_{1+})}{|M_{1+}|^2}$ parameter from M_{1+} dominance fit for Q^2=6.36 GeV^2.

production and pQCD limits. In particular, if the ratio changes sign in the range 3-10 GeV or so, this may indicate an influence from the perturbative limit.[5]

The information obtained in the differential cross sections coupled with similar fitting methods to those demonstrated in this brief report can allow the extraction of both the multipole amplitudes in the Δ resonance region[c] and the transition form factor G_M^*. The data discussed will be used to extract all possible multipole amplitudes and related quantities with both M_{1+} dominance fits and phenomenological fits.

Acknowledgements

The authors acknowledge the Jefferson Lab technical and accelerator staff for making the experiment possible. This work is partially supported by the U.S. National Science Foundation and the U.S. Department of Energy. The Southeastern Universities Research Association operates the Thomas Jefferson National Accelerator Facility under the U.S. Department of Energy contract DEAC05-84ER40150.

[c]The extraction of multipole amplitudes from this data alone requires some assumptions since both the phase and magnitude of the amplitudes cannot be determined from only unpolarized data. Including this data set in the world's data and then making appropriate phenomenological fits is the most powerful way to accurately extract the parameters in the most model independent way possible.

References

1. A. Afanasev, I. Akushevich, V. Burkert, and K. Joo. Qed radiative corrections in processes of exclusive pion electroproduction. *Phys. Rev.*, D66:074004, 2002.
2. O. Keith Baker et al. The high momentum spectrometer drift chambers in hall c at cebaf. *Nucl. Instrum. Meth.*, A367:92–95, 1995.
3. V. D. Burkert and T. S. H. Lee. Electromagnetic meson production in the nucleon resonance region. *Int. J. Mod. Phys.*, E13:1035–1112, 2004.
4. Carl E. Carlson. Electromagnetic n - delta transition at high q**2. *Phys. Rev.*, D34:2704, 1986.
5. Carl E. Carlson and Nimai C. Mukhopadhyay. Approach to perturbative results in the n delta transition. *Phys. Rev. Lett.*, 81:2646–2649, 1998.
6. Norman Dombey. Scattering of polarized leptons at high energy. *Rev. Mod. Phys.*, 41:236–246, 1969.
7. D. Drechsel and L. Tiator. Threshold pion photoproduction on nucleons. *J. Phys.*, G18:449–497, 1992.
8. R. Ent et al. Radiative corrections for (e,e-primep) reactions at gev energies. *Phys. Rev.*, C64:054610, 2001.
9. V. V. Frolov et al. Electroproduction of the delta(1232) resonance at high momentum transfer. *Phys. Rev. Lett.*, 82:45–48, 1999.
10. G. Peter Lepage and Stanley J. Brodsky. Exclusive processes in perturbative quantum chromodynamics. *Phys. Rev.*, D22:2157, 1980.
11. D. H. Lyth. Exclusive electroproduction processes. In G. Shaw A. Donnachie, editor, *Electromagnetic Interactions of Hadrons*, volume 1, pages 159–217. Plenum Press, New York, 1978.
12. L. C. Maximon and J. A. Tjon. Radiative corrections to electron proton scattering. *Phys. Rev.*, C62:054320, 2000.
13. Vladimir Pascalutsa and Marc Vanderhaeghen. New large-N_c relations among the nucleon and nucleon-to- Δ GPDs. hep-ph/0611050, 2006.
14. P. Stoler. Baryon form-factors at high q**2 and the transition to perturbative qcd. *Phys. Rept.*, 226:103–171, 1993.
15. Paul Stoler. Connection between the elastic g(e)(p)/g(m)(p) and p -¿ delta form factors. *Phys. Rev. Lett.*, 91:172303, 2003.
16. L. Tang, C. Yan, and Ed. V. Hungerford. Systematic resolution study of the cebaf hall c spectrometers - hms and sos. *Nucl. Instrum. Meth.*, A366:259–262, 1995.

EXCLUSIVE ELECTROPRODUCTION OF η MESONS IN THE $S_{11}(1535)$ RESONANCE REGION AT HIGH MOMENTUM TRANSFER

M. M. DALTON for the JLab E01-002 Collaboration

University of the Witwatersrand, Johannesburg, South Africa
E-mail: dalton@jlab.org

The differential cross section for the process $p(e, e'p)\eta$ has been measured for centre-of-mass energies from threshold to 1.8 GeV, encompassing the $S_{11}(1535)$ resonance which dominates the channel. This was at $Q^2 = 5.7$ and 7.0 $(\text{GeV/c})^2$, the largest momentum transfer measurement of this exclusive process to date. The photocoupling amplitude for $\gamma p \to S_{11}$ is extracted from the data, and the possible onset of scaling in this reaction is observed.

Keywords: Baryon; Eta; Electroproduction.

1. Introduction

Photoproduction and electroproduction of baryons allows the measurement of transition form factors. This is a rich testing ground for models of hadronic structure in addition to that contributed by the measurement of elastic form factors. The leptonic part of the interaction is well understood and thus such measurements are cleaner than hadron-hadron interactions. The difficulty is in isolating any of the multitude of wide and overlapping resonant states.

The $S_{11}(1535)$ is a baryon resonance that can be accessed relatively easily. Although there are many overlapping states in its mass region, it is very strongly excited over all Q^2 and is the only resonance with a large branching fraction to η mesons, causing it to dominate the $p(e, e'p)\eta$ channel.

It is expected from helicity conservation in pQCD that at sufficiently high Q^2 the photocoupling amplitude will begin to scale[1] as $1/Q^3$, or equivalently the quantity $Q^3 A_{1/2}$ will flatten. The observation of such scaling is thus a possible signal of the transition to the dominance of hard processes. Our data appears to show this behaviour, beginning at about $Q^2 \sim 5$

GeV^2, although it has been pointed out that such scaling may have a non-perturbative explanation.[2,3]

2. Overview

The data described in this article were taken concurrently, in Jefferson Lab's Hall C, with the $p(e, e'p)\pi^0$ data described elsewhere in this volume.[4] The remainder of this section summarises the more detailed descriptions given there.

In the one photon exchange approximation, the 5-fold laboratory differential cross section can be factorised into the centre of mass differential cross section multiplied by a virtual photon flux factor. This centre of mass cross section is then written in terms of response functions corresponding to the virtual photon polarisation states and their interference terms. These can be expanded in terms of multipole amplitudes in the centre on mass.

Timing considerations were used to select protons that originated in the same physical scattering as the associated coincident electron. The data was taken with an electron beam of 5.5 GeV at two distinct settings of the electron spectrometer, namely 1.74 GeV at 47.5° and 1.04 GeV at 70.0°. This corresponds to two distinct data sets, which at the $S_{11}(1535)$ resonance mass, have $Q^2 \sim 5.7$ (GeV/c)^2 and 7.0 (GeV/c)^2 respectively.

3. Multipion Background

Since only correlated protons and electrons are detected, the channel of interest is selected by requiring that the square of the missing mass quantity, $m_x^2 \equiv (q + p - p')^2$, is that for an η particle.

Unlike the π^0 case,[4] the major background for the $p(e, e'p)\eta$ is from multiple pion production. For events where the exit channel has two or more undetected particles, the reconstructed missing mass squared doesn't correspond to any physical mass, and thus forms a continuous background from $m_x^2 \sim 0.08$ GeV^2, under the missing mass peaks, Fig. 1.

The multipion background is subtracted by simulating the spectrum of the background using a Monte Carlo generator, which produces events from various channels that involve multiple pions, as input to an accurate spectrometer acceptance model. The simulation was verified to describe the data well, with only one parameter for normalisation, and was subtracted from the data. Other sources of background and error are much smaller and not discussed here.

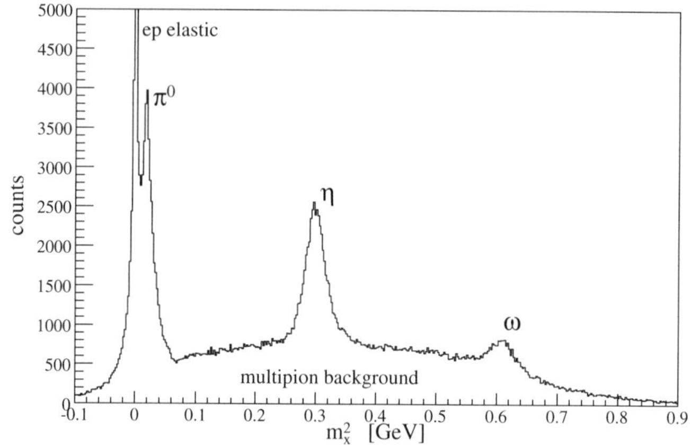

Fig. 1. Missing mass squared, m_x^2, for all data taken at $\theta_e = 47.5°$.

4. Differential Cross Sections

The differential cross sections for the process were obtained in 12 $\cos\theta_\eta$ bins and 8 ϕ_η for 30 MeV wide bins in W from threshold to $W \sim 1.85$ GeV. The angular acceptance begins to decrease significantly for $W \gtrsim 1.6$ GeV since a telescopic magnetic spectrometer was used to acquire the data.

$$\frac{d^2\sigma}{d\Omega^*} = A + B\ \cos\theta^* + C\ \cos^2\theta^* + D\ \sin\theta^*\cos\phi^* \\ +E\ \cos\theta^*\sin\theta^*\cos\phi^* + F\ \sin^2\theta^*\cos2\phi^* \qquad (1)$$

Equation (1) is the parametrisation of the the virtual photon cross section in terms of its angular dependence. The preliminary extracted differential cross section was fitted with Eq. (1), for the lowest W bins where the is sufficient angular acceptance for a fit, and is plotted in Fig. 2. The parameters extracted from the fit are plotted in Fig. 3. These parameters A–F can be written in terms of electric, magnetic and scalar multipoles of S, P, D-wave etc. excitations, but for the purposes of this article it is merely pointed out that the isotropic component dominates the fit results near threshold.

5. Helicity Amplitude

Under the assumption that the $S_{11}(1535)$ resonance completely dominates the η channel near threshold, contributions from other resonances and the nonresonant amplitude can be ignored. Further, the longitudinal component

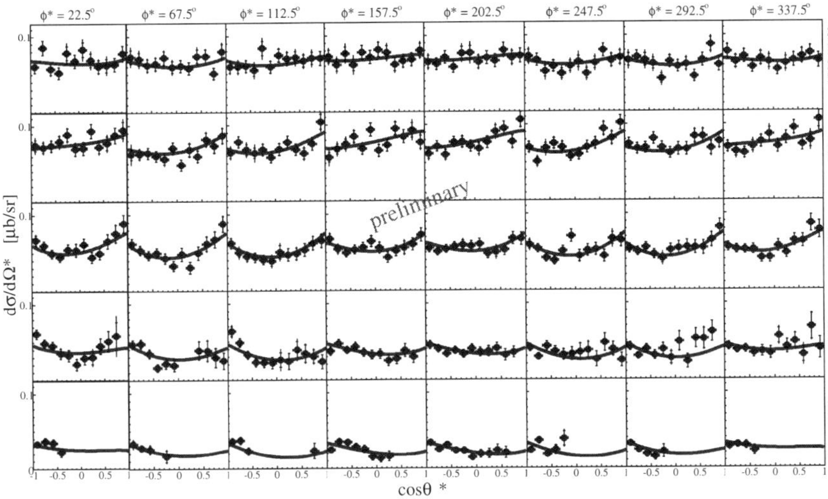

Fig. 2. Extracted differential cross sections for the first 5 W bins of the $\theta_e = 47.5°$ data set. The curve is a fit of Eq. (1) to each W bin.

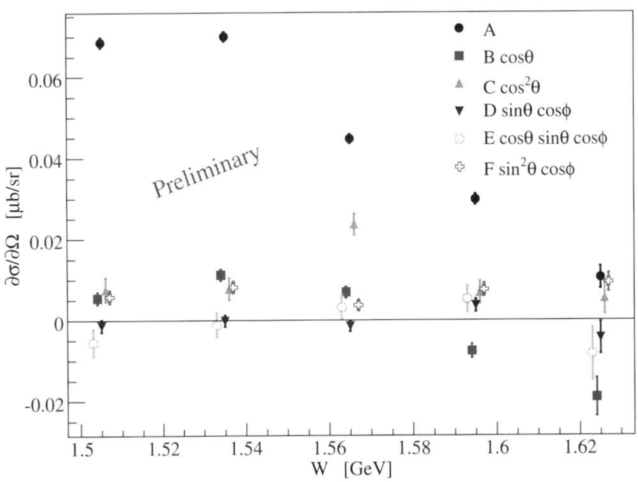

Fig. 3. Parameters from fit of Eq. (1) to angular dependence of virtual photon differential cross section, given in Fig. 2.

to the cross section for the $S_{11}(1535)$ resonance has been shown to be small[5] at low Q^2, and the cross section can thus be written as depending only on the dominant E_{0+} multipole in the simple form of Eq. (2).

$$\frac{d^2\sigma}{d\Omega^*_\eta} = \frac{|\boldsymbol{p}^*_\eta|W}{m_p K}|E_{0+}|^2 \tag{2}$$

Equation (3) relates $A_{1/2}$, the transverse photocoupling amplitude for total γN helicity of $\frac{1}{2}$, to the contribution of the $S_{11}(1535)$ resonance to the E_{0+} multipole at $W = W_R$.[6]

$$A_{1/2} = \left[2\pi \frac{|\boldsymbol{p}^*_\eta|_R W_R}{m_p K} \frac{W_R}{m_p} \frac{\Gamma_R}{b_\eta}\right]^{1/2} |E_{0+}(W_R)| \tag{3}$$

The combination of Eqs. (2) and (3) yields the helicity amplitude as a function of $\sigma_R \equiv \sigma(W_R)$, the total cross section of the $S_{11}(1535)$ resonance, measured at the resonance mass itself.

$$A_{1/2}(Q^2) = \sqrt{\frac{W_R \Gamma_R}{2 m_p b_\eta} \sigma_R(Q^2)} \tag{4}$$

The value of σ_R was obtained by fitting a relativistic Breit-Wigner to the total cross section and evaluating it at the resonance mass. $A_{1/2}$ was then determined using Eq. (4), consistently for this work and other published data[7–9] using $\Gamma_R = 150$ MeV, $b_\eta = 0.55$ and $W_R = 1535$ MeV. The uncertainties in $A_{1/2}$ do not include uncertainties in W_R, b_η or Γ_R. Figure 4 is a plot of $Q^3 A_{1/2}$, showing that the amplitude $A_{1/2}$ appears to begin scaling like $1/Q^3$ at a photon momentum transfer of $Q^2 \sim 5$ GeV^2. It is interesting to note that the magnetic form-factor of the proton G^p_M, begins scaling as $1/Q^4$ at a very similar value of Q^2.[10]

6. Conclusion

This work presents some of the preliminary results of a precise, high statistics measurement of the differential cross section for the $ep \to e'p\eta$ exclusive process, at a significantly higher momentum transfer than available previously. S-wave multipoles are observed to dominate the channel near threshold, with other multipoles contributing less than 15%. The helicity-conserving transition amplitude to the S_{11} resonance, $A_{1/2}$, is extracted from the data, and the possible onset of $1/Q^3$ scaling in this reaction is observed to begin at about $Q^2 \sim 5$ GeV^2.

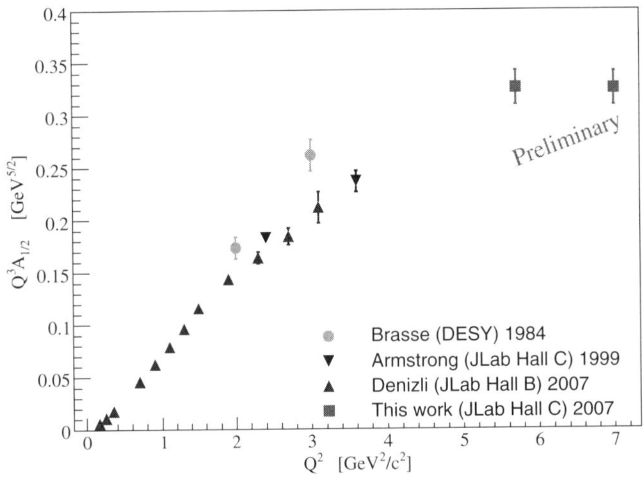

Fig. 4. Values for $Q^3 A_{1/2}$, determined as described in text, with data from other experiments.[7–9]

Acknowledgements

This work is supported in part by research grants from the U.S. Department of Energy, the U.S. National Science Foundation and the South African National Research Foundation. The Southeastern Universities Research Association operates the Thomas Jefferson National Accelerator Facility under the U.S. Department of Energy contract DEAC05-84ER40150.

References

1. C. E. Carlson and J. L. Poor, *Phys. Rev.* **D38**, p. 2758 (1988).
2. A. V. Radyushkin, *Nucl. Phys.* **A527**, 153c (1991).
3. N. Isgur and C. H. Llewellyn Smith, *Phys. Rev. Lett.* **52**, p. 1080 (1984).
4. A. Villano *et al.*, *elsewhere in these proceedings* (2007).
5. H. Breuker *et al.*, *Phys. Lett.* **B74**, 409 (1978).
6. Particle Data Group, *Rev. Mod. Phys.* **48**, p. S157 (1976).
7. C. S. Armstrong, P. Stoler *et al.*, *Phys. Rev.* **D60**, 052004(September 1999).
8. F. W. Brasse *et al.*, *Z. Phys.* **C22**, 33 (1984).
9. H. Denizli *et al.*, *Phys. Rev.* **C76**, p. 015204 (2007).
10. R. G. Arnold *et al.*, *Phys. Rev. Lett.* **57**, p. 174 (1986).

TWO-PHOTON EXCHANGE IN ELECTRON-PROTON ELASTIC SCATTERING: THEORY UPDATE

ANDREI V. AFANASEV

Department of Physics, Hampton University, Hampton, VA 23668, USA
and
Theory Center, Jefferson Lab, Newport News, VA 23606, USA

Recent theoretical developments in the studies of two-photon exchange effects in elastic electron-proton scattering are reviewed. Two-photon exchange mechanism is considered a likely source of discrepancy between polarized and unpolarized experimental measurements of the proton electric form factor at momentum transfers of several GeV^2. This mechanism predicts measurable effects that are currently studied experimentally.

Keywords: Electron scattering; nucleon form factors; two–photon exchange; QED radiative corrections

Measurements of elastic nucleon form factors reached a new level of accuracy, with separation of electric and magnetic contributions made possible at high transferred momenta. At Jefferson Lab, due to a 100% duty factor of the electron beam and implementation of nucleon spin polarization techniques, electric nucleon form factors were measured up to 4-momentum transfers Q^2= 5.6 GeV^2 for the proton[1] and Q^2= 1.45 GeV^2 for the neutron.[2] Extension of the measurements up to Q^2= 9 GeV^2 via recoil proton polarimetry is underway.[3]

Polarization-based results, however, appeared to be in conflict with earlier unpolarized cross section measurements at SLAC.[4] In high-Q^2 kinematics, the difference between the measured values of the proton electric/magetic form factor ratio, G_{Ep}/G_{Mp}, was as large as a factor of five, resulting in important theoretical and phenomenological implications, *c.f.* Ref. [5]. The observed discrepancy between unpolarized and polarized experimental techniques prompted new cross section measurements at Jefferson Lab.[6] These later measurements also appeared to be in conflict with polarization data, confirming a systematic difference between the data from the different experimental techniques.

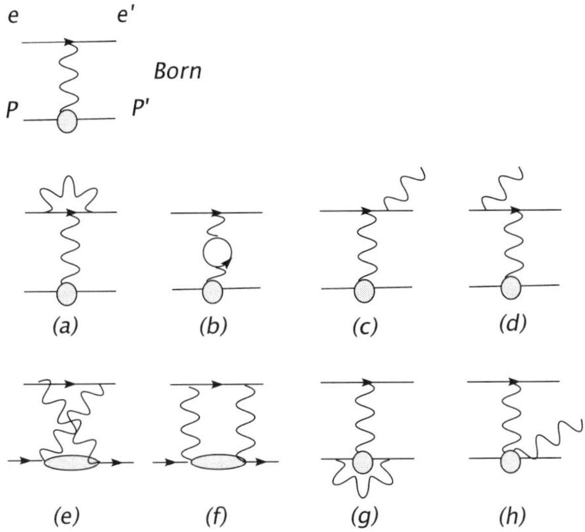

Fig. 1. QED radiative corrections for electron-proton scattering: (a) Electron vertex correction; (b) Vacuum polarization; (c,d) Electron bremsstrahlung; (e,f) Two-photon exchange; (g) Proton vertex correction; and (f) Proton bremsstahlung (virtual Compton scattering).

The resolution of this conflict was suggested[9] to be due to a higher-order electromagnetic effect of two-photon exchange not accounted for in the experimental analysis. Model calculations[10] and[11] lead to similar conclusions, attributing over a half of experimental discrepancy to two-photon exchange corrections. A detailed account of the status of theory and experiment in two-photon exchange can be found in the recent reviews, Refs.[7,8]. Here, I present an update on the status of the two-photon exchange problem and its implications, and highlight related theoretical issues.

Let us start with reviewing a full set of higher-order electromagnetic corrections to electron-proton scattering Fig.1. Contributions of the diagrams Fig.1a,c,d can be calculated using standard QED techniques. They are enhanced by large logarithmic factors $log\frac{Q^2}{m_e^2}$, resulting in radiative corrections of the order of tens per cent that in addition depend on details of experimental cuts in the phase space of the radiated photon and electron scattering angles (at fixed Q^2). Vacuum polarization, Fig.1b, albeit has model uncertainties due to hadronic loop contributions, does not alter angular dependence of cross section at *fixed* momentum transfer Q^2, and hence it has no impact on Rosenbluth separation. Subprocesses with an additional photon coupling only to the proton Fig.1g,f show negligible angular dependence when constrained by kinematic cuts of elastic scattering.

The bremsstrahlung correction of Fig.1c,d was calculated in Ref.[13] in soft-photon approximation, and this result was applied in data analysis in Ref.[4]. If this contribution is calculated fully including also hard-photon emission, for example, according to Ref.[14] or Ref.[15], it leads to about 1 per cent additional absolute correction[16] to the experimental[4] Rosenbluth slope at Q^2=6 GeV^2. This additional correction accounts for about one fifth of the discrepancy between Rosenbluth[4] and polarization data[16] when missing mass cuts on the radiated photon are chosen to match experimental ones. The choice of kinematic cuts is essential since the magnitude of bremsstrahlung correction strongly depends on them. For example, if one uses a generic energy cut parameter for all electron scattering angles (*e.g.*, c=0.97 as in Ref.[17]) the extracted Rosenbluth slope reduces by about 5% at Q^2=6 GeV^2, thereby seemingly 'resolving' disagreement between Rosenbluth and polarization data. It is therefore very important that all refined calculations of bremsstrahlung corrections are also as accurate in the choice of kinematic cuts when compared with specific experimental analysis.

Let us take a closer look at the two-photon exchange process, Fig.1e,f. In the approach developed by Tsai,[13] these contributions were calculated in a limit when one of the exchanged photons carries a negligible 4-momentum. This contribution to the scattering amplitude is (infra-red) divergent, and the divergence is cancelled at the cross-section level by adding interference between the bremsstrahhlung diagrams Fig.1c,d and f. It therefore also depends on the details of experimental cuts on the radiated photon kinematics. The good news is that the calculation with soft second photon exchange does not require additional knowledge on the nucleon structure: It can be done in terms of one-photon exchange contribution times a 'soft' factor that is independent on nucleon structure.[13]

As opposed to bremsstrahlung and vertex corrections, two-photon exchange is not enhanced by large logarithms. It is instructive to see the effect of the soft-photon-exchange portion on the Rosenbluth plot. In Fig.2, its effect on the cross section is shown for the kinematics of SLAC experiment[4] at $Q^2 = 6$ GeV^2. The correction is angular-dependent, varying between about -5% for backward scattering and 0 for forward scattering angles. The Rosenbluth slope measured at SLAC[4] at $Q^2 = 6$ GeV^2 was close to 5% with the above correction *included*. It emphasizes importance of two-photon exchange: Without this correction included, the extracted value of electric proton form factor[4] would be about a factor of $\sqrt{2}$ larger!

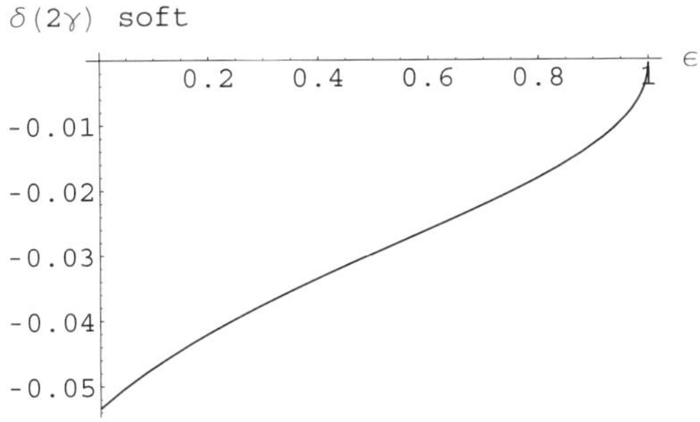

Fig. 2. 'Soft' two-photon exchange correction combined with interference of electron and proton bremsstrahlung according to Ref. [13]. The relative radiative correction to the cross section is plotted aganist the standard kinematic variable ϵ for $Q^2 = 6$ GeV^2.

Let us estimate the effect of terms neglected in the two-photon exchange in Ref.[[13]]. Using a well-established QED results for a structureless spin-1/2 (muon) target,[18] we find the correction for backward-angle electron scattering on a quark with a mass m_q and a charge e_q:

$$\delta^{2\gamma} = -\frac{e_q \alpha}{2\pi} \log^2 \frac{s}{m_q^2}, \tag{1}$$

where s is a Mandelstam variable, and α is a fine structure constant. The corresponding correction is zero for forward electron scattering. Note that the correction is negative for positive-charge quarks, and it grows logarithmically with beam energy; numerically, it is a few per cent for relevant kinematics, if a constituent quark mass of $m_q = 300$ MeV is taken for the estimate. Therefore this correction has the proper sign, magnitude and angular dependence to mimic a contribution of electric form factor to the cross section of electron-proton scattering.

It motivates more detailed studies of two-photon exchange, especially at a partonic level. Such a partonic approach was developed in Refs.[[11,12]], where two-photon exchange amplitude was factorized into a hard subprocess of electron-quark scattering and a soft subprocess described by generalized parton distributions (GPD). A representative result is shown in Fig.3, where it can be seen that linear ϵ-dependence of the cross section is modified by a non-linear contribution from two-photon exchange. A dotted line labeled '1γ' is an expectation from a pure one-photon exchange mechanism with a proton electric form factor taken from polarization measurements.[1] Noticing that at the same time two-photon exchange does not significantly alter interpretation of polarization data, we conclude that within the con-

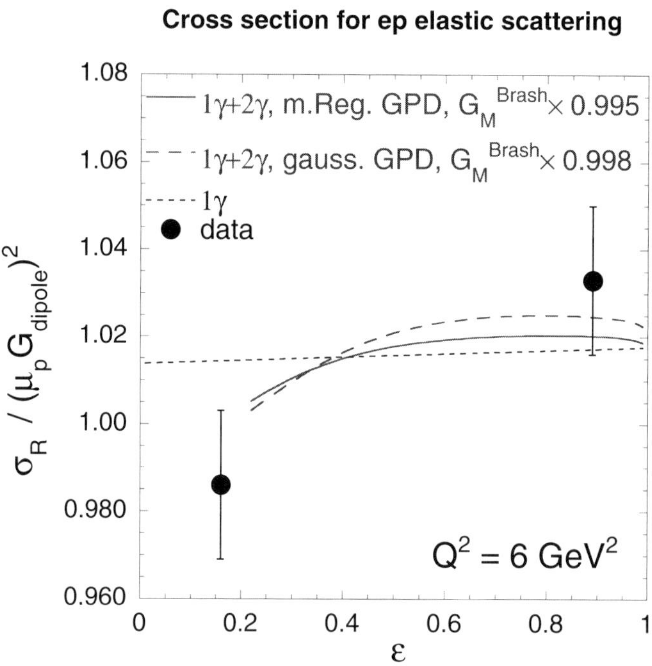

Fig. 3. Reduced ep-scattering cross section at $Q^2 = 6$ GeV^2. Data points are from Ref.[4]. The dotted line shows an expected result from one-photon exchange using G_{Ep} fit to polarization data;[1] solid and dashed curves have the two-photon exchange mechanism included within partonic approach[11,12] using different models of GPD.

sidered model, this mechanism partially reconciles results of experimental techniques using polarized and unpolarized scattering.

In a different approach,[10,19] the virtual Compton amplitude entering the two-photon exchange mechanism was approximated with nucleon pole diagrams with on-shell form factors substituted in photon-nucleon vertices. Despite of different dynamical models for the nucleon Compton amplitudes, the conclusions of Refs.[10,19] and Refs.[11,12] are in qualitative agreement. Addition of Δ-excitation mechanism[20] to the approach of Refs.[10,19] somewhat reduced the predicted magnitude of the two-photon effect. Higher nucleon resonances are estimated Ref.[21] to contribute about an order of magnitude less than nucleon and Δ.

Clearly, the problem of two-photon exchange, especially the real part of the amplitude, is challenging because 4-dimensional momentum integration in the box (and cross-box) diagrams Fig.1e,f requires knowledge of the nucleon Compton amplitude over a broad (infinite, to be exact!) range of kinematic variables not available from experiment. On the other hand, theoretical models applied so far are valid only within certain kinematic regions. One may also try a dispersive approach that takes advantage of analyticity and unitarity of the two-photon amplitude. In the kinematics of forward electron scattering, it is possible to reduce model uncertainties by using inelastic electroproduction structure functions measured experimentally.[22] A different category of papers attempt to evaluate the two-photon exchange effect using the experimentally observed difference between Rosenbluth and polarization data,[9,23–25]

The two-photon exchange mechanism also contributes to parity-violation studies of electron scattering through interference with Z-boson exchange, as was pointed out in Ref.[26]. The effect was evaluated in Ref.[26] in GPD framework at about 2% for backward angles and large Q^2. For smaller momentum transfers the two-photon effect is less significant, but Q^2 dependence of a γZ box contribution was found to be essential[27] for extraction of strange-quark effects in the proton neutral weak current. Authors of Ref.[27] used a hadronic model[10,19] and found that a combined effect of 2γ and γZ exchange on the values of $G_E^s + \beta G_M^s$ extracted in recent experiments can be as large as -40% in certain kinematics. A similar calculation was also presented in Ref.[28].

A comprehensive series of experiments are either underway or in preparation at Jefferson Lab with a purpose to study two-photon exchange effects in electron-proton scattering. Since two-photon exchange correction to electron scattering observables is proportional to an odd power of the electron charge, it can be measured directly by comparing electron and positron scattering. This method will be used in JLab experiment E-07-005, with a tertiary beam obtained from photoproduction of electron-positron pairs. Another JLab experiment, E-05-017, analyzes non-linearity of Rosenbluth plot caused two-photon exchange. Angular dependence of double-spin observables is also affected by two-photon exchange at a few per cent level,[12] and it is being looked for in polarization-transfer measurements (JLab experiment E-04-019). A single-spin target asymmetry caused by two-photon exchange will be studied in JLab experiment E-05-015 on a polarized ^{3}He target.

So far, the only definitive experimental observation of two-photon exchange effects came from the measurements of normal beam asymmetry at MIT-Bates,[29] MAMI,[30] and JLab.[31,32] The observations appear to be in reasonable agreement with theoretical calculations at lower energies,[33,34]

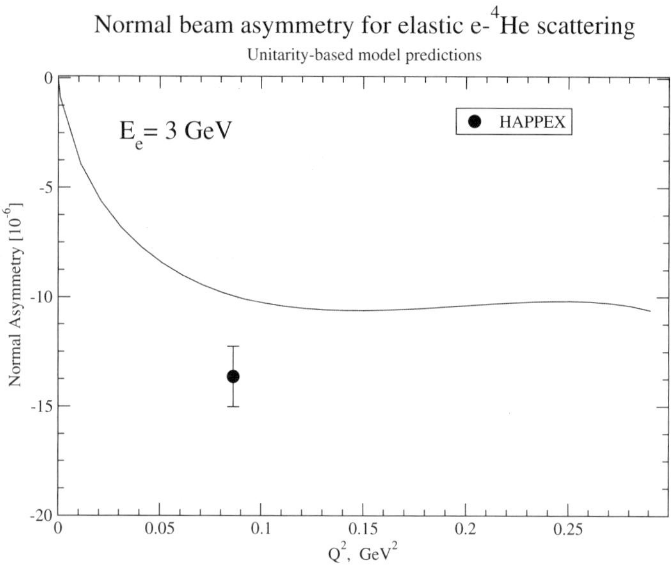

Fig. 4. Single-spin normal beam asymmetry on a ^{4}He target in units of parts per million. The curve is a prediction of a unitarity-based model[35] extended to a nuclear target, with total photoproduction cross section and Compton t-slope on ^{4}He used for input. Experimental data point is from Ref.[32]. Contribution of Coulomb distortion is below a few parts per *billion* in the shown kinematics.

nucleon resonance region[34] and above the resonance region.[22,35,36] Note that a unitarity-based approach[35] applied for a nuclear ^{4}He target agrees both in sign and magnitude with recent measuments from JLab HAPPEX collaboration,[32] see Fig.4; while the prediction in the kinematics of upcoming PREX measurement (JLAB E06-002) on Pb-target is about -5 ppm. At the same time, this asymmetry appears to be several orders of magnitude larger than predictions from a known mechanism of Coulomb distortion for small-angle electron scattering kinematics.[37]

We conclude that the two-photon exchange effect stands as a possible source of the difference between Rosenbluth and polarization techniques for proton electric form factor measurements. It has to be included in the analysis of other precision experiments. The two-photon exchange mechanism leads to new effects that can be studied experimentally.

Notice: Authored by Jefferson Science Associates, LLC under U.S. DOE Contract No. DE-AC05-06OR23177.

References

1. M. K. Jones *et al.*, Phys. Rev. Lett. **84**, 1398 (2000); O. Gayou *et al.*, Phys. Rev. Lett. **88**, 092301 (2002); Phys. Rev. C **64**, 038202 (2001).
2. B. Plaster *et al.*, Phys. Rev. C **73**, 025205 (2006).
3. Jefferson Lab Experiment E-04-108, "Measurement of G_{Ep}/G_{Mp} to Q^2=9 GeV^2 via Recoil Polarization", Spokepersons: E. Brash *et al.*
4. L. Andivahis *et al.*, Phys. Rev. D **50**, 5491 (1994).
5. C. E. Hyde-Wright and K. de Jager, Ann. Rev. Nucl. Part. Sci. **54**, 217 (2004).
6. M. E. Christy *et al.* [E94110 Collaboration], Phys. Rev. C **70**, 015206 (2004); I. A. Qattan *et al.*, Phys. Rev. Lett. **94**, 142301 (2005).
7. C. E. Carlson and M. Vanderhaeghen, Annu. Rev. Nucl. Part. Sci. **57**, 171 (2007).
8. J. Arrington, C. D. Roberts and J. M. Zanotti, J. Phys. G **34**, S23 (2007).
9. P. A. M. Guichon and M. Vanderhaeghen, Phys. Rev. Lett. **91**, 142303 (2003)
10. P. G. Blunden, W. Melnitchouk and J. A. Tjon, Phys. Rev. Lett. **91**, 142304 (2003)
11. Y. C. Chen, A. Afanasev, S. J. Brodsky, C. E. Carlson and M. Vanderhaeghen, Phys. Rev. Lett. **93**, 122301 (2004).
12. A. V. Afanasev, S. J. Brodsky, C. E. Carlson, Y. C. Chen and M. Vanderhaeghen, Phys. Rev. D **72**, 013008 (2005).
13. Y.S. Tsai, Phys. Rev. **122**, 1898 (1961).
14. A. Afanasev, I. Akushevich and N. Merenkov, Phys. Rev. D **64**, 113009 (2001).
15. A. V. Afanasev, I. Akushevich and N. P. Merenkov, Phys. Rev. D **65**, 013006 (2002).
16. A. Afanasev, Talk presented at Workshop of Precision Electroweak Interactions, Williamsburg, VA, USA, Aug. 15-17, 2005 (unpublished).
17. Yu. M. Bystritskiy, E. A. Kuraev and E. Tomasi-Gustafsson, Phys. Rev. C **75**, 015207 (2007).
18. V. G. Gorshkov, V. N. Gribov, L. N. Lipatov, and G. V. Frolov, Sov. J. Nucl. Phys. **6**, 95 (1968); in V. B. Berestetskii, E. M. Lifshitz, and L. P. Pitaevskii, Quantum Electrodynamics, Course of Theoretical Physics (Pergamon, Oxford, 1982), Vol. 4, Second edition, pp. 616.
19. P. G. Blunden, W. Melnitchouk and J. A. Tjon, Phys. Rev. C **72**, 034612 (2005).
20. S. Kondratyuk, P. G. Blunden, W. Melnitchouk and J. A. Tjon, Phys. Rev. Lett. **95**, 172503 (2005).
21. S. Kondratyuk and P. G. Blunden, Phys. Rev. C **75**, 038201 (2007).
22. M. Gorchtein, Phys. Lett. B **644**, 322 (2007).
23. J. Arrington, Phys. Rev. C **71**, 015202 (2005).
24. P. Jain, S. D. Joglekar and S. Mitra, Eur. Phys. J. C **52**, 339 (2007).
25. M. A. Belushkin, H. W. Hammer and U. G. Meissner, arXiv:0705.3385 [hep-ph].
26. A. V. Afanasev and C. E. Carlson, Phys. Rev. Lett. **94**, 212301 (2005).

27. H. Q. Zhou, C. W. Kao and S. N. Yang, arXiv:0708.4297 [hep-ph].
28. J. A. Tjon and W. Melnitchouk, arXiv:0711.0143 [nucl-th].
29. S. P. Wells *et al.* [SAMPLE collaboration], Phys. Rev. C **63**, 064001 (2001).
30. F. E. Maas *et al.*, Phys. Rev. Lett. **94**, 082001 (2005).
31. D. S. Armstrong *et al.* [G0 Collaboration], Phys. Rev. Lett. **99**, 092301 (2007).
32. L.J. Kaufman, Eur. Phys. J. A **32**, 501 (2007).
33. A. Afanasev, I. Akushevich and N. P. Merenkov, arXiv:hep-ph/0208260.
34. B. Pasquini and M. Vanderhaeghen, Phys. Rev. C **70**, 045206 (2004).
35. A. V. Afanasev and N. P. Merenkov, Phys. Rev. D **70**, 073002 (2004); Phys. Lett. B **599**, 48 (2004) [arXiv:hep-ph/0407167v2].
36. M. Gorchtein, Phys. Rev. C **73**, 035213 (2006).
37. E. D. Cooper and C. J. Horowitz, Phys. Rev. C **72**, 034602 (2005)

TWO-PHOTON EXCHANGE CONTRIBUTIONS TO ELASTIC *ep* SCATTERING IN THE NON-LOCAL FIELD FORMALISM

PANKAJ JAIN, SATISH D. JOGLEKAR AND SUBHADIP MITRA

Physics Department, IIT Kanpur, Kanpur - 208016, India

We construct a non-local gauge invariant Lagrangian to model the electromagnetic interaction of proton. The Lagrangian includes all allowed operators with dimension up to five. We compute the two-photon exchange contribution to elastic electron-proton scattering using this effective non-local Lagrangian. The calculation uses experimentally extracted electromagnetic form factors and depends on one unknown constant. The final result explains the difference between the form factor ratio G_E/G_M extracted by Rosenbluth at SLAC and the polarization transfer technique.

Keywords: electromagnetic form factors, two photon exchange, non-local field theory

1. Introduction

The proton electromagnetic form factor ratio $\mu_p G_E/G_M$, extracted experimentally using the Rosenbluth separation technique,[1] shows the behaviour,[2–4] $\frac{\mu_p G_E}{G_M} \approx 1$ upto momentum transfer $Q^2 \approx 6$ GeV^2. This ratio can also be extracted using the polarization transfer technique[5] which shows that it decreases[6–8] with Q^2. Hence the results obtained in these two different extractions of the form factors are very different.[9,10] It has been suggested that the difference may be explained by the two-photon exchange diagram.[11–14] The two-photon exchange diagram is included in the radiative corrections but only in the infrared limit.[15–18] Several estimates show that this diagram may give a sizeable contribution.[19,20] Here we employ an effective non-local model in order to estimate the two-photon exchange contributions.

2. Effective Nonlocal Lagrangian Model

We are interested in formulating a model which incorporates the electromagnetic form factors at tree level, while preserving gauge invariance. Furthermore we restrict the Lagrangian to incorporate up to dimension 5 operators. This is justified since for the momentum transfers of interest, the higher dimension operators are expected to contribute much less.[21] A Lagrangian model which satisfies these constraints is given by,

$$\mathcal{L} = \overline{\psi}\left(i\partial\!\!\!/ - ef_1'\left[\frac{\partial^2}{\Lambda^2}\right]A\!\!\!/ - M_p\right)\psi + \frac{a''}{2M_p}\overline{\psi}\left(\sigma_{\mu\nu}f_2'\left[\frac{\partial^2}{\Lambda^2}\right]F^{\mu\nu}\right)\psi$$
$$+\frac{b''}{2M_p}\overline{\psi}\widetilde{D}^2\psi \qquad (1)$$

where A_μ and ψ are the photon and proton fields respectively, a" and b" are constants, $i\widetilde{D} = i\partial - ef_1'\left[\frac{\partial^2}{\Lambda^2}\right]A$ is the *non-local covariant* derivative and Λ is the scale of non-locality. The Lagrangian is exactly valid as long as the proton is on-shell, irrespective of the value of the momentum transfer q^2. In the limit $\Lambda \to \infty$ we reproduce the local field theory model for a proton with an anomalous magnetic moment. We point out that the form factors f_1' and f_2' are to be extracted directly from experiments.

This lagrangian is invariant under a non-local form of gauge transformations:[22,23]

$$\delta A_\mu = -\partial_\mu\alpha\,(x)\,;\psi(x) \to e^{ief_1'\left[\frac{\partial^2}{\Lambda^2}\right]\alpha(x)}\psi(x), \overline{\psi}(x) \to \overline{\psi}(x)e^{-ief_1'\left[\frac{\partial^2}{\Lambda^2}\right]\alpha(x)} \qquad (2)$$

This gauge-invariance leads to a non-local form of the Ward-Takahashi[21] identity.

The lagrangian, eq. 2, leads[21] to three two photon exchange diagrams. These are the standard box and cross box diagrams, with on-shell electromagnetic form factor of proton, and an additional two-photon exchange diagram arising from the term,

$$-\left(\frac{\bar{a}^2\bar{b}^2}{16M_p^3}\right)\overline{\psi}'\left(\sigma_{\mu\nu}f_2'\left[\frac{\partial^2}{\Lambda^2}\right]F^{\mu\nu}\right)\left(i\widetilde{D\!\!\!/} - M_p\right)\left(\sigma_{\mu\nu}f_2'\left[\frac{\partial^2}{\Lambda^2}\right]F^{\mu\nu}\right)\psi',$$

where $\bar{a}$ is related to the proton anomalous magnetic moment, κ_p, by the relation, $\bar{a} = e\kappa_p/2$, and $\bar{b}$ is an unknown constant.

3. Calculation and Results

In order to compute the two-photon exchange diagrams we require models for the form factors both in the space-like and time-like regions. The dominant contribution is expected to come from the unphysical region $0 < q^2 < 4M_p^2$ where the form factors are very large. In this region they can be extracted[24,25] using the dispersion relations.[26,27] Our model for the form factors is given in Jain *et al.*[21]

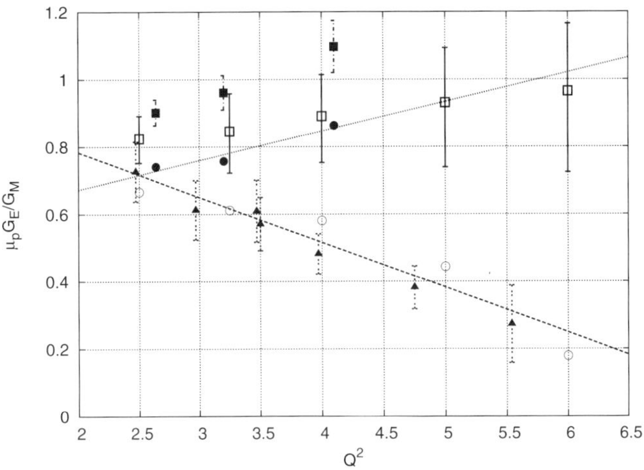

Fig. 1. The ratio, $\mu_p G_E/G_M$ obtained by polarization transfer technique at JLAB (filled triangles) and Rosenbluth separation technique at SLAC (unfilled squares) and JLAB (filled squares). The ratio after correcting for the two-photon exchange contribution is also shown. The filled circles are the corrected JLAB Rosenbluth data and the dotted line is the best fit through these points. The unfilled circles are the corrected SLAC Rosenbluth data and the dashed is line the best fit through these points.

In Fig. 1 we show how the experimental results are modified due to two-photon exchange contribution. The figure shows the ratio $\mu G_E/G_M$ extracted from the JLAB polarization transfer experiment as well as the Rosenbluth separation experiment performed at SLAC and JLAB. The Rosenbluth result after subtracting the two-photon exchange contribution computed in this paper is also shown both for JLAB and SLAC data. We find that the two-photon exchange explains a significant part of the difference between the two experimental results. The SLAC Rosenbluth data is in agreement with the JLAB polarization transfer data after correcting for the two photon exchange contributions. However our corrected JLAB result is still far from the polarization transfer data. The predicted dependence of the reduced cross section, $\sigma_{\rm R}$, on ϵ is slightly non-linear but within the experimental limits.[28]

4. Conclusions

We have constructed a non-local Lagrangian to model the electromagnetic interaction of proton. We use this model to compute the two-photon exchange diagrams contributing to elastic scattering of electron with proton. The result explains the discrepancy between the form factors extracted using the polarization transfer technique and those extracted at SLAC using the Rosenbluth separation technique.

References

1. M. N. Rosenbluth, Phys. Rev. **79**, 615 (1950).
2. R. C. Walker *et al.*, Phys. Rev. **D 49**, 5671 (1994).
3. L. Andivahis *et al.*, Phys. Rev. **D 50**, 5491 (1994).
4. I. A. Qattan *et al*, Phys. Rev. Lett. **94**, 142301 (2005); nucl-ex/0410010.
5. A. I. Akhiezer, L. N. Rosentsweig, I. M. Shmushkevich, Sov. Phys. JETP **6**, 588 (1958); J. Scofield, Phys. Rev. **113**, 1599 (1959); *ibid* **141**, 1352 (1966); N. Dombey, Rev. Mod. Phys. **41**, 236 (1969); A. I. Akhiezer and M. P. Rekalo, Sov. J. Part. Nucl. **4**, 277 (1974); R. G. Arnold, C. E. Carlson and F. Gross, Phys. Rev. **C 23**, 363 (1981).
6. M. K. Jones *et al*, Phys. Rev. Lett. **84**, 1398 (2000).
7. O. Gayou *et al*, Phys. Rev. Lett. **88**, 092301, (2002).
8. V. Punjabi *et al*, Phys. Rev. **C 71**, 055202 (2005); Phys. Rev. **C 71**, 069902(E) (2005).
9. J. Arrington, C. D. Roberts, J. M. Zanotti, nucl-th/0611050.
10. C. F. Perdrisat, V. Punjabi, M. Vanderhaeghen, hep-ph/0612014.
11. P. A. M. Guichon and M. Vanderhaeghen, Phys. Rev. Lett. **91**, 142303 (2003).
12. J. Arrington, Phys. Rev. **C 71**, 015202 (2005); hep-ph/0408261.
13. M. P. Rekalo and E. Tomasi-Gustafsson, Eur. Phys. J. **A 22**, 331 (2004).
14. M. A. Belushkin, H. W. Hammer and U. G. Meissner, hep-ph/0705.3385.
15. R. Ent *et al*, Phys. Rev. **C 64**, 054610-1 (2001).
16. Y. S. Tsai, Phys. Rev. **122**, 1898 (1961).
17. L. M. Mo and Y. S. Tsai, Rev. Mod. Phys. **41**, 205 (1969).
18. L. C. Maximon and J. A. Tjon, Phys. Rev. **C 62**, 054320 (2000); nucl-th/0002058.
19. P. G. Blunden, W. Melnitchouk and J. A. Tjon, Phys. Rev. Lett. **91**, 142304 (2003), nucl-th/0306076; P. G. Blunden, W. Melnitchouk and J. A. Tjon, Phys. Rev. **C 72**, 034612 (2005), nucl-th/0506039.
20. A. V. Afanasev, S. J. Brodsky, C. E. Carlson, Yu-Chun Chen and M. Vanderhaeghen, Phys. Rev. **D 72**, 013008 (2005), hep-ph/0502013.
21. P. Jain, S. D. Joglekar and S. Mitra, to appear in EPJC.
22. G. Kleppe, and R. P. Woodard, Nucl. Phys. **B 388**, 81 (1992).
23. J. W. Moffat, Phys. Rev. **D 41**, 1177 (1990);
24. R. Baldini et al, Eur. Phys. J. **C 11**, 709 (1999).
25. R. Baldini et al, Nucl. Phys. **A 755**, 286 (2005).

26. P. Mergell, U. G. Meissner, and D. Drechsel, Nucl. Phys. **A 596**, 367 (1996).
27. M. A. Belushkin, H. W. Hammer and U. G. Meissner, Phys. Rev. **C 75**, 035202 (2007).
28. I. A. Qattan, nucl-ex/0610006.

BEYOND THE BORN APPROXIMATION: A PRECISE COMPARISON OF POSITRON-PROTON AND ELECTRON-PROTON ELASTIC SCATTERING IN CLAS

J.LACHNIET*, A.AFANASEV, J.ARRINGTON, W.K. BROOKS, K.JOO, M. MOTEABBED, M. NIROULA, B.A. RAUE, M. UNGARO, L.B. WEINSTEIN, *et al.*
(THE CLAS COLLABORATION)

**Department of Physics, Old Dominion University, Norfolk, VA 23529, USA*
lachniet@physics.odu.edu

Recent polarization transfer measurements of the ratio of elastic form factors G_E/G_M on the proton are not in agreement with the results of Rosenbluth-type experiments. A two photon exchange contribution to the elastic cross section is the leading theoretical candidate to explain the discrepancy. A definitive test of this theory is the direct comparison of e^+p and e^-p elastic scattering. A proposed measurement of this sort using a novel mixed e^+e^- lepton beam and the CEBAF Large Acceptance Spectrometer (CLAS) at Jefferson Lab is described. Preliminary results from an engineering test run showing a clear signal of elastic e^+p and e^-p are presented.

Keywords: Two Photon Exchange; Form Factors; Elastic Scattering; Proceedings; World Scientific Publishing.

1. Introduction

Understanding the internal structure of the nucleon represents one of the fundamental problems of hadronic physics. The nucleon electromagnetic form factors are the most basic set of observables describing the internal structure of the nucleon and their evolution with Q^2 characterizes the distribution of charge and magnetization within the nucleon.

The nucleon elastic form factors are defined through the matrix elements of the nucleon electromagnetic current $J_\mu = \overline{\psi}\gamma_\mu\psi$ as

$$\langle N(P')|J_\mu(0)|N(P)\rangle = \overline{u}(P')\left(\gamma_\mu F_1(Q^2) + \frac{i\sigma_{\mu\nu}q^\nu\kappa}{2M}F_2(Q^2)\right)u(P) \quad (1)$$

where P and P' are the initial and final nucleon momenta, $q = P - P'$, $Q^2 = -q^2$, M is the nucleon mass, κ is the anomalous magnetic moment, and F_1

and F_2 are scalar functions of Q^2 that characterize the internal structure of the nucleon. These are the Dirac and Pauli form factors, respectively. The differential cross section for elastic electron-nucleon scattering can then be calculated in the laboratory frame as[1]

$$\frac{d\sigma}{d\Omega} = \sigma_{Mott}\left[\left(F_1^2 + \frac{\kappa^2 Q^2}{4M^2}F_2^2\right) + \frac{Q^2}{2M^2}\left(F_1 + \kappa F_2\right)^2 \tan^2\left(\frac{\theta}{2}\right)\right] \quad (2)$$

where θ is the electron scattering angle, κ is the nucleon anomalous magnetic moment, and σ_{Mott} is

$$\sigma_{Mott} = \frac{\alpha^2 E' \cos^2(\frac{\theta}{2})}{4E^3 \sin^4(\frac{\theta}{2})}. \quad (3)$$

It is convenient to define different electromagnetic form factors that are related to the charge and magnetization density of the nucleon in the appropriate kinematics. These so-called Sachs form factors are defined as

$$G_E = F_1 - \frac{\kappa Q^2}{4M^2}F_2 \qquad G_M = F_1 + \kappa F_2 \quad (4)$$

so Equation 2 can be written as

$$\frac{d\sigma}{d\Omega} = \frac{\sigma_{Mott}}{\epsilon}\left(\epsilon G_E^2 + \tau G_M^2\right)\left(\frac{1}{1+\tau}\right) \quad (5)$$

where

$$\tau = \frac{Q^2}{4M^2} \qquad \text{and} \qquad \epsilon = \frac{1}{1 + 2(1+\tau)\tan^2(\frac{\theta}{2})}. \quad (6)$$

Equation 5 shows us that G_E^2 and G_M^2 can be determined by a series of cross section measurements holding Q^2 fixed and varying ϵ by using a range of incident beam energies. Until the 1990s, most proton form factor measurements were made using the Rosenbluth separation method.

The proton form factors have also been measured in polarization transfer experiments using $p(\vec{e}, e'\vec{p})$. The ratio of the transverse and longitudinal polarization components of the recoil proton are proportional to the ratio G_E/G_M. As shown in Fig. 1, the polarization transfer measurements show a significant decrease in G_E relative to G_M as Q^2 increases, a trend not seen in the Rosenbluth results. The two techniques disagree by a factor of around 3 at $Q^2 = 5.6$ GeV^2.

Two photon exchange (TPE) contributions to the cross section, of the type illustrated in diagrams (e) and (f) in Fig. 2, may be able to explain the discrepancy[11,12] seen in Fig. 1. Because the contribution of G_E to the cross section is kinematically suppressed at large Q^2, a small ϵ dependent

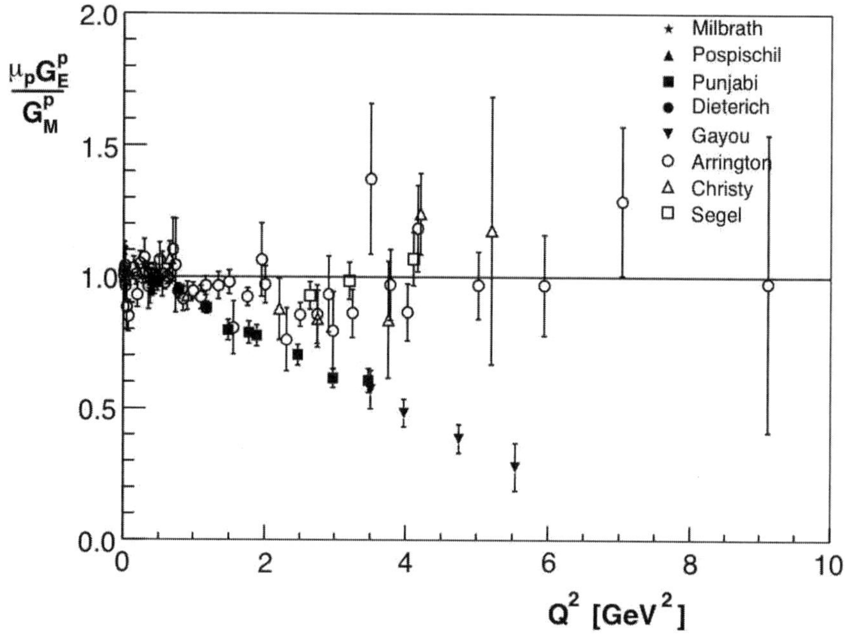

Fig. 1. The $\mu_p G_E/G_M$ ratio of the proton from Rosenbluth and polarization transfer measurements. The open symbols are Rosenbluth data[2,3] and reanalyzed SLAC data from Arrington.[9] The solid symbols are polarization transfer data.[4–8] The figure is from Hyde-Wright and de Jager.[10]

correction (5-8%) from TPE processes can produce the large difference in the observed $\mu_p G_E/G_M$ ratio.[11,13] The dominant TPE effect on the cross section comes from the interference between the one and two photon exchange diagrams. Since the sign of the interference term changes when the charge of the lepton is changed, a comparison of the cross section for e^-p and e^+p elastic scattering is sensitive to the magnitude of the two photon amplitude:

$$\sigma(e^{\pm}) = \sigma_{Born}(1 \mp \delta_{2\gamma}) \tag{7}$$

where $\delta_{2\gamma}$ is a TPE correction. This yields a cross section ratio of:

$$R = \frac{\sigma(e^+)}{\sigma(e^-)} \approx 1 - 2\delta_{2\gamma} \tag{8}$$

A measurement of the ratio of the e^+p to e^-p cross section provides direct access to the TPE term of the elastic cross section.

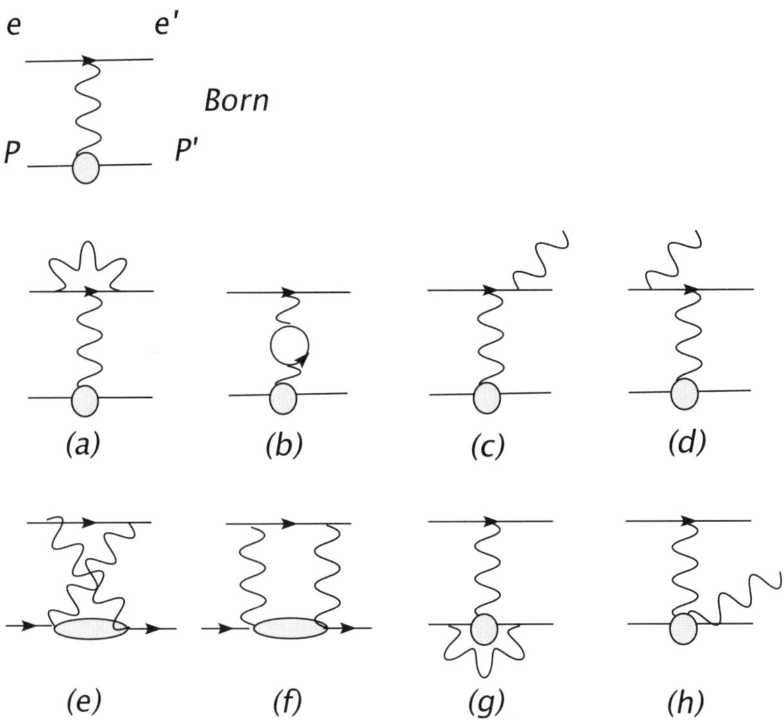

Fig. 2. Feynman diagrams for elastic lepton-proton scattering including first-order radiative corrections. Diagrams (e) and (f) show the TPE terms.

2. Hall-B lepton beamline

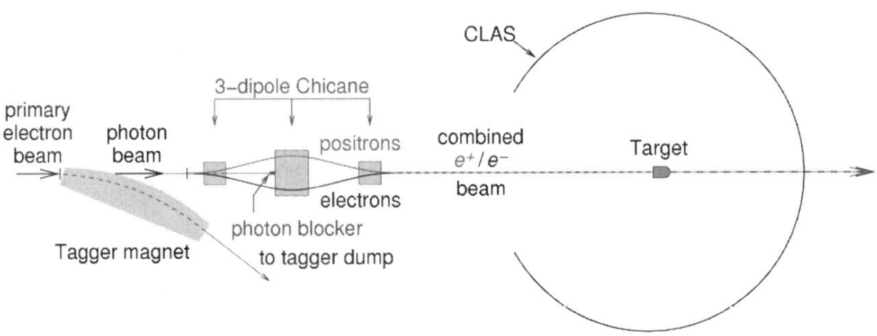

Fig. 3. Schematic illustration of CLAS lepton beamline.

The electron beamline in Hall-B at Jefferson Lab can be reconfigured to produce a mixed electron/positron beam. A simplified schematic of the mixed lepton beamline is shown in Fig. 3. The primary electron beam impinges on a thin radiator and is used to create a bremsstrahlung photon beam. The CLAS tagger magnet is used to deflect the primary electron beam to a beam dump below the Hall-B floor. The bremsstrahlung photon beam strikes a convertor foil, creating electrons and positrons from pair production. The lepton pairs and the unconverted photons enter a 3-dipole magnetic chicane. The first dipole spatially separates the pairs from the unconverted photons by bending them away from the beam axis. The unconverted photons are then stopped by a tungsten photon blocker at the center of the second dipole. The second and third dipoles recombine the lepton pairs which then proceed down the beamline to a hydrogen target at the center of CLAS. CLAS is then used to detect the recoil protons and scattered leptons.

This technique creates nearly identical e^+ and e^- beams with a continuous energy distribution, and the large acceptance of CLAS allows simultaneous measurement of the e^+p and e^-p cross section over a wide range of Q^2 and ϵ. The overdetermined elastic kinematics provide efficient background rejection and allow determination of the incident lepton energy.

3. October 2006 engineering run

In October 2006, an engineering test was conducted in Hall-B at Jefferson Lab. Because the mixed lepton beam is a tertiary beam, the current of the primary beam had to be increased to an unusually large value for CLAS. Previous measurements and detailed GEANT4 simulations indicated that the materials around the tagger vacuum box exit were the primary source of background in CLAS. Extensive modifications to the CLAS photon tagger system were made prior to the 2006 test run, simplifying the vacuum box exit region and adding extensive shielding. The tagger improvements reduced the background rate by a factor of approximately twenty.

The mixed lepton beamline described in Sec. 2 was installed and using a 3.2 GeV primary electron beam successfully created and transported both electrons and positrons into the CLAS target. An opposite sector trigger was used to preselect elastic candidates. Some results from an analysis of the test run data using a preliminary calibration of the CLAS detector systems are shown in Fig. 4 and Fig. 5. Elastic events are selected using a set of cuts that require the event satisfies the kinematic constraints of elastic scattering. The cuts combine information regarding the agreement between the incident lepton energy determined from the total momentum of final state particles and the incident lepton energy determined from the observed proton and lepton scattering angles:

$$\Delta E = E(P_1, P_2) - E(\theta_1, \theta_2). \tag{9}$$

A cut was also placed on the angle, θ_B, between the total final state momentum and the beamline. Both ΔE and θ_B are expected to be zero for elastic scattering.

Fig. 4 shows the distribution of the difference in azimuthal angle, $\Delta\phi$, for events with two positive tracks (candidate e^+p events), with and without elasticity cut. The distribution peaks at $\Delta\phi \approx 180$ degrees, as expected for elastic scattering.

Fig. 5 shows the W distribution for events with two positive tracks. The distribution is peaked near the proton mass. Together, these two figures give a clear indication that elastic e^+p scattering has been observed at CLAS for the first time. Results of similar quality were obtained for the e^-p channel.

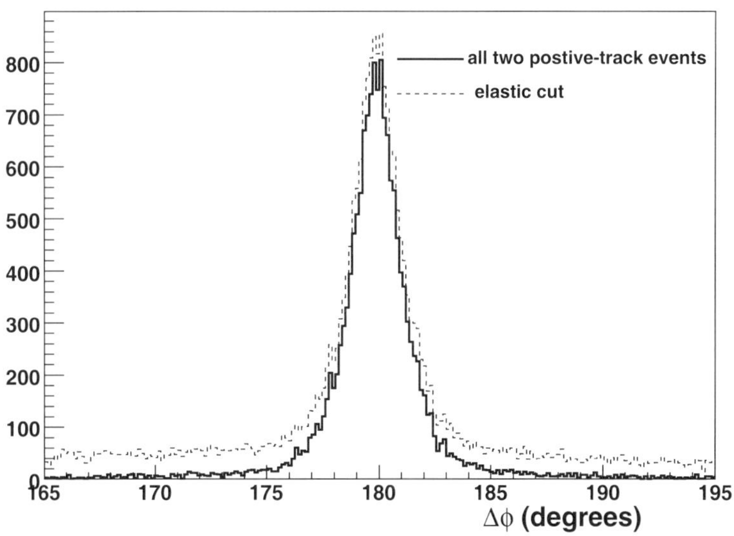

Fig. 4. $\Delta\phi$ distribution for candidate e^+p events, with and without a cut on elasticity.

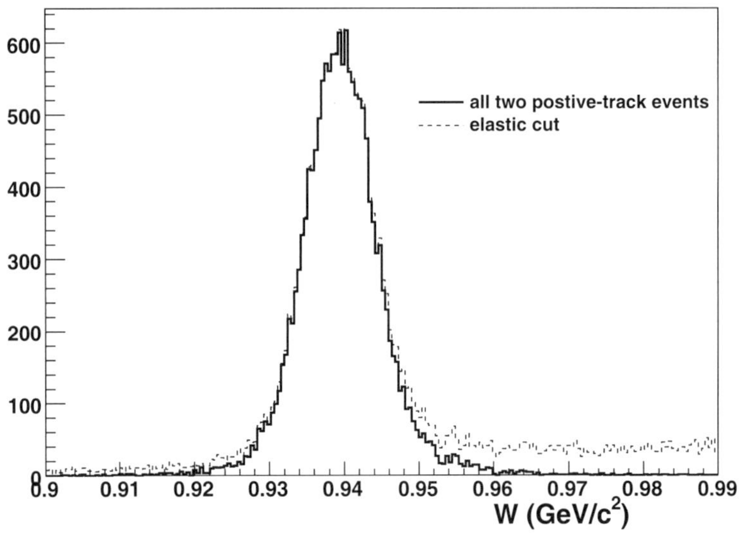

Fig. 5. W distribution for candidate e^+p events, with and without a cut on elasticity.

4. Future measurements

Based on the results obtained in the October 2006 test run, JLab PAC31 approved 30 days of running time for a TPE measurement in Hall-B using a 5.7 GeV primary electron beam.[14] This will allow a direct measurement of the TPE contribution to the elastic cross sector over a Q^2 range from 0.5 to 2.0 GeV^2 and ϵ range from approximately 0.1 to 0.9. In this region we will make a precise determination of the ϵ dependence of the TPE contribution, which can be used to test theoretical models of TPE effects. For Q^2 up to 3 GeV^2 we will measure the Q^2 dependence of the TPE contribution and an approximate ϵ dependence. This data will allow model independent TPE correction to be made to existing Rosenbluth data up to $Q^2 \approx 3$ GeV^2, and will allow us to determine if TPE effects can fully explain the observed discrepancy between the Rosenbluth and polarization transfer data. Two outcomes are possible:

(1) We confirm that TPE effects are responsible for the observed difference between Rosenbluth and polarization transfer measurements. Our measurements will be used to constrain theoretical calculations of TPE. In addition, measurements of the Rosenbluth/polarization transfer dis-

crepancy itself can then be used to extend these studies of the two-photon effects to larger Q^2.

(2) We determine that the TPE corrections do not fully explain the discrepancy. Then the explanation for the difference between Rosenbluth and polarization transfer measurements will need to be found elsewhere.

In either case, this measurement will have interesting results.

References

1. Rosenbluth, M.N. *Phys. Rev.* 79:615 (1950)
2. Qattan I.A., et al. *Phys. Rev. Lett.* 94:142301 (2005)
3. Christy M.E., et al. *Phys. Rev.* C 70:015206 (2004)
4. Milbrath B., et al. *Phys. Rev. Lett.* 80:452 (1998); erratum, *Phys. Rev. Lett.* 82:2221 (1999)
5. Pospischil T., et al. *Eur. Phys. Jour.* A 12:125 (2001)
6. Punjabi V., et al. *Phys. Rev.* C 71:055202 (2005); Jones MK, et al. *Phys. Rev. Lett.* 84:1398 (2000)
7. Gayou O., et al. *Phys. Rev. Lett.* 88:092301 (2002)
8. Dieterich S., et al. *Phys. Lett.* B 500:47 (2001)
9. Arrington J. *Phys. Rev.* C 68:034325 (2003)
10. Hyde-Wright, C.E. and K. de Jager *Ann. Rev. Nuc. Part. Sci.* 54:217-267 (2004)
11. Guichon, P.A and Vandehaeghen, M. *Phys. Rev. Lett.* 91:142303 (2003)
12. Arrington, J. *Phys. Rev.* C 71:015202 (2005)
13. Arrington, J. *Phys. Rev.* C 69:022201 (2004)
14. Afanasev, A., et al. Jefferson Lab experiment E07005 (2007)

MESON FORM FACTORS IN THE SPACE–LIKE REGION

D. GASKELL

Thomas Jefferson National Accelerator Facility,
Newport News, VA 23606, USA
E-mail: gaskelld@jlab.org

Experimental studies of meson form factors provide a unique and powerful tool for understanding bound quark systems due to the meson's simple $q\bar{q}$ valence structure. However, experiments in the space–like regime ($Q^2 = -q^2 > 0$) are made challenging by the lack of "free meson" targets. Recent measurements at Jefferson Lab have significantly improved the quality of measurements of the charged pion form factor up to $Q^2 = 2.5$ GeV^2, and future measurements will extend the database to 6 GeV^2.

1. Meson Form Factors in QCD

The experimental study of meson form factors is particularly attractive in that it provides access to the simplest, bound QCD system: a quark, anti–quark pair. In addition, the fact that the asymptotic value of the meson form factor can be calculated with no free parameters in perturbative Quantum Chromodynamics means that such experiments, performed over a large enough kinematic range, provide access to the transition between effective degrees of freedom at low Q^2 to a description in terms of quarks and gluons at large Q^2. Understanding this transition is perhaps one of the more intriguing questions remaining in the study of hadronic systems.

As an example of the asymptotic expression for meson form factors, we briefly discuss the case for the charged pion. In general, the form factor can be described in terms of a hard scattering piece and a convolution over initial and final pion wave functions;

$$F_\pi(Q^2) = \int_0^1 dxdy\phi_\pi(y,\mu^2)T(y,x,Q^2,\mu^2)\phi_\pi(x,\mu^2), \qquad (1)$$

where $\phi_\pi(x,\mu^2)$ is the pion wave function and describes a quark with fraction x of the pion momentum and $T(x,y,Q^2,\mu^2)$ is the amplitude describing the parton hard scattering process at a scale μ^2.[1] The hard scattering function can be expressed as an expansion in the strong coupling constant α_s while the pion wave function can be described using a power series in terms of Gegenbauer polynomials. In the limit of infinite Q^2 only the leading terms for both T and ϕ_π survive leading to,[2]

$$F_\pi(Q^2)\xrightarrow[Q^2\to\infty]{}\frac{16\pi\alpha_s(Q^2)f_\pi^2}{Q^2}, \tag{2}$$

where $f_\pi \approx 92$ MeV is the pion decay constant. A similar result holds for other mesons, *e.g.* the charged kaon, replacing f_π with f_K.

The implications of the result in Eq. 2 should not be underestimated. In this case, there is a model–independent statement about not only the large Q^2 behavior of the charged pion form factor, but its absolute magnitude as well. One can make reasonable arguments that Eq. 2 is truly an asymptotic limit, perhaps unreachable experimentally. However, the mere fact that this limit is well defined is already more information than is typically available in other cases (the proton form factor being a case in point).

2. Measurement of Meson Form Factors at $Q^2 > 0$

While experimental studies of meson form factors are clearly of interest on theoretical grounds, the experiments are unfortunately quite challenging in practice. The lack of a "free meson" target limits the Q^2 range accessible in direct measurements of charge form-factors. For example, the charged pion and kaon form factors have been extracted using elastic scattering of pion and kaon beams from atomic electrons at CERN.[3–5] However, even for pion beams of 300 GeV, the largest accessible Q^2 was 0.25 GeV^2 (for kaons, the beam momentum of 250 GeV resulted in a maximum accessible Q^2 of 0.1 GeV^2). Access to larger values of Q^2 can only be achieved by making use of the virtual meson cloud of the nucleon. However, this method is not without its difficulties, both theoretical and experimental. In contrast to the charged pion and kaon form factors, however, the $\gamma\gamma^* \to \pi^0$ form factor is relatively straightforward to access experimentally in e^+e^- collisions.

2.1. $\gamma\gamma^* \to \pi^0$ *Form Factor Measurements*

The $\gamma\gamma^* \to \pi^0$ transition form factor has been measured in e^+e^- collisions at the PETRA storage ring at DESY by the CELLO Collaboration[6] and at the Cornell Electron Storage Ring (CESR) by the CLEO Collaboration.[7] In these experiments, one of the electrons is detected at large angles, yielding a virtual photon with large Q^2. The second electron is scattered at small angles and is undetected, yielding a nearly real photon. The process $\gamma\gamma^* \to \pi^0$ then contains the π^0 structure information.

The $\gamma\gamma^* \to \pi^0$ form factor is advantageous in that the lowest order diagram is a pure QED process (see Fig. 1.a). This form factor is then, in principle, a good test of the pion wave function (or distribution amplitude) used at finite Q^2. Indeed, this can be seen in Fig. 2 where the CLEO and CELLO measurements of the $\gamma\gamma^* \to \pi^0$ form factor are compared to calculations using different forms of the pion distribution amplitude.[8] It is interesting to note that the asymptotic form of the pion wave function does a decent job describing the experimental data above $Q^2 = 4$ GeV^2, although the calculation is systematically smaller than the data. This relatively good agreement has motivated some to argue that the asymptotic form of the pion distribution, with some inclusion of effects like intrinsic transverse momentum of the bound quarks, is sufficient to describe the $\gamma\gamma^* \to \pi^0$ form factor at virtually all Q^2 (see for example [9–11]). Regardless of one's point of view on this particular issue, the main point of interest here is that one can, in principle, use the simplicity of the lowest order diagram contributing to the $\gamma\gamma^* \to \pi^0$ form factor to provide information on the appropriate pion wave function to be used in the more complicated charged pion case.

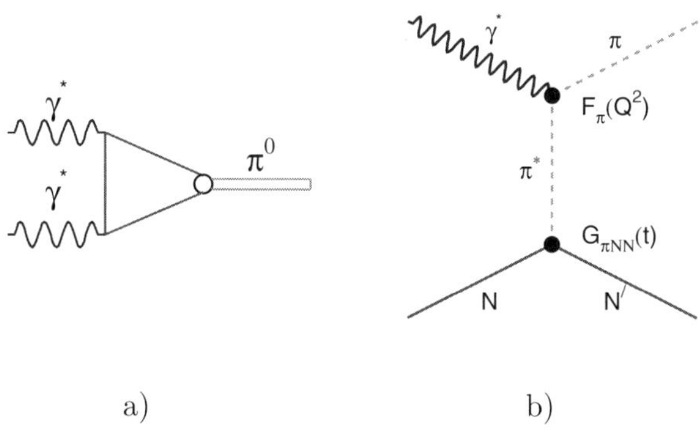

a) b)

Fig. 1. a) Lowest order diagram for the $\gamma\gamma^* \to \pi^0$ form factor. Note that there are no gluons exchanged at lowest order. b) Pole diagram for the pion electroproduction process. $g_{\pi NN}$ describes the probability to emit a virtual pion with four–momentum t, while $F_\pi(Q^2)$ describes the coupling of the virtual photon to the pion.

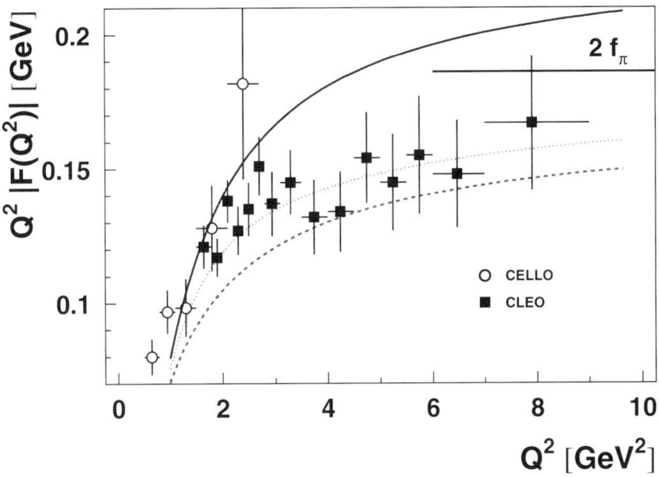

Fig. 2. Measurements of the $\gamma\gamma^* \rightarrow \pi^0$ form factor[6,7] compared to calculations from Ref. [8]. The dashed curve uses the asymptotic form of the pion distribution amplitude while the solid curve uses the so-called "CZ" distribution amplitude[12] derived using QCD sum rules. The dotted curve uses a new pion distribution amplitude that attempts to incorporate non-perturbative contributions from the QCD vacuum. Also shown (solid line upper right corner) is the asymptotic limit of the $\gamma\gamma^* \rightarrow \pi^0$ form factor; $Q^2F \rightarrow 2f_\pi$.

2.2. *Charged Pion (and Kaon) Form Factor Measurements*

As noted earlier, direct measurements of the charged pion and kaon form factors are limited to very low values of Q^2. In order to access the form factor at larger Q^2 one must make use of the virtual meson cloud of the nucleon. In the case of the charged pion form factor, this technique has been applied for all measurements of the form factor above $Q^2 = 0.3$ GeV^2.

At low values of $-t$ (the four-momentum transferred to the nucleon), the pion pole term (see Fig. 1.b) dominates the longitudinal part (σ_L) of the pion electroproduction cross section. Assuming pole dominance, the (longitudinal) cross section can be written in a Born term model,

$$\frac{d\sigma_L}{dt} \propto \frac{-tQ^2}{(t - m_\pi^2)} g_{\pi NN}^2(t) F_\pi^2(Q^2, t), \tag{3}$$

where $g_{\pi NN}(t)$ denotes the probability to find a virtual pion with four-momentum, t, and $F_\pi^2(Q^2, t)$ is the pion form factor. Using this expression, one can extract the "form factor" as a function of t and extrapolate to the physical pole ($t = m_\pi^2$). In practice, this technique is difficult in that the functional form of the extrapolation is not clear and the experimental uncertainties are amplified by the need to extrapolate from experimentally accessible t values ($t < 0$ for electroproduction), through zero to the pion pole. In addition, the Born term expression in Eq. 3 assumes no contributions from other, non-pole processes and it is unclear how one would account for the presence such backgrounds when performing the extrapolation. A more robust technique involves the use of a pion electroproduction model that incorporates the pion form factor. In this case, the extrapolation to the pole is implicit, but one has the advantage of only fitting the model to the data in the region measured.

Fig. 3 shows the existing data on F_π. In this figure, all results above $Q^2 = 0.3$ GeV^2 have been extracted using the electroproduction technique. Note that in this figure, only electroproduction results from true Rosenbluth separations have been included - earlier experiments performed at Cornell[13–16] either performed Rosenbluth separations from two different experiments resulting in poorly controlled systematic errors, or extracted the longitudinal cross section from unseparated cross sections using a model of the transverse contributions.

A few representative calculations of the charged pion form factor using a variety of models are also shown in Fig. 3. These include a calculation using QCD Sum Rules,[17] a relativistic treatment of bound quarks using the Bethe-Salpether equation,[18] as well as a very recent calculation using an Anti-de Sitter/Conformal Field Theory approach.[19] Of particular interest here is the perturbative QCD calculation[8] which employs the same pion distribution amplitude that described the $\gamma\gamma^* \to \pi^0$ transition form factor very well in Fig. 2. The "hard" curve dramatically under-predicts the data and reasonable agreement is only achieved after including an explicit "soft" contribution, in this case described using local duality.

The points at $Q^2 > 1$ GeV^2 are results from the Jefferson Lab pion form factor program in Hall C.[20,21] Since the extraction of the pion form factor requires isolation of the longitudinal cross section via a Rosenbluth separation, control of point–to–point systematic uncertainties, high luminosity, and isolation of the $H(e, e'\pi^+)n$ channel are crucial. The excellent properties of the CW beam at Jefferson Lab, combined with the easy–to–understand acceptance and reproducibility of the magnetic spectrometers in Hall C are crucial to achieving high precision measurements of the pion form factor. The maximum Q^2 accessible at JLab (≈ 2.5 GeV^2) is presently limited by the 1) maximum available beam energy of 6 GeV and 2) the minimum angle achievable by the pion arm spectrometer - 10.5 degrees. Both of these issues will be addressed by the JLab 12 GeV Upgrade. The maximum beam energy of 11 GeV available to the existing experimental halls, combined with the reduction of the minimum angle of the pion spectrometer (the new Super High Momentum Spectrometer) to 5.5 degrees will allow measurements of the pion form factor up to $Q^2 = 6$ GeV^2.[22]

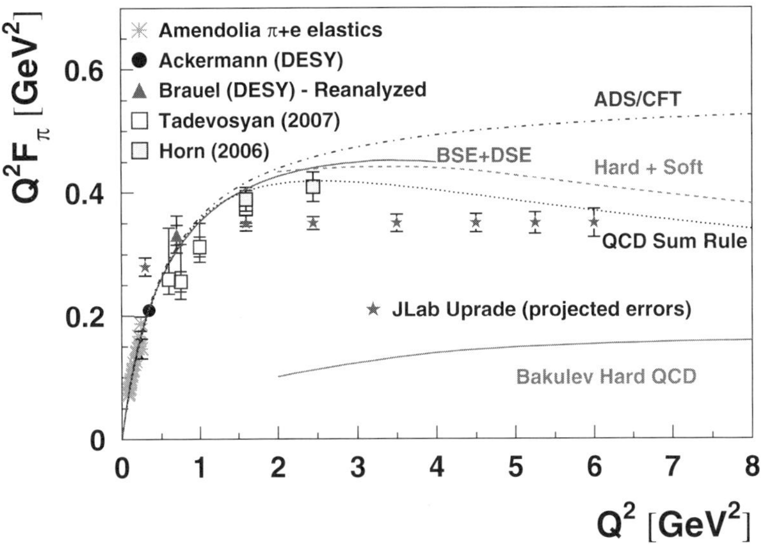

Fig. 3. Charged pion form factor from $\pi + e$ elastic scattering[4] and pion electroproduction data.[20,21,23,24] Several calculations are also shown; the lower, solid and dashed curves from Ref. [8], the dotted curve from Ref. [17], the upper, solid curve from Ref. [18], and the dashed–dotted curve is from Ref. [19]. Also shown are projected errors for an approved measurement to be carried at JLab after the 12 GeV upgrade .[22]

It is important to check experimentally that the form factor extracted using the electroproduction technique corresponds to that from the physical pion. The first, and most straightforward method is to simply measure the form factor using the electroproduction technique in kinematics that overlap the region that has been measured using elastic $\pi + e$ scattering. As shown in Fig. 4, the electroproduction measurement by Ackermann *et al*[23] at $Q^2 = 0.35$ GeV^2 is consistent with the extrapolation of the pion scattering data from Amendolia *et al*[4] (maximum $Q^2 = 0.25$ GeV^2). In addition, as part of the pion form factor measurement at JLab at 12 GeV, data will be taken at $Q^2 = 0.3$ GeV^2 to further verify the applicability of the electroproduction technique.

An additional, although less direct method to test the validity of the electroproduction method as a means to extract the pion form factor is to

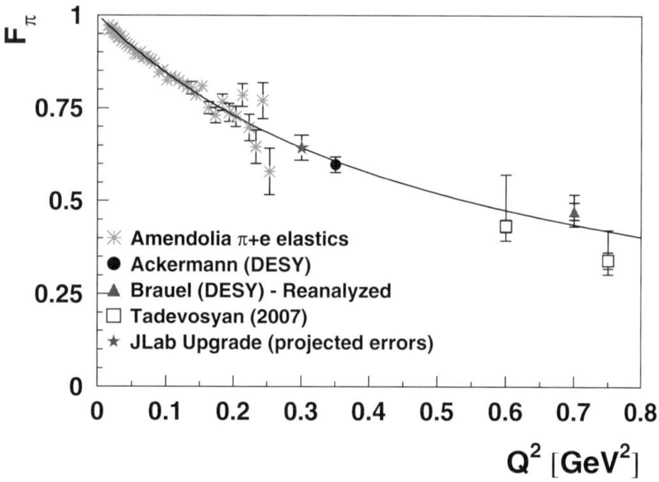

Fig. 4. Charged pion form factor measurements at low Q^2. The existing electroproduction point at $Q^2 = 0.35$ GeV^2 is consistent with the extrapolation of the low Q^2, $\pi + e$ data.

make measurements at different values of t_{min} but same Q^2, *i.e.* various separations from the physical pion pole, $t = m_\pi^2$. While the longitudinal cross section, σ_L depends on three Lorentz invariant quantities, t, W, and Q^2, the minimum accessible value of $-t$ is correlated with W at fixed Q^2. Hence, measurements at smaller values of $-t_{min}$ require larger values of W and are only practical at larger beam energies. The JLab F_π program has already performed such a test where the pion form factor at $Q^2 = 1.6$ GeV^2 was extracted during both the first [20] and second [21] phases of the program. The second measurement was 30% closer to the physical pion pole than the first, $-t_{min} = 0.093$ GeV^2 vs. $-t_{min} = 0.15$ GeV^2 (see Fig. 3). Additional measurements at $Q^2 = 1.6$ GeV^2, even closer to the pion pole at $-t_{min} = 0.03$ GeV^2, will be performed as part of the JLab 12 GeV F_π program.[22]

Finally, it is worth noting that there is some speculation that the kaon form factor at large Q^2 can also be accessed via the electroproduction technique using the $H(e, e'K^+)\Lambda$ reaction. In this case, the measurements are potentially less interpretable since the kinematically accessible value of $-t_{min}$ is even further from the physical kaon mass. In addition, it is not clear to what extent the pole process dominates the K^+ production mechanism in the longitudinal channel. Nonetheless, JLab experiment 93–0118 has made precision measurements of σ_L for K^+ electroproduction and demonstrated that reliable measurements of the $-t$ dependence of the cross section are feasible.[25] In addition, at $Q^2 = 1$ GeV^2 a "Chew–Low" extrapolation to the physical kaon mass was attempted, and the resulting form factor was found to be consistent with extrapolations of parameterizations constrained by the low Q^2 $K + e$ data.

Acknowledgments

This work is supported in part by the United States Department of Energy; Jefferson Science Associates, LLC operates the Thomas Jefferson National Accelerator Facility under Department of Energy contract DE-AC05-06OR23177.

References

1. G. Sterman and P. Stoler, *Ann. Rev. Nucl. Part. Sci.* **47**, 193 (1997).
2. G. P. LePage and S. J. Brodsky, *Phys. Lett. B* **87**, 359 (1979).
3. S. R. Amendolia *et al.*, *Phys. Lett. B* **146**, 116 (1984).
4. S. R. Amendolia *et al.*, *Nucl. Phys. B* **277**, 168 (1986).
5. S. R. Amendolia *et al.*, *Phys. Lett. B* **178**, 435 (1986).
6. H. J. Behrend *et al.*, *Z. Phys. C.* **49**, 401 (1991).
7. J. Gronberg *et al.*, *Phys. Rev. D* **57**, 33 (1998).
8. A. P. Bakulev *et al.*, *Phys. Rev. D* **70**, 033014 (2004).
9. M. Vanderhaeghen, P. A. M. Guichon and M. Guidal, *Phys. Rev. D* **60**, 094017 (1999).
10. R. Jakob, P. Kroll and M. Raulfs, *J. Phys. G* **22**, 45 (1996).
11. P. Kroll and M. Raulfs, *Phys. Lett. B* **387**, 848 (1996).
12. V. L. Chernyak and A. R. Zhitnitsky, *Phys. Rep.* **112**, 173 (1984).
13. C. N. Brown *et al.*, *Phys. Rev. D* **8**, 92 (1973).
14. C. J. Bebek *et al.*, *Phys. Rev. D* **9**, 1229 (1974).
15. C. J. Bebek *et al.*, *Phys. Rev. D* **13**, 25 (1976).
16. C. J. Bebek *et al.*, *Phys. Rev. D* **17**, 1693 (1978).
17. V. A. Nesterenko and A. V. Radyushkin, *Phys. Lett. B* **115**, 410 (1982).
18. P. Maris and P. Tandy, *Phys. Rev. C* **62**, 055204 (2000).
19. S. J. Brodsky and G. F. de Teramond, *arXiv:hep-th/0702205* (2007).
20. V. Tadevosyan *et al.*, *Phys. Rev. C* **75**, 055205 (2007).
21. T. Horn *et al.*, *Phys. Rev. Lett.* **97**, 192001 (2006).
22. D. Gaskell and G. M. Huber (spokespersons), *Measurement of the Charged Pion Form Factor to High Q^2*, Jefferson Lab Experiment E12-06-101 (2006).
23. H. Ackermann *et al.*, *Nucl. Phys. B* **137**, 294 (1978).
24. P. Brauel *et al.*, *Z. Phys. C* **3**, 101 (1979).
25. G. Niculescu, First measurement of the longitudinal and transverse cross sections in $H(e, e'K^+)\Lambda$, PhD thesis, Hampton University, (VA, USA, 1998), pp. 207–214.

PION-NUCLEON DISTRIBUTION AMPLITUDES

ANDREAS PETERS
Institut für Theoretische Physik,
Universität Regensburg,
D-93040 Regensburg, Germany
E-mail: andreas.peters@physik.uni-regensburg.de

This is a short presentation of the results for the pion-nucleon distribution amplitudes which are expressed in terms of the nucleon distribution amplitudes with the help of current algebra. Everything is considered to be at threshold.

Keywords: distribution amplitudes, chiral symmetry, threshold electroproduction

1. Introduction

In recent years there has been increasing attention to hard exclusive processes involving emission of soft pions in the final state. In the near future new experimental data will emerge and their interpretation requires information about the nucleon wave function. Such processes are attractive as they provide us new insights in the hadron (nucleon) structure.

At this point the nucleon distribution amplitudes (DAs) play an important role, as they contain direct information about the wave function. Our aim will be to treat the outgoing nucleon and produced π as a $N\pi$ final state and describe the $N\pi$ DAs in terms of nucleon DAs. The main physical tool we use to calculate such $N\pi$ DAs is the well known soft-pion theorem. Everything is considered to be at threshold.

2. Distribution Amplitudes (DAs)

2.1. *Nucleon DAs of twist-3*

Let us first consider the leading twist nucleon DAs in some detail to become acquainted with the method. As introductory remarks we first want to repeat some notations from [1].

The object of interest is the hadron-to-vacuum matrix element of a trilocal operator built of quark and gluon fields at light-like separations.

$$\langle 0|\, \epsilon^{ijk} u_\alpha^{i'}(a_1 z)\, [a_1 z, a_0 z]_{i',i}\, u_\beta^{j'}(a_2 z)\, [\ldots]_{j',j}\, d_\gamma^{k'}(a_3 z)\, [a_3 z, a_0 z]_{k',k}\, |p(P,\lambda)\rangle\;, \tag{1}$$

where $|p(P,\lambda)\rangle$ denotes the proton state with momentum P, $P^2 = M^2$ and helicity λ. u, d are the quark-field operators. The Greek letters α, β, γ stand for Dirac indices, the Latin letters i, j, k refer to color. z is an arbitrary light-like vector, $z^2 = 0$, the a_i are real numbers. The gauge-factors $[x, y]$ can be seen in [1, 2] and keep the matrix element in (1) gauge-invariant. In what follows we spare on writing the gauge factors explicitely, but keep them always as present in mind.

In order to fullfill Lorentz covariance, spin and parity conservation of the nucleon it is convenient to write Eq. (1) in terms of 24 invariant functions [1]. In the leading twist approximation only three amplitudes are relevant. In a shorthand notation it reads

$$\begin{aligned} &4\,\langle 0|\, \varepsilon^{ijk} u_\alpha^i(a_1 z) u_\beta^j(a_2 z) d_\gamma^k(a_3 z)\, |p(P,\lambda)\rangle_{\text{twist}-3} = \\ &= V_1^p\,(v_1)_{\alpha\beta,\gamma} + A_1^p\,(a_1)_{\alpha\beta,\gamma} + T_1^p\,(t_1)_{\alpha\beta,\gamma} \end{aligned} \tag{2}$$

where

$$\begin{aligned} (v_1)_{\alpha\beta,\gamma} &= (\not{p} C)_{\alpha\beta}\, (\gamma_5 N^+)_\gamma \\ (a_1)_{\alpha\beta,\gamma} &= (\not{p}\gamma_5 C)_{\alpha\beta}\, N_\gamma^+ \\ (t_1)_{\alpha\beta,\gamma} &= (i\sigma_{\perp p} C)_{\alpha\beta}\, (\gamma^\perp \gamma_5 N^+)_\gamma \end{aligned} \tag{3}$$

stand for the Lorentz structures.

The amplitudes V_1^p, A_1^p, T_1^p can be written as

$$F(a_i p\cdot z) = \int \mathcal{D}x\, e^{-ipz\sum_i x_i a_i} F(x_i)\,, \tag{4}$$

where the functions $F(x_i)$ depend on the dimensionless variables x_i, $0 < x_i < 1, \sum_i x_i = 1$ which correspond to the longitudinal momentum fractions carried by the quarks inside the nucleon. The integration measure is defined as

$$\int \mathcal{D}x = \int_0^1 \mathrm{d}x_1 \mathrm{d}x_2 \mathrm{d}x_3\, \delta(x_1 + x_2 + x_3 - 1)\,. \tag{5}$$

Applying the set of Fierz transformations

$$\begin{aligned} (v_1)_{\gamma\beta,\alpha} &= \frac{1}{2}\,(v_1 - a_1 - t_1)_{\alpha\beta,\gamma} \\ (a_1)_{\gamma\beta,\alpha} &= \frac{1}{2}\,(-v_1 + a_1 - t_1)_{\alpha\beta,\gamma} \\ (t_1)_{\gamma\beta,\alpha} &= -\,(v_1 + a_1)_{\alpha\beta,\gamma} \end{aligned} \tag{6}$$

one ends up with the condition

$$2T_1^p(1,2,3) = [V_1^p - A_1^p](1,3,2) + [V_1^p - A_1^p](2,3,1)\,, \tag{7}$$

which allows to express the tensor DA of the leading twist in terms of the vector and axial vector distributions. Since the latter have different symmetry, they can be combined together to define the single independent leading twist-3 proton DA

$$\Phi_3^p(x_1,x_2,x_3) = [V_1^p - A_1^p](x_1,x_2,x_3) \tag{8}$$

which is well known and received a lot of attention in the literature. The neutron leading twist DA $\Phi_3^n(x_1,x_2,x_3)$ can readily be obtained by the interchange of u and d quarks in the defining Eq. (2). For all invariant functions $F = V, A, T$ proton and neutron DAs differ by an overall sign:

$$F^p(1,2,3) = -F^n(1,2,3)\,, \tag{9}$$

as follows from the isospin symmetry. This property is retained for all twists. A complete collection of the results for the higher twist DAs and the x^2-corrections can be seen in [1, 3].

2.2. *Pion-Nucleon DAs of twist-3*

We want to recall the three-quark DAs of a $N\pi$ pair from [2] in the limit that the momentum of the pion relative to that of the nucleon is small. The central idea is to use the well kown soft-pion theorem and current algebra, similar to the ideas developed in [4]. To this end we define for the $N\pi$-system:

$$\begin{aligned} &4\,\langle 0|\,\varepsilon^{ijk} u_\alpha^i(a_1 z) u_\beta^j(a_2 z) d_\gamma^k(a_3 z)\,|N(P,\lambda)\pi\rangle_{\mathrm{tw-3}} = \\ &\qquad (\gamma_5)_{\gamma\delta}\frac{-i}{f_\pi}\left[V_1^{N\pi}(v_1)_{\alpha\beta,\delta} + A_1^{N\pi}(a_1)_{\alpha\beta,\delta} + T_1^{N\pi}(t_1)_{\alpha\beta,\delta}\right]. \end{aligned} \tag{10}$$

An extra γ_5 is needed to conserve parity. Similar to the proton case, the symmetry of the two u-quarks implies that the DAs V and T are symmetric, and A is antisymmetric to the exchange of the first two arguments, respectively. Here $f_\pi = 93$ MeV is the pion decay constant.

On the other hand one can calculate (10) explicitely. To this end one has to evaluate (10) with the help of the well known soft-pion theorem at threshold. For a detailed description, see [2]. After some calculation one ends up with the results in leading twist [2]:

$$V_1^{n\pi^+}(1,2,3) = \frac{1}{\sqrt{2}}\Big\{V_1^n(1,3,2) + V_1^n(1,2,3) + V_1^n(2,3,1) + A_1^n(1,3,2) + A_1^n(2,3,1)\Big\},$$

$$A_1^{n\pi^+}(1,2,3) = -\frac{1}{\sqrt{2}}\Big\{V_1^n(3,2,1) - V_1^n(1,3,2) + A_1^n(2,1,3) + A_1^n(2,3,1) + A_1^n(3,1,2)\Big\},$$

$$T_1^{n\pi^+}(1,2,3) = \frac{1}{2\sqrt{2}}\Big\{A_1^n(2,3,1) + A_1^n(1,3,2) - V_1^n(2,3,1) - V_1^n(1,3,2)\Big\} \tag{11}$$

and

$$F_1^{p\pi^0}(1,2,3) = \frac{1}{2}F_1^p(1,2,3)\,. \tag{12}$$

The isospin relation similar to (7) is not valid any more, since the pion-nucleon pair can have both isospin 1/2 and 3/2. Calculation of higher twists can be made in a complete analogy, but is much more extensive. A short review of the results is given in the Appendix.

3. Conclusions

We have shown the results for the $N\pi$ DAs. The tools we used were the light-cone formalism, the soft-pion theorem at threshold and current algebra. In this way it was possible to express the $N\pi$ DAs in terms of nucleon DAs.

Acknowledgements

I would like to thank V. Braun, D.Yu Ivanov and A. Lenz for collaboration and G. Peters for interesting discussion. Special thanks are to the organizers of the *Exclusive Reactions at High Momentum Transfer 2007* workshop at Jefferson Laboratory for their hospitality and enlightening debates. This work was supported by the Studienstiftung des deutschen Volkes.

Appendix A. $n\pi^+$ and $p\pi^0$ DAs in higher twist

For $n\pi^+$ we obtain in twist-4:

$$V_2^{n\pi^+}(1,2,3) = -\frac{1}{\sqrt{2}}\left\{\frac{1}{2}\Big(\Big\{A_3^n(1,3,2) - P_1^n(1,3,2) + S_1^n(1,3,2) - T_3^n(1,3,2) - T_7^n(1,3,2) - V_3^n(1,3,2)\Big\} + \Big\{1 \leftrightarrow 2\Big\}\Big)\right\},$$

$$
\begin{aligned}
A_2^{n\pi^+}(1,2,3) = -\frac{1}{\sqrt{2}}\Bigg\{ \frac{1}{2}\Big(\Big\{ & A_3^n(1,3,2)+P_1^n(1,3,2)-S_1^n(1,3,2)+T_3^n(1,3,2) \\
& +T_7^n(1,3,2)-V_3^n(1,3,2)\Big\}-\Big\{1\leftrightarrow 2\Big\}\Big)\Bigg\}, \\
V_3^{n\pi^+}(1,2,3) = \frac{1}{\sqrt{2}}\Bigg\{ \frac{1}{2}\Big(\Big\{ & A_2^n(1,3,2)-P_1^n(1,3,2)+S_1^n(1,3,2)+T_3^n(1,3,2) \\
& +T_7^n(1,3,2)+V_2^n(1,3,2)\Big\}+\Big\{1\leftrightarrow 2\Big\}\Big)\Bigg\}, \\
A_3^{n\pi^+}(1,2,3) = \frac{1}{\sqrt{2}}\Bigg\{ \frac{1}{2}\Big(\Big\{ & -A_2^n(1,3,2)-P_1^n(1,3,2)+S_1^n(1,3,2)+T_3^n(1,3,2) \\
& +T_7^n(1,3,2)-V_2^n(1,3,2)\Big\}-\Big\{1\leftrightarrow 2\Big\}\Big)\Bigg\}, \\
S_1^{n\pi^+}(1,2,3) = -\frac{1}{\sqrt{2}}\Bigg\{ \frac{1}{4}\Big(\Big\{ & A_2^n(1,3,2)+A_3^n(1,3,2)+P_1^n(1,3,2)+S_1^n(1,3,2) \\
& +2T_2^n(1,3,2)+T_3^n(1,3,2)-T_7^n(1,3,2)-V_2^n(1,3,2) \\
& +V_3^n(1,3,2)\Big\}-\Big\{1\leftrightarrow 2\Big\}\Big)\Bigg\}, \\
P_1^{n\pi^+}(1,2,3) = -\frac{1}{\sqrt{2}}\Bigg\{ \frac{1}{4}\Big(\Big\{ & -A_2^n(1,3,2)-A_3^n(1,3,2)+P_1^n(1,3,2)+S_1^n(1,3,2) \\
& +2T_2^n(1,3,2)+T_3^n(1,3,2)-T_7^n(1,3,2)+V_2^n(1,3,2) \\
& -V_3^n(1,3,2)\Big\}-\Big\{1\leftrightarrow 2\Big\}\Big)\Bigg\}, \\
T_2^{n\pi^+}(1,2,3) = -\frac{1}{\sqrt{2}}\Bigg\{ \frac{1}{2}\Big(\Big\{ & P_1^n(1,3,2)+S_1^n(1,3,2)-T_3^n(1,3,2)+T_7^n(1,3,2)\Big\} \\
& +\Big\{1\leftrightarrow 2\Big\}\Big)\Bigg\},
\end{aligned}
\tag{A.1}
$$

$$T_3^{n\pi^+}(1,2,3)=\frac{1}{\sqrt{2}}\left\{\frac{1}{4}\left(\left\{A_2^n(1,3,2)-A_3^n(1,3,2)-P_1^n(1,3,2)-S_1^n(1,3,2)\right.\right.\right.$$
$$+2T_2^n(1,3,2)-T_3^n(1,3,2)+T_7^n(1,3,2)-V_2^n(1,3,2)$$
$$\left.\left.\left.-V_3^n(1,3,2)\right\}+\left\{1\leftrightarrow 2\right\}\right)\right\},$$

$$T_7^{n\pi^+}(1,2,3)=\frac{1}{\sqrt{2}}\left\{\frac{1}{4}\left(\left\{A_2^n(1,3,2)-A_3^n(1,3,2)+P_1^n(1,3,2)+S_1^n(1,3,2)\right.\right.\right.$$
$$-2T_2^n(1,3,2)+T_3^n(1,3,2)-T_7^n(1,3,2)-V_2^n(1,3,2)$$
$$\left.\left.\left.-V_3^n(1,3,2)\right\}+\left\{1\leftrightarrow 2\right\}\right)\right\}. \tag{A.2}$$

And similar for $p\pi^0$ in twist-4

$$S_1^{p\pi^0}(1,2,3) = -\frac{1}{2}\left\{2P_1^p(1,3,2) - S_1^p(1,2,3)\right\},$$
$$P_1^{p\pi^0}(1,2,3) = -\frac{1}{2}\left\{2S_1^p(1,3,2) - P_1^p(1,2,3)\right\},$$
$$V_2^{p\pi^0}(1,2,3) = \frac{1}{2}V_2^p(1,2,3)\,, \qquad V_3^{p\pi^0}(1,2,3) = \frac{1}{2}V_3^p(1,2,3)\,,$$
$$A_2^{p\pi^0}(1,2,3) = \frac{1}{2}A_2^p(1,2,3)\,, \qquad A_3^{p\pi^0}(1,2,3) = \frac{1}{2}A_3^p(1,2,3)\,,$$
$$T_2^{p\pi^0}(1,2,3) = -\frac{1}{2}T_2^p(1,2,3)\,,$$
$$T_3^{p\pi^0}(1,2,3) = \frac{1}{2}\left\{T_3^p(1,2,3) + 2T_7^p(1,2,3)\right\},$$
$$T_7^{p\pi^0}(1,2,3) = \frac{1}{2}\left\{T_7^p(1,2,3) + 2T_3^p(1,2,3)\right\}. \tag{A.3}$$

The expressions for twist-5 and twist-6 DAs are identical to those for twist-4 and twist-3, respectively, with some substitutions in the DAs, see [2].

References

1. V. Braun, R. J. Fries, N. Mahnke and E. Stein, Nucl. Phys. B **589** (2000) 381 [Erratum-ibid. B **607** (2001) 433].
2. V. M. Braun, D. Y. Ivanov, A. Lenz and A. Peters, Phys. Rev. D **75** (2007) 014021 [arXiv:hep-ph/0611386].
3. V. M. Braun, A. Lenz and M. Wittmann, Phys. Rev. D **73** (2006) 094019 [arXiv:hep-ph/0604050].
4. P. V. Pobylitsa, M. V. Polyakov and M. Strikman, Phys. Rev. Lett. **87** (2001) 022001.

$\pi N \rightarrow$ MULTI-πN SCATTERING IN THE $1/N_c$ EXPANSION

HERRY J. KWEE

Physics Department, Arizona State University,
Tempe, AZ 85287-1504, USA
E-mail: Herry.Kwee@asu.edu

We extend the $1/N_c$ meson-baryon scattering formalism to $\pi N \rightarrow$ multi-πN case. We first show that the leading-order large N_c processes proceed through resonant intermediate states (e.g., ρN or $\pi\Delta$). We find that the pole structure of baryon resonances can be uniquely identified by their (non)appearance in ηN or mixed partial-wave $\pi\Delta$ final states.

Keywords: Large N_c QCD; pion nucleon scattering

1. Introduction

This talk mainly focuses on $\pi N \rightarrow$ multi-πN processes in the $1/N_c$ scattering formalism.[1] Recent development of the original scattering formalism[2–8] can be reviewed in Refs. [9]. The central idea rests upon symmetries that emerge for QCD in the large N_c limit. The symmetries relate scattering amplitudes in channels of different I, J, and other quantum numbers and consequently impose degeneracies among poles that occur within them.

However, this scattering formalism depends upon a single incoming source and a single outgoing source scattering from the baryon. Therefore, a meaningful way to constrain such multipion processes must be considered. Also, a standard N_c counting shows that the generic scattering amplitude for $\pi N \rightarrow \pi N$ is $O(N_c^0)$, while that for $\pi N \rightarrow \pi\pi N$ [Fig. 1(a)] is $O(N_c^{-1/2})$. Nevertheless, circumstances exist in which processes that *eventually* produce two (or more) pions nevertheless appear with amplitudes at leading order, $O(N_c^0)$. In particular, the Δ is stable for sufficiently large N_c; its width scales as $1/N_c^2$. Therefore, as in Fig. 1(b), the $\pi N \rightarrow \pi\pi N$ process may be cut (indicating an on-shell state) at the intermediate stage, $\pi N \rightarrow \pi\Delta$. Of course, we live in the $N_c = 3$ world where Γ_Δ is over 100 MeV; even so, Γ_Δ is considered sufficiently small that researchers regularly extract $\pi N \rightarrow \pi\Delta$ partial widths. Similarly, processes such as in Fig. 1(c) ($\pi N \rightarrow \rho N$ followed

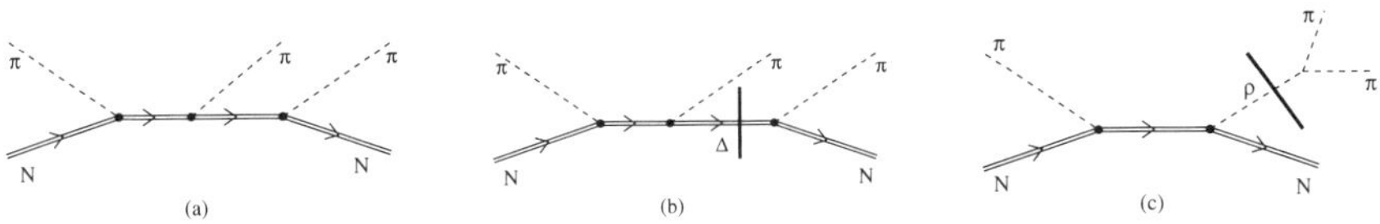

Fig. 1. Diagrams for $\pi N \to \pi\pi N$ scattering. (a) Nonresonant scattering (1 of 6 diagrams); (b) $\pi N \to \pi\Delta$ (Δ on shell), followed by $\Delta \to \pi N$; (c) $\pi N \to \rho N$, (ρ on shell) followed by $\rho \to \pi\pi$.

by $\rho \to \pi\pi$) may be analyzed in the two-body formalism since $\pi N \to \rho N$ partial widths have been tabulated.

2. Group Theory Preliminaries

The derivation of the meson-baryon scattering amplitudes formalism[10–12] is done by considering only baryons lying in the ground-state band of large N_c, whose nonstrange members have spin = isospin $R = \frac{1}{2}, \frac{3}{2}, \ldots, \frac{N_c}{2}$. This restriction is not physically constraining since all observed meson-baryon scattering processes fit into this category. The basic process is

$$m + B \to m' + B', \tag{1}$$

where $m\,(m')$ is a meson of spin $s\,(s')$ and isospin $i\,(i')$, in a state of relative orbital angular momentum L (L') with a baryon $B\,(B')$ of spin = isospin $R\,(R')$ in the ground-state multiplet, and the total spin angular momentum of the meson and baryon is denoted $S\,(S')$. The intermediate state is labeled by quantum numbers I and J, giving the full partial wave $S_{LL'SS'IJ}$. Abbreviating the multiplicity $2X+1$ of an SU(2) representation of quantum number X by $[X]$, one finds[2,12]

$$S_{LL'SS'IJ} = \sum_{K,\tilde{K},\tilde{K}'} [K]([R][R'][S][S'][\tilde{K}][\tilde{K}'])^{1/2} \times \begin{Bmatrix} L & i & \tilde{K} \\ S & R & s \\ J & I & K \end{Bmatrix} \begin{Bmatrix} L' & i' & \tilde{K}' \\ S' & R' & s' \\ J & I & K \end{Bmatrix} \tau_{K\tilde{K}\tilde{K}'LL'}. \tag{2}$$

The remaining symbols, K, $\tilde{K}$, and $\tilde{K}'$, are intermediate quantum numbers, with $\mathbf{K} \equiv \mathbf{I} + \mathbf{J}$, $\tilde{\mathbf{K}} \equiv \mathbf{i} + \mathbf{L}$, and $\tilde{\mathbf{K}}' \equiv \mathbf{i}' + \mathbf{L}'$ (so that $\mathbf{K} = \tilde{\mathbf{K}} + \mathbf{s} = \tilde{\mathbf{K}}' + \mathbf{s}'$). The factors in braces are $9j$ coefficients, and $\tau_{K\tilde{K}\tilde{K}'LL'}$ are universal amplitudes (*reduced* or K amplitudes) that are independent of I, J, R, R', i, i', s, and s'. The linear relations among the scattering amplitudes can be seen from the structure of Eq. (2); the point is there are more $S_{LL'SS'IJ}$ amplitudes than $\tau_{K\tilde{K}\tilde{K}'LL'}$ amplitudes.

3. Amplitude Tables

Here we present the transition amplitude for ηN, $\eta\Delta$, πN, $\pi\Delta$, ωN, and ρN final states of spin $\frac{1}{2}$, $I=\frac{1}{2}$ and both positive-$\mathcal{P}$ and negative-$\mathcal{P}$ parity resonances in Tables 1 and 2. The notation for $\pi N \to m'B'$ partial waves is $LL'^{(\pi N)(m'B')_{S'}}_{2I,2J}$; if $L'=L$ then the label L' is suppressed, while if s' (the m' spin) is zero, then S' equals the spin of baryon B' and is suppressed.

Table 1. Partial-wave amplitudes for positive-parity $N_{1/2}$ resonances in multipion processes (the πN final state is included for comparison). Expansions are given in terms of K amplitudes.

State	Poles	Partial Wave, K-Amplitudes		
$N^+_{1/2}$	$K=0,1$	$P_{11}^{(\pi N)(\eta N)}$	$=$	$-\frac{\sqrt{2}}{\sqrt{3}}\tau_{11111}$
		$P_{11}^{(\pi N)(\pi N)}$	$=$	$\frac{1}{3}\tau_{00011}+\frac{2}{3}\tau_{11111}$
		$P_{11}^{(\pi N)(\pi\Delta)}$	$=$	$\frac{\sqrt{2}}{3}\tau_{00011}-\frac{\sqrt{2}}{3}\tau_{11111}$
		$P_{11}^{(\pi N)(\omega N)_1}$	$=$	$\frac{1}{3}\tau_{00111}+\frac{2}{3}\tau_{11111}$
		$P_{11}^{(\pi N)(\omega N)_3}$	$=$	$\frac{\sqrt{2}}{3}\tau_{00111}-\frac{\sqrt{2}}{3}\tau_{11111}$
		$P_{11}^{(\pi N)(\rho N)_1}$	$=$	$\frac{\sqrt{2}}{3\sqrt{3}}\tau_{00111}-\frac{\sqrt{2}}{9}\tau_{11011}+\frac{2\sqrt{10}}{9}\tau_{11211}$
		$P_{11}^{(\pi N)(\rho N)_3}$	$=$	$-\frac{1}{3\sqrt{3}}\tau_{00111}-\frac{4}{9}\tau_{11011}$ $+\frac{1}{\sqrt{3}}\tau_{11111}+\frac{\sqrt{5}}{9}\tau_{11211}$

Table 2. Partial-wave amplitudes for negative-parity $N_{1/2}$ resonances in multipion processes.

State	Poles	Partial Wave, K-Amplitudes		
$N^-_{1/2}$	$K=1$	$S_{11}^{(\pi N)(\eta N)}$	$=$	0
		$S_{11}^{(\pi N)(\pi N)}$	$=$	τ_{11100}
		$SD_{11}^{(\pi N)(\pi\Delta)}$	$=$	$-\tau_{11102}$
		$S_{11}^{(\pi N)(\omega N)_1}$	$=$	τ_{11000}
		$SD_{11}^{(\pi N)(\omega N)_3}$	$=$	$-\tau_{11202}$
		$S_{11}^{(\pi N)(\rho N)_1}$	$=$	$\sqrt{\frac{2}{3}}\tau_{11100}$
		$SD_{11}^{(\pi N)(\rho N)_3}$	$=$	$\frac{1}{\sqrt{6}}\tau_{11102}+\frac{1}{\sqrt{2}}\tau_{11202}$

4. Phenomenological Results

The association of resonances with poles—as determined by presence of absence of certain decay channels—seems robust. In particular, Eq. (2) can be employed in a straightforward fashion to show that $\pi N \to \eta N$ contains a single K amplitude [with $K=L$], and the mixed partial wave $\pi N(L) \to$

$\pi\Delta(L')$ contains a single K amplitude [with $K = \frac{1}{2}(L + L')$].[3] For given I, J, and $\mathcal{P}$ these two amplitudes always probe distinct K. The following is an example of brief analysis for one of the two channels.

(1) $\boldsymbol{N^{+}_{1/2}}$ $\boldsymbol{(P_{11})}$: The two well-established resonances in this channel are $N(1440)$ (the Roper) and $N(1710)$, while our transition amplitude calculations provide two distinct pole structures, $K = 0$ and $K = 1$. The $N(1440)$ has a very small, $(0{\pm}1)\%$, ηN BR while $N(1710)$ has a small but nonnegligible ηN BR, $(6.2 \pm 1.0)\%$. Comparing this observation to our tabulated result for the $\pi N \to \eta N$ transition amplitude suggests that the Roper is a $K=0$ pole and the $N(1710)$ is a $K=1$ pole. This assignment agrees very well with the assumption of the Roper as a radial excitation of ground-state N, which is a (nonresonant) $K=0$ state.

Acknowledgments

The authors thank Tom Cohen for valuable discussions. This work was supported by the NSF under Grant No. PHY-0456520.

References

1. H. J. Kwee and R. F. Lebed, Phys. Rev. D **75**, 016002 (2007).
2. T.D. Cohen and R.F. Lebed, Phys. Rev. D **68**, 056003 (2003).
3. T.D. Cohen and R.F. Lebed, Phys. Rev. D **74**, 036001 (2006).
4. T.D. Cohen and R.F. Lebed, Phys. Rev. Lett. **91**, 012001 (2003); Phys. Rev. D **67**, 096008 (2003).
5. T.D. Cohen and R.F. Lebed, Phys. Lett. B **578**, 150 (2004); Phys. Rev. D **70**, 096015 (2004); Phys. Lett. B **619**, 115 (2005); Phys. Rev. D **72**, 056001 (2005); Phys. Rev. D **74**, 056006 (2006).
6. T.D. Cohen, D.C. Dakin, A. Nellore, and R.F. Lebed, Phys. Rev. D **69**, 056001 (2004); Phys. Rev. D **70**, 056004 (2004).
7. T.D. Cohen, D.C. Dakin, R.F. Lebed, and D.R. Martin, Phys. Rev. D **71**, 076010 (2005).
8. R.F. Lebed, Phys. Lett. B **639**, 68 (2006).
9. R.F. Lebed, `hep-ph/0601022`, published in *NStar 2005*, edited by S. Capstick *et al.*, World Scientific, Singapore (2006).
10. A. Hayashi, G. Eckart, G. Holzwarth, H. Walliser, Phys. Lett. B **147**, 5 (1984).
11. M.P. Mattis and M. Karliner, Phys. Rev. D **31**, 2833 (1985); M.P. Mattis and M.E. Peskin, Phys. Rev. D **32**, 58 (1985); M.P. Mattis, Phys. Rev. D **39**, 994 (1989); Phys. Rev. Lett. **63**, 1455 (1989); Phys. Rev. Lett. **56**, 1103 (1986).
12. M.P. Mattis and M. Mukerjee, Phys. Rev. Lett. **61**, 1344 (1988).

e^+e^- ANNIHILATIONS INTO QUASI-TWO-BODY FINAL STATES AT 10.58 GEV

KAI YI

Stanford Linear Accelerator Center
Menlo Park, California 94025, USA
**E-mail: yik@slac.stanford.edu*

We report the first observation of e^+e^- annihilations into hadronic states of positive C-parity, $\rho^0\rho^0$ and $\phi\rho^0$. The angular distributions support two-virtual-photon annihilation production. We also report the observations of $e^+e^- \to \phi\eta$ and a preliminary result on $e^+e^- \to \rho^+\rho^-$.

Keywords: Exclusive; High momentum; C parity; Helicity amplitude.

1. Introduction

The large datasets collected by the B factories provide unique opportunities for studying rare processes and discovering new states. We report several observations of e^+e^- annihilations into quasi-two-body hadronic final states with $C = \pm 1$ at *BABAR* [1–3]. A new avenue for the study of hadron production mechanisms is opened with these observations, and a testing ground for QCD at the amplitude level is provided.

2. $e^+e^- \to \rho^0\rho^0, \phi\rho^0$

The process $e^+e^- \to$ hadrons at center-of-mass (c.m.) energy $\sqrt{s}$ far below the Z^0 mass is dominated by annihilation via a single virtual photon, thus yielding final state charge-conjugation parity $C = -1$. The Two-Virtual-Photon-Annihilation (TVPA) process, depicted in Fig. 1, with positive final state C parity, has been ignored in incorporating the total hadronic cross section in e^+e^- annihilations into calculations [4] of muon g–2, and the running of the QED coupling constant, α.

The present analysis uses a 205 fb^{-1} data sample collected at the $\Upsilon(4S)$ resonance, and 20 fb^{-1} collected at c.m. energy 40 MeV lower, using the

BABAR detector at the SLAC PEP-II asymmetric-energy e^+e^- collider. The *BABAR* detector is described in detail elsewhere [5].

Events with four well-reconstructed charged tracks and net charge zero are selected. The χ^2 probability of the fitted four track vertex is required to exceed 0.1%, and two oppositely charged tracks must be identified as pions; the other pair must be identified as two pions or two kaons. We accept events with four-particle invariant mass within 170 MeV/c^2 of the nominal c.m. energy. Loose signal regions are defined by the mass ranges $0.5 < m_{\pi^+\pi^-} < 1.1$ GeV/c^2 and $1.008 < m_{K^+K^-} < 1.035$ GeV/c^2. The extracted $\rho^0\rho^0$ and $\phi\rho^0$ yields in these intervals are 1243 ± 43 and 147 ± 13 events, respectively.

The efficiency-corrected production angular distributions are shown in Fig. 2, where θ^* is defined as the angle between the ρ^0_f (ϕ) direction and the e^- beam direction in the c.m. frame. The observed sharply peaking $|\cos\theta^*|$ distributions are consistent with the TVPA expectation [6], which is approximated by:

$$\frac{d\sigma}{d\cos\theta^*} \propto \frac{1+\cos^2\theta^*}{1-\cos^2\theta^*}. \tag{1}$$

For the signal mass regions defined above, and $|\cos\theta^*| < 0.8$, we obtain the following results for the TVPA cross sections near $\sqrt{s} = 10.58$ GeV:

$$\sigma_{\text{fid}}(e^+e^- \to \rho^0\rho^0) = 20.7 \pm 0.7(\text{stat}) \pm 2.7(\text{syst})\ \text{fb}$$
$$\sigma_{\text{fid}}(e^+e^- \to \phi\rho^0) = 5.7 \pm 0.5(\text{stat}) \pm 0.8(\text{syst})\ \text{fb}.$$

The measured cross sections are in good agreement with the calculations [6, 7]. The Standard Model calculations of the anomalous magnetic moment of the muon and of the QED coupling constant rely on measurements of low-energy e^+e^- hadronic cross sections, which are assumed to be entirely due to single-photon exchange. We have estimated the effect due to the TVPA processes [6] and find it to be small compared with the current precision [4].

3. $e^+e^- \to \phi\eta$

The process $e^+e^- \to J/\psi\eta_c$ and other double charmonium processes are observed [8] at rates approximately ten times larger than expected from QCD-based models [9]. Various theoretical efforts to understand this have been made recently [10]. An alternate avenue of investigation is provided by the process $e^+e^- \to \phi\eta$, which also involves a vector – pseudoscalar (VP) final state. Different models predict different s dependences for the cross

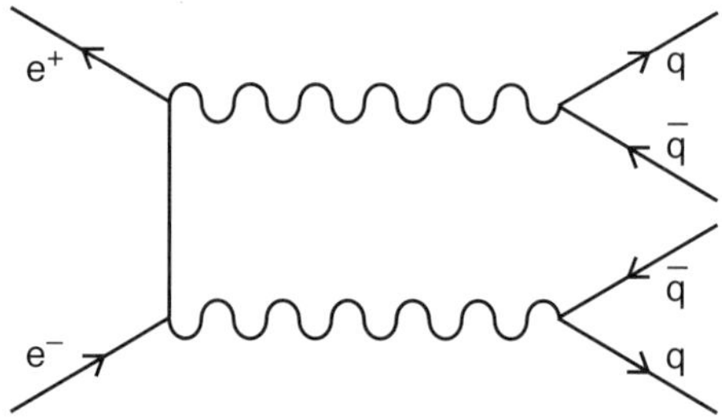

Fig. 1. The two-virtual-photon annihilation diagram.

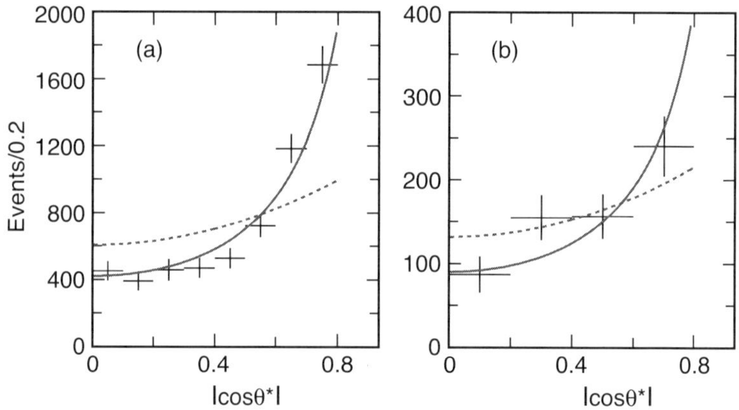

Fig. 2. Production angle distributions, after correction for efficiency, for a) $\rho^0\rho^0$ and b) $\phi\rho^0$. The solid and dashed lines are the normalized $\frac{1+\cos^2\theta^*}{1-\cos^2\theta^*}$ and $1+\cos^2\theta^*$ distributions, respectively.

section, and so it is interesting to investigate this by comparing a measurement at $\sqrt{s} = 10.58$ GeV to the CLEO measurement at $\sqrt{s} = 3.67$ GeV [11].

This analysis uses 204 fb^{-1} of e^+e^- colliding beam data collected on the $\Upsilon(4S)$ resonance at $\sqrt{s} = 10.58$ GeV and 20 fb^{-1} collected 40 MeV below. Events with exactly two well-reconstructed, oppositely charged kaon tracks and at least two well-identified photons are selected. We fit the two tracks to a common vertex, and require the χ^2 probability to exceed 0.1%. Each photon candidate is required to have a minimum laboratory energy of 500 MeV. Events with a reconstructed $K^+K^-\gamma\gamma$ invariant mass within 230 MeV/c^2 of the e^+e^- c.m. energy are accepted for further study.

We define the ϕ mass window as $1.008 < m_{KK} < 1.035$ GeV/c^2, and extract 24 ± 5 $\phi\eta$ signal events in the ϕ mass window, with $\eta \to \gamma\gamma$. The significance is estimated to be 6.5 sigma.

The final radiation-corrected cross section for $1.008 < m_\phi < 1.035$ GeV/c^2 within $|\cos\theta^*| < 0.8$ near $\sqrt{s} = 10.58$ GeV is:

$$\sigma_{\text{fid}}(e^+e^- \to \phi\eta) = 2.1 \pm 0.4(\text{stat}) \pm 0.1(\text{syst})\ \text{fb}.$$

The cross section, extended to the full range of $\cos\theta^*$ by assuming a $1 + \cos^2\theta^*$ distribution, is:

$$\sigma(e^+e^- \to \phi\eta) = 2.9 \pm 0.5(\text{stat}) \pm 0.1(\text{syst})\ \text{fb}.$$

There is currently no direct prediction for the cross section of this process at this energy, but the $e^+e^- \to$ VP cross section is expected to have a $1/s^2$ [12] or $1/s^4$ [13,14] dependence in QCD-based models. A comparison between our result and that of CLEO, ($\sigma = 2.1^{+1.9}_{-1.2} \pm 0.2$ pb) at $\sqrt{s} = 3.67$ GeV (continuum) [11], favors a $1/s^3$ dependence (Fig. 3).

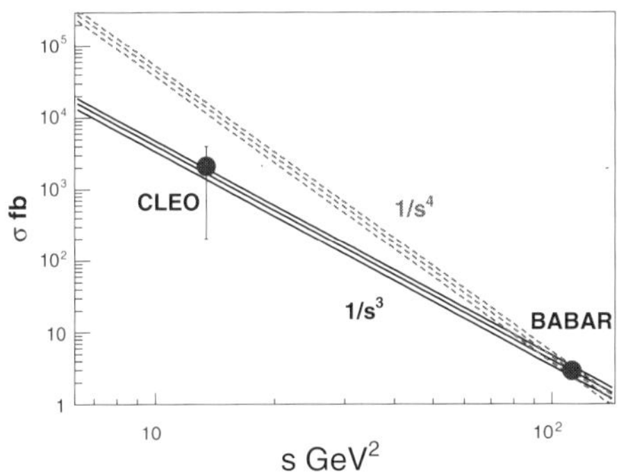

Fig. 3. Cross section extrapolations based on *BABAR*'s measurement at $\sqrt{s} = 10.58$ GeV assuming a $1/s^3$ (lower curve) or $1/s^4$ (upper curve) energy dependence. The bands show one standard deviation uncertainties in the extrapolations. The CLEO measurement at $\sqrt{s} = 3.67$ GeV is also shown.

4. $e^+e^- \to \rho^+\rho^-$ (Preliminary result)

Since charged ρ's are involved, the $e^+e^- \to \rho^+\rho^-$ process is unlikely to occur through TVPA [1,6,7], unless there is significant final quark recombination between the products of the two virtual photons, or unless there is significant final state interaction ($e^+e^- \to \rho^0\rho^0 \to \rho^+\rho^-$) [15]. Assuming a one-photon production mechanism, this VV ($\rho^+\rho^-$) final state is described

BABAR Preliminary

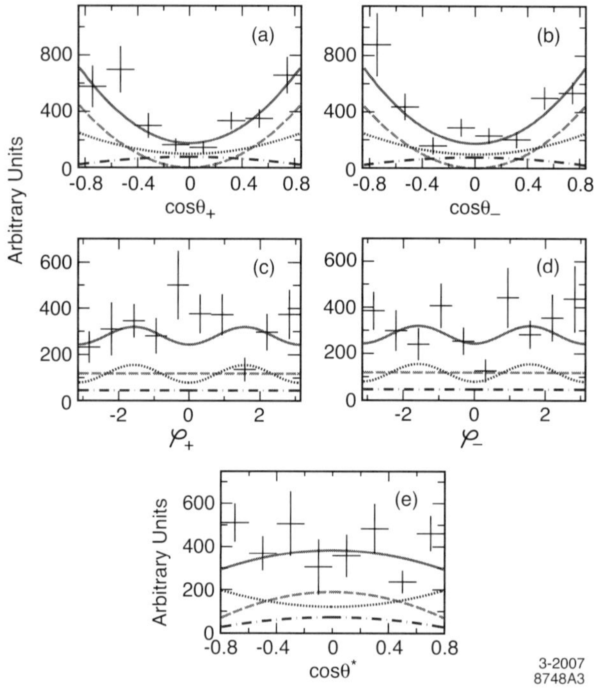

Fig. 4. The s-weighted and efficiency corrected a) $\cos\theta_+$ b) $\cos\theta_-$ c) φ_+ d) φ_- e) $\cos\theta^*$ distributions for $e^+e^- \to \rho^+\rho^-$. The dashed curves show the contributions from F_{00}, the dotted curves are F_{10}, the dashed-dotted curves are F_{11}, and the solid curves show the total result.

by three helicity amplitudes. A study of this reaction can then provide an experimental test of QCD at the amplitude level [13,14] through investigation of the final states angular correlations.

This analysis uses 343 fb^{-1} of e^+e^- colliding beam data collected on the $\Upsilon(4S)$ resonance at $\sqrt{s} = 10.58$ GeV and 36 fb^{-1} collected 40 MeV lower. Events with exactly two well-reconstructed, oppositely charged tracks identified as pions and at least two well-reconstructed π^0s are selected. We fit the charged tracks to a common vertex, and require the χ^2 probability to exceed 0.1%. Each π^0 is reconstructed through its $\gamma\gamma$ decay channel by requiring the two photon invariant mass to be within the range [0.1,0.16] GeV/c^2, and then constraining its mass to the nominal value. We accept events with $|m_{\pi^+\pi^0\pi^-\pi^0} - E_{cm}| < 0.28$ GeV and $|\Delta p| < 0.2$ GeV/c, where E_{cm} is the total c.m. energy, and Δp is the momentum difference between the $\pi^+\pi^0\pi^-\pi^0$ system and the e^+e^- system.

We define the $\rho^\pm$ mass intervals as [0.5,1.1] GeV/c^2, and extract 308$\pm$25 $\rho^+\rho^-$ signal events in the defined mass region. The significance is estimated to be 9.5 sigma.

Assuming $\rho^+\rho^-$ is produced through one photon or $\Upsilon(4S)$, there are three independent helicity amplitudes ($F_{\mu\nu}$, μ/ν is the helicity of ρ^+/ρ^-), F_{00}, F_{10}, and F_{11} ($F_{10} = F_{-10} = F_{0\pm1}$, $F_{11} = F_{-1-1}$) [16]. The one-dimensional projections for the decay angles involved can be expressed as:

$$\frac{dN}{d\cos\theta^*} \propto (\sin^2\theta^*|F_{00}|^2 + 2(1+\cos^2\theta^*)|F_{10}|^2 + 2\sin^2\theta^*|F_{11}|^2) \tag{2}$$

$$\frac{dN}{d\cos\theta_\pm} \propto (\cos^2\theta_\pm|F_{00}|^2 + (1+\cos^2\theta_\pm)|F_{10}|^2 + \sin^2\theta_\pm|F_{11}|^2) \tag{3}$$

$$\frac{dN}{d\varphi_\pm} \propto (|F_{00}|^2 + (4-\cos 2\varphi_\pm)|F_{10}|^2 + 2|F_{11}|^2) \tag{4}$$

where θ^* is the ρ production angle, $\theta_\pm$ ($\varphi_\pm$) is the helicity (azimuthal) angle of the pion from ρ decay. From the two dimensional mass fit ($\pi^+\pi^0$ and $\pi^-\pi^0$), we can calculate a $\rho^+\rho^-$ signal sWeight [17] for each event (including those events outside the defined $\rho^\pm$ mass window) and use it to produce signal angular distributions. We fit the five angular distributions to Equs. 2, 3 and 4 simultaneously by minimizing χ^2. The correlations among the five angles are neglected; this is justified by means of fits to events generated according the assumed PDFs (toy MC). We normalize the amplitudes such that $|F_{00}|^2 + 4|F_{10}|^2 + 2|F_{11}|^2 = 1$ since we have 1 F_{00}, 4 F_{10} and 2 F_{11} amplitude contributions. The normalized amplitudes from the fit are found to be in the ratio: $|F_{00}|^2 : |F_{10}|^2 : |F_{11}|^2 = 0.51 \pm 0.14(\text{stat}) \pm 0.02(\text{syst}) : 0.10 \pm 0.04(\text{stat}) \pm 0.01(\text{syst}) : 0.04 \pm 0.03(\text{stat}) \pm 0.00(\text{syst})$, and $|F_{00}|^2$ deviates from 1 with significance more than 3 sigma. This disagrees with a QCD prediction [14], and suggests that either the decay is not dominated by single-photon exchange as naively expected, or that the QCD prediction does not apply to data in our energy region. The final radiation-corrected cross section for $0.5 < m_{\rho^\pm} < 1.1$ GeV/c^2, and within $|\cos\theta^*| < 0.8$, $|\cos\theta_\pm| < 0.85$, at near $\sqrt{s} = 10.58$ GeV (assuming only one-photon production) is:

$$\sigma_{\text{fid}}(e^+e^- \to \rho^+\rho^-) = 8.5 \pm 0.7(\text{stat}) \pm 1.5(\text{syst}) \text{ fb}.$$

We extend the cross section calculation from our acceptance region to the full phase space using the fitted amplitude values, and find 20.0$\pm$1.6(stat)$\pm$

3.6(syst) $\pm$ 1.7(ampl) fb; the third uncertainty is due to the amplitude uncertainties.

5. Conclusion

We report the first observation of e^+e^- annihilations into hadronic states of positive C-parity, $\rho^0\rho^0$ and $\phi\rho^0$. We also report the observation of the process $e^+e^- \to \phi\eta$, and obtain preliminary results on $e^+e^- \to \rho^+\rho^-$. The measured helicity amplitude magnitudes from $e^+e^- \to \rho^+\rho^-$ contradict a QCD prediction at a significance of more than 3 sigma.

Acknowledgments

The author wishes to thank his *BABAR* colleagues for their support.

References

1. B. Aubert *et al.*, Phys. Rev. Lett. **97**, 112002 (2006).
2. B. Aubert *et al.*, Phys. Rev. D (R) **74**, 111103(R) (2006).
3. F. Blanc for *BABAR* Collaboration, Morind QCD 2006.
4. M. Davier, S. Eidelman, A. Hoecker, Z. Zhang, Eur. Phys. Jour. C 31, 503 (2003).
5. B. Aubert *et al.*, Nucl. Instrum. Methods, Sect. A **479**, 1 (2002).
6. M. Davier, M. Peskin, and A. Snyder, hep-ph/0606155.
7. G. T. Bodwin *et al.*, Phys. Rev. D **74**, 074014 (2006).
8. K. Abe *et al.*, Phys. Rev. Lett. **89**, 142001 (2002); B. Aubert *et al.*, Phys. Rev. D **72**, 031101 (2005).
9. E. Braaten and J. Lee, Phys. Rev. D **67**, 054007 (2003), [Erratum-ibid. D **72**, 099901 (2005)]; K. Y. Liu, Z. G. He and K. T. Chao, Phys. Lett. B **557**, 45 (2003).
10. A. E. Bondar and V. L. Chernyak, Phys. Lett. B **612**, 215 (2005); G. T. Bodwin, D. Kang and J. Lee, hep-ph/0603185; Y. Zhang, Y. Gao, and K. T. Chao, Phys. Rev. Lett. **96**, 092001 (2006); G. T. Bodwin, D. Kang, T. Kim, J. Lee, and C. Yu, hep-ph/0611002.
11. G. S. Adams *et al.* (CLEO Collaboration), Phys. Rev. D **73**, 012002 (2006).
12. J. M. Gerard and G. Lopez Castro, Phys. Lett. B **425**, 365 (1998).
13. V. Chernyak, hep-ph/9906387.
14. V. L. Chernyak and A. R. Zhitnitsky, JETP Lett. **25**, 510 (1977); [Pisma Zh. Eksp. Teor. Fiz. **25**, 544 (1977)]; G. P. Lepage and S. J. Brodsky, Phys. Lett. B **87**, 359 (1979); G. P. Lepage and S. J. Brodsky, Phys. Rev. D **22**, 2157 (1980); S. J. Brodsky and G. P. Lepage, Phys. Rev. D **24**, 2848 (1981).
15. M. Gell-Mann, D. Sharp, and W. G. Wagner, Phys. Rev. Lett. **8**, 261 (1962); P. Kroll, hep-ph/0112274.
16. M. Jacob, and G. C. Wick, Ann. Phys.. (N. Y.) **7**, 404 (1959); S. U. Chung, Phys. Rev. D **57**, 431 (1998).
17. M. Pivk and F. R. Le Diberder, Nucl. Instrum. Meth. A **555**, 356 (2005).

TRANSITION DISTRIBUTION AMPLITUDES

J.P. LANSBERG[1]*, B. PIRE[1] and L. SZYMANOWSKI[2,3]

[1] *Centre de Physique Théorique, École Polytechnique, CNRS, 91128, Palaiseau, France*
[2] *Fundamental Interactions in Physics and Astrophysics, Université de Liège, Belgique*
[3] *Soltan Institute for Nuclear Studies, Warsaw, Poland*

We give a brief overview of the theoretical status of the Transition Distribution Amplitudes and discuss their experimental near future.

Keywords: Hadronic Physics, Exclusive Processes, Factorisation

1. Introduction

According to a now well-established framework,[1] the Bjorken limit of near forward exclusive reactions with a hard probe allows factorisation of the leading twist amplitudes into a perturbatively calculable sub-process at quark and gluon level on the one hand and hadronic matrix elements of light-cone non-local operators describing the transition from the baryon target to a final baryon on the other hand. Those matrix elements can be expressed through hadron distribution amplitudes (DAs) and generalised parton distributions (GPDs). Experimental data[2] from DESY and JLab are now confirming this framework, and they seem to show that its applicability is quite precocious in terms of Q^2. The harmonic analysis[3] of spin asymmetries which are particularly sensitive to the interference between the deeply-virtual-Compton scattering (DVCS) and Bethe-Heitler processes is very relevant for that purpose.

Investigations on GPDs are very important since they are new QCD objects which carry much information on the hadronic structure. A further generalisation of the GPD concept has been proposed[4] in cases where the initial and final states are different hadronic states. If those new hadronic objects are defined through a quark-antiquark operator (meson to meson or meson to photon transition), we call them *mesonic* transition distribution

*lansberg@cpht.polytechnique.fr

amplitudes (TDA),[5] if they are defined through a three quark operator (baryon to meson or baryon to photon transition), we call them *baryonic* transition distribution amplitudes.[6]

Both appear in the description of hard exclusive processes, where a highly off-shell photon provides a hard scale permitting, on the one hand, to treat perturbatively the interaction between the photon and the quarks off the target, on the other hand to advocate the factorisation of the amplitude as a convolution of a hard amplitude M_h with the *universal* TDAs describing the non-perturbative transition between two hadronic states and DAs describing the formation of another hadron. This is illustrated in Fig. 1 for three cases.The first is the process $\gamma^\star\gamma \to A\pi$ at small t, where A is a meson. There appear the mesonic $\gamma \to \pi$ TDAs. The second is the backward electroproduction of a pion and the third is a crossed process $\bar{p}p \to \gamma^\star\pi^0$. In those two last processes, there appear the baryonic $p \to \pi^0$ TDAs.

2. The mesonic TDAs

2.1. *Mesonic TDAs vs. GPDs*

For definiteness, let us consider the $\gamma \to \pi^-$ TDAs. Those are defined from the correlators[a b]

$$\int \frac{dz^-}{2\pi} e^{ixP^+z^-} \langle \pi^-(p_{\pi^-})|\bar{d}(-\frac{z}{2})(\Gamma)u(\frac{z}{2})|\gamma(p_\gamma,\varepsilon)\rangle\Big|_{z^+=0,\, z_T=0}, \tag{1}$$

where Γ is one of the Dirac forms, γ^μ,$\gamma^\mu\gamma^5$ or $\sigma^{\mu\nu}$. The first two enable to respectively define the vector TDA $V^{\gamma\pi^-}(x,\xi,\Delta^2)$ and the axial-vector one $A^{\gamma\pi^-}(x,\xi,\Delta^2)$, going along with their respective Lorentz structures.[5] The latter Dirac form, $\sigma^{\mu\nu}$, is related to two chiral-odd tensorial TDAs $T_1^{\gamma\pi^-}(x,\xi,\Delta^2)$ and $T_2^{\gamma\pi^-}(x,\xi,\Delta^2)$.

The mesonic TDAs have much in common with meson GPDs. They are defined from matrix elements of the same quark - antiquark operator and thus obey the same QCD evolution equations. Sum rules may be derived for the photon to meson TDAs. Since the local matrix elements appear in radiative weak decays, we can relate the TDAs to the vector and axial form factors F_V and F_A in the $\pi^\pm$ case and to $F_{\pi^0\gamma^\star\gamma}$ in the π^0 case. Those form factors are well measured.[7]

---

[a]Working in light-like (or axial) gauge both for the gluons and the photons, we omitted to include the QCD and QED Wilson line between the quark fields.

[b]If p and p' are respectively the momenta of the initial and final particles in the TDAs, we have $(p'-p) = \Delta$, $t = \Delta^2$, $P = 1/2(p+p')$ and $2\xi = -\Delta^+/P^+$. We also have $\xi \approx x_B/(2-x_B)$.

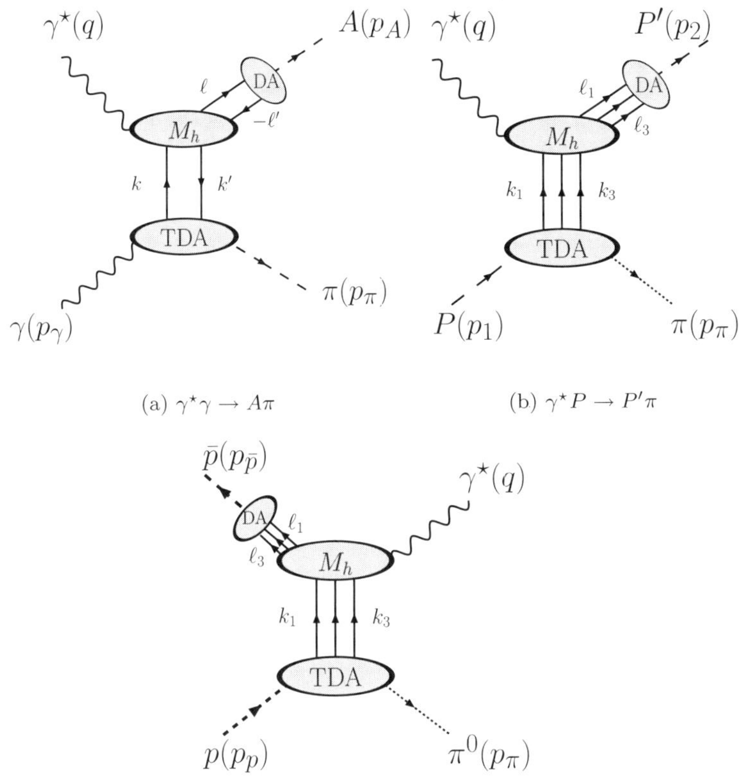

(a) $\gamma^\star\gamma \to A\pi$

(b) $\gamma^\star P \to P'\pi$

(c) $\bar{p}p \to \gamma^\star\pi^0$

Fig. 1. (a) Illustration of the factorisation for $\gamma^\star\gamma \to A\pi$ at small transfer momentum. (b) Idem for backward electroproduction of a pion. (c) Idem for $\bar{p}p \to \gamma^\star\pi^0$.

Moreover, the TDAs satisfy similar polynomiality conditions and they may be constructed from a spectral decomposition, in analogy with the construction of GPDs through double distributions.[8] The x and ξ dependence of the TDAs is then given as

$$\int_{-1}^{1} d\beta \int_{-1+|\beta|}^{1-|\beta|} d\alpha\, \delta(x-\beta-\xi\alpha) f(\beta,\alpha).$$

In the GPD case, $f(\beta,\alpha) = q(\beta)h(\beta,\alpha)$ with $q(\beta)$ the forward quark distribution and $h(\beta,\alpha)$ a profile function. Note however that TDAs possess different properties than GPDs with respect to time reversal since initial and final states are different. A consequence of this is the appearance of odd-powers of ξ in their moments in x.

In the aforementioned case of $\gamma \to \pi^-$, the leading-twist decomposition of the matrix element differs from the π^- GPD one mainly because of the presence of the photon polarisation vectors $\varepsilon(p_\gamma)$ – a similar situation would occur for the $\pi \to \rho$ transition. This enables to build 4 independent leading-twist Lorentz structures, and thus to define 4 TDAs, whereas there are only 2 leading-twist pion GPDs.

2.2. *A perturbative limit*

An interesting perturbative limit of a mesonic TDA may be derived[9] by considering the Born order amplitude $\mathcal{A}$ for the process

$$\gamma^\star_L(q')\,\gamma^\star_L(q) \to \rho^0_L(k')\,\rho^0_L(k).$$

in the region where $Q'^2 \gg Q^2$ and in the forward kinematics ($t = t_{min}$).

One can see that the amplitude factorises and the $\gamma^\star_L \to \rho_L$ vector $V(x,\xi,t)$ TDA[c] is obtained at leading order as

$$V(x,\xi,t_{min}) = 3\left[\Theta(1 \geq x \geq \xi)\Phi\left(\frac{x-\xi}{1-\xi}\right) - \Theta(-\xi \geq x \geq -1)\Phi\left(\frac{1+x}{1-\xi}\right)\right],$$

where Φ is the ρ DA. This TDA describes the transition between the *least off-shell* photon (Q^2) and the longitudinally polarised ρ to which it is collinear. Note that this perturbative expression vanishes in the ERBL region.

2.3. *Models*

We have already at our disposal several models for the vector and axial-vector TDAs describing the pion to photon transition. In Refs 10,11, those were modelled via double-distributions and more recently in the spectral quark model.[12] Other approaches[13] used to construct pion GPDs could provide us with reasonable modelling of the mesonic TDAs. Let us cite for instance the Nambu-Jona Lasinio Model.[14] Finally, lattice QCD, which has been recently applied to extract moments of pion GPDs,[15] could also be applied to the TDA case.

---

[c]Defining n_1 and n_2 such that $q = \frac{-Q^2}{(1+\xi)s}n_1 + (1+\xi)n_2$ and $k = (1-\xi)n_2$ and considering the quark antiquark operator along the light-cone vector $n = 2n_1/s, s = 2n_1.n_2$, $V(x,\xi,t)$ is defined as (omitting the QED Wilson line)

$$\int \frac{dz^-}{2\pi} e^{ix(P.z)} \langle \rho^q_L(k)|\bar{q}(-\frac{z}{2})\not{n} q(\frac{z}{2})|\gamma^\star(q)\rangle = \frac{eQ_q\epsilon_\nu(q)}{P^+Q^2}[(1+\xi)n_2^\nu + \frac{Q^2 n_1^\nu}{s(1+\xi)}]V(x,\xi,t).$$

2.4. *Experimental possibilities*

The introduction of the $\gamma \rightarrow$meson TDAs completes the kinematical domain of understanding of the reactions $\gamma\gamma^\star \rightarrow M_1 M_2$ in the framework of QCD factorisation, supplementing the near threshold kinematical domain described by generalised distribution amplitudes (GDAs)[16] and the fixed (or large) angle domain historically described by the Brodsky-Lepage factorisation.[17] Data have been collected at LEP and CLEO on these reactions, mostly in the GDA domain for $\rho\rho$ final states, with some phenomenological success.[18] More data are obviously needed and are eagerly waited for in the TDA region, and much hope comes from the high luminosity electron colliders. This requires in general the exclusive detection of two mesons, when tagging simultaneously one outgoing electron.

Another way to access the mesonic TDAs is to study DVCS on virtual pion target, which may be studied[19] at Hermes and JLab in the reaction $\gamma^\star p \rightarrow \gamma\pi^+ n$, when the transition $p \rightarrow n$ is dominated by the pion pole and the π^+ flies in the direction of the $\gamma^\star$ in the $\gamma\pi^+$ CMS.

3. The baryonic TDAs

3.1. *Definitions*

Let us concentrate here on the $p \rightarrow \pi$ TDAs. The leading twist TDAs for the $p \rightarrow \pi^0$ transition are defined from the correlator :

$$\langle \pi^0(p_\pi) | \, \epsilon^{ijk} u^i_\alpha(z_1\, n) u^j_\beta(z_2\, n) d^k_\gamma(z_3\, n) \, | p(p_p, s_p) \rangle$$

These TDAs are matrix elements of the same operator that appears in baryonic distribution amplitudes. The known evolution equations of this operator lead to derive evolution equations which have different forms in different regions; one defines one ERBL and two DGLAP regions much in the same spirit as in the GPD case, so that those equations in momentum space depend on the signs of the quark momentum fractions x_i.

As for DAs, an asymptotic solution for this evolution equation exists but the phenomenological study of electromagnetic form factors leads us to strongly doubt that it is of any phenomenological relevance. In some sense, this is not a surprise since the corresponding asymptotic solution ($\delta(x)$) for parton distribution functions is far from a realistic description of DIS data. We thus do not propose to take an asymptotic TDA as a realistic input for phenomenology.

On the other hand, there exists an interesting soft limit[20] when the emerging pion momentum is small, which allows to relate proton $\rightarrow$ pion

TDAs to proton DAs. The soft pion theorems indeed allow to write:

$$\langle \pi^a(p_\pi)|\mathcal{O}|P(p_1,s_1)\rangle \to -\frac{i}{f_\pi}\langle 0|[Q_5^a,\mathcal{O}]|P(p,s)\rangle$$

when $\xi \to 1$ ($E_\pi \to 0$) ; the neglected nucleon pole term, which does not contribute at threshold but is likely to be important for ξ significantly different from 1, may also be taken into account. One then gets relations between the nucleon DAs[21] A^p, V^p and T^p on the one hand and the $p \to \pi$ TDAs $V_1^{p\pi^0}$, $A_1^{p\pi^0}$ and $T_1^{p\pi^0}$ on the other hand :

$$V_1^{p\pi^0}(x_1,x_2,x_3,1,M^2) = \frac{1}{4}V^p\Big(\frac{x_1}{2},\frac{x_2}{2},\frac{x_3}{2}\Big),$$

$$A_1^{p\pi^0}(x_1,x_2,x_3,1,M^2) = \frac{1}{4}A^p\Big(\frac{x_1}{2},\frac{x_2}{2},\frac{x_3}{2}\Big),$$

$$T_1^{p\pi^0}(x_1,x_2,x_3,1,M^2) = \frac{3}{4}T^p\Big(\frac{x_1}{2},\frac{x_2}{2},\frac{x_3}{2}\Big).$$

To conclude, let us mention that the proton to photon TDAs, entering the description of backward DVCS, have been defined in Ref. 22.

3.2. *Experimental situation*

As we mentioned above, $p \to \pi$ baryonic TDAs appear in the description of backward electroproduction of a pion on a proton target. In terms of angle, in the $\gamma^\star p$ center of momentum (CM) frame, the angle between the $\gamma^\star$ and the pion, $\theta_\pi^\star$, is close to 180°. We then have $|u| \ll s$ and $t \simeq -(s+Q^2)$, in contrast to the fixed angle regime $u \simeq t \simeq -(s+Q^2)/2$ ($\theta_\pi^\star \simeq 90°$) and the forward (GPD) one $|t| \ll s$ and $u \simeq -(s+Q^2)$ ($\theta_\pi^\star \simeq 0°$).

The TDAs appear also in similar electroproduction processes such as $ep \to e\ (p,\Delta^+)\ (\eta,\rho^0)$, $ep \to e\ (n,\Delta)\ (\pi^+,\rho^+)$, $ep \to e\ \Delta^{++}\ (\pi^-,\rho^-)$. Those processes have already been analysed, at backward angles, at JLab in the resonance region, *i.e.* $\sqrt{s_{\gamma^\star p}} = W < 1.8$ GeV, in order to study the baryonic transition form factors in the π channel[23] or in the η channel.[24,25] Data are being extracted in some channels above the resonance region. The number of events seems large enough to expect to get cross section measurements for $\Delta_T^2 < 1$ GeV^2, which is the region described in terms of TDAs. Hermes analysis[26] for forward electroproduction may also be extended to larger values of $-t$. It has to be noted though that present studies are limited to Q^2 of order a few GeV^2, which gives no guarantee to reach the TDA regime yet. Higher-Q^2 data may be obtained at JLab-12 GeV and in muoproduction at Compass within the next few years. Besides comparisons with forthcoming experimental data, one may also consider results from global Partial Wave Analysis (*e.g.* SAID[27]).

Crossed reactions involving TDAs in proton-antiproton annihilation (GSI-FAIR[28]), with time-like photons (*i.e.* di-leptons) can also be studied with other mesons than a pion, *e.g.* $\bar{p}p \to \gamma^\star\ (\eta, \rho^0)$, or on a different target than proton $\bar{p}N \to \gamma^\star\pi$. Finally, one may also consider associated J/ψ production with a pion $\bar{p}p \to \psi\ \pi^0$ or another meson $\bar{p}p \to \psi\ (\eta, \rho^0)$, which involve the *same* TDAs as with an off-shell photon or in backward electroproduction. They will serve as very strong tests of the universality of the TDAs in different processes.

3.3. *Models and applications*

The first application of baryonic TDAs was centered on backward electroproduction of a pion.[20] In that case, the hard contribution which consists in the scattering of the hard photon with three quarks is known at leading order. Extrapolating the limiting value of the TDAs obtained from the soft pion theorems to the large-ξ region, we obtained a first evaluation of the unpolarised cross section for backward electroproduction.This estimate, which is unfortunately reliable only in a restricted kinematical domain (large-ξ), shows an interesting sensitivity to the underlying model for the proton DA. This study will soon be extended to hard exclusive production of a $\gamma^\star\pi^0$ pair in $\bar{p}p$ annihilation at GSI-FAIR.[29]

In the near future, a modelling inspired from the soft-pion limit including the nucleon pole term will be applied to backward electroproduction at $\theta^\star_\pi \neq 180°$. Application of the Pion-cloud Model[30] to $p \to \pi$ transitions are also expected to be available soon.[31] Specific non-perturbative approaches such as lattice studies, instanton-based models, chiral perturbation theory,[32] AdS-CFT correspondence,[33] can also provide us with realistic expression for the TDAs to be checked experimentally and are therefore eagerly waited for.

Acknowledgments

L.Sz. acknowledges the support by the Polish Grant 1 P03B 028 28. L.Sz. is a Visiting Fellow of FNRS (Belgium).This work was partly supported by the Joint Research Activity "Generalised Parton Distributions" of the European I3 program Hadronic Physics, contract RII3-CT-2004-506078.

References

1. D. Müller *et al.*, Fortschr. Phys. **42**, 101 (1994); X. D. Ji, Phys. Rev. Lett. **78**, 610 (1997); A. V. Radyushkin, Phys. Rev. **D56**, 5524 (1997).

2. C. Munoz Camacho *et al.* [Jefferson Lab Hall A Collaboration], Phys. Rev. Lett. **97** (2006) 262002; S. Chen *et al.* [CLAS Collaboration], Phys. Rev. Lett. **97** (2006) 072002; A. Airapetian *et al.* [HERMES Collaboration], arXiv:0707.0222 [hep-ex]; L. Schoeffel, arXiv:0705.2925 [hep-ph].
3. M. Diehl *et al.* Phys. Lett. B **411** (1997) 193; A. V. Belitsky *et al.*, Nucl. Phys. B **593** (2001) 289.
4. L. L. Frankfurt *et al.* Phys. Rev. D **60** (1999) 014010; T. Feldmann and P. Kroll, Eur. Phys. J. C **12** (2000) 99.
5. B. Pire and L. Szymanowski, Phys. Rev. D **71** (2005) 111501.
6. B. Pire and L. Szymanowski, Phys. Lett. B **622** (2005) 83 and PoS **HEP2005** (2006) 103.
7. W. M. Yao *et al.* [Particle Data Group], J. Phys. G **33** (2006) 1.
8. A. V. Radyushkin, Phys. Rev. D **59** (1999) 014030.
9. B. Pire *et al.* Phys. Lett. B **639** (2006) 642.
10. J. P. Lansberg, B. Pire and L. Szymanowski, Phys. Rev. D **73** (2006) 074014.
11. B. C. Tiburzi, Phys. Rev. D **72** (2005) 094001.
12. W. Broniowski and E. R. Arriola, Phys. Lett. B **649** (2007) 49.
13. A. E. Dorokhov and L. Tomio, Phys. Rev. D **62** (2000) 014016 ; F. Bissey, *et al.*, Phys. Lett. B **547** (2002) 210; F. Bissey, *et al.*, Phys. Lett. B **587** (2004) 189.
14. A. Courtoy and S. Noguera, arXiv:0707.3366 [hep-ph].
15. D. Brommel *et al.* [QCDSF Collaboration], arXiv:0708.2249 [hep-lat].
16. M. Diehl *et al*, Phys. Rev. Lett. **81**, 1782 (1998) and Phys. Rev. D **62**, 073014 (2000).
17. G. P. Lepage and S. J. Brodsky, Phys. Rev. D **22**, 2157 (1980).
18. I. V. Anikin, B. Pire and O. V. Teryaev, Phys. Rev. D **69**, 014018 (2004) and Phys. Lett. B **626** (2005) 86.
19. D. Amrath, J.P. Lansberg and M. Diehl, in preparation.
20. J. P. Lansberg, B. Pire and L. Szymanowski, Phys. Rev. D **75** (2007) 074004.
21. V. L. Chernyak and A. R. Zhitnitsky, Phys. Rept. **112**, 173 (1984);
22. J. P. Lansberg, B. Pire and L. Szymanowski, Nucl. Phys. A **782** (2007) 16.
23. K.J. Park *et al.*, [CLAS collaboration], *Cross sections and beam asymmetries for* $\vec{e}p \to en\pi^+$ *in the nucleon resonance region for* $1.7 < Q^2 < 4.5\ GeV^2$
24. C. S. Armstrong *et al.* [Jefferson Lab E94014 Collaboration], Phys. Rev. D **60** (1999) 052004.
25. H. Denizli *et al.* [CLAS Collaboration], Phys. Rev. C **76** (2007) 015204.
26. A. Airapetian *et al.* [HERMES Collaboration], arXiv:0707.0222 [hep-ex].
27. R. A. Arndt, W. J. Briscoe, I. I. Strakovsky and R. L. Workman, AIP Conf. Proc. **904** (2007) 269; R. A. Arndt *et al.*, work in progress.
28. P. Spiller and G. Franchetti, Nucl. Instrum. Meth. A **561** (2006) 305; see also `http://www.gsi.de/fair`.
29. J. P. Lansberg, B. Pire and L. Szymanowski, in preparation.
30. B. Pasquini and S. Boffi, Phys. Rev. D **73** (2006) 094001.
31. B. Pasquini, private communication.
32. V. Bernard and U. G. Meissner, Ann. Rev. Nucl. Part. Sci. **57** (2007).
33. S. J. Brodsky and G. F. de Teramond, Phys. Lett. B **582** (2004) 211.

NOVEL QCD EFFECTS FROM INITIAL AND FINAL STATE INTERACTIONS

STANLEY J. BRODSKY

Stanford Linear Accelerator Center
Stanford University, Stanford, California 94309
E-mail: sjbth@slac.stanford.edu

Initial-state and final-state interactions which are conventionally neglected in the parton model, have a profound effect in QCD hard-scattering reactions. These effects, which arise from gluon exchange between the active and spectator quarks, cause leading-twist single-spin asymmetries, diffractive deep inelastic scattering, diffractive hard hadronic reactions, and the breakdown of the Lam-Tung relation in Drell-Yan reactions. Diffractive deep inelastic scattering also leads to nuclear shadowing and non-universal antishadowing of nuclear structure functions through multiple scattering reactions in the nuclear target. Factorization-breaking effects are particularly important for hard hadron interactions since both initial-state and final-state interactions appear. Related factorization breaking effects can also appear in exclusive electroproduction reactions and in deeply virtual Compton scattering. None of the effects of initial-state and final-state interactions are incorporated in the light-front wavefunctions of the target hadron computed in isolation.

1. Introduction

Deep inelastic lepton scattering provides a direct window to the fundamental quark and gluon structure of nucleons and the nucleus.

In the conventional description of of deep inelastic lepton scattering, the final-state interactions of the struck quark can be systematically neglected at leading order in $1/Q^2$. This intuitive picture, which is based on the quark-parton model, is reinforced by the argument that the Wilson line which describes the gauge interactions of the outgoing colored quark current can be set to unity simply by choosing the light cone gauge $A^+ = A^0 + A^3 = 0$. In this intuitive picture, the leading-twist structure functions of the target hadron or nucleus can be computed as a probability distribution in x defined from the square of its light-front wavefunctions. The Bjorken variable x_{bj} can be identified at leading twist with the light-cone fraction $x^+ = k^+/P^+ = (k^0 + k^3)/(P^0 + P^3)$ of the struck quark.

Surprising, this simple picture is not actually correct in QCD. The effects of final-state gluonic interactions of the struck quark cannot be neglected in any gauge even at very high Q^2. There are a number of areas of phenomenology where the effects of final-state interactions become manifest:

(1) in the case of diffractive deep inelastic scattering (DDIS), experiments at HERA,[1] have shown the proton remains intact and separated by a large rapidity in nearly 15% of the DIS events. This remarkable phenomena can be understood as due to the exchange of gluons in the final state which neutralize the color separation,[2] an effect which persists at high s and Q^2 in any gauge. This is illustrated in fig. 1. The net effect of the FSI is in fact not even unitary, so that standard interpretation of structure functions as probability distributions is not accurate. The Wilson line can also be interpreted as the final-state phase induced by a noncausal gauge induced in the DIS reaction. Thus the structure functions measured in DIS cannot be computed solely from the wavefunctions of a hadron in isolation. This picture also contradicts models based on an intrinsic pomeron component of the proton.

(2) A new understanding of nuclear shadowing and antishadowing has emerged based on the presence of multi-step coherent reactions involving leading twist diffractive reactions[3,4] as illustrated in fig. 2. Thus the nuclear shadowing of structure functions is a consequence of the lepton-nucleus collision; it is not an intrinsic property of the nuclear wavefunction. The same analysis shows that antishadowing is *not universal*, but it depends in detail on the flavor of the quark or antiquark constituent.[4]

(3) The final-state interactions illustrated in fig. 3 which cause DDIS also produce the pseudo-T-odd $i\, \vec{S}_p \cdot \vec{p}_H \times \vec{q}$ Sivers correlation between the production plane of the hadron or jet produced in semi-inclusive deep inelastic scattering (DIS) and the target nucleon spin. These interactions produce the Sivers effect at leading twist[5] with different signs in semi-inclusive deep inelastic scattering and the Drell-Yan reaction.[6] Double initial-state interactions[7] also produce anomalous angular effects, including the breakdown of the Lam-Tung relation[8] in the Drell-Yan process. As recently noted by Collins and Qiu,[9] the traditional factorization formalism of perturbative QCD for high transverse momentum hadron production in hadron collisions also fails in detail because of initial- and final-state gluonic interactions.

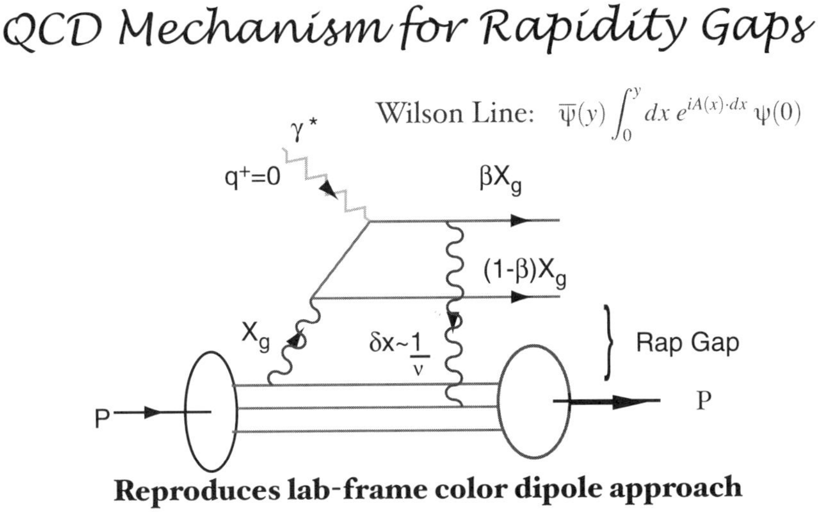

Fig. 1. Final-state interactions in QCD lead to diffractive deep inelastic scattering $\ell p \to \ell' p' + X$ at leading twist.

These examples of unconventional wisdom highlight the need for a fundamental understanding the dynamics of hadrons in QCD at the amplitude level. This is essential for understanding phenomena such as the quantum mechanics of hadron formation and the origins of diffractive phenomena, as well and single-spin asymmetries.

2. Diffractive Deep Inelastic Scattering

A remarkable feature of deep inelastic lepton-proton scattering at HERA is that up to 15% of the events are diffractive:[10,11] the target proton remains intact, and there is a large rapidity gap between the proton and the other hadrons in the final state. These diffractive deep inelastic scattering (DDIS) events can be understood most simply from the perspective of the color-dipole model: the $q\bar{q}$ Fock state of the high-energy virtual photon diffractively dissociates into a diffractive dijet system. The exchange of multiple gluons between the color dipole of the $q\bar{q}$ and the quarks of the target proton neutralizes the color separation and leads to the diffractive final state. The same multiple gluon exchange also controls diffractive vector meson electroproduction at large photon virtuality.[12] This observation

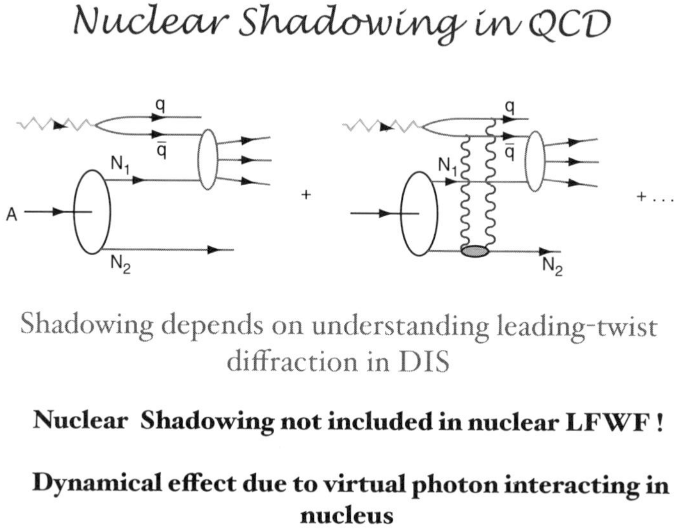

Fig. 2. Relation of nuclear shadowing of structure functions to leading-twist diffractive deep inelastic scattering.

presents a paradox: if one chooses the conventional parton model frame where the photon light-front momentum is negative $q+ = q^0 + q^z < 0$, the virtual photon interacts with a quark constituent with light-cone momentum fraction $x = k^+/p^+ = x_{bj}$. Furthermore, the gauge link associated with the struck quark (the Wilson line) becomes unity in light-cone gauge $A^+ = 0$. Thus the struck "current" quark apparently experiences no final-state interactions. Since the light-front wavefunctions $\psi_n(x_i, k_{\perp i})$ of a stable hadron are real, it appears impossible to generate the required imaginary phase associated with pomeron exchange, let alone large rapidity gaps.

This paradox was resolved by Hoyer, Marchal, Peigne, Sannino and myself.[2] Consider the case where the virtual photon interacts with a strange quark—the $s\bar{s}$ pair is assumed to be produced in the target by gluon splitting. In the case of Feynman gauge, the struck s quark continues to interact in the final state via gluon exchange as described by the Wilson line. The final-state interactions occur at a light-cone time $\Delta\tau \simeq 1/\nu$ shortly after the virtual photon interacts with the struck quark. When one integrates over the nearly-on-shell intermediate state, the amplitude acquires an imaginary part. Thus the rescattering of the quark produces

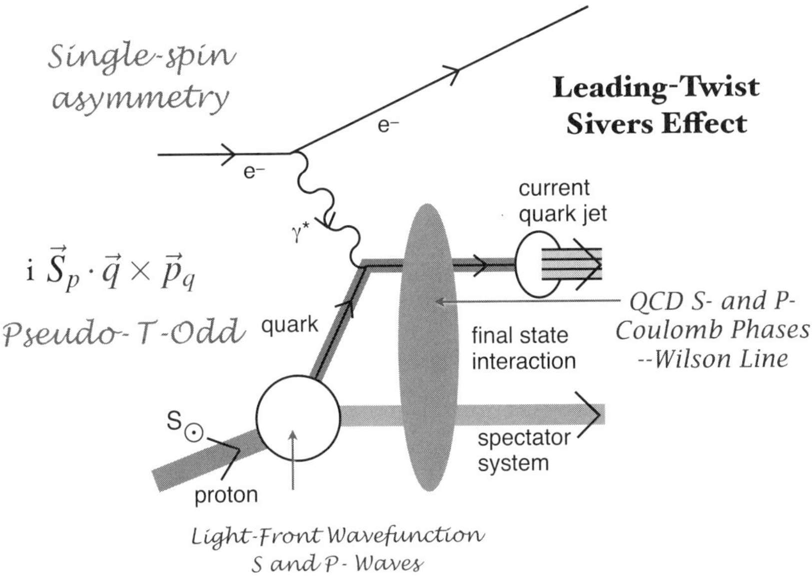

Fig. 3. Final-state interactions in QCD and the physics of the leading-twist Sivers single-spin asymmetry in semi-inclusive deep inelastic lepton-proton scattering.

a separated color-singlet $s\bar{s}$ and an imaginary phase. In the case of the light-cone gauge $A^+ = \eta \cdot A = 0$, one must also consider the final-state interactions of the (unstruck) $\bar{s}$ quark. The gluon propagator in light-cone gauge $d^{\mu\nu}_{LC}(k) = (i/k^2 + i\epsilon)\left[-g^{\mu\nu} + (\eta^\mu k^\nu + k^\mu \eta^\nu/\eta \cdot k)\right]$ is singular at $k^+ = \eta \cdot k = 0$. The momentum of the exchanged gluon k^+ is of $\mathcal{O}(1/\nu)$; thus rescattering contributes at leading twist even in light-cone gauge. This is illustrated in fig. 4. The net result is gauge invariant and is identical to the color dipole model calculation. The calculation of the rescattering effects on DIS in Feynman and light-cone gauge through three loops is given in detail for an Abelian model in reference [2]. The result shows that the rescattering corrections reduce the magnitude of the DIS cross section in analogy to nuclear shadowing.

A new understanding of the role of final-state interactions in deep inelastic scattering has thus emerged. The multiple scattering of the struck parton via instantaneous interactions in the target generates dominantly imaginary diffractive amplitudes, giving rise to an effective "hard pomeron" exchange.

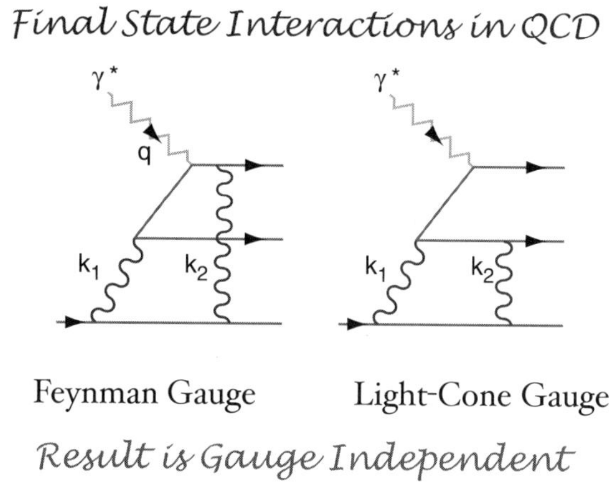

Fig. 4. Final-state interactions in QCD are nonzero even in light-cone gauge.

The presence of a rapidity gap between the target and diffractive system requires that the target remnant emerges in a color-singlet state; this is made possible in any gauge by the soft rescattering. The resulting diffractive contributions leave the target intact and do not resolve its quark structure; thus there are contributions to the DIS structure functions which cannot be interpreted as parton probabilities;[2] the leading-twist contribution to DIS from rescattering of a quark in the target is a coherent effect which is not included in the light-front wave functions computed in isolation. One can augment the light-front wave functions with a gauge link corresponding to an external field created by the virtual photon $q\bar{q}$ pair current.[13,14] Such a gauge link is process dependent,[6] so the resulting augmented LFWFs are not universal.[2,13,15] We also note that the shadowing of nuclear structure functions is due to the destructive interference between multi-nucleon amplitudes involving diffractive DIS and on-shell intermediate states with a complex phase. In contrast, the wave function of a stable target is strictly real since it does not have on-energy-shell intermediate state configurations. The physics of rescattering and shadowing is thus not included in the nuclear light-front wave functions, and a probabilistic interpretation of the nuclear DIS cross section is precluded.

Rikard Enberg, Paul Hoyer, Gunnar Ingelman and I[16] have shown that the quark structure function of the effective hard pomeron has the same form as the quark contribution of the gluon structure function. The hard pomeron is not an intrinsic part of the proton; rather it must be considered as a dynamical effect of the lepton-proton interaction. Our QCD-based picture also applies to diffraction in hadron-initiated processes. The rescattering is different in virtual photon- and hadron-induced processes due to the different color environment, which accounts for the observed non-universality of diffractive parton distributions. This framework also provides a theoretical basis for the phenomenologically successful Soft Color Interaction (SCI) model[17] which includes rescattering effects and thus generates a variety of final states with rapidity gaps.

3. Shadowing and Antishadowing of Nuclear Structure Functions

One of the novel features of QCD involving nuclei is the *antishadowing* of the nuclear structure functions which is observed in deep inelastic lepton scattering and other hard processes. Empirically, one finds $R_A(x, Q^2) \equiv \big(F_{2A}(x, Q^2)/(A/2)F_d(x, Q^2)\big) > 1$ in the domain $0.1 < x < 0.2$; *i.e.*, the measured nuclear structure function (referenced to the deuteron) is larger than than the scattering on a set of A independent nucleons.

The shadowing of the nuclear structure functions: $R_A(x, Q^2) < 1$ at small $x < 0.1$ can be readily understood in terms of the Gribov-Glauber theory. Consider a two-step process in the nuclear target rest frame. The incoming $q\overline{q}$ dipole first interacts diffractively $\gamma^* + N_1 \to (q\overline{q})N_1$ on nucleon N_1 leaving it intact. This is the leading-twist diffractive deep inelastic scattering (DDIS) process which has been measured at HERA to constitute approximately 10% of the DIS cross section at high energies. The $q\overline{q}$ state then interacts inelastically on a downstream nucleon $N_2 : (q\overline{q})N_2 \to X$. The phase of the pomeron-dominated DDIS amplitude is close to imaginary, and the Glauber cut provides another phase i, so that the two-step process has opposite phase and destructively interferes with the one-step DIS process $\gamma^* + N_2 \to X$ where N_1 acts as an unscattered spectator. The one-step and-two step amplitudes can coherently interfere as long as the momentum transfer to the nucleon N_1 is sufficiently small that it remains in the nuclear target; *i.e.*, the Ioffe length[18] $L_I = 2M\nu/Q^2$ is large compared to the inter-nucleon separation. In effect, the flux reaching the interior nucleons is diminished, thus reducing the number of effective nucleons and $R_A(x, Q^2) < 1$.

There are also leading-twist diffractive contributions $\gamma^* N_1 \to (q\bar{q}) N_1$ arising from Reggeon exchanges in the t-channel.[3] For example, isospin–non-singlet $C = +$ Reggeons contribute to the difference of proton and neutron structure functions, giving the characteristic Kuti-Weisskopf $F_{2p} - F_{2n} \sim x^{1-\alpha_R(0)} \sim x^{0.5}$ behavior at small x. The x dependence of the structure functions reflects the Regge behavior $\nu^{\alpha_R(0)}$ of the virtual Compton amplitude at fixed Q^2 and $t = 0$. The phase of the diffractive amplitude is determined by analyticity and crossing to be proportional to $-1 + i$ for $\alpha_R = 0.5$, which together with the phase from the Glauber cut, leads to *constructive* interference of the diffractive and nondiffractive multi-step nuclear amplitudes. Furthermore, because of its x dependence, the nuclear structure function is enhanced precisely in the domain $0.1 < x < 0.2$ where antishadowing is empirically observed. The strength of the Reggeon amplitudes is fixed by the fits to the nucleon structure functions, so there is little model dependence.

As noted above, the Bjorken-scaling diffractive contribution to DIS arises from the rescattering of the struck quark after it is struck (in the parton model frame $q^+ \leq 0$), an effect induced by the Wilson line connecting the currents. Thus one cannot attribute DDIS to the physics of the target nucleon computed in isolation.[2] Similarly, since shadowing and antishadowing arise from the physics of diffraction, we cannot attribute these phenomena to the structure of the nucleus itself: shadowing and antishadowing arise because of the $\gamma^* A$ collision and the history of the $q\bar{q}$ dipole as it propagates through the nucleus.

Ivan Schmidt, Jian-Jun Yang, and I[4] have extended the Glauber analysis to the shadowing and antishadowing of all of the electroweak structure functions. Quarks of different flavors will couple to different Reggeons; this leads to the remarkable prediction that nuclear antishadowing is not universal; it depends on the quantum numbers of the struck quark. This picture implies substantially different antishadowing for charged and neutral current reactions, thus affecting the extraction of the weak-mixing angle θ_W. We find that part of the anomalous NuTeV result[19] for θ_W could be due to the non-universality of nuclear antishadowing for charged and neutral currents. Detailed measurements of the nuclear dependence of individual quark structure functions are thus needed to establish the distinctive phenomenology of shadowing and antishadowing and to make the NuTeV results definitive. Schmidt, Yang, and I have also identified contributions to the nuclear multi-step reactions which arise from odderon exchange and hidden color degrees of freedom in the nuclear wavefunction. There are other ways in which this new view of antishadowing can be tested; antishadowing can also depend on the target and beam polarization.

4. Single-Spin Asymmetries from Initial- and Final-State Interactions

Among the most interesting polarization effects are single-spin azimuthal asymmetries in semi-inclusive deep inelastic scattering, representing the correlation of the spin of the proton target and the virtual photon to hadron production plane: $\vec{S}_p \cdot \vec{q} \times \vec{p}_H$. Such asymmetries are time-reversal odd, but they can arise in QCD through phase differences in different spin amplitudes. In fact, final-state interactions from gluon exchange between the outgoing quarks and the target spectator system lead to single-spin asymmetries in semi-inclusive deep inelastic lepton-proton scattering which are not power-law suppressed at large photon virtuality Q^2 at fixed x_{bj}.[5] In contrast to the SSAs arising from transversity and the Collins fragmentation function, the fragmentation of the quark into hadrons is not necessary; one predicts a correlation with the production plane of the quark jet itself. Physically, the final-state interaction phase arises as the infrared-finite difference of QCD Coulomb phases for hadron wave functions with differing orbital angular momentum. This is illustrated in fig. 3. The same proton matrix element which determines the spin-orbit correlation $\vec{S} \cdot \vec{L}$ also produces the anomalous magnetic moment of the proton, the Pauli form factor, and the generalized parton distribution E which is measured in deeply virtual Compton scattering. Thus the contribution of each quark current to the SSA is proportional to the contribution $\kappa_{q/p}$ of that quark to the proton target's anomalous magnetic moment $\kappa_p = \sum_q e_q \kappa_{q/p}$.[5,20] The HERMES collaboration has recently measured the SSA in pion and kaon electroproduction using transverse target polarization.[21] The Sivers and Collins effects can be separated using planar correlations; both contributions are observed to contribute, with values not in disagreement with theory expectations .[21,22] The larger Sivers effect seem for K^+ production compared to $\pi+$ production at small x_{bj} suggests a role for polarized sea quarks.

The final-state interaction mechanism provides an appealing physical explanation within QCD of single-spin asymmetries. Remarkably, the same matrix element which determines the spin-orbit correlation $\vec{S} \cdot \vec{L}$ also produces the anomalous magnetic moment of the proton, the Pauli form factor, and the generalized parton distribution E which is measured in deeply virtual Compton scattering. Physically, the final-state interaction phase arises

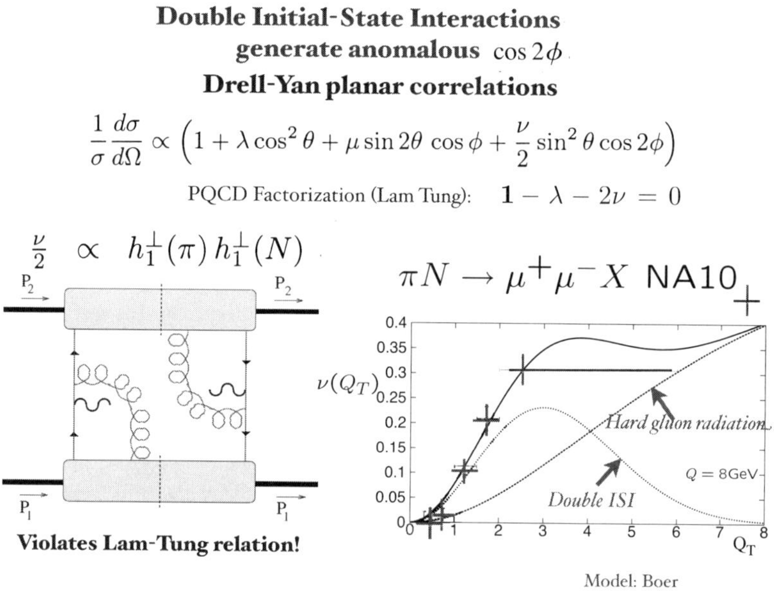

Fig. 5. Double initial-state interactions in QCD violate the Lam-Tung relation for Drell-Yan massive lepton pair production.

as the infrared-finite difference of QCD Coulomb phases for hadron wave functions with differing orbital angular momentum. An elegant discussion of the Sivers effect including its sign has been given by Burkardt.[20] As shown recently by Gardner and myself,[25] one can also use the Sivers effect to study the orbital angular momentum of gluons by tagging a gluon jet in semi-inclusive DIS. In this case, the final-state interactions are enhanced by the large color charge of the gluons.

The final-state interaction effects can also be identified with the gauge link which is present in the gauge-invariant definition of parton distributions.[14] Even in light-cone gauge, a transverse gauge link is required which reflects the external conditions of electroproduction. Thus the parton amplitudes need to be augmented by an additional eikonal factor incorporating the final-state interaction and its phase.[13,26] This procedure allows one to formally define transverse momentum dependent parton distribution functions which contain the effect of the QCD final-state interactions. However, the physics of final state interactions is not contained in the wavefunction of a hadron in isolation.

5. The Exclusive Sivers Effect

It would also be interesting to study the Sivers effect in exclusive electroproduction reactions. For example, there should be a $i\vec{S}_p \cdot \vec{q} \times \vec{p}_\pi$ correlation in pion electroproduction $\gamma^* p_\updownarrow \to \pi^+ n$. This could be an ideal experiment for the 12 GeV program at JLab.

A central uncertainty in the analysis of B decays is the unknown nature and magnitude of the strong phase. It would thus be interesting to make a connection between the final-state hadronic phases which cause the Sivers effect in exclusive electroproduction and the strong interaction phases which appear in exclusive B decays. The final-state QCD phase in such hard processes would be expected to be diminished because of color transparency as the momentum transfer squared t to the meson increases.

6. The Sivers Effect in General Inclusive Reactions

A related analysis also predicts that the initial-state interactions from gluon exchange between the incoming quark and the target spectator system lead to leading-twist single-spin asymmetries in the Drell-Yan process $H_1 H_2^\updownarrow \to \ell^+ \ell^- X$.[6,23] The SSA in the Drell-Yan process is the same as that obtained in SIDIS, with the appropriate identification of variables, but with the opposite sign. There is no Sivers effect in charged-current reactions since the W only couples to left-handed quarks.[24]

If both the quark and antiquark in the initial state of the Drell-Yan subprocess $q\bar{q} \to \mu^+ \mu^-$ interact with the spectators of the other incident hadron, one finds a breakdown of the Lam-Tung relation, which was formerly believed to be a general prediction of leading-twist QCD. These double initial-state interactions also lead to a $\cos 2\phi$ planar correlation in unpolarized Drell-Yan reactions.[7] More generally one must consider subprocesses involving initial-state gluons such as $ngq\bar{q} \to \ell\bar{\ell}$ as well as subprocesses with extra final-state gluons. This is illustrated in fig. 5.

The situation becomes more complicated in the case of hard hadron interactions where both initial and final state interactions are present. An example involving heavy quark production is shown in fig. 6. As noted by Collins and Qiu[9] the combination of such effects endanger the standard arguments for factorization in general hadroproduction processes. In addition, the final-state interactions which produce diffractive deep inelastic scattering and the Sivers effect in leptoproduction at leading twist will also affect the intermediate quark line in the virtual Compton amplitude, thus correcting the handbag approximation to DVCS.

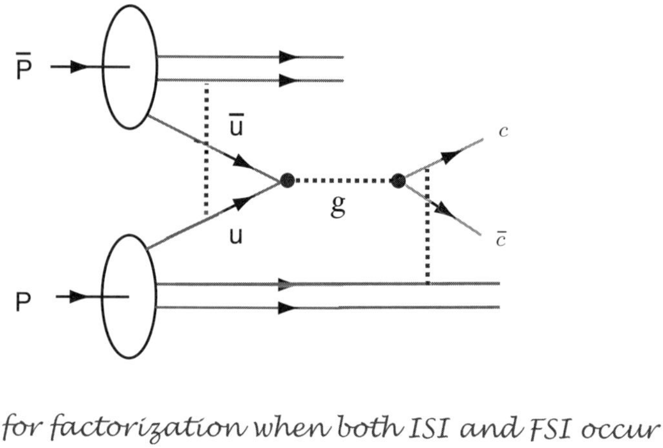

Fig. 6. Initial-state and final-state interactions in QCD both contribute to massive heavy quark production.

7. Summary

Initial- and final-state interactions from gluon-exchange, which are neglected in the parton model, have a profound effect in QCD hard-scattering reactions. These effects cause leading-twist single-spin asymmetries, diffractive deep inelastic scattering, diffractive hard hadronic reactions, and the breakdown of the Lam Tung relation in Drell-Yan reactions. Diffractive deep inelastic scattering leads to nuclear shadowing and non-universal antishadowing. Related effects can appear in exclusive electroproduction reactions and in deeply virtual Compton scattering. None of the effects of initial or final state interactions are incorporated in the light-front wavefunctions of the target hadron computed in isolation.

Acknowledgments

The results in this talk are based on collaborations with Daniel Boer, Rikard Enberg, Susan Gardner, Paul Hoyer, Dae Sung Hwang, Gunnar Ingelman, Hung Jung Lu, Nils Marchal, Stephane Peigne, Francesco Sannino, and Ivan Schmidt. This work was supported in part by the Department of Energy, contract No. DE-AC02-76SF00515.

References

1. M. Derrick *et al.* [ZEUS Collaboration], Phys. Lett. B **315**, 481 (1993).
2. S. J. Brodsky, P. Hoyer, N. Marchal, S. Peigne and F. Sannino, Phys. Rev. D **65**, 114025 (2002) [arXiv:hep-ph/0104291].
3. S. J. Brodsky and H. J. Lu, Phys. Rev. Lett. **64**, 1342 (1990).
4. S. J. Brodsky, I. Schmidt and J. J. Yang, Phys. Rev. D **70**, 116003 (2004) [arXiv:hep-ph/0409279].
5. S. J. Brodsky, D. S. Hwang and I. Schmidt, Int. J. Mod. Phys. A **18**, 1327 (2003) [Phys. Lett. B **530**, 99 (2002)] [arXiv:hep-ph/0201296].
6. J. C. Collins, Phys. Lett. B **536**, 43 (2002) [arXiv:hep-ph/0204004].
7. D. Boer, S. J. Brodsky and D. S. Hwang, Phys. Rev. D **67**, 054003 (2003) [arXiv:hep-ph/0211110].
8. C. S. Lam and W. K. Tung, Phys. Rev. D **21**, 2712 (1980).
9. J. Collins and J. W. Qiu, Phys. Rev. D **75**, 114014 (2007) [arXiv:0705.2141 [hep-ph]].
10. C. Adloff *et al.* [H1 Collaboration], Z. Phys. C **76**, 613 (1997) [arXiv:hep-ex/9708016].
11. J. Breitweg *et al.* [ZEUS Collaboration], Eur. Phys. J. C **6**, 43 (1999) [arXiv:hep-ex/9807010].
12. S. J. Brodsky, L. Frankfurt, J. F. Gunion, A. H. Mueller and M. Strikman, Phys. Rev. D **50**, 3134 (1994) [arXiv:hep-ph/9402283].
13. A. V. Belitsky, X. Ji and F. Yuan, Nucl. Phys. B **656**, 165 (2003) [arXiv:hep-ph/0208038].
14. J. C. Collins and A. Metz, Phys. Rev. Lett. **93**, 252001 (2004) [arXiv:hep-ph/0408249].
15. J. C. Collins, Acta Phys. Polon. B **34**, 3103 (2003) [arXiv:hep-ph/0304122].
16. S. J. Brodsky, R. Enberg, P. Hoyer and G. Ingelman, Phys. Rev. D **71**, 074020 (2005) [arXiv:hep-ph/0409119].
17. A. Edin, G. Ingelman and J. Rathsman, Phys. Lett. B **366**, 371 (1996) [arXiv:hep-ph/9508386].
18. B. L. Ioffe, Phys. Lett. B **30**, 123 (1969).
19. G. P. Zeller *et al.* [NuTeV Collaboration], Phys. Rev. Lett. **88**, 091802 (2002) [Erratum-ibid. **90**, 239902 (2003)] [arXiv:hep-ex/0110059].
20. M. Burkardt, Nucl. Phys. Proc. Suppl. **141**, 86 (2005) [arXiv:hep-ph/0408009].
21. A. Airapetian *et al.* [HERMES Collaboration], Phys. Rev. Lett. **94**, 012002 (2005) [arXiv:hep-ex/0408013].
22. H. Avakian and L. Elouadrhiri [CLAS Collaboration], AIP Conf. Proc. **698**, 612 (2004).
23. S. J. Brodsky, D. S. Hwang and I. Schmidt, Nucl. Phys. B **642**, 344 (2002) [arXiv:hep-ph/0206259].
24. S. J. Brodsky, D. S. Hwang and I. Schmidt, Phys. Lett. B **553**, 223 (2003) [arXiv:hep-ph/0211212].
25. S. J. Brodsky and S. Gardner, Phys. Lett. B **643**, 22 (2006) [arXiv:hep-ph/0608219].
26. X. d. Ji and F. Yuan, Phys. Lett. B **543**, 66 (2002) [arXiv:hep-ph/0206057].

PARTON DISTRIBUTIONS AND SPIN-ORBITAL CORRELATIONS

FENG YUAN

RIKEN/BNL Research Center, Brookhaven National Laboratory, Upton, NY 11973
Nuclear Science Division, Lawrence Berkeley National Laboratory, Berkeley, CA 94720

In this talk, I summarize a recent study showing that the large-x parton distributions contain important information on the quark orbital angular momentum of nucleon. This contribution could explain the conflict between the experimental data and the theory predictions for the polarized quark distributions. Future experiments at JLab shall provide further test for our predictions.

Keywords: Quark Orbital Angular Momentum; Large-x Parton Distributions

1. Introduction

Power-counting rules for the large-x parton distributions were derived many years ago based on perturbative quantum chromodynamics (pQCD) combined with a S-wave quark model of hadrons.[1–4] The basic argument is that when the valence quark carries nearly all of the longitudinal momentum of the hadron, the relevant QCD configurations in the hadronic wave function become far off-shell and can be treated in pQCD. The power-counting rule has also been generalized to sea quarks, gluons, helicity-dependent distributions,[5,6] and generalized parton distributions.[7]

The leading pQCD diagrams associated with the leading Fock state of the proton wave function predict that the positive helicity (quark spin aligned with the proton spin) quark distribution $q^+(x)$ scales as $(1-x)^3$, whereas the negative helicity (quark spin anti-aligned with the proton spin) quark distribution $q^-(x)$ is suppressed by $(1-x)^2$ relative to the positive helicity one, scaling as $(1-x)^5$ at large x.[3] The direct consequence of these power laws for the quark distributions is that the ratio of polarized quark distribution $\Delta q(x) = q^+(x) - q^-(x)$ over the unpolarized quark distribution $q(x) = q^+(x) + q^-(x)$ approaches 1 in the limit $x \to 1$; i.e., at large x, q^+ dominates over q^-. When this prediction is compared to the experimental data,[8–11] it is interesting to observe that, for the up quark the ratio increases

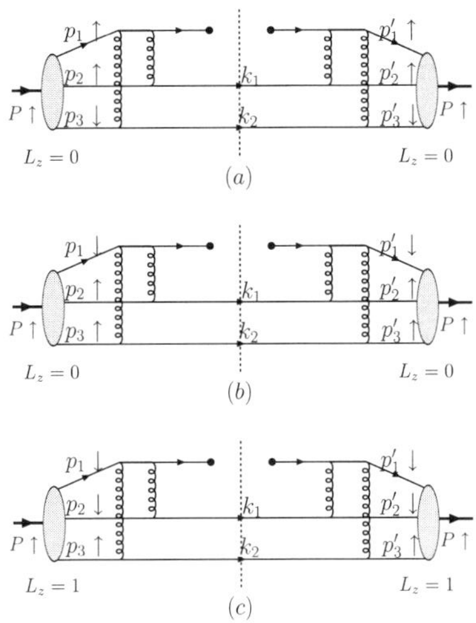

Fig. 1. *Examples of Feynman diagrams which contribute to the $q^{\pm}$ quark distributions at large x: (a) for q^+ with power contribution of $(1-x)^3$; (b) for q^- with $(1-x)^5$; (c) for q^- with $(1-x)^5\log^2(1-x)$. The wave functions at the left and right sides of the cut line in (a) and (b) represent the leading Fock state expansion with zero quark orbital angular momentum, whereas those for (c) represent the valence Fock state with one-unit of quark orbital angular momentum.*

with x, and seems to approach 1 at large x. However, the ratio for the down quark is still far below 1, and remains negative for a wide range of $x \leq 0.6$.[8]

In our recent study,[12] we have reexamined the large-x quark helicity distributions in the perturbative QCD framework,[3,4] and found that for the negative helicity distribution q^-, there exist large logarithmic enhancements from the $|L_z| = 1$ Fock states. With this large logarithmic modification, we can explain the discrepancy between the power-counting rule and experimental data.

2. Large-x q^- distribution

In the previous analysis, one only considered the contributions from the leading Fock state of the proton with zero quark orbital angular momentum. As we show the typical Feynman diagrams in Fig. 1(a) and (b) from this contribution, the positive helicity distribution q^+ scales as $(1-x)^3$, whereas the negative helicity distribution q^- scales as $(1-x)^5$.[3–6] In general, the contributions from the higher Fock states and the valence Fock states with nonzero quark orbital angular momentum will introduce additional suppression in (1-x).[4,6] However, the nonzero-quark-orbital-angular-momentum Fock state can provide large logarithmic enhancement to the helicity flip amplitudes. In Fig. 1(c), we show an example of a contribution from the $L_z = 1$ Fock state of proton. Because the quark orbital angular momentum contributes one unit of the proton spin, we will have difference between the total quark spin and the proton spin. If the two spectator quarks are in the spin-0 configuration, this will enhance the power-counting in the hard factor. On the other hand, in order to get a nonzero contribution, we have to perform the intrinsic transverse momentum expansion for the hard partonic scattering amplitudes,[13] which will introduce an additional suppression factor in $(1-x)$.[6] One intrinsic transverse momentum expansion comes from the propagator of momentum $(p_3 - k_2)$ will be,

$$\begin{aligned}\frac{1}{(p_3-k_2)^2} &= \frac{1}{(y_3P - k_2 + p_{3\perp})^2} \\ &\approx \frac{\beta(1-x)}{y_3k_{2\perp}^2}\left(1 - \frac{\beta(1-x)}{y_3k_{2\perp}^2}2p_{3\perp}\cdot k_{2\perp}\right) ,\end{aligned} \tag{1}$$

where β is the longitudinal momentum fraction of the spectator carried by k_2, and we have kept the linear dependence on $p_{3\perp}$ in the above expansion. Only this linear term will contribute when integrating over $p_{i\perp}$: $\int k_{2\perp}\cdot p_{3\perp}(p_1^x+ip_1^y)\tilde{\psi}^{(3)} \propto (k_2^x+ik_2^y)y_3\Phi_4(y_1,y_2,y_3)$, where Φ_4 is one of the twist-4 quark distribution amplitudes of the proton.[13,14] From the above expansion, we find that this term will introduce additional factor of $(1-x)/y_3$ in the hard factor. Similarly, we have to do the expansion in intrinsic transverse momentum associated with the wave function at the right side of the cut line, and again the expansion of the gluon propagator with momentum of $p_3' - k_2$ will introduce another suppression factor of $(1-x)/y_3'$ in the hard factor. Thus the total suppression factor from the above two expansions will be $(1-x)^2/y_3y_3'$, which gives the same power counting contribution to q^- as that from the leading Fock state with $L_z = 0$ in the above.

We thus find the contributions from $L_z = 1$ Fock state of the proton do not change the power counting for the q^- quark distribution at large x. However, the additional factor $1/y_3y_3'$ from the intrinsic transverse momentum expansions will lead to a large logarithm when integrating over y_i and y_i'. This is because, combining the above two factors with all other factors from the propagators shown in Fig. 1(c), the total dependence on y_i and y_i' for the hard factor will be

$$\sim \frac{1}{y_2y_3^2(1-y_2)y_2'y_3'^2(1-y_2')} , \tag{2}$$

where we have y_3^2 and $y_3'^2$ in the denominator. On the other hand, we expect the twist-4 quark distribution amplitude to have the following behavior at the end point region: $y_3\Phi_4(y_1, y_2, y_3) \propto y_1y_2y_3$ and $y_3'\Phi_4(y_1', y_2', y_3') \propto y_1'y_2'y_3'$.[14] Thus we will have logarithmic divergences for the integrations over y_3 and y_3', for which we can regularize in terms of $\log(1-x)$ as indicated in the above propagator expansion. This will lead to a double logarithmic contribution $\log^2(1-x)$ in addition to the power term $(1-x)^5$ to the q^- quark distribution at large x.

In summary, for the negative helicity distribution q^-, the leading Fock state with zero quark orbital angular momentum $L_z = 0$ contributes to a power term $(1-x)^5$, whereas the valence Fock state with $|L_z| = 1$ contributes to a double logarithmical enhanced term $(1-x)^5\log^2(1-x)$. So, in the limit $x \to 1$, the q^- distribution will be dominated by the contributions from $L_z = 1$ Fock state of the proton, scaling as $(1-x)^5\log^2(1-x)$. In the intermediate x range, the sub-leading terms can also be important. For example in Ref. [5], the quark helicity distributions were parameterized by the leading and sub-leading power terms and fit to the experimental data. This was later updated to account for the latest data in Ref. [15]. Thus, as a first step towards a comprehensive phenomenology, we follow the parameterizations for q^+ and q^- in Ref. [5] by adding the newly discovered double logarithms enhanced contributions,

$$
\begin{aligned}
u^+(x) &= \frac{1}{x^\alpha}\left[A_u(1-x)^3 + B_u(1-x)^4\right] \\
d^+(x) &= \frac{1}{x^\alpha}\left[A_d(1-x)^3 + B_d(1-x)^4\right] \\
u^-(x) &= \frac{1}{x^\alpha}\left[C_u(1-x)^5 + C_u'(1-x)^5\log^2(1-x)\right. \\
&\qquad \left. + D_u(1-x)^6\right] \\
d^-(x) &= \frac{1}{x^\alpha}\left[C_d(1-x)^5 + C_d'(1-x)^5\log^2(1-x)\right. \\
&\qquad \left. + D_d(1-x)^6\right] \ ,
\end{aligned}
\tag{3}
$$

where the additional two parameters C_u' and C_d' come from the logarithmic modifications to the q^- quark distribution at large x, and all other parameters refer to [5]. In the following, we will fit to the current experimental data at large x region with the above parameterizations for the valence up and down quarks.

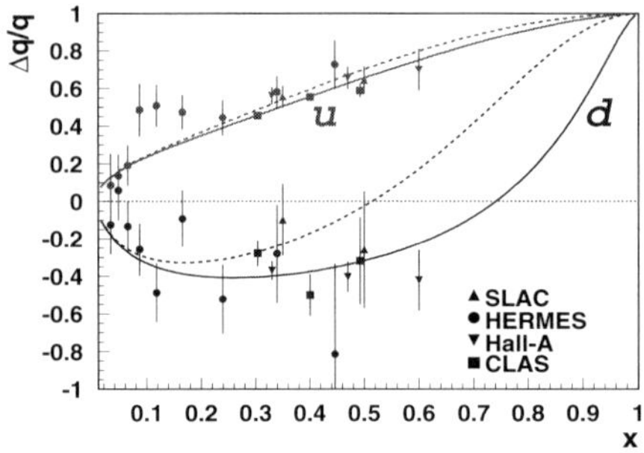

Fig. 2. *Comparison of the quark helicity distributions Eq. (3) with the experimental data, plotted as functions of x for up (the upper curves) and down (the lower curves) quarks. The circles are for HERMES data,[10] the triangles up for SLAC,[11] the triangles down for JLab Hall-A data,[8] the filled squares for CLAS.[9] The dashed curves are the predictions from [15], and the solid ones are our fit results (only the large-x (> 0.3) experimental data were used in the fit).*

3. Phenomenological applications

In order to demonstrate the importance of the new scaling behavior for the negative helicity distributions for the valence up and down quarks, we analyze the latest experimental data from SLAC, HERMES and Jefferson Lab, including Hall A and Hall B data.[8–11] We will keep the original fit values for other parameters[15] except the two new parameters: C'_u and C'_d. We only use the experimental data in the large-x region, i.e., $x > 0.3$, where the sea contribution is not significant. From our fit, we find the following values for C'_u and C'_d,[12]

$$C'_u = 0.493 \pm 0.249, \qquad C'_d = 1.592 \pm 0.378 \,, \tag{4}$$

The minimum of the functional χ^2 is achieved at $\chi^2 = 11.4$ and $\chi^2/DOF = 1.14$. We further notice that the additional two terms in Eq. (3) do not change significantly the sum rules for the up and down quarks, such as the Bjorken and momentum sum rule, which are essential for constraining the parameters in Refs. [5,15].

In Fig. 2, we show the above fit. We plot the ratios of the polarized quark distributions Δq over the unpolarized quark distributions q as functions of x for both up and down quarks, compared with the experimental data. From these comparisons, we find that the ratio for the up quark $\Delta u/u$ can still be described by the parameterization based on the original power counting rule for u^+ and u^-.[15] However, for the down quark we have to take into account a large contribution from the newly discovered term for the negative helicity distribution d^-; the difference between our result and the original parameterization[15] becomes significant at large x. Another important prediction of our fit is that the ratio of $\Delta d/d$ will approach 1 at extremely large x, and it will cross zero at $x \approx 0.75$. It will be interesting to check this prediction in future experiments, such as the 12 GeV upgrade of Jefferson Lab.

We thank Harut Avakian, Stan Brodsky, and Alexandre Deur for the collaborations and many important conversations. This work was supported in part by the Department of Energy, contract No. DE-AC02-76SF00515, and by the Jefferson Science Associates (JSA) operates the Thomas Jefferson National Accelerator Facility for the U. S. DOE under contract DE-AC05-060R23177. F.Y. is grateful to RIKEN, Brookhaven National Laboratory and the U.S. DOE (contract number DE-AC02-98CH10886) for providing the facilities essential for the completion of this contribution.

References

1. J. F. Gunion, Phys. Rev. D **10**, 242 (1974).
2. R. Blankenbecler, S. J. Brodsky, Phys. Rev. D **10**, 2973 (1974).
3. G. R. Farrar, D. R. Jackson, Phys. Rev. Lett. **35**, 1416 (1975).
4. G. P. Lepage, S. J. Brodsky, Phys. Rev. D **22**, 2157 (1980).
5. S. J. Brodsky, M. Burkardt and I. Schmidt, Nucl. Phys. B **441**, 197 (1995).
6. S. J. Brodsky, F. Yuan, Phys. Rev. D **74**, 094018 (2006).
7. F. Yuan, Phys. Rev. D **69**, 051501 (2004).
8. X. Zheng *et al.*, Phys. Rev. Lett. **92**, 012004 (2004); Phys. Rev. C **70**, 065207 (2004).
9. K.V. Dharmawardane *et al.*, Phys. Lett. **B641**,11 (2006).
10. A. Airapetian *et al.*, Phys. Rev. D **71**, 012003 (2005).
11. K. Abe *et al.*, Phys. Lett. B **405**, 180 (1997).
12. H. Avakian, S. J. Brodsky, A. Deur and F. Yuan, Phys. Rev. Lett. **99**, 082001 (2007).
13. A. V. Belitsky, X. Ji and F. Yuan, Phys. Rev. Lett. **91**, 092003 (2003).
14. V. Braun, R. J. Fries, N. Mahnke and E. Stein, Nucl. Phys. B **589**, 381 (2000).
15. E. Leader, A. V. Sidorov and D. B. Stamenov, Int. J. Mod. Phys. A **13**, 5573 (1998).

TRANSVERSE MOMENTUM DEPENDENCES OF DISTRIBUTION AND FRAGMENTATION FUNCTIONS

DAE SUNG HWANG[a] and DONG SOO KIM[b]

[a] *Department of Physics, Sejong University*
Seoul 143-747, South Korea
E-mail: dshwang@sejong.ac.kr

[b] *Department of Physics, Kangnung National University*
Kangnung 210-702, South Korea
E-mail: dongskim@kangnung.ac.kr

We study the transverse momentum dependent distribution functions of the up and down quarks inside the proton using the spectator model by Jakob et al. We calculate the double spin asymmetry for the setups of COMPASS, HERMES, and JLab. We investigate the differences between the spectator model and the model based on factorization ansatz.

1. Introduction

In recent years the role of the transverse momentum of the parton has been more important in the field of the hadron physics since, for example, it provides time-odd distribution and fragmentation functions, and makes the single-spin asymmetries in hadronic processes possible.[1,2] The transverse momentum of the parton inside the proton is also related to the orbital angular momentum carried by the parton, which is an important subject since it is considered as a part of the spin contents of the proton.

We investigate the flavor dependence of the transverse momentum distribution functions of the up and down quarks inside the proton. As an explicit model, we use the model of Jakob et al.[3] which is based on the scalar and axial-vector diquark models of the nucleon. We will call this model the spectator model. The important character of the spectator model is that the longitudinal momentum fraction x and the transverse momentum $k_\perp$ of the parton are intimately correlated with each other, since the spectator model is based on Lorentz invariant Feynman diagram. The transverse momentum distributions of the up and down quarks inside the proton are different, since for the proton the up quark is composed of a linear combination of the scalar and axial-vector diquark components and the down quark is only composed of the axial-vector diquark component.

In a lot of researches on the transverse momentum dependent distribution (fragmentation) functions, the ansatz which factorizes x (z) and $k_\perp$ ($p_\perp$) is adopted. For example, Ref. [4] investigated the double spin asymmetry A_{LL} by using such factorized distribution and fragmentation functions. We study the differences of the distribution and fragmentation functions of the factorized model and those of the spectator model. We also investigate the consequence of these differences in the double spin asymmetry A_{LL} in SIDIS, which is a measurable physical quantity.

2. Distribution and Fragmentation Functions

Let us consider the spectator model by Jakob, Mulders and Rodrigues,[3] which was constructed starting from Lorentz invariant amplitude:

$$f_{1R}(x,\mathbf{k}_\perp)=N_R\,\frac{(xM+m)^2+\mathbf{k}_\perp^2}{(\mathbf{k}_\perp^2+\lambda_R^2)^{2\alpha}}\,,\qquad g_{1R}(x,\mathbf{k}_\perp)=N_R\;a_R\,\frac{(xM+m)^2-\mathbf{k}_\perp^2}{(\mathbf{k}_\perp^2+\lambda_R^2)^{2\alpha}}\,, \tag{1}$$

where $\lambda_R^2(x) = (1-x)\Lambda^2 + xM_R^2 - x(1-x)M^2$ and $a_s = 1$, $a_a = -\frac{1}{3}$ for a_R, with $\Lambda = 0.5$ GeV, $M = 0.94$ GeV, $M_s = 0.6$ GeV, $M_a = 0.8$ GeV for M_R. We take $\alpha = 2$ in this paper. The normalization constant N_R is fixed by the normalization condition of $f_{1R}(x,\mathbf{k}_\perp)$.

From the $SU(4)$ wave function of the proton, we have (also for g_1^q)[3]

$$f_1^u = \frac{3}{2}f_{1s} + \frac{1}{2}f_{1a}\,,\quad f_1^d = f_{1a}\,. \tag{2}$$

Then, Eqs. (1) and (2) give

$$\begin{aligned}
f_1^u(x,\mathbf{k}_\perp) &= \frac{3}{2}N_s\frac{(xM+m)^2+\mathbf{k}_\perp^2}{(\mathbf{k}_\perp^2+\lambda_s^2)^{2\alpha}} + \frac{1}{2}N_a\frac{(xM+m)^2+\mathbf{k}_\perp^2}{(\mathbf{k}_\perp^2+\lambda_a^2)^{2\alpha}}, \\
g_1^u(x,\mathbf{k}_\perp) &= \frac{3}{2}N_s\frac{(xM+m)^2-\mathbf{k}_\perp^2}{(\mathbf{k}_\perp^2+\lambda_s^2)^{2\alpha}} - \frac{1}{6}N_a\frac{(xM+m)^2-\mathbf{k}_\perp^2}{(\mathbf{k}_\perp^2+\lambda_a^2)^{2\alpha}}, \\
f_1^d(x,\mathbf{k}_\perp) &= N_a\frac{(xM+m)^2+\mathbf{k}_\perp^2}{(\mathbf{k}_\perp^2+\lambda_a^2)^{2\alpha}}, \\
g_1^d(x,\mathbf{k}_\perp) &= -\frac{1}{3}N_a\frac{(xM+m)^2-\mathbf{k}_\perp^2}{(\mathbf{k}_\perp^2+\lambda_a^2)^{2\alpha}}.
\end{aligned} \tag{3}$$

The distribution functions given in (3) are plotted in Figs. 1 and 2. Fig. 3 presents the widths of the distribution functions in $\mathbf{k}_\perp$ as functions of x.

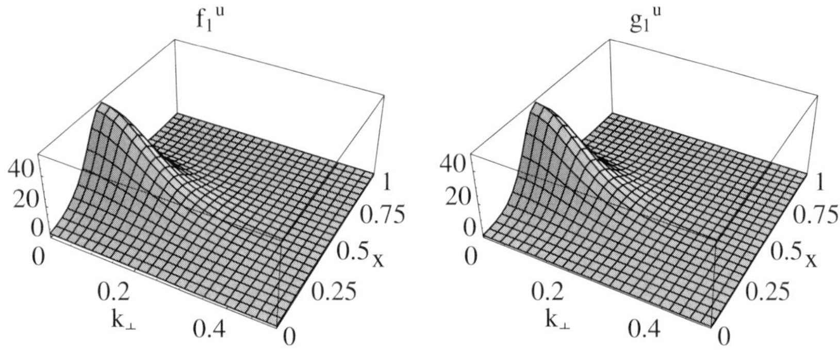

Fig. 1. The distribution functions of u quark $f_1^u(x, \mathbf{k}_\perp)$ (left) and $g_1^u(x, \mathbf{k}_\perp)$ (right) given in (3).

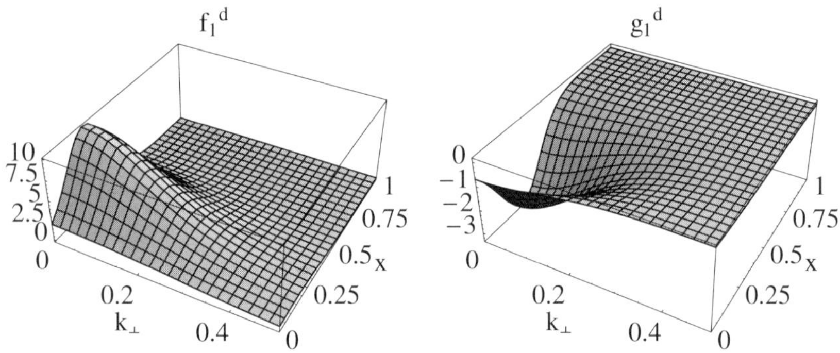

Fig. 2. The distribution functions of d quark $f_1^d(x, \mathbf{k}_\perp)$ (left) and $g_1^d(x, \mathbf{k}_\perp)$ (right) given in (3).

We use for both u and d quarks the fragmentation function given in Ref. [5], which is plotted in Fig. 4:

$$D_1(z, \mathbf{p}_\perp) = \frac{1}{z}\frac{g^2}{16\pi^3}\frac{\mathbf{p}_\perp^2 + m^2}{(\mathbf{p}_\perp^2 + m^2 + \frac{1-z}{z^2}m_\pi^2)^2} , \tag{4}$$

where m_π is pion mass and $m = 0.3$ GeV.

3. Double Spin Asymmetry

The double spin asymmetry is given by [4]

$$A_{LL}(x, y, z, P_{hT}) = \frac{\Delta\sigma_{LL}}{\sigma_0} , \tag{5}$$

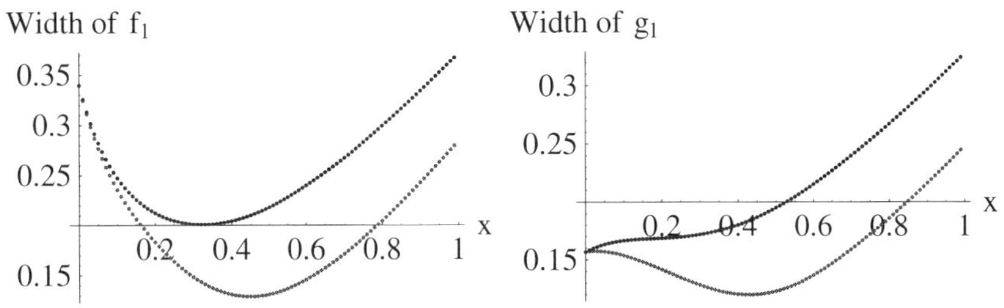

Fig. 3. The width in $\mathbf{k}_\perp$, which is defined as the value of $\mathbf{k}_\perp$ which satisfies $f_1(g_1)(x,\mathbf{k}_\perp) = \frac{1}{2} f_1(g_1)(x,\mathbf{0}_\perp)$ for u quark (lower line) and d quark (upper line).

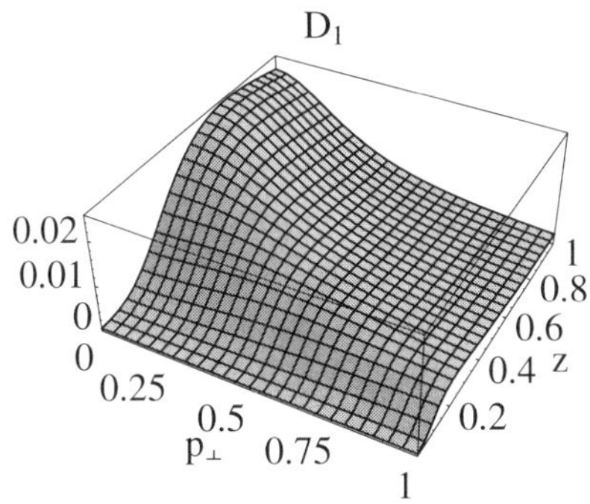

Fig. 4. The fragmentation function $D_1(z,\mathbf{p}_\perp)$ given in (4).

where

$$\Delta\sigma_{LL} = \frac{\pi}{xy^2}\left[\, y(2-y)\,\right] \Sigma_q e_q^2 \int d^2\mathbf{k}_\perp\; g_1^q(x,\mathbf{k}_\perp)\; D_q^h(z,\mathbf{P}_{hT} - z\mathbf{k}_\perp)\;, \quad (6)$$

$$\sigma_0 = \frac{\pi}{xy^2}\left[\, 1 + (1-y)^2\,\right] \Sigma_q e_q^2 \int d^2\mathbf{k}_\perp\; f_1^q(x,\mathbf{k}_\perp)\; D_q^h(z,\mathbf{P}_{hT} - z\mathbf{k}_\perp)\;.$$

We study A_{LL} with the model of Ref. [3] by using $f_1^q(x,\mathbf{k}_\perp)$ and $g_1^q(x,\mathbf{k}_\perp)$ given in (3), and $D_q^h(z,\mathbf{p}_\perp)$ given in (4) for both u and d quarks, and we ignore the contributions from antiquarks. In this paper we study the π^0 production by adding u and d quark contributions. The results for A_{LL} are presented in Fig. 5.

We integrate both the numerator and denominator of Eq. (5) over the following ranges covered by the setups of the experiments,[4] and the results are presented in Fig. 6. COMPASS: $0.1 < x < 0.6$, $0.5 < y < 0.9$, and

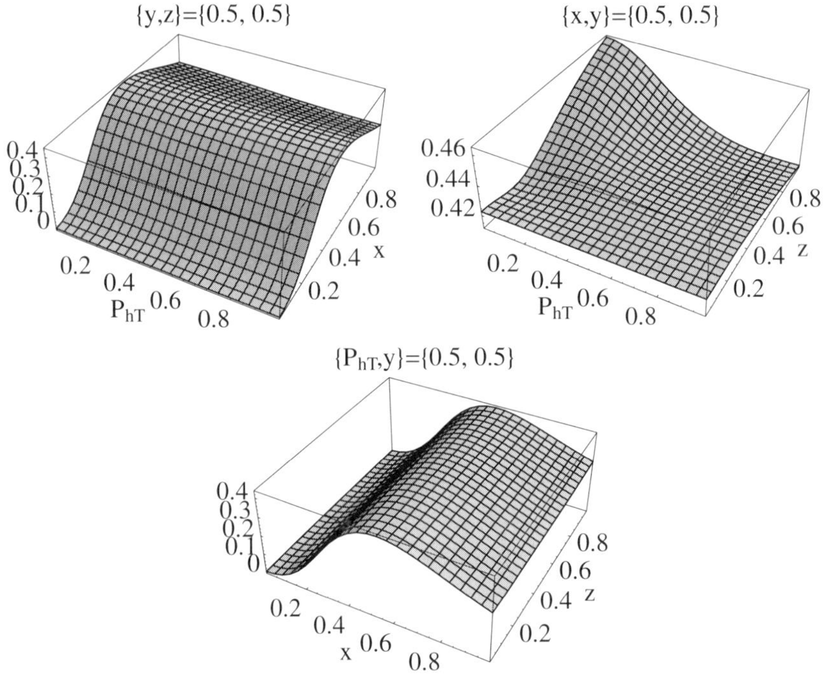

Fig. 5. A_{LL} of π^0 production as a function of P_{hT} and x (up-left) with fixed $y = 0.5$ and $z = 0.5$, that of P_{hT} and z (up-right) with fixed $x = 0.5$ and $y = 0.5$, and that of x and z (down) with fixed $P_{hT} = 0.5$ and $y = 0.5$.

$0.4 < z < 0.9$; HERMES: $0.1 < x < 0.6$, $0.45 < y < 0.85$, and $0.4 < z < 0.7$; JLab: $0.2 < x < 0.6$, $0.4 < y < 0.85$, and $0.4 < z < 0.7$.

4. Discussions

For the transverse momentum dependent distribution (fragmentation) functions, the ansatz which factorizes x (z) and $k_\perp$ ($p_\perp$) is often adopted. For example, Ref. [4] used the factorized functions given by

$$f_1^q(x, \mathbf{k}_\perp) = f_1^q(x) \frac{1}{\pi \mu_0^2} \exp\left(-\frac{\mathbf{k}_\perp^2}{\mu_0^2} \right), \tag{7}$$

$$g_1^q(x, \mathbf{k}_\perp) = g_1^q(x) \frac{1}{\pi \mu_2^2} \exp\left(-\frac{\mathbf{k}_\perp^2}{\mu_2^2} \right),$$

$$D_q^h(z, \mathbf{p}_\perp) = D_q^h(z) \frac{1}{\pi \mu_D^2} \exp\left(-\frac{\mathbf{p}_\perp^2}{\mu_D^2} \right).$$

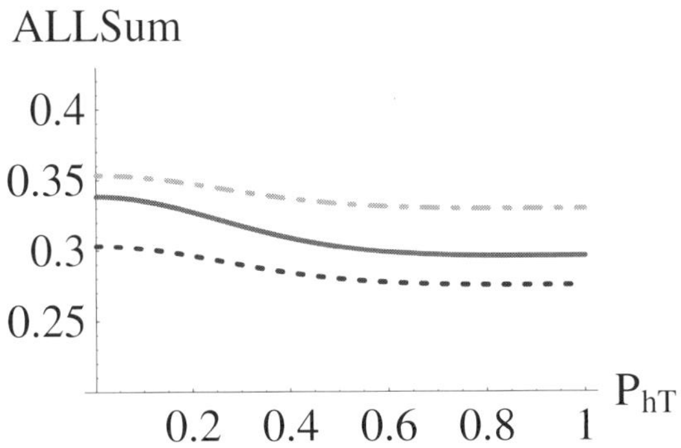

Fig. 6. A_{LL} of π^0 production for the integration over the ranges of (x, y, z) for the setups of the experiments of COMPASS(solid), HERMES(dotted), and JLab(dash-dotted line).

Then, one has

$$\Delta\sigma_{LL} = \frac{y(2-y)}{xy^2}\frac{1}{\mu_D^2+z^2\mu_2^2}\exp\left(-\frac{\mathbf{P}_{hT}^2}{\mu_D^2+z^2\mu_2^2}\right)\ \Sigma_q e_q^2 g_1^q(x) D_q^h(z)\ , \quad (8)$$

$$\sigma_0 = \frac{1+(1-y)^2}{xy^2}\frac{1}{\mu_D^2+z^2\mu_0^2}\exp\left(-\frac{\mathbf{P}_{hT}^2}{\mu_D^2+z^2\mu_0^2}\right)\ \Sigma_q e_q^2 f_1^q(x) D_q^h(z)\ .$$

Ref.[4] used $\mu_0^2 = 0.25$ (GeV^2), $\mu_D^2 = 0.20$, and $\mu_2^2 = 0.10,\ 0.17,\ 0.25$. The results are presented in their FIG. 1, which shows that the curves are very dependent on the values of μ_2^2.

It is important to have informations on how the distribution and fragmentation functions are dependent on the transverse momenta of partons. At first, it should be useful to discriminate the spectator model and the model based on the factorization ansatz. We found that A_{LL} presented in Fig. 6 obtained by (6) is characteristically different from that obtained by (8), which can be used as an experimental discrimination of the two models.

Acknowledgments

We thank Harut Avagyan and Stan Brodsky for useful discussions. This work was supported in part by the International Cooperation Program of the KICOS (Korea Foundation for International Cooperation of Science & Technology), and in part by the 2007 research fund from Kangnung National University.

References

1. P.J. Mulders and R.D. Tangerman, Nucl. Phys. B 461, 197 (1996); B 484, 538(E) (1997).
2. S.J. Brodsky, D.S. Hwang, and I. Schmidt, Phys. Lett. B 530, 99 (2002).
3. R. Jakob, P.J. Mulders, and J. Rodrigues, Nucl. Phys. A 626, 937 (1997).
4. M. Anselmino, A. Efremov, A. Kotzinian, and B. Parsamyan, Phys. Rev. D 74, 074015 (2006).
5. D. Amrath, A. Bacchetta, and A. Metz, Phys. Rev. D 71, 114018 (2005).

FLAVOR DEPENDENCE OF THE BOER-MULDERS FUNCTION AND ITS INFLUENCE ON AZIMUTHAL AND SINGLE-SPIN ASYMMETRIES IN SEMI-INCLUSIVE DIS

LEONARD P. GAMBERG

Department of Physics and Astronomy, Penn State University-Berks, Reading, PA 19610, USA
E-mail: lpg10@psu.edu

GARY R. GOLDSTEIN

Department of Physics and Astronomy, Tufts University Medford, MA 02155, USA
E-mail: gary.goldstein@tufts.edu

MARC SCHLEGEL

Theory Center, Thomas Jefferson National Accelerator Facility Newport News, VA 23608, USA
E-mail: schlegel@jlab.org

The chiral-odd and time reversal-odd transverse momentum dependent (TMD) parton distribution $h_1^{\perp}$, the so-called Boer-Mulders function, is calculated in the diquark spectator model for both an active up- and down-quark. In particular the signs of $h_1^{\perp(u)}$ and $h_1^{\perp(d)}$ are of interest as they were predicted in other works, and the result of this analysis is consistent with these predictions. Both flavors of the Boer-Mulders function contribute to observables such as the azimuthal $\cos(2\phi)$ asymmetry in semi-inclusive DIS. Using the results of this analysis for $h_1^{\perp(u)}$ and $h_1^{\perp(d)}$ and recent calculations of the Collins fragmentation function in a similar spectator model, phenomenological predictions for the $\cos(2\phi)$ asymmetry are presented. In addition, the chiral-odd but time-reversal even TMD $h_{1L}^{\perp}$ is computed in the diquark spectator model, and is used to present predictions for the $\sin(2\phi)$ single-spin asymmetry for a longitudinally polarized target in SIDIS.

Keywords: Boer-Mulders function; Diquark; Azimuthal asymmetry

1. Introduction

Naive time reversal-odd ("T-odd") transverse momentum dependent (TMD) parton distributions (PDFs) have gained considerable attention in recent years. Theoretically they can account for non-trivial transverse spin and momentum correlations such as single spin asymmetries (SSA) in hard scattering processes when transverse momentum scales are on the order of that of quarks in hadrons, namely $P_T \sim k_\perp \ll \sqrt{Q^2}$. A prominent example is the Sivers function $f_{1T}^{\perp}$[1] which explains the observed SSA in semi-inclusive deep inelastic scattering (SIDIS) for a transversely polarized proton target by the HERMES collaboration.[2] It describes correlations of the intrinsic quark transverse momentum and the transverse nucleon spin. The corresponding SSA on a deuteron target measured by COMPASS[3] vanishes, indicating a flavor dependence of the Sivers function. Another leading twist "T-odd" parton distribution, the chiral-odd Boer-Mulders function $h_1^{\perp}$[4] correlates the transverse spin of a quark with its transverse momentum within the nucleon.

Theoretically, twist two T-odd PDFs are of particular interest as they formally emerge from the gauge link structure of the color gauge invariant definition of the quark-qluon-quark correlation function.[5–7] This gauge link not only ensures a color gauge invariant definition of correlation functions in general, but also describes final state interactions which are necessary to generate SSA.[7–9] Assuming factorization of leading twist SIDIS spin observables in terms of the T-even[10] and T-odd TMDs and fragmentaion functions[4] shows how spin observables in SIDIS can be expressed in terms of convolutions of TMD parton distributions and fragmentation functions. Formal proofs of factorization of leading twist SIDIS spin observables were presented later in Refs. [11–13]. We focus on the flavor dependence of these "T-odd" functions where for example $h_1^{\perp}$ is important for the analysis of the azimuthal $\cos(2\phi)$ asymmetry in unpolarized SIDIS and Drell-Yan.[14,15] We also consider the flavor dependence of the "T-even" function $h_{1L}^{\perp}$, which is of interest in the transverse momentum and quark spin correlations in a longitudinally polarized target.[16]

Considerable understanding of TMDs and fragmentation functions (FF) have been gained from model calculations using the spectator framework.[8,9,14,15,17,18]

2. T-even parton distributions

In this formalism we start (cf. [19]) from the definition of the unintegrated color gauge invariant quark-quark correlator which contains the gauge link indicated by the (future-pointing) Wilson line, $\mathcal{W}[a \mid b]$, and work in Feynman gauge in which the transverse Wilson line vanishes.[6] In the diquark model the sum over the complete set of intermediate on-shell states in the definition of the correlator is represented by a single one-particle diquark state $|\,dq;\, p_{dq}, \lambda\rangle$, where p_{dq} is the diquark momentum and λ its polarization. The diquark is "built" from two valence quarks which can be scalar-spin 0 or axial vector-spin 1. The unintegrated correlator is then given by

$$\begin{aligned}\Phi_{ij}(p; P, S) = \sum_{\lambda} \frac{\delta((P-p)^2 - m_s^2)\Theta(P^0 - p^0)}{(2\pi)^3} \times \\ \langle P, S|\, \bar{\psi}_j(0)\, \mathcal{W}[0\,|\,\infty, 0, \vec{0}_T]\,|\, dq;\, P-p, \lambda\rangle \times \\ \langle\, dq;\, P-p, \lambda|\, \mathcal{W}[\infty, 0, \vec{0}_T\,|\,0]\, \psi_i(0)\,|P, S\rangle. \end{aligned} \quad (1)$$

The essence of the diquark spectator model is to calculate the matrix elements in Eq. (1) by the introduction of effective nucleon-diquark-quark vertices $\Upsilon_s(N)$ and $\Upsilon^{\mu}_{ax}(N)$ represented in Fig. 1 (a). In the following the

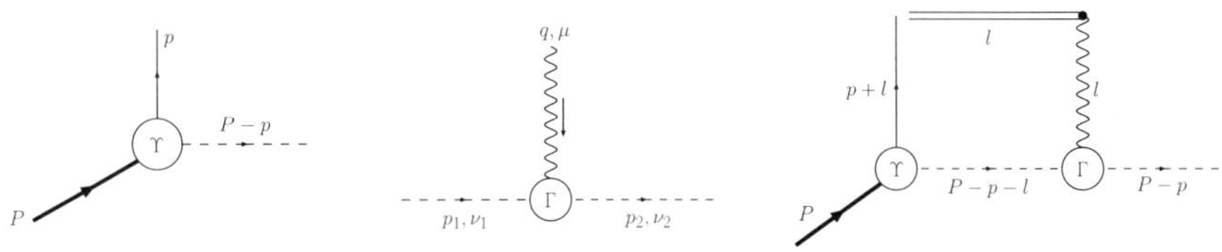

Fig. 1. Different vertices for the axial-vector diquark and contribution of the gauge link in the one-gluon approximation. Left: Nucleon-Quark-Diquark vertex, Middle: Diquark-Gluon vertex, Right: Box-graph.

nucleon-quark-diquark vertices of Ref.[19] are used. In such a way the matrix elements in Eq. (1) then read for a scalar diquark

$$\langle sdq;\, P-p|\, \psi_i(0)\,|P, S\rangle = i g_{sc}(p^2) \frac{[(\not{p} + m_q) u(P, S)]_i}{p^2 - m_q^2 + i0}, \quad (2)$$

and for the axial vector diquark

$$\begin{aligned}\langle adq;\, P-p; \lambda|\, \psi_i(0)\,|P, S\rangle = i \frac{g_{ax}(p^2)}{\sqrt{3}} \varepsilon^*_{\mu}(P-p; \lambda) \times \\ \frac{\left[(\not{p} + m_q)\gamma_5 \left[\gamma^{\mu} - R_g \frac{P^{\mu}}{M}\right] u(P, S)\right]_i}{p^2 - m_q^2 + i0}, \end{aligned} \quad (3)$$

where the polarization vector of the axial-vector diquark is ε_{μ}, $u(P, S)$ denotes the nucleon spinor and M and m_q are nucleon and quark masses, respectively. The unpolarized TMD f_1 is obtained by inserting these expressions into Eq. (1) and projecting from the quark-quark correlator

$$f_1(x, \vec{p}_T^2) = \frac{1}{4} \int dp^- \left(\mathrm{Tr}\left[\gamma^+ \Phi(p; P, S)\right] + \mathrm{Tr}\left[\gamma^+ \Phi(p; P, -S)\right] \right) \Big|_{p^+ = xP^+}, \tag{4}$$

We have also calculated the distribution of transversely polarized quarks in a longitudinally polarized target, $h_{1L}^{\perp}$ by replacing γ^+ with $\gamma^+\gamma^i\gamma_5$ and S by S_L, the projection of the spin 4-vector in longitudinal direction. Their analytic expressions in the scalar and axial vector diquark sectors are given in [20].

3. T-odd parton distributions

In the spectator framework the "T-odd" TMDs[8] are generated by the gauge link.[9,14,15] The leading contribution, arising from the interference between tree- and box graph which contains an imaginary part necessary for "T-odds", is represented in Fig. 1 (c) in which the double line is an eikonal line, and l is the loop momentum.

In the following the calculational method is shown for an axial-vector diquark, the much simpler case of a scalar diquark is analogous. For details see Ref. [20]. For an axial-vector diquark we model the composite nature of the diquark through an anomalous magnetic moment κ.[21] In the notation of Fig. 1 (b) the gluon-diquark axial diquark vertex is

$$\begin{aligned} \Gamma_{ax}^{\mu\nu_1\nu_2} = & -ie_{dq}\left[g^{\nu_1\nu_2}(p_1+p_2)^{\mu} + (1+\kappa)\left(g^{\mu\nu_2}(p_2+q)^{\nu_1}\right.\right. \\ & \left.\left. + g^{\mu\nu_1}(p_1-q)^{\nu_2}\right)\right]. \end{aligned} \tag{5}$$

For $\kappa = -2$ the vertex Γ_{ax} reduces to the standard γWW-vertex. We express the matrix elements including the gauge link, $\langle dq; P - p|\, \mathcal{W}[\infty, 0, \vec{0}_T \,|\, 0]\, \psi_i(0)\, |P, S\rangle$ in the one gluon approximation.[20] Projecting the Boer-Mulders function,

$$\begin{aligned} \epsilon_T^{ij} p_T^j h_1^{\perp}(x, \vec{p}_T^2) = & \frac{M}{4} \int dp^- \left(\mathrm{Tr}\left[\Phi_{\mathrm{unpol}}(p, S) i\sigma^{i+}\gamma_5\right] \right. \\ & \left. + \mathrm{Tr}\left[\Phi_{\mathrm{unpol}}(p, -S) i\sigma^{i+}\gamma_5\right] \right), \end{aligned} \tag{6}$$

where $\epsilon_T^{ij} \equiv \epsilon^{-+ij}$ the axial-vector diquark contribution is given by the expression,

$$
\begin{aligned}
\epsilon_T^{ij} p_T^j h_1^{\perp,ax}(x,\vec{p}_T^2) = -\frac{e_q e_{dq}}{8(2\pi)^3}\frac{1}{\vec{p}_T^2+\tilde{m}^2}\frac{M}{P^+} \times \\
\int \frac{d^4l}{(2\pi)^4}\left\{\frac{1}{3} g_{ax}\left((l+p)^2\right) g_{ax}^*\left(p^2\right)\times\right. \\
\mathcal{D}_{\rho\eta}(P-p-l)\left(\sum_\lambda \varepsilon_\sigma^*(P-p;\lambda)\varepsilon_\mu(P-p;\lambda)\right)\times \\
\frac{\left[g^{\sigma\rho}\, v\cdot(2P-2p-l)+(1+\kappa)\left(v^\sigma\,(P-p+l)^\rho+v^\rho\,(P-p-2l)^\sigma\right)\right]}{[l\cdot v+i0]\,[l^2-\lambda^2+i0]\left[(l+p)^2-m_q^2+i0\right]}\times \\
\mathrm{Tr}\Big[(\not{P}+M)\left(\gamma^\mu - R_g\frac{P^\mu}{M}\right)(\not{p}-m_q)\times \\
\left.\gamma^+\gamma^i\,(\not{l}+\not{p}+m_q)\left(\gamma^\eta+R_g\frac{P^\eta}{M}\right)\gamma_5\Big]\right\} + \text{ h.c. }. \qquad (7)
\end{aligned}
$$

$\mathcal{D}(P-p-l)$ denotes the propagator of the axial-vector diquark and is given by

$$
\mathcal{D}^{\mu\nu}(k) = \frac{-i\left(g^{\mu\nu}-\frac{1}{m_s^2}k^\mu k^\nu\right)}{[k^2-m_s^2+i0]}. \qquad (8)
$$

It is emphasized that the term $\frac{1}{m_s^2}k^\mu k^\nu$ was not taken into account in a previous similar analysis[17] and leads to major differences.

Since the numerator in Eq. (7) contains at most the loop momentum to the fourth power we can write it in the following manner, $\sum_{i=1}^4 N^{(i)}_{\alpha_1\ldots\alpha_i} l^{\alpha_1}\ldots l^{\alpha_i} + N^{(0)}$. The (real) coefficients (tensors) $N^{(i)}_{\alpha_1\ldots\alpha_i}$ depend only on external momenta and can be computed in a straight-forward way. The integration over the light cone components, l^+ and l^-, are easily performed first by contour integration (l^-) and then by using the fact that the addition of the complex conjugated box-graph integral leads to a δ-function $\delta(l^+)$; however, calculating the l^+-integral results in an integral that is potentially ill-defined. This happens when the form factor $g(p^2)$ is a holomorphic function in p^2 and at least one of the Minkowski indices is light-like in the minus direction, e.g. $\alpha_1 = -$, $\alpha_2, ..., \alpha_i \in \{+,\perp\}$ resulting in an integral of the form $\int dl^+\,\delta(l^+)\Theta(-l^+)$, implying that $l^+ = 0$ and $l^- = \infty$. This signals the existence of a light cone divergence in Ref. [18]. It is pointed out that this light cone divergence is a consequence of the $k^\mu k^\nu$-term of the axial-vector diquark propagator in Eq. (8). Since such a term is missing for a scalar diquark, we encounter no light cone divergence in that case.

From the standpoint of phenomenology one can regard these light cone divergences as model artefacts (for the axial-vector diquark model). It was shown in Ref. [18] how to regularize these light cone divergences by introducing non-light-like Wilson line. It was pointed out that one can also handle the light cone divergences by introducing phenomenological form factors with additional poles,[20]

$$g_{ax}(p^2) = \frac{(p^2 - m_q^2) f(p^2)}{[p^2 - \Lambda^2 + i0]^n}. \tag{9}$$

For $n \geq 3$ there are enough powers of l^+ to eliminate this divergence. $f(p^2)$ is chosen to bea covariant Gaussian,[20] and Λ is an arbitrary mass scale fixed by fitting f_1 to data. The Gaussian is in particular convenient as it regularizes a UV-divergence arising in the integration over the transverse loop momentum (again, the UV-divergence is a consequence of the axial-vector diquark propagator, and missing for a scalar diquark) as well as it suppresses large transverse parton momenta. After having performed the integration over the light cone components of the loop momenta, the integration over the transverse momentum leads for the choice of a Gaussian form factor in (9) to a result which can be expressed in terms of incomplete Gamma-functions.[20]

Similarly, the Sivers-function is projected from the trace of the quark-quark correlator (1) (see e.g. [22]),

$$2S_T^i \epsilon_T^{ij} p_T^j f_{1T}^{\perp}(x, \vec{p}_T^{\,2}) = \frac{M}{2} \int dp^- \left(\mathrm{Tr}\left[\gamma^+ \Phi(p; P, S_T)\right] \right.$$
$$\left. - \mathrm{Tr}\left[\gamma^+ \Phi(p; P, -S_T)\right] \right)\Big|_{p^+ = xP^+}. \tag{10}$$

It is well-known[14] that in the scalar diquark approximation the $h_1^{\perp}$ and $f_{1T}^{\perp}$ coincide. By contrast the different Dirac structure for the chiral even $f_{1T}^{\perp}$ and chiral odd $h_1^{\perp}$ in the axial-vector diquark sector, Eq. (6) and (10) respectively, lead to different coefficients in the decomposition $N^{(i)}_{\alpha_1 \ldots \alpha_i}$.[20]

4. Phenomenology

We fix most of the model parameters such as masses and normalizations by comparing the model result for the unpolarized f_1 for u and d quarks to the low-scale ($\mu^2 = 0.34$GeV) data parameterization of GRV[23] (see Fig. 2). Note that PDFs for u and d quarks are given by linear combinations of PDFs for an axial vector and scalar diquark, $u = \frac{3}{2} f^{sc} + \frac{1}{2} f^{ax}$ and $d = f^{ax}$.[17,19] For "T-odd" PDFs we fix the sign and the strength of the final state interactions, the product of the charges of the diquark and quark, and the parameter κ by comparing $f_{1T}^{\perp}$ for u and d quarks in the diquark model with the existing data parameterizations (see Ref. [24]). The "one-half" moments $q_T^{(1/2)}$ (where $q = u, d$) of the Sivers and Boer-Mulders functions

$$\frac{1}{2}q_T^{(1/2)}(x) = f_{1T}^{\perp(q,1/2)}(x) = \int d^2p_T \, \frac{|\vec{p}_T|}{2M} \, f_{1T}^{\perp(q)}(x, \vec{p}_T^{\,2}),$$

are displayed along with the unpolarized u and d quark distributions in Fig. 2. The "one-half" and first moments[20] of the up and down quark Sivers functions are negative and positive respectively while the up and down quark Boer-Mulders functions are *both negative* over the full range in Bjorken-x. These results are in agreement with the large N_c predictions[25] , Bag Model results reported in [26] , impact parameter distortion picture of Burkardt[27] and recent studies of nucleon transverse spin structure in lattice QCD.[28] We also note that the u-quark Sivers function and Boer-Mulders function are nearly equal, even with the inclusion of the axial vector spectator diquark.

Having explored the flavor dependence of the $h_1^{\perp}$ we are now in a position to extend early phenomenological work on "T-odd" contributions to azimuthal asymmetries in SIDIS.[15] We consider the spin independent double "T-odd" $\cos 2\phi$ asymmetry for π^+ and π^- production. We focus on the important contributions to the cross section for unpolarized SIDIS[22]

$$\frac{d\sigma}{dx\,dy\,dz\,d\phi_h\,dP_{h\perp}^2} \approx \frac{2\pi\alpha^2}{xyQ^2}\Big[\left(1 - y + \frac{1}{2}y^2\right) F_{UU} + (1-y)\cos(2\phi_h)\, F_{UU}^{\cos 2\phi_h}\Big], \tag{11}$$

where the structure function $F_{UU}^{\cos 2\phi_h}$ involves a convolution of the Boer-Mulders and Collins fragmentation function

$$F_{UU}^{\cos 2\phi_h} = \mathcal{C}\Big[-\frac{2\hat{\boldsymbol{h}}\cdot\boldsymbol{k}_T\hat{\boldsymbol{h}}\cdot\boldsymbol{p}_T - \boldsymbol{k}_T\cdot\boldsymbol{p}_T}{MM_h} h_1^{\perp} H_1^{\perp}\Big], \tag{12}$$

$\mathcal{C}$ is the convolution integral. Our input for the Collins functions is based on recent work in [29] where the Collins function was calculated in the spectator framework. It was assumed that $H_1^{\perp(dis-fav)} \approx -H_1^{\perp(fav)}$ in the pion sector, thereby satisfying the Schaefer-Teryaev sum rule[30] locally. We estimate the azimuthal asymmetry $A_{UU}^{\cos 2\phi}$ (cf. Eq. (12)), where $A_{UU}^{\cos 2\phi} \equiv \int \cos 2\phi \; d\sigma / \int \; d\sigma$. In Fig. 3 we display the $A_{UU}^{\cos 2\phi}(P_T)$ in the range of HERMES[2] and future JLAB kinematics[31] as well as x dependence in the range $0.5 < P_T < 1.5$ GeV/c.

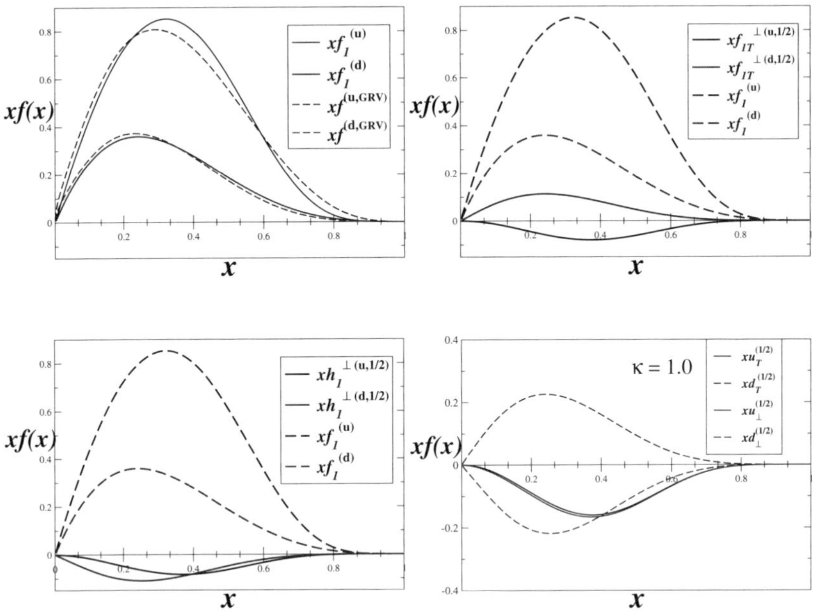

Fig. 2. The unpolarized up and down quark distributions functions (first line, left) versus x compared to the low scale parameterization of GRV .[23] The half moment of the Sivers function (first line, right) and Boer-Mulders function (second line, left) and the unpolarized up and down quark distributions ($\kappa = 1.0$). The half-moments (second line, right) of the Boer-Mulders and Sivers functions versus x fitted to extractions from data were presented in Ref. [24].

Having calculated the chiral-odd but "T-even" parton distribution $h_{1L}^{\perp}$ we use this result together with the result of Ref. [29] for the Collins function to give a prediction for the $\sin(2\phi)$ single spin asymmetry A_{UL} for a longitudinally polarized target. A decomposition into structure functions of the cross section of semi-inclusive DIS for a longitudinally polarized target reads (see e.g. [22])

$$\frac{d\sigma_{UL}}{dx\,dy\,dz\,d\phi_h\,dP_{h\perp}^2} \approx \frac{2\pi\alpha^2}{xyQ^2}\,S_\parallel\left[(1-y)\sin(2\phi_h)\,F_{UL}^{\sin(2\phi)} + (2-y)\sqrt{1-y}\,\sin(\phi_h)\,F_{UL}^{\sin\phi}\right], \tag{13}$$

$S_\parallel$ is the projection of the spin vector on the direction of the virtual photon. In a partonic picture the structure function $F_{UL}^{\sin(2\phi)}$ is a leading twist object (while $F_{UL}^{\sin\phi}$ is sub-leading) and given by a convolution of the TMD $h_{1L}^{\perp}$

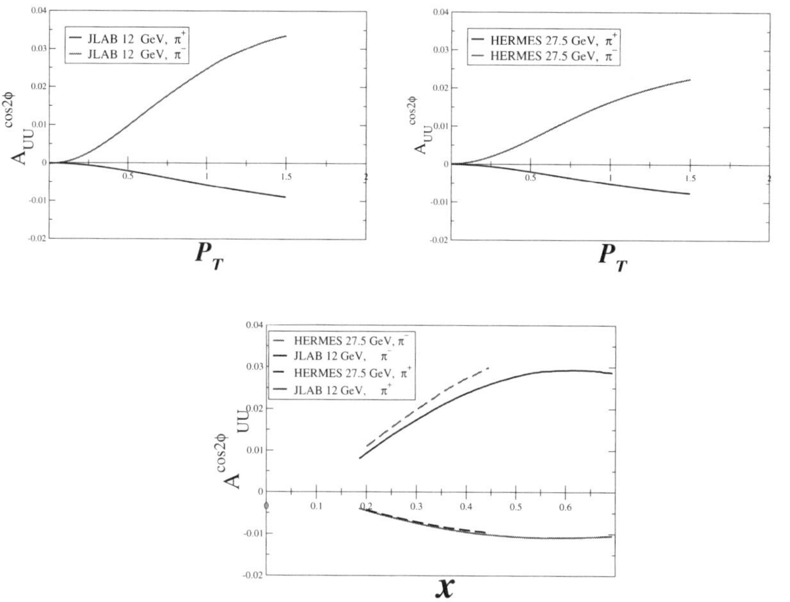

Fig. 3. The $\cos 2\phi$ asymmetry for $\pi^{\pm}$ as a function of P_T at JLAB 12GeV and HERMES (center) kinematics, and the $\cos 2\phi$ asymmetry for $\pi^{\pm}$ as a function of x at JLAB 12GeV and HERMES kinematics.

and the Collins function (cf. [22])

$$F_{UL}^{\sin(2\phi)} = \mathcal{C}\left[-\frac{2\hat{\boldsymbol{h}}\cdot\boldsymbol{k}_T\,\hat{\boldsymbol{h}}\cdot\boldsymbol{p}_T - \boldsymbol{k}_T\cdot\boldsymbol{p}_T}{MM_h}\, h_{1L}^{\perp} H_1^{\perp}\right]. \tag{14}$$

We display the results for the single spin asymmetry $A_{UL}^{\sin(2\phi)}$ in Fig. 4 using the kinematics of the upcoming JLab 12 GeV upgrade. We note that the π^- asymmetry is large and positive due to the model assumption $H_1^{\perp(dis-fav)} \approx -H_1^{\perp(fav)}$. This asymmetry has been measured at HERMES for longitudinally polarized protons[32] and deuterons.[33] The data show that for the proton target at HERMES 27.5 GeV kinematics both π^+ and π^- asymmetries are consistent with 0 down to a sensitivity of about 0.01. These asymmetries could be non-zero, but with magnitudes less than 0.01 or 0.02. These results are considerably smaller than our predictions for the JLab upgrade. For the deuteron target the results are consistent with 0 for π^+ and π^-. This SIDIS data for polarized deuterons could reflect the near cancellation of u- and d-quark $h_{1L}^{\perp}$ functions and/or the large unfavored Collins function contributions. There is also CLAS preliminary data[34] at 5.7 GeV

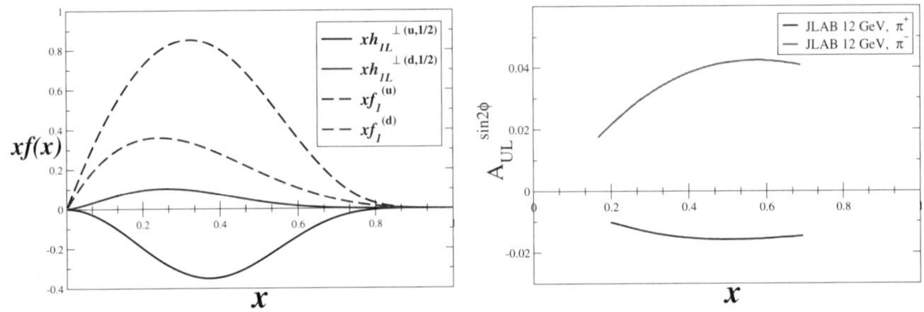

Fig. 4. Left Panel: The half-moment of $xh_{1L}^{\perp(1/2)}$ versus x compared to the unpolarized up and down quark distribution functions. Right Panel: The $\sin 2\phi$ asymmetry for π^+ and π^- as a function of x at JLAB 12GeV kinematics.

that shows slightly negative asymmetries for π^+ and π^- and leads to the extraction of a negative $xh_{1L}^{\perp(u)}$. This suggests that the unfavored Collins function (for $d \to \pi^+$) is not contributing much. Data from the upgrade should help resolve these phenomenological questions.

We have performed calculations of transverse momentum dependent parton distributions, including the Boer-Mulders function $h_1^\perp$, the Sivers function $f_{1T}^\perp$ along with $h_{1L}^\perp$ in the framework of an axial-vector and a scalar diquark spectator model. The calculation of these functions in both sectors allowed us to explore their flavor dependence, i.e. to compute $h_1^\perp$, $f_{1T}^\perp$ and $h_{1L}^\perp$ for a u-quark and a d-quark. We used these results along with the Collins fragmentation function $H_1^\perp$. to estimate the azimuthal asymmetries $A_{UU}^{(\cos(2\phi))}$ and $A_{UL}^{(\sin(2\phi))}$ in SIDIS. In summary, a refined diquark spectator model, including axial vector di-quarks leads to both u- and d-quark "T-odd" TMDs and provides the ingredients for predicting a range of asymmetries for future experiments.

Acknowledgments

Notice: Authored by Jefferson Science Associates, LLC under U.S. DOE Contract No. DE-AC05-06OR23177. This work is supported in part by the U.S. Department of Energy under contracts, DE-FG02-07ER41460 (LG), and DE-FG02-92ER40702 (GRG).

References

1. D. W. Sivers, *Phys. Rev.* **D41**, p. 83 (1990).
2. A. Airapetian *et al.*, *Phys. Rev. Lett.* **94**, p. 012002 (2005).

3. V. Y. Alexakhin *et al.*, *Phys. Rev. Lett.* **94**, p. 202002 (2005).
4. D. Boer and P. J. Mulders, *Phys. Rev.* **D57**, 5780 (1998).
5. J. C. Collins, *Phys. Lett.* **B536**, 43 (2002).
6. A. V. Belitsky, X. Ji and F. Yuan, *Nucl. Phys.* **B656**, 165 (2003).
7. D. Boer, P. J. Mulders and F. Pijlman, *Nucl. Phys.* **B667**, 201 (2003).
8. S. J. Brodsky, D. S. Hwang and I. Schmidt, *Phys. Lett.* **B530**, 99 (2002).
9. X.-d. Ji and F. Yuan, *Phys. Lett.* **B543**, 66 (2002).
10. P. J. Mulders and R. D. Tangerman, *Nucl. Phys.* **B461**, 197 (1996).
11. X.-d. Ji, J.-p. Ma and F. Yuan, *Phys. Rev.* **D71**, p. 034005 (2005).
12. X.-d. Ji, J.-P. Ma and F. Yuan, *Phys. Lett.* **B597**, 299 (2004).
13. J. C. Collins and A. Metz, *Phys. Rev. Lett.* **93**, p. 252001 (2004).
14. G. R. Goldstein and L. Gamberg, *hep-ph/0209085* (2002).
15. L. P. Gamberg, G. R. Goldstein and K. A. Oganessyan, *Phys. Rev.* **D67**, p. 071504 (2003).
16. A. M. Kotzinian and P. J. Mulders, *Phys. Lett.* **B406**, 373 (1997).
17. A. Bacchetta, A. Schaefer and J.-J. Yang, *Phys. Lett.* **B578**, 109 (2004).
18. L. P. Gamberg, D. S. Hwang, A. Metz and M. Schlegel, *Phys. Lett.* **B639**, 508 (2006).
19. R. Jakob, P. J. Mulders and J. Rodrigues, *Nucl. Phys.* **A626**, 937 (1997).
20. L. P. Gamberg, G. R. Goldstein and M. Schlegel, *arXiv:0708.0324 [hep-ph]* (2007).
21. G. R. Goldstein and J. Maharana, *Nuovo Cim.* **A59**, p. 393 (1980).
22. A. Bacchetta, M. Diehl, K. Goeke, A. Metz, P. Mulders and M. Schlegel, *JHEP* **02**, p. 093 (2007).
23. M. Gluck, E. Reya and A. Vogt, *Eur. Phys. J.* **C5**, 461 (1998).
24. M. Anselmino *et al.*, *hep-ph/0511017* (2005).
25. P. V. Pobylitsa, *hep-ph/0301236* (2003).
26. F. Yuan, *Phys. Lett.* **B575**, 45 (2003).
27. M. Burkardt, *Phys. Rev.* **D72**, p. 094020 (2005).
28. M. Gockeler *et al.*, *Phys. Rev. Lett.* **98**, p. 222001 (2007).
29. A. Bacchetta, L. Gamberg, G. Goldstein and A. Mukherjee.
30. A. Schafer and O. V. Teryaev, *Phys. Rev.* **D61**, p. 077903 (2000).
31. H. Avakian *et al.*, *approved JLab proposal PR 12-06-112.* (2006).
32. A. Airapetian *et al.*, *Phys. Rev. Lett.* **84**, 4047 (2000).
33. A. Airapetian *et al.*, *Phys. Lett.* **B562**, 182 (2003).
34. H. Avakian, P. E. Bosted, V. Burkert and L. Elouadrhiri, *AIP Conf. Proc.* **792**, 945 (2005).

SYMMETRIC SPIN-DEPENDENT STRUCTURE FUNCTION IN DEEP INELASTIC PROCESSES

DAE SUNG HWANG, JONG HYUN KIM, AND SEYONG KIM

Department of Physics, Sejong University
Seoul 143-747, South Korea

We investigate the single-spin asymmetries in deep inelastic electron-proton scattering and electron-positron annihilation. We show that it is possible to measure the single-spin asymmetry in the inclusive Λ production at the present B-factories.

1. Introduction

There have been important progresses in exploring the structure of the proton recently. The Jeffereson Lab Hall A Collaboration[1] obtained the ratio of the proton's elastic electromagnetic form factors G_E/G_M by the measuring the polarization transfers. The BABAR Collaboration[2] measured, using the initial state radiation, the $p\bar{p}$ mass dependence of the ratio of the time-like proton electric and magnetic form factors. These progresses are valuable for improving our understanding of the nucleon structure.

In this paper we investigate the single-spin asymmetries in deep inelastic electron-proton scattering and electron-positron annihilation. In the decomposition of hadronic tensor, a Lorentz symmetric spin-dependent term is generally present. The existence of such a term implies single-spin asymmetries in these processes. We show that it is possible to measure the single-spin asymmetry in the inclusive Λ production at the present B-factories. This will lead to the first measurement of the symmetric spin-dependent structure function in the electron-positron annihilation process.

2. Space-like Process ($e^- p \to e^- X$)

The structure of the proton is probed by the inelastic scattering $e^- p \to e^- X$ in which the scattered electron is detected at a fixed energy and angle, and

X indicates all possible states. The structure functions are defined by

$$\begin{aligned} W_{\mu\nu} &= \Big(-g_{\mu\nu} + \frac{q_\mu q_\nu}{q^2}\Big) W_1(q^2,\nu) \\ &+ \frac{1}{M^2}\Big(P_\mu - \frac{P\cdot q}{q^2}q_\mu\Big)\Big(P_\nu - \frac{P\cdot q}{q^2}q_\nu\Big) W_2(q^2,\nu) \\ &+ \frac{1}{M^3}\Big[\Big(P_\mu - \frac{P\cdot q}{q^2}q_\mu\Big)\epsilon_{\nu abc}s^a P^b q^c \\ &+ \Big(P_\nu - \frac{P\cdot q}{q^2}q_\nu\Big)\epsilon_{\mu abc}s^a P^b q^c\Big] G_3(q^2,\nu), \end{aligned} \tag{1}$$

in which M is the proton mass and $q = k - k'$ where k and k' are the momenta of the initial and final electrons, respectively. $M\nu = P\cdot q$ and we use $\epsilon_{0123} = 1$. $-q^2 > 0$ and $1 < \omega \equiv \frac{2M\nu}{|q^2|} < \infty$ for the space-like processes.

W_1 and W_2 are usual spin-independent structure functions, and G_3 is a symmetric spin-dependent structure function which measures a violation of the time reversal invariance in electromagnetic interactions.[3,4] From the product of the hadronic tensor $W_{\mu\nu}$ in (1) and the symmetric part of the leptonic tensor, we get the differential cross section in the rest frame of the target proton

$$\begin{aligned} \frac{d\sigma}{d\Omega dE'} &= \frac{4\alpha^2 E'^2}{|q^2|^2}\sin^2\frac{\theta}{2}\Big[\ 2W_1(q^2,\nu)\ +\ \cot^2\frac{\theta}{2}W_2(q^2,\nu) \\ &+ (\vec{s}\cdot\hat{y})\ \frac{4(E+E')EE'}{M\,|q^2|}\ |\sin\theta|\ G_3(q^2,\nu)\ \Big], \end{aligned} \tag{2}$$

where E ($\hat{k}$) and E' ($\hat{k}'$) are the initial and final energies (momentum directions) and θ the scattering angle of the electron, and $\hat{y} = (\hat{k}\times\hat{k}')/|\hat{k}\times\hat{k}'|$. If there exists the angular distribution given by the third term in (2), it would imply the T-violation.

3. Time-like Process ($e^+e^- \to p$(or Λ)X)

The structure of the proton is probed by the inelastic scattering $e^+e^- \to pX$ in which the produced proton is detected at a fixed energy and angle, and X indicates all possible states. The structure functions are defined by the same formula as Eq. (1) with $W_1(q^2,\nu)$ and $W_2(q^2,\nu)$ being replaced by $\bar{W}_1(q^2,\nu)$ and $\bar{W}_2(q^2,\nu)$ following the notation in Ref. [5]. In the time-like process $e^+e^- \to pX$, $q = k + k'$ where k and k' are the momenta of the initial electron and positron, respectively. $M\nu = P\cdot q$ and we use $\epsilon_{0123} = 1$. $q^2 > 0$ and $0 < \omega \equiv \frac{2M\nu}{q^2} < 1$ for the time-like processes.

$\bar{W}_1$ and $\bar{W}_2$ are usual spin-independent structure functions,[5] and G_3 is a symmetric spin-dependent structure function which has its origin in the strong-interaction phases.[4,6] In this section we study the single-spin asymmetry which would be induced by G_3 in the time-like deep inelatic process.

From the product of the hadronic tensor $W_{\mu\nu}$ and the symmetric part of the leptonic tensor, we get the differential cross section in the center of mass frame of the initial electron and positron

$$\begin{aligned}\frac{d^2\sigma}{dE d\cos\theta} &= \frac{2\pi\alpha^2 M}{(q^2)^{\frac{3}{2}}}\Big(\frac{2M\nu}{q^2}\Big)\Big(1-\frac{q^2}{\nu^2}\Big)^{\frac{1}{2}} \\ &\times \Big(\Big[2\bar{W}_1(q^2,\nu) + \Big(\frac{2M\nu}{q^2}\Big)\Big(1-\frac{q^2}{\nu^2}\Big)\sin^2\theta\,\frac{\nu\bar{W}_2(q^2,\nu)}{2M}\Big] \\ &+ (\vec{s}\cdot\hat{y})\,\frac{(q^2)^{\frac{3}{2}}}{4M^3}\Big(\frac{2M\nu}{q^2}\Big)^2\Big(1-\frac{q^2}{\nu^2}\Big)\sin 2\theta\, G_3(q^2,\nu)\Big)\,,\end{aligned} \tag{3}$$

where E is the energy of the produced proton, $\hat{k}$ and $\hat{p}$ the initial electron and produced proton momentum directions, and θ the angle between $\hat{k}$ and $\hat{p}$, and $\hat{y} = (\hat{k}\times\hat{p})/|\hat{k}\times\hat{p}|$. From (3) the single-spin asymmetry is given by

$$\mathcal{P}_y = \frac{\frac{(q^2)^{\frac{3}{2}}}{4M^3}\Big(\frac{2M\nu}{q^2}\Big)^2\Big(1-\frac{q^2}{\nu^2}\Big)\sin 2\theta\, G_3(q^2,\nu)}{2\bar{W}_1(q^2,\nu) + \Big(\frac{2M\nu}{q^2}\Big)\Big(1-\frac{q^2}{\nu^2}\Big)\sin^2\theta\,\frac{\nu\bar{W}_2(q^2,\nu)}{2M}}\,. \tag{4}$$

We also have

$$\frac{1}{2}\Big(\frac{d^2\sigma(\vec{s}=\hat{y})}{dE d\cos\theta} - \frac{d^2\sigma(\vec{s}=-\hat{y})}{dE d\cos\theta}\Big) = \frac{\pi\alpha^2}{2M^2}\Big(\frac{2M\nu}{q^2}\Big)^3\Big(1-\frac{q^2}{\nu^2}\Big)^{\frac{3}{2}}\sin 2\theta\, G_3(q^2,\nu)\,. \tag{5}$$

When we follow Ref. [5], in the Bjorken limit for the time-like q^2 we have

$$M\bar{W}_1 = -F_1(\omega)\,, \qquad \nu\bar{W}_2 = F_2(\omega)\,, \qquad \text{and}\;\; F_1(\omega) = \frac{\omega}{2}F_2(\omega)\,. \tag{6}$$

Then, the differential cross section is given by

$$\begin{aligned}\frac{d^2\sigma}{dE d\cos\theta} &= \frac{2\pi\alpha^2 M}{(q^2)^{\frac{3}{2}}}\,\omega\Big(1-\frac{4M^2}{q^2}\frac{1}{\omega^2}\Big)^{\frac{1}{2}} \\ &\times \Big(\frac{2}{M}\Big(-F_1(\omega)\Big)\Big[1-\frac{1}{2}\Big(1-\frac{4M^2}{q^2}\frac{1}{\omega^2}\Big)\sin^2\theta\Big] \\ &+ (\vec{s}\cdot\hat{y})\,\frac{(q^2)^{\frac{3}{2}}}{4M^3}\,\omega^2\Big(1-\frac{4M^2}{q^2}\frac{1}{\omega^2}\Big)\sin 2\theta\, G_3(\omega)\Big)\,,\end{aligned} \tag{7}$$

and the single-spin asymmetry is given by

$$\mathcal{P}_y = \frac{G_3(\omega)\ \frac{(q^2)^{\frac{3}{2}}}{4M^3}\ \omega^2 \left(1 - \frac{4M^2}{q^2}\frac{1}{\omega^2}\right)}{\left(-F_1(\omega)\right)\ \frac{2}{M}} \frac{\sin 2\theta}{\left[1\ -\ \frac{1}{2}\left(1 - \frac{4M^2}{q^2}\frac{1}{\omega^2}\right)\ \sin^2\theta\right]}\ . \quad (8)$$

We present the graph of Eq. (8) in Fig. 1, where we used the proton mass for M, and $q^2 = (10.6\ \text{GeV})^2$ which is the value of BABAR and BELLE. The existence of $G_3(q^2, \nu)$ can also be investigated by the $\sin 2\theta$ distribution in the differential cross section given in Eq. (3) or (7).

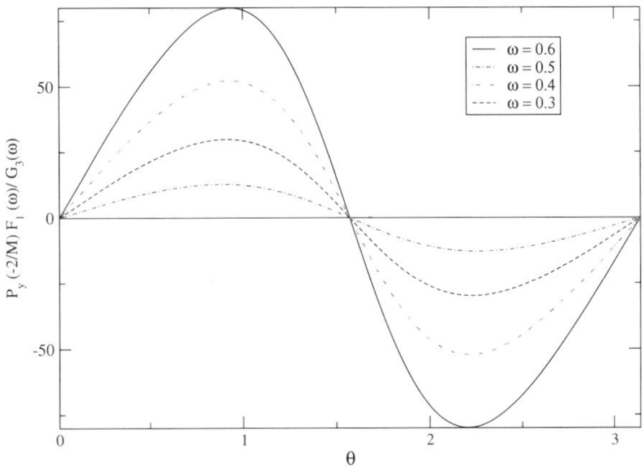

Fig. 1. Single-spin asymmetry, $\mathcal{P}_y\ (-\frac{2}{M})\ \frac{F_1(\omega)}{G_3(\omega)}$ from Eq. (8).

4. Conclusion

A symmetric spin-dependent structure function, $G_3(q^2, \nu)$, is allowed in the deep inelastic electron-positron annihilation process by the strong phases which are made from the final-state interactions. We showed that it is possible to measure the single-spin asymmetry in the inclusive Λ production at the present B-factories. This will lead to the first measurement of the symmetric spin-dependent structure function of in the deep inelastic electron-positron annihilation process.

Acknowledgments

This work was supported in part by the International Cooperation Program of the KICOS (Korea Foundation for International Cooperation of Science & Technology).

References

1. Jefferson Lab Hall A Collaboration, M.K. Jones et al., Phys. Rev. Lett. 84 (2000) 1398.
2. BABAR Collaboration, B. Aubert et al., Phys. Rev. D 73 (2006) 012005.
3. N. Christ and T.D. Lee, Phys. Rev. 143 (1966) 1310.
4. M. Gourdin, Nucl. Phys. B 38 (1972) 418.
5. S.D. Drell, D.J. Levy, and T.-M. Yan, Phys. Rev. 187 (1969) 2159.
6. R. Gatto and I. Vendramin, Nucl. Phys. B 78 (1974) 413; B 153 (1979) 546(E).

HERMES RECOIL DETECTOR

R. PEREZ-BENITO*
On behalf of the HERMES Collaboration
II Physikalisches Institut, University of Giessen,
35392 Giessen, Germany
**E-mail: perez@mail.desy.de*

To study hard exclusive processes which provide access to Generalized Parton Distributions (GPDs) and hence to the orbital angular momentum of the quarks, a new Recoil Detector surrounding the internal gas target of the HERMES experiment at DESY was installed. The Recoil Detector improves the selection of exclusive events by a direct measurement of the momentum and track direction of the recoiling particle as well as by rejecting non-exclusive background. This detector is an ideal tool to combine energy and position measurements for charged particles in an intermediate momentum range of 0.1 to 1.4 GeV. The HERMES Recoil Detector consists of three main components located inside a solenoidal magnet which provides a 1 T longitudinal magnetic field: a silicon strip detector surrounding the target cell inside the vacuum, a scintillating fiber tracker and a photon detector.

1. Introduction

Since 1995, the HERMES experiment has been in operation. The main goal is the experimental investigation of the spin structure of the nucleon. For this purpose, the 27.6 GeV longitudinal polarized electron/positron beam of HERA is used, in combination with a polarized or unpolarized gas target internal to the storage ring.
The HERMES spectrometer[1] is a forward angle instrument of conventional design, with a dipole magnet providing an integrated field of 1.3 Tm. It is symmetric about a central, horizontal shielding plate in the magnet. The acceptance is limited at small angles by an iron plate in the beam plane, which shields the positron and proton beam from the magnetic field of the spectrometer magnet. Particles with scattering angles within ± 170 mrad in the horizontal direction and between ±(40-140) mrad in the vertical direction are accepted.

The energy resolution for the electrons/positrons and photons of the existing HERMES spectrometer is not sufficient to select a clean sample of hard exclusive processes. These hard exclusive processes provide access to the generalized parton distributions (GPDs).[2] GPDs offer a possibility to determine the orbital angular momentum of the quarks, and can be accessed by studing Deeply Virtual Compton Scattering (DVCS).[3] In the reaction $ep \rightarrow e'\gamma p$, the DVCS process interferes with the Bethe-Heitler process where the incoming or outgoing electron/positron radiates a hard real photon in the Coulomb field of the nucleon. This interference leads to measurable beam spin and beam charge asymmetries around the virtual photon direction.

2. The HERMES Recoil Detector

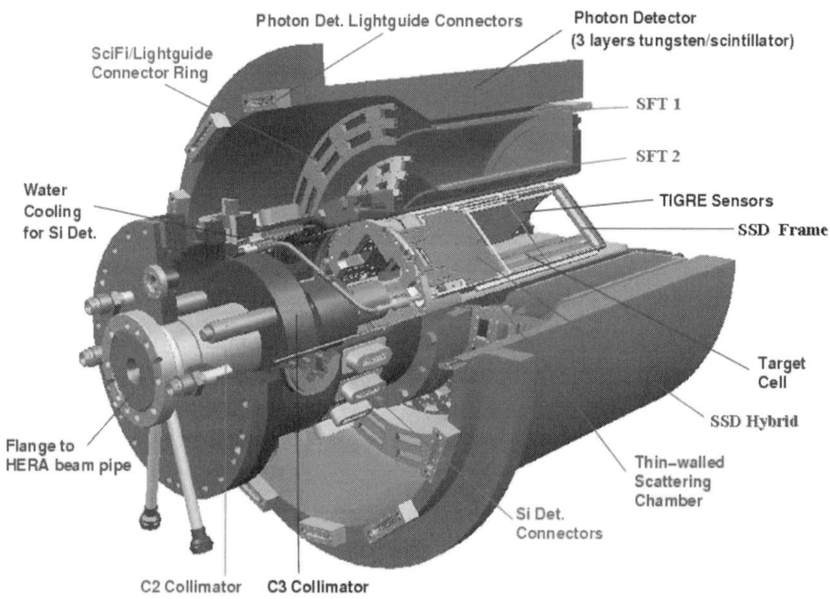

Fig. 1. Schematic overview of the HERMES Recoil Detector.

The design requirements on the Recoil Detector[4] are defined by the kinematics and the nature of the particles involved in the exclusive reactions, as well as by those of the expected background reactions. The Recoil Detector shown in fig.1 consists of three active detector parts: a silicon strip detector surrounding the target cell inside the beam vacuum, a scintillating fiber tracker, a the photon detector consisting of three layers of tungsten and scintillator bars. All three detectors are placed in a longitudinal magnetic field of 1 T generated by a superconducting solenoidal magnet. The fiber tracker and the silicon detectors are optimized to detect recoiling protons from hard exclusive processes. They cover a momentum range around 0.1 GeV to 1.4 GeV and provide the particle identification properties to discriminate protons from pions. The photon detector is used for particle identification of high momentum particles and for the background rejection of neutral particles. All three detectors provide a measurement of the deposited energy of the traversing recoil particles. Momentum determination for low momenta particles stopping inside the silicon detector is performed using the total energy deposition in both layers of the silicon in combination with the reconstructed track direction. The high momenta particles are determined by their bending in the longitudinal magnetic field.

3. Performance and Statistics

The recoil detector was install at HERMES in January 2006 and was taking data until the HERA end-of-running in July 2007. During this period data were collected with an electron beam, where only the scintillating fiber tracker was operational and with a positron beam where all subdetectors were fully operational. Table 1 shows the collected statistics of Deep Inelastic Scattering (DIS) and DVCS data from HERMES with the recoil detector operational.

Table 1. Collected statistic (preliminary) with recoil detector.

	H_2	D_2
Electron beam (only scintillating fibre tracker operational)	5k DVCS 3Mio. DIS	1k DVCS 0.8Mio. DIS
Positron beam (all subdetectors fully operational)	42k DVCS 28Mio. DIS	10k DVCS 7Mio. DIS

Figure 2 shows a particle identification in the recoil detector only using the total energy deposition in both layers of the silicon in combination with the reconstructed track direction.

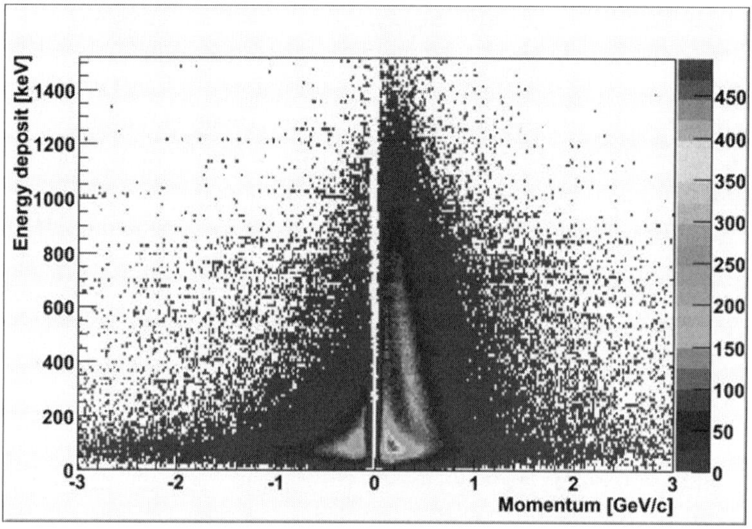

Fig. 2. Energy loss of the particles through the whole Silicon strip detector.

In Figure 2 the negative momentum correspond to negative particles and the positive momentum to positive particles. Positive and negative particles are identified according to their bending in the longitudinal magnetic field. We can see clearly the separation obtained by the Silicon strip detector between p and π^+. For higher momentum particles a similar energy loss plot is obtained from the signals in the scintillating fibre tracker.

References

1. K. Ackerstaff *et al.* [HERMES Collaboration], Nucl. Instrum. Meth. A **417**, 230 (1998)
2. X. D. Ji, Phys. Rev. D **55**, 7114 (1997)
3. V. A. Korotkov and W. D. Nowak, Nucl. Phys. A **711**, 175 (2002)
4. A. Airapetian *et al.* [HERMES Collaboration], DESY PRC 02-01,

INNER CALORIMETER IN CLAS/DVCS EXPERIMENT

RUSTAM NIYAZOV

Hall-B, Jefferson Lab
Newport News, VA 23606, USA
E-mail: rustam@jlab.org

The new Inner Calorimeter (IC) was developed specifically for dedicated CLAS/DVCS experiment.[1] IC consists of 424 $PbWO_4$ tapered crystals readout by avalanche photodiodes (APD) followed by low-noise preamplifiers. It is a high precision calorimeter with energy resolution better than 3% at 2-5 GeV and has an angle resolution of 3 to 4 mrad.

1. Introduction

The CEBAF large acceptance spectrometer[2] (CLAS) allows the detection of multi-particle final states in reactions induced by electrons or photons with high efficiency. The exploration of the nucleon partonic structure through the so-called Generalized Parton Distributions (GPD) requires the measurement of exclusive electroproduction reactions $ep \to ep/\gamma/\pi/\eta/\rho/\cdots$ in a kinematical regime where cross sections become very small. The study of Deeply Virtual Compton Scattering (DVCS) is of primary importance for the determination of GPDs. A high luminosity and acceptance is required. Also, in order to be fully exclusive, the detection of all three particles (electron, proton, photon) in the final state is preferable.

The standard CLAS set-up allows the detection of photons in the forward electromagnetic calorimeter[3] (EC), down to angles of about 10°. The photons from the DVCS process, however, are emitted at smaller angles, and the optimal angular coverage for this study is about 5° to 25° with the largest possible azimuthal acceptance.

2. CLAS/DVCS Experiment

The CLAS/DVCS Experiment used 5.77 GeV polarized electrons (with average polarization of 79.4%) incident on a 2.5 cm long liquid-hydrogen target. Outgoing electron and proton were detected in CLAS and the photons produced in reactions $ep \to ep/\gamma/\pi/\eta/\rho/$ were detected in EC and IC. The Inner Calorimeter (IC) was installed 55 cm downstream from the target. The IC was built out of 424 $PbWO_4$ tapered crystals, read-out by APDs and followed by specially designed low-noise preamplifiers. This calorimeter was designed to operate in conjunction with the superconducting solenoidal magnet that allowed us to shield or trap, along the beam direction, the numerous low energy electrons coming from Møller scattering on the electrons from the liquid hydrogen target. The solenoid, the target and the IC were inserted inside the CLAS bore. The temperature stabilization system was built to keep the temperature inside the calorimeter stable to within 0.1°. A Laser Monitoring System that allowed us to monitor gains in the calorimeter channels was installed as well.

3. Energy Calibration

The photons from $\pi^0 \to \gamma\gamma$ decay were used to calibrate gains of the calorimeter channels.[4] We selected events in the reaction $ep \to ep\pi^0$ with two photons hitting the IC, represented by two reconstructed clusters. For each counter in the fiducial region the distribution of invariant mass of two photons was obtained for the calibrated counter that detected the most energy in the cluster with second cluster's central energy absorbed in any other counter. Each distribution was fitted and the position of π^0 mass peak was found. Then the gain in these channels was corrected with respect to the found peak position. The final results were obtained after several iterations of the gain correction in each channel. The results of the calibration showing the final invariant mass $M_{\gamma\gamma}$ distribution for one out of six calibration runs are presented in Figure 1. Results for six calibration runs are similar. Two peaks corresponding to π^0 and η are fitted with Gaussian and a polynomial function.

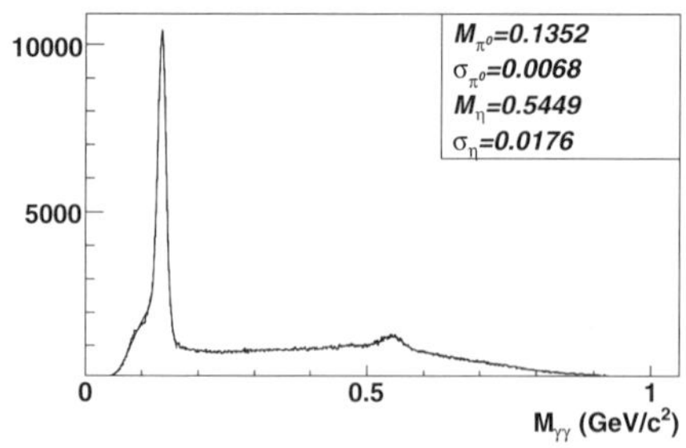

Fig. 1. Calibration results. Invariant mass of two reconstructed photons $M_{\gamma\gamma}$ integrated over all calorimeter channels for one of the calibration runs. Fit results for π^0 and η peaks are shown in the box.

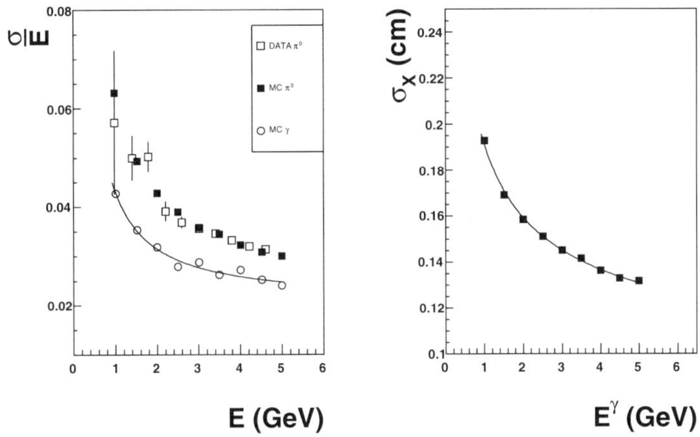

Fig. 2. a) IC energy resolution σ_E/E as a function of energy. Open squares are π^0 data. Closed squares are π^0 MC simulation. Open circles are single photon simulation. Line is a three parameter fit. b) IC position resolution from MC.

4. Energy and Position Resolution

The IC energy resolution was extracted using the reaction $ep \to ep\pi^0$. As a resolution the width of the distrubution of difference between missing energy for $ep \to epX$, measured by the CLAS and energy of π^0s reconstructed in IC was used. Several cuts were applied to eliminate background. Monte Carlo (MC) simulation for π^0 and single photons absorbed by IC was also studied. The energy resolution as a function of energy is shown in Figure 2a. The π^0 data and simulation results have very good agreement. From single photon simulation we extract the following resolution:

$$\frac{\sigma_E}{E} = \frac{0.024}{E} \oplus \frac{0.029}{\sqrt{E}} \oplus 0.02,$$

where the first term corresponds to the preamplifier noise, the second term includes the photostatistics fluctuations and the third term represents the fluctuations of the leakage of energy from the back of the crystals.

Position resolution as a function of energy for single photons from simulation is shown in Figure 2b.

5. CLAS and IC Measurements

In order to check compatibility of the CLAS and IC we studied missing mass distribution of $e\pi^0$ in the reaction $ep \to ep\gamma\gamma$, see Figure 3.

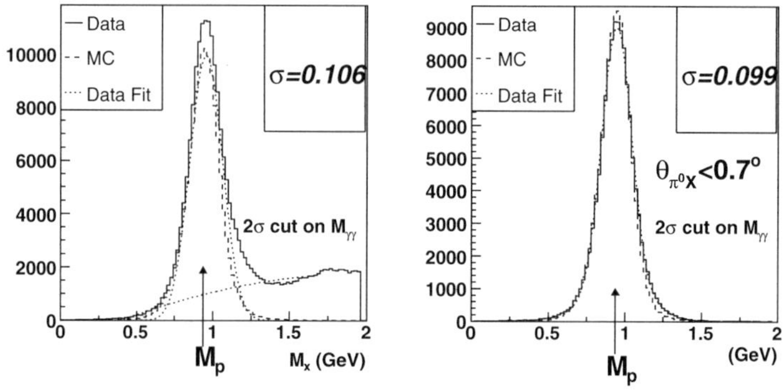

Fig. 3. Missing mass distribution for the reaction $ep \rightarrow e\pi^0 X$. Data – solid line; MC simulation – dashed line; fit to Data – dotted line. a) Both Data and MC have 2σ cut on $M_{\gamma\gamma}$. b) Both Data and MC have 2σ cut on $M_{\gamma\gamma}$ and on direction of π^0. Width of the peak from the fit is shown in the box.

Figure 3a shows Data and MC with cut on $M_{\gamma\gamma}$. One can clearly see the missing proton mass peak and the background in the data. Background can be elliminated by applying a cut on the direction of the π^0 detected in IC with respect to direction of the π^0 inferred from the detected electron and proton. Figure 3b shows clean proton peak after applying such a cut. The peaks in Figures 3a and 3b have almost the same widths.

6. Summary

Inner Calorimeter performed well during the CLAS/DVCS run. This is a first functional calorimeter with APD readout used in experiment. The energy resolution was determined to be 3% at 2-5 GeV. The data analysis[5] confirmed that around 90% of all DVCS events are within IC acceptance. Inner Calorimeter will be used in the second part of CLAS/DVCS experiment as well as in three other approved experiments at 6 GeV.

References

1. V. Burkert, L. Elouadrhiri, M. Garçon, S. Stepanyan *et al.*, CEBAF experiments 01-113 and 06-003.
2. B.A. Mecking *et al.*, Nucl. Instr. and Meth. A **503**, 513 (2003).
3. M. Amarian *et al.*, Nucl. Instr. and Meth. A **460**, 239 (2001).
4. R. Niyazov, S. Stepanyan, CLAS-NOTE 2005-021.
5. F.X, Girod, R.A. Niyazov, H.Avakian *et al.*, DVCS Beam-Spin Asymmetries, 2007, In CLAS review. To be published in PRL.

FROZEN-SPIN HD AS A POSSIBLE TARGET FOR ELECTRO-PRODUCTION EXPERIMENTS

A. M. SANDORFI, M. LOWRY,
C. E. THORN, A. CARACAPPA
Physics Dept., Brookhaven National Lab, Upton, NY 11973

T. KAGEYEVA
Physics Dept., Virginia Polytechnic Institute & State University, Blacksburg, VA 24061

X. WEI, S. HOBLIT
Physics Dept., University of Virginia, Charlottesville, VA 22901

C. S. WHISNANT
Physics Dept., James Madison University, Harrisonbury, VA 22807

There are many potentially interesting 12 GeV-era electron experiments that require transversely polarized H or D. The high fields associated with conventional Dynamically Polarized targets lead to challenging background conditions if oriented transverse to an electron beam. Frozen-spin targets that are maintained by low holding fields could greatly mitigate such problems. There is at least indirect evidence that frozen-spin HD could be quite radiation resistant and might maintain D polarization with modest electron currents.

1. Introduction

A diverse range of potentially exciting electro-production experiments require transversely polarized hydrogen or deuterium targets. Generalized Parton Distributions (GPD) hold the ultimate promise of producing a $2+1$ dimensional picture of the nucleon, provided the complete set of four GPDs can be mapped out. Of these, the $\tilde{E}$ GPD can only be accessed with transverse target polarization. Transverse asymmetries of hadron fragmentation in semi-inclusive-DIS (the Collins function) probe the transverse quark orbital angular momentum, and single-spin asymmetries (the Sivers function) can provide information on $u - d$ separation in the nucleon. The g_2 and A_2 Parton-Distribution Functions measured through inclusive-DIS contain information on the color-polarizability of the gluon field, and N^* transition form factors extracted with transverse target polarization can constrain models of baryon structure.

While measurements with transversely polarized targets have the potential for making a major impact in many areas, the magnetic holding fields accompanying such targets can deflect an electron beam and create challenging background conditions. A magnetic chicane can be installed upstream of the target and arranged in such a way that the target's magnetic field bends the electron beam back onto the detector axis. However, bremsstrahlung created in the target material will be peaked along the direction of the incoming electrons, which will then be at several degrees to the detector axis. Generally, one can arrange to have either the electron beam or the target bremsstrahlung centered at $0°$, but not both.

A transversely polarized target in a frozen-spin state that required only small holding fields would greatly mitigate such background problems. Butanol targets such as FROST, now under development at JLab, and HDice developed at LEGS in Brookhaven, have been used quite successfully in photon beam experiments. Two factors potentially limit their use with a charged particle beam: heating and the creation of free radicals. Butanol, C_4H_9OH (and similar materials), are polarized dynamically by first polarizing the electrons of paramagnetic impurities in a large external field and then using microwaves to transfer polarization to the hydrogen nuclei. This is typically carried out above 1/2 K. Since the spin relaxation (decay) time is a steep function of temperature, a frozen-spin state can be reached by turning off the microwave power and quickly lowering the target temperature into the 50-100 mK region. With holding fields of 0.5 tesla ($\sim 1/10$th of the polarizing field), relaxation times of days -to- weeks can be achieved in this way. However, heating in an electron beam can easily raise the target temperature out of the frozen-spin regime. In addition, bremsstrahlung in the target material will create ionized nuclei with randomly oriented spins. Although their absolute number will be small, such free radicals can couple to nearest neighbor spins, which can then couple to next-to-nearest neighbors, and so create polarization sinks within the target. The electron currents that would depolarized frozen-spin butanol are estimated to be on the order of pAmps, too small to be of general use.

The HDice target is migrating from BNL to JLab and will first be used in E06-101 in Hall-B. The polarization physics associated with frozen-spin

HD is very different from other targets and there is at least very suggestive evidence that it could withstand useful electron currents.

Considering first the H_2 molecule, its ground state consists of two identical fermions with opposed nuclear spins ($I = 0$) that cannot couple to a magnetic field. The symmetry of the molecular wavefunction requires the first rotational state ($J = 1$) to have aligned nuclear spins ($I = 1$). Since the energy of this state corresponds to an equivalent temperature of 172 K, its population is saturated at room temperature, while in the solid state (below 16 K) the $J = 1$ state is metastable since transitions to the ground state require the flip of a nuclear spin. The molecular $J = 1$ level is not spherically symmetric and is accompanied by a weak magnetic dipole generated by the rotation of the molecule and a relatively strong electric quadrupole field that couples to lattice vibrations. Similar considerations apply to molecular D_2 (although with boson symmetry requirements). As a result, the first rotational $J = I = 1$ levels of H_2 and D_2 fulfill two basic polarization requirements, the capability to couple to an applied magnetic field and a means to dissipate (through phonons) the heat generated by aligning spins. While these states polarize rapidly at high field and low temperature, they also depolarize rapidly when the field is lowered and the temperature raised.

Since the HD molecule has no requirements of symmetry, its rotational states are not metastable. Upon freezing, all population quickly reverts to the $J = 0$ ground state which is spherically symmetric, decoupled from its lattice and unable to dissipate the heat from spin alignment. However, if HD is prepared with small ($\sim 10^{-3}$) H_2 and D_2 impurities, their $J = I = 1$ levels can be polarized and can then transfer polarization to HD through the interaction between the magnetic dipole field of the rotating H_2 or D_2 molecule and the spin dipoles of HD. High polarizations can be reached by lowering the sample to the $\sim$ 15 mK range in a dilution refrigerator equipped with a strong (15 Tesla) magnet. An equilibrium polarization is reached after some days, limited by the time constants for heat removal (which is facilitated by a mesh of 50 μm aluminum wires imbedded in the target that accounts for $\sim$ 20% of the mass). Were the magnetic field to be lowered, or the temperature raised at this point, the HD polarization would rapidly relax by the same spin-exchange mechanisms. However, the metastable $J = 1$ H_2 and D_2 impurities decay to their magnetically inert $J = 0$ ground states with 1/e time constants of 6.3 days and 18.6 days, respectively. Thus by simply maintaining the high-field/low-temperature polarizing conditions for many time constants (in practice, a few months), relaxation through $J = 1$ impurities is switched off and the HD is left in a frozen-spin state that can be maintained at much lower fields (< 1 tesla) and higher temperatures (0.2 - 0.5 K).

At BNL, HD target polarizations of 60% H and 35% D have been used in photon experiments with spin-relaxation times in excess of a year, and polarizations are expected to be higher with the smaller diameter cells that will be used at JLab. The deuterium polarization is particularly stable; spin-relaxation times of 2 months have been measured with only 0.01 tesla (100 gauss) holding field and 0.2 K. The projected D-decay time for a 0.04 tesla saddle coil, 0.12 m in length ($B \cdot dL = 0.005$ tesla-m), is ~ 7 months. Comparable H relaxation times require higher fields but should be possible with $B \cdot dL = 0.050$ tesla-m, which is still about 30 times less than a dynamically polarized ammonia target.

The beam heating expected from 5 nAmps of 10 GeV electrons traversing a 2 cm HD target is ~ 5 mW (as calculated with GEANT). This is about the cooling power of the existing BNL In-Beam-Cryostat (IBC) at 0.5 K and will be significantly increased in the CLAS-IBC now under design for E06-101. Beam heating is considerably less in HD, as compared to Butanol, due to the lower Z and, unlike butanol, HD relaxation times are not so strong functions of temperature so that long life-times are achievable up to about 0.7 K.

Free radicals generated by electron bremsstrahlung will have randomly oriented polarizations. While their absolute number is small, they can generate polarization sinks within the target if the spin-diffusion time is short. This time constant has been indirectly measured at BNL by using RF to punch a local polarization hole within a highly polarized target. The rate at which this hole heals after the RF power is lowered reflects the in-diffusion of spin from other regions of the target. At 2 K, this measured spin-diffusion is ~ 1 day for H but unmeasurably long for D (greater than a year). Whether or not the H performance improves at lower temperatures is a matter for further study, but the extremely slow spin-diffusion for D already suggests that frozen-spin HD could maintain its deuterium polarization during electron experiments.

Frozen-spin HDice may provide an attractive alternative for electron experiments with at least transversely polarized deuterons in CLAS12. Problems associated with beam deflection are virtually eliminating by the small holding fields and this potentially allows the target to be located in the center of the detector, thus dramatically increasing the acceptance. In addition, the target has almost no dilution. The only unpolarizable nucleons are associated with the target cell and these can be sampled and subtracted in conventionally empty-cell measurements. At the same time, the low Z results in a long radiation length and comparatively few bremsstrahlung photons.

The factors affecting target polarization are complex and intertwined; a direct test of the performance of polarized HD with electrons is essential. This will be carried out during the course of the E06-101 run.

This work has been supported by the United States Department of Energy under contract DE-AC02-98CH10886.

PARTICIPANTS LIST

Afanasev, Andrei
Hampton University
afanas@jlab.org
Physics Department
Hampton University
Hampton, VA 23668

Ahmad, Saeed
Univ of Virginia
sa8y@virginia.edu
Physics (Univ of Virginia)
382 McCormick Road
Charlottesville, VA 22904 USA

Alexandrou, Constantia
University of Cyprus
alexand@ucy.ac.cy
Department of Physics
P.O. Box 20537
Nicosia, 1678 Cyprus

Ashery, Daniel
Tel Aviv University
ashery@tauphy.tau.ac.il
Ramat Aviv
Tel Aviv, Israel

Avagyan, Harut
JLab
avakian@jlab.org

Baldini, Rinaldo
INFN-LabNazFrascati
baldini@lnf.infn.it
Roma, 00186 ITALY

Balewski, Jan
IUCF
jtbalews@indiana.edu
Bloomington, IN 47408

Bedir, Islam
ODU
bedir@physics.odu.edu
12000 Jefferson Avenue,
CEBAF Center
Newport News, VA 23606

Braun, Vladimir
Regensburg University
vladimir.braun@physik.uni-regensburg.de
Universitaet Regensburg, Institut fuer Theor.Physik,
Universitaetsstrasse 31
Regensburg, D-93053 Germany

Brodsky, Stanley
SLAC, Stanford University
sjbth@slac.stanford.edu
2575 Sand Hill Road
Menlo Park, CA 94025

Burkardt, Matthias
NMSU
burkardt@nmsu.edu

Department of Physics, MSC 3D
New Mexico State University
Las Cruces, NM 88011

Burkert, Volker
Jefferson Lab
burkert@jlab.org
12000 Jefferson Avenue
Newport News, VA 23606

Camsonne, Alexandre
Jefferson Laboratory
camsonne@jlab.org
12000 Jefferson Avenue
Cebaf Center A120
Newport News, VA 23606

Canan, Mustafa
Old Dominion University
mcanan@odu.edu

Cardman, Lawrence
JLab
cardman@jlab.org
12000 Jefferson Ave
Newport News, VA 23606

Carlson, Carl
William & Mary
carlson@physics.wm.edu

Chen, Jian-ping
Jefferson Lab
jpchen@jlab.org

Chirilli, Giovanni
JLAB/ODU
chirilli@jlab.org
4600 Elkhorn ave.
Norfolk, VA 23529

Christy, Michael
Hampton University
christy@jlab.org
Hampton, VA 23668

Chudakov, Eugene
JLab
gen@jlab.org
12000 Jeffereson ave
Newport News, VA 23606

Commisso, Melissa
UVA
mac5cx@virginia.edu

D'Hose, Nicole
CEA- Scalay
nicole.dhose@cea.fr

Dalton, Mark
University of the Witwatersrand
dalton@jlab.org
School of Physics
Private Bag 3
Johannesburg, 2192 South Africa

Danagoulian, Samuel
North Carolina A&T
State University
danagu@jlab.org
Department of Physics
1601 E. Market St.
Greensboro, NC 27411

Day, Donal
University of Virginia
dbd@virginia.edu
Department of Physics
McCormick Road
Charlottesville, VA 22903

de Jager, Kees
Jefferson Lab
kees@jlab.org
12000 Jefferson Avenue
Newport News, VA 23606

Dubnicka, Stanislav
SAS
fyzidubn@savba.sk
Inst.of Physics, SAS, Bratislava
Dubravska cesta
Bratislava, SK 84311
Slovak Republic

Dubnickova, Anna
Comenius University
dubnickova@fmph.uniba.sk
Dept.of Theor.Phys., FMFI UK,
Mlynska dolina
Bratislava, SK 84245
Slovak Republic

Ellinghaus, Frank
University of Colorado
Frank.Ellinghaus@desy.de
Brookhaven National Lab,
Physics Department
Building 510 C
Upton, NY 11973

Elouadrhiri, Latifa
Jefferson Lab
latifa@jlab.org

Ent, Rolf
Jefferson Lab
lcarlson@jlab.org

Feuerbach, Robert
College of William and Mary
feuerbac@jlab.org

Gamberg, Leonard
Penn State Berks
lpg10@psu.edu
Reading, PA 19610

Gaskell, Dave
JLab
gaskelld@jlab.org

Gilfoyle, Gerard
University of Richmond
gilfoyle@jlab.org
Physics Department,
University of Richmond
28 Westhampton Way
Richmond, VA 23173

Gilman, Ronald
Rutgers/JLab
gilman@jlab.org

Girod, Francois-Xavier
Jefferson Lab
fxgirod@jlab.org

Goldstein, Gary
Tufts University
gary.goldstein@tufts.edu
Department of Physics
Tufts University
Medford, MA 02155

Gothe, Ralf
University of South Carolina
gothe@sc.edu
Department of Physics
and Astronomy
712 Main Street
Columbia, SC 29208

Griffioen, Keith
College of William and Mary
griff@jlab.org

Grosse Perdekamp, Matthias
UIUC and RBRC
mgp@uiuc.edu
1110 W. Green Str.
Urbana, IL

Guidal, Michel
IPN Orsay

guidal@ipno.in2p3.fr
rue Georges Clemenceau
Orsay, 91405 France

Hafidi, Kawtar
Argonne National Laboratory
kawtar@anl.gov
9700 S Caa Av
Argonne, IL 60439

Hansen, Ole
Jefferson Lab
ole@jlab.org
12000 Jefferson Avenue
Suite 4
Newport News, VA 23606

Hayrapetyan, Avetik
University of Michigan
Avetik.Airapetian@desy.de
HERMES/DESY
Notkestrasse 85
Hamburg, 22607 Germany

Honkanen, Heli
UVa
hh9e@virginia.edu

Horn, Tanja
JLab
hornt@jlab.org

Hoyer, Paul
University of Helsinki
paul.hoyer@helsinki.fi

Hwang, Dae Sung
Sejong University
dshwang@sejong.ac.kr
Gwangjin-Gu, Gunja-Dong 98
Seoul, 143-747 South Korea

Hyde, Charles
Old Dominion University and
University Blaise Pascal
hyde@jlab.org
Laboratoire de Physique
Corpusculaire CNRS/IN2P3
24 avenue des Landais
AUBIERE, cedex 63177 France

Itard, Florian
University of Blaise Pascal
flosourcier@yahoo.fr

Jain, Pankaj
IIT Kanpur
pkjain@iitk.ac.in
Physics Department
IIT
Kanpur, UP 208016 India

Ji, Chueng
North Carolina State University
ji@ncsu.edu
Dept of Physics
North Carolina State University
Raleigh, NC 27695-8202

Jiang, Xiaodong
Rutgers University
jiang@jlab.org

Jones, Mark
JLab
jones@jlab.org

Khetarpal, Puneet
Rensselaer Polytechnic Institute
puneetk@jlab.org
JR Science Center Room 1C25
110 Eighth Street
Troy, NY 12180

Kroll, Peter
Universitaet Wuppertal
kroll@physik.uni-wuppertal.de

Universitaet Wuppertal
Physik
Wuppertal, D-42097 Germany

Kubarovsky, Valery
RPI/Jlab
vpk@jlab.org
Jefferson Lab
12000 Jefferson av., suite 5, B105
Newport News, VA 23606

Kuchina, Elena
Rutgers University
kuchina@jlab.org

Kwee, Herry
Arizona State University
herry.kwee@asu.edu
Physics Department,
Bateman Physical Science F-Wing
Arizona State University
Tempe, AZ 85287

Lachniet, Jeffrey
ODU
lachniet@physics.odu.edu
Dept. of Physics
Old Dominion University
Norfolk, VA 23529

Laget, Jean-Marc
JLab
laget@jlab.org

Lansberg, Jean-Philippe
Ecole Polytechnique
Jean-Philippe.Lansberg@
cpht.polytechnique.fr
Ecole Polytechnique - CPHT
Palaiseau, 91128 France

LeRose, John
JLab
lerose@jlab.org
JLab 12000 Jefferson Ave
Suite 4, MS 12H3
Newport News, Va 23188

Liuti, Simonetta
University of Virginia
sl4y@virginia.edu
382, McCormick Rd.
Charlottesville, VA 22904-4714

Matevosyan, Hrayr
JLab
hrayr@jlab.org
Theory Center, MS-12H2, Suite 1
12000 Jefferson Avenue
Newport News, VA 23606

Mazouz, Malek
Faculte des sciences de Monastir
mazouz@jlab.org
Departement de physique
Monastir, 5000 Tunisia

Mbianda Njencheu, Georgie
Wits University/Jlab
mbianda@jlab.org
School of Physics, Private Bag 3,
Wits University
Johannesburg, Gauteng 2050
South Africa

Miller, Gerald
Physics Dep't UW
miller@phys.washington.edu
University of Washington,
Physics Dep't
Box 351560
Seattle, WA 98195-1560

Mueller, Dieter
University Bochum

dieter.mueller@physik.uni-regensburg.de
Inst. f. Theoretische Physik II
University Bochum
Bochum, 44780 Germany

Nathan, Alan
University of Illinois
a-nathan@uiuc.edu
Department of Physics
1110 W. Green Street
Urbana, IL 61821

Niyazov, Rustam
JLAB
rustam@jlab.org
B120 12000 Jefferson Ave
Newport News, VA 23606

Pentchev, Lubomir
The College of William and Mary
pentchev@jlab.org
Jefferson Lab, 12000 Jefferson Ave.
12H5/F318
Newport News, VA 23606

Perdrisat, Charles
William and Mary
perdrisa@jlab.org
College of William and Mary
Williamsburg, VA 23187

Perez-Benito, Roberto
DESY Hermes
perez@mail.desy.de

Peters, Andreas
Regensburg University
andreas.peters@physik.uni-regensburg.de
Universitaetsstrasse 31,
Regensburg, Bavaria
93053 Germany

Petratos, Makis
Kent State University
gpetrato@kent.edu
Physics Department
Kent State University
Kent, OH 44236

Psaker, Ales
Hampton University/JLab
apsaker@physics.odu.edu

Punjabi, Vina
Norfolk State University
punjabi@jlab.org
Physics Department
Norfolk State University
Norfolk, VA 23504

Quinn, Brian
Carnegie Mellon University
bquinn@cmu.edu
Pittsburgh, PA 15213

Radyushkin, Anatoly
ODU/JLab
radyush@jlab.org
12000 Jefferson Ave
MS-12H2, Suite 1
Newport News, VA 23606

Richards, David
Jefferson Lab
dgr@jlab.org

Roberts, Craig
Argonne National Laboratory
cdroberts@anl.gov
Physics Division
Argonne National Laboratory
Argonne, IL 60439

Roche, Julie
Ohio University

rochej@ohio.edu
206 Accelerator Laboratory
Department of Physics
& Astronomy Athens, OH 45701

Rohrwild, Jurgen
Universitat Regensburg
juergen.rohrwild@physik.uni-r.de
Institut fur theoretische Physik
Regensburg, 93040 Germany

Rostomyan, Armine
DESY
armine.rostomyan@desy.de
Notkestrasse 85
Hamburg, 22607 Germany

Sabatie, Franck
CEA Saclay
fsabatie@cea.fr
Dapnia/SPhN
Gif sur Yvette, 91191 France

Saha, Arun
JLAB
saha@jlab.org

Sandorfi, Andrew
Brookhaven National Laboratory
sandorfi@bnl.gov
Physics Department
Building 510
Upton, NY 11973

Santoro, Joseph
CUA/JLab
santoro@jlab.org
12000 Jefferson Ave
Newport News, VA 23602

Sargsian, Misak
Florida International University
sargsian@fiu.edu
Department of Physics
Florida International University
Miami, FL 33199

Schierholz, Gerrit
DESY
Gerrit.Schierholz@desy.de

Schlegel, Marc
Jefferson Lab
schlegel@jlab.org
Theory Center, MS 12H2
12000 Jefferson Avenue
Newport News, VA 23606

Seth, Kamal
Northwestern University
kseth@northwestern.edu
Physics Department,
Northwestern University
Evanston, IL 60201

Smith, Elton
JLab
elton@jlab.org
12000 Jefferson Ave
Newport News, Va 23606

Stoler, Paul
RPI
stoler@rpi.edu
Physics Dept
RPI
Troy, NY 12180

Taneja, Swadhin
Ecole Polytechnique
Swadhin.Taneja@
cpht.polytechnique.fr
CPHT, Ecole Polytechnique
Palaiseau, 91128 France

Thomas, Anthony
JLab

awthomas@jlab.org
Theory Center, MS-12H2, Suite 1
12000 Jefferson Avenue
Newport News, VA 23606

van der Nat, Paul
NIKHEF
travel@nikhef.nl
Kruislaan 409
AMSTERDAM, 1098 SJ
Netherlands

Vanderhaeghen, Marc
William &Mary/JLab
marcvdh@jlab.org
College of William and Mary
Dept Physics
PO Box 8795
Williamsburg, VA 23187-8795

Villano, Anthony
RPI
villaa@jlab.org

Weiss, Christian
Jefferson Lab
weiss@jlab.org
Theory Center
Jefferson Lab
Newport News, VA 23606

Wojtsekhowski, Bogdan
TJNAF
bogdanw@jlab.org

Yi, Kai
SLAC
yik@slac.stanford.edu
2575 sand hill road
Menlo Park, Ca 94025

Yuan, Feng
Brookhaven National Lab
fyuan@quark.phy.bnl.gov
RIKEN/BNL Research Center
Brookhaven National Laboratory
Upton, NY 11973